Karger-Kocsis, Fakirov
**Nano- and Micromechanics of
Polymer Blends and Composites**

József Karger-Kocsis
Stoyko Fakirov

Nano- and Micro-mechanics of Polymer Blends and Composites

HANSER

Hanser Publishers, Munich   Hanser Publications, Cincinnati

The Editors:

Prof. Dr.-Ing. József Karger-Kocsis, Department of Polymer Engineering, Faculty of Mechanical Engineering, Budapest University of Technology and Economics, 1111 Budapest, Hungary

Prof. Stoyko Fakirov, Ph.D., The University of Auckland, Department of Mechanical Engineering, Private bag 92019, Auckland, New Zealand

Distributed in the USA and in Canada by
Hanser Publications
6915 Valley Avenue, Cincinnati, Ohio 45244-3029, USA
Fax: (513) 527-8801
Phone: (513) 527-8896 or 1-800-950-8977
www.hanserpublications.com

Distributed in all other countries by
Carl Hanser Verlag
Postfach 86 04 20, 81631 München, Germany
Fax: +49 (89) 98 48 09
www.hanser.de

The use of general descriptive names, trademarks, etc., in this publication, even if the former are not especially identified, is not to be taken as a sign that such names, as understood by the Trade Marks and Merchandise Marks Act, may accordingly be used freely by anyone.

While the advice and information in this book are believed to be true and accurate at the date of going to press, neither the authors nor the editors nor the publisher can accept any legal responsibility for any errors or omissions that may be made. The publisher makes no warranty, express or implied, with respect to the material contained herein.

Library of Congress Cataloging-in-Publication Data

Nano- and micromechanics of polymer blends and composites / edited by Jszsef Karger-Kocsis, and Stoyko Fakirov.
    p. cm.
 Includes bibliographical references.
  Summary: „The aim of this book is to give a state-of-art overview on aspects of micro- and nanomechanics of polymers, polymeric blends and composites"--Provided by publisher.
 ISBN 978-1-56990-435-0
 1. Polymeric composites--Mechanical properties. 2. Micromechanics. 3. Microstructure. 4. Nanostructured materials. I. Fakirov, Stoyko. II. Karger-Kocsis, J. (Jszsef)
  TA455.P58N355 2009
  620.1'920423--dc22

                                    2009010703

Bibliografische Information Der Deutschen Bibliothek
Die Deutsche Bibliothek verzeichnet diese Publikation in der Deutschen Nationalbibliografie; detaillierte bibliografische Daten sind im Internet über <http://dnb.d-nb.de> abrufbar.

ISBN 978-3-446-41323-8

All rights reserved. No part of this book may be reproduced or transmitted in any form or by any means, electronic or mechanical, including photocopying or by any information storage and retrieval system, without permission in wirting from the publisher.

© Carl Hanser Verlag, Munich 2009
Production Management: Steffen Jörg
Coverconcept: Marc Müller-Bremer, Rebranding, München, Germany
Coverdesign: Stephan Rönigk
Printed and bound by Druckhaus "Thomas Müntzer" GmbH, Bad Langensalza
Printed in Germany

Preface

The deduction of *structure-property relationships* is a major target of materials scientists. Unfortunately, these relationships are hardly ever universal, but mostly of specific nature. On the one hand, this is due to the fact that there is no correlation between the structural units in different materials, such as metals, ceramics and polymers. On the other hand, the materials have a very complex structural build-up, the constituents of which are strongly interrelated. This note holds especially for polymers.

Depending on the size of the volume elements which contain all structural "inhomogeneities" in a way that *via* their combination the related material can be treated as "homogenous" (representative volume element concept), one may distinguish between macro-, micro- and nanocomposites. The size of the representative volume element in *macrocomposites* is in the range of millimeters and above. Such polymer macrocomposites, as those with various textile architecture (woven, braided), resin infiltrated open cell metallic and ceramic foams *etc.*, are beyond the scope of this book. The term polymer *microcomposite* covers very different systems, accepting the definition that at least one dimension of the constituents is in the micrometer range. So, unidirectional aligned carbon (diameter 6–7 µm) or glass (diameter 10–15 µm) fiber reinforced composites, neat polymers with injection molding-induced skin-core structure (oriented skin layer thickness is usually less then 200 µm), polymer blends and reinforced/filled polymers having microscale inclusions (at least in one dimension again!) – all belong to this category. In the case of *nanocomposites*, at least one dimension of the constituents should be on the nanoscale (such as clay or carbon nanotubes). As mentioned above, the structure of polymers may be very complex and multiscaled. For example, fiber-reinforced microcomposites with a nanomodified matrix or semi-crystalline polymers themselves (just consider the mean sizes of spherulites and crystalline lamellae) are on the borderline between micro- and nanocomposites. This may be the reason why polymer scientists often use the term morphology instead of micro- and nano-scale structures for neat polymers.

The structural elements of *one-component multiphase systems* (homopolymers and copolymers) are in the same size-range as nano- and micro-sized additives and reinforcements of the *multicomponent systems* (blends and composites). This fact makes possible the application of the same characterization techniques and modeling approaches to these rather different, from material point of view, systems. What is more, the structural elements in the homopolymers and copolymers as well as the constituents of the blends and composites are distinguished by their individual mechanical characteristics and each of them makes its own contribution to the overall mechanical behavior of the whole system. At the same time, multicomponent systems, as a rule, can be considered multiphase systems as well. Only in rare cases, when a particular property obeys the additivity law, the complex system can be easily predicted with respect of this property. In the rest of the cases, empirical approaches and modeling have to be applied for the deduction of structure-properties relationships.

With the appearance of polymeric nanocomposites, the structural build-up of polymeric materials, including polymeric microcomposites, has become even more complex. This has forced the researchers to check whether or not the knowledge acquired with traditional microcomposites can be transferred to nanocomposites, produced by different methods. As the reader will notice, this is sometimes the case for structural polymeric micro- and nanocomposites. On the other hand, polymer nanocomposites may show peculiar properties that are not present, not even in analogy, in microcomposites. To find the cause of such behavior is a very challenging task, which requires a *multiscale approach*. The latter covers in-depth structural investigations as well as molecular modeling using different approaches and techniques.

Our aim with this book is to demonstrate that the *multiscale approach* is the right tool for a deeper understanding of the structure-property relationships in polymeric micro- and nanocomposites. The term "mechanics" in the title is foreseen to demonstrate that emphasis has been put on structural instead of functional composites. Note that the mechanical behavior of structural composites is of great practical relevance and the driving force of their development nowadays. The book contributions are grouped in 5 sections. Part I is devoted to polymers (Galeski and Regnier and Ginzburg *et al.*). Part II highlights selected aspects of composite production (Zhang *et al.*, Bokobza, and Bhattacharyya and Fakirov). Part III considers interphase aspects (Kalfus and Jancar). Part IV gives an overview of characterization methods and properties, including *in situ* deformation (Stribeck), creep and fatigue (Pegoretti), deformation and fracture properties (Tjong, Dasari *et al.* and Karger-Kocsis), and hardness (Fakirov). The book ends with Part V dealing with modeling of rubber and polymer nanocomposites (Mark *et al.* and Spencer and Sweeney).

Our intention was to cover all kinds of polymeric materials, *viz.* thermoplastics, thermosets, and rubbers, including their different types and combinations.

We are thankful to our contributors for their high quality chapters and their timely delivery. We strongly hope that you, the readers, will find this book useful for your work.

A.m.D.g., March, 2009 *József Karger-Kocsis*
Budapest, Kaiserslautern, Pretoria,

Stoyko Fakirov
Auckland

Content

PART I   POLYMERS

Chapter 1   **Nano- and Micromechanics of Crystalline Polymers**
             A. Galeski, G. Regnier

1.1. Introduction	3
1.2. Tensile deformation of crystalline polymers	4
1.3. Cavitation in tensile deformation	4
1.4. Tensile deformation of polyethylene and polypropylene	8
1.5. Deformation micromechanisms in crystalline polymers	13
1.6. Molecular mechanisms at a nanometer scale	16
1.7. Dislocations in crystal plasticity	23
1.8. Generation of dislocations	25
1.9. Competition between crystal plasticity and cavitation	34
1.10. Micromechanics modeling in semicrystalline polymers	35
1.10.1. Microstructure and mechanical properties	35
1.10.2. The micromechanical models	36
1.10.3. Idealizing the microstructure of semicrystalline polymers	38
1.10.4. Elastic behavior prediction	40
1.11. Large deformations and bottlenecks	45
1.12. Phenomenological models of polymer deformation under tensile and compressive stresses	45
1.13. Conclusions	47
References	48

Chapter 2   **Modeling Mechanical Properties of Segmented Polyurethanes**

V. V. Ginzburg, J. Bicerano, C. P. Christenson, A. K. Schrock, A. Z. Patashinski

2.1. Introduction 59
2.2. Predicting Young's modulus of segmented polyurethanes 63
   2.2.1. Relationship between Young's modulus and formulation – experimental observations 63
   2.2.2. Theory 64
   2.2.3. Young's modulus: comparing theory with experiments 72
2.3. Modeling tensile stress-strain behavior 76
2.4. Linear viscoelasticity 82
2.5. Non-equilibrium factors and their influence on mechanical properties 84
2.6. Conclusions and Outlook 84
   Acknowledgment 85
   References 85

PART II   NANOCOMPOSITES: INFLUENCE OF PREPARATION

Chapter 3   **Nanoparticles/Polymer Composites: Fabrication and Mechanical Properties**

M. Q. Zhang, M. Z. Rong, W. H. Ruan

3.1. Introduction 93
3.2. Dispersion-oriented manufacturing of nanocomposites 95
   3.2.1. Conventional two-step manufacturing 95
   3.2.2. Specific two-step manufacturing 107
   3.2.3. One-step manufacturing 118
3.3. Dispersion and filler/matrix interaction-oriented manufacturing of nanocomposites 120
   3.3.1. Two-step manufacturing in terms of in situ reactive compatibilization 120
   3.3.2. One-step manufacturing in terms of in situ graft and crosslinking 124
3.4. Dispersion, filler/filler interaction and filler/matrix interaction-oriented manufacturing of nanocomposites 129
3.5. Conclusions 135
   Acknowledgements 136
   References 136

Chapter 4 Rubber Nanocomposites: New Developments, New Opportunities

L. Bokobza

4.1.	Introduction	141
4.2.	General considerations on elastomeric composites	142
4.3.	Spherical in situ generated reinforcing particles	144
4.4.	Carbon nanotube-filled rubber composites	153
4.5.	Conclusions	161
	References	162

Chapter 5 Organoclay, Particulate and Nanofibril Reinforced Polymer-Polymer Composites: Manufacturing, Modeling and Applications

D. Bhattacharyya, S. Fakirov

5.1.	Introduction	167
5.2.	Polypropylene/organoclay nanocomposites: experimental characterisation and modeling	169
5.2.1.	Peculiarities of polymer/clay nanocomposites	169
5.2.2.	Parametric study and associated properties of PP/organoclay nanocomposites	171
5.2.3.	Evaluation of the experimental data by means of Taguchi and Pareto ANOVA methods	174
5.2.4.	Materials, manufacturing and characterisation of nanocomposites	178
5.2.5.	Analytical models for composites	179
5.2.6.	Comparisons of experimental results with the calculated values	182
5.3.	The dispersion problem in the case of polymer-polymer nanocomposites	185
5.3.1.	Manufacturing of nanofibrillar polymer-polymer composites	187
5.3.2.	Nanofibrillar *vs.* microfibrillar polymer-polymer composites and their peculiarities	188
5.4.	Directional, thermal and mechanical characterisation of polymer-polymer nanofibrillar composites	190
5.4.1.	Directional state of NFC as revealed by wide-angle X-ray scattering	190
5.4.2.	Thermal characterization of NFC	192
5.4.3.	Mechanical properties of NFC	193
5.5.	Potentials for application of nanofibrillar composites and the materials developed from neat nanofibrils	196

5.6.	Conclusions and outlook	199
	Acknowledgments	200
	References	201

PART III   NANO- AND MICROCOMPOSITES: INTERPHASE

Chapter 6   Viscoelasticity of Amorphous Polymer Nanocomposites with Individual Nanoparticles

J. Kalfus

6.1.	Introduction	209
6.2.	Brief physics of amorphous polymer matrices	210
6.2.1.	Equilibrium structure of amorphous chains	210
6.2.2.	Microscopic relaxation modes and segmental mobility	212
6.2.3.	Entropy vs. energy driven mechanical response	214
6.3.	Basic aspects of amorphous polymer nanocomposites	216
6.3.1.	Structure of surface adsorbed chains	217
6.3.2.	Segmental immobilization of chains in the presence of solid surfaces	219
6.4.	Reinforcement of amorphous nanocomposite below and above matrix T_g	222
6.5.	Strain induced softening of amorphous polymer nanocomposites	228
6.6.	Relaxation of chains in the presence of nanoparticles	233
6.7.	Conclusions and outlook	235
	Acknowledgements	236
	References	236

Chapter 7   Interphase Phenomena in Polymer Micro- and Nanocomposites

J. Jancar

7.1.	Introduction	241
7.2.	Micro-scale interphase in polymer composites	246
7.3.	Nano-scale interphase	250
7.4.	Chain immobilization on the nano-scale	252
7.5.	Characteristic length-scale in polymer matrix nanocomposites	255
7.6.	Conclusions and outlook	257
	Acknowledgement	258
	References	258

PART IV   NANO- AND MICROCOMPOSITES: CHARACTERIZATION

Chapter 8   Deformation Behavior of Nanocomposites Studied by X-Ray Scattering: Instrumentation and Methodology

N. Stribeck

8.1. Introduction 269
8.2. Scattering theory and materials structure 272
    8.2.1. Relation between a CDF and IDFs 275
8.3. Analysis options derived from scattering theory 276
    8.3.1. Completeness – a preliminary note 276
    8.3.2. Analysis options 276
    8.3.3. Parameters, functions and operations 277
8.4. The experiment 278
    8.4.1. Principal design 278
    8.4.2. Engineering solutions 279
    8.4.3. Scattering data and its evaluation 284
8.5. Techniques: Dynamic *vs.* stretch-hold 286
8.6. Advanced goal: Identification of mechanisms 286
8.7. Observed promising effects from stretch-hold experiments 289
    8.7.1. Orientation of nanofibrils in highly oriented polymer blends by means of USAXS 289
    8.7.2. USAXS studies on undrawn and highly drawn PP/PET blends 291
8.8. Choosing experiments 293
    8.8.1. Experiments with a macrobeam 293
    8.8.2. Experiments with a microbeam 294
8.9. Conclusion and outlook 295
    References 296

Chapter 9   Creep and Fatigue Behavior of Polymer Nanocomposites

A. Pegoretti

9.1. Introduction 301
9.2. Generalities on the creep behavior of viscoelastic materials ... 302
9.3. Generalities on the fatigue resistance of polymeric materials . 306
9.4. Creep behavior of polymer nanocomposites 309
    9.4.1. Creep response of PNCs containing one-dimensional nanofillers 309

    9.4.2. Creep response of PNCs containing
          two-dimensional nanofillers 315
    9.4.3. Creep response of PNCs containing
          three-dimensional nanoparticles 317
9.5. Fatigue resistance of polymer nanocomposites 321
    9.5.1. Fatigue behavior of PNCs containing
          one-dimensional nanofillers 322
    9.5.2. Fatigue behavior of PNCs containing
          two-dimensional nanofillers 326
    9.5.3. Fatigue behavior of PNCs containing
          three-dimensional nanoparticles 332
9.6. Conclusions and outlook 334
    References 335

Chapter 10   Deformation Mechanisms of Functionalized Carbon Nanotube Reinforced Polymer Nanocomposites

S. C. Tjong

10.1. Introduction 341
10.2. Deformation characteristics 343
    10.2.1. CNT/glassy thermoplastic nanocomposites 345
    10.2.2. CNT/semicrystalline thermoplastic nanocomposites 356
    10.2.3. CNT/epoxy nanocomposites 362
    10.2.4. CNT/elastomer nanocomposites 369
10.3. Conclusions 371
    References 371

Chapter 11   Fracture Properties and Mechanisms of Polyamide/Clay Nanocomposites

A. Dasari, S.-H. Lim, Z.-Z. Yu, Y.-W. Mai

11.1. Introduction 377
11.2. Dispersion of clay in polymers 378
11.3. Crystallization behavior 384
11.4. Fracture properties and mechanisms 387
    11.4.1. Improved toughness in polymer/clay nanocomposites 387
    11.4.2. Brittleness of polymer/clay nanocomposites 393
    11.4.3. Approaches to improve fracture toughness
           of polymer/clay nanocomposites 399
11.5. Conclusions and future work 414
    Acknowledgements 415
    References 415

Content   xiii

Chapter 12   On the Toughness of "Nanomodified" Polymers and Their Traditional Polymer Composites

J. Karger-Kocsis

12.1. Introduction 425
12.2. Toughness assessment 427
12.3. Nanomodified thermoplastics 428
    12.3.1. Amorphous polymers 428
    12.3.2. Semicrystalline polymers 432
12.4. Nanomodified thermosets 444
    12.4.1. (Neat) Resins 444
    12.4.2. Toughened and hybrid resins 453
12.5. Nanomodified traditional composites 456
    12.5.1. Thermoplastic matrices 457
    12.5.2. Thermoset matrices 457
12.6. Outlook and future trends 460
    Acknowledgements 460
    References 461

Chapter 13   Micromechanics of Polymer Blends: Microhardness of Polymer Systems Containing a Soft Component and/or Phase

S. Fakirov

13.1. Introduction 471
13.2. The peculiarity of polymer systems containing
    a soft component and/or phase 472
13.3. Comparison between measured and computed microhardness
    values for various systems 477
    13.3.1. Two-component multiphase systems comprising
    soft phase(s) (blends of semicrystalline homopolymers) 477
    13.3.2. One-component multiphase systems
    containing soft phase(s) (polyblock copolymers) 478
    13.3.3. Two-component one-phase systems
    (miscible blends of amorphous polymers) 482
    13.3.4. Two-component two-phase amorphous systems
    containing a soft phase 484
    13.3.5. One-component two-phase systems
    (semicrystalline polymers with T_g below room temperature) 487
13.4. Main factors determining the microhardness of polymer systems
    containing a soft component and/or phase 489

13.4.1. Importance of the ratio hard/soft components (or phases) 489
13.4.2. Crystalline or amorphous solids 490
13.4.3. Copolymers vs. polymer blends 492
13.4.4. New data on the relationship between H and T_g
of amorphous polymers 493
13.4.5. Modified additivity law for systems
containing soft component and/or phase 495
13.5. Microhardness on the interphase boundaries in polymer blends
and composites and doubly injection molding processing 495
13.5.1. Microhardness on the interphase boundaries
in polymer blends 495
13.5.2. Microhardness on the interphase boundaries
in polymers after double injection molding processing 502
13.6. Conclusions and outlook 510
Acknowledgements 512
References 512

PART V   NANOCOMPOSITES: MODELING

Chapter 14   Some Monte Carlo Simulations on Nanoparticle Reinforcement of Elastomers

J. E. Mark, T. Z. Sen, A. Kloczkowski

14.1. Introduction 519
14.2. Description of simulations 520
14.2.1. Rotational isomeric state theory
for conformation-dependent properties 520
14.2.2. Distribution functions 520
14.2.3. Applications to unfilled elastomers 521
14.2.4. Applications to filled elastomers 522
14.3. Spherical particles 522
14.3.1. Particle sizes, shapes, concentrations, and arrangements 522
14.3.2. Distributions of chain end-to-end distances 523
14.3.3. Stress-strain isotherms 525
14.3.4. Effects of arbitrary changes in the distributions 526
14.3.5. Some preliminary results on physisorption 528
14.3.6. Relevance of cross linking in solution 530
14.3.7. Detailed descriptions of conformational changes
during chain extension 534

14.4. Ellipsoidal particles	534
14.4.1. General features	534
14.4.2. Oblate ellipsoids	536
14.5. Aggregated particles	537
14.5.1. Real systems	537
14.5.2. Types of aggregates for modeling	537
14.5.3. Deformabilities of aggregates	538
14.6. Potential refinements	538
14.7. Conclusions	538
Acknowledgments	538
References	539

Chapter 15 Modeling of Polymer Clay Nanocomposites for a Multiscale Approach

P. E. Spencer, J. Sweeney

15.1. Introduction	545
15.2. Sequential multiscale modeling	547
15.3. Representative volume element	548
15.3.1. Effective elastic material properties	549
15.3.2. Statistical ensemble	550
15.3.3. Periodic boundary conditions	551
15.4. Generating RVE geometry	553
15.4.1. Number of platelets	553
15.4.2. Generation of platelet configurations	554
15.5. Periodic finite element mesh	556
15.6. Numerical solution process	558
15.6.1. Finite element analysis of boundary value problem	558
15.6.2. Ensemble averaged elastic properties	560
15.6.3. Automation	561
15.7. Elastic RVE numerical results	562
15.7.1. Fully exfoliated straight platelets	565
15.7.2. Effect of platelet orientation	567
15.7.3. Curved platelets	569
15.7.4. Multi-layer stacks of intercalated platelets	572
15.8. Conclusions	574
Acknowledgements	576
References	576
Acknowledgements to previous publishers	579
Author Index	591
Subject Index	597

Contributors

Bhattacharyya, D., Prof. Dr.
Centre for Advanced Composite Materials (CACM),
Department of Mechanical Engineering,
The University of Auckland,
Private Bag 92019, Auckland, New Zealand
Fax: +64 9 3737479, e-mail: d.bhattacharyya@auckland.ac.nz

Bicerano, J., Dr.
Bicerano & Associates Consulting, Inc.
1208 Wildwood Street, Midland, Michigan 48642, USA
Fax: +1-(480)-247-4754; e-mail: bicerano@polymerexpert.biz

Bokobza, L., Prof. Dr.
ESPCI,
Laboratoire PPMD,
10 rue Vauquelin, 75231 Paris Cedex, France
Fax: +33 1 40 79 46 86; e-mail: liliane.bokobza@wanadoo.fr

Christenson, Ch. P., Dr.
The Dow Chemical Company,
Building B1608, Freeport, Texas 78155, USA
Fax: +1-(979)-238-0957; e-mail: cpchristenson@dow.com

Dasari, A., Dr.
Centre for Advanced Materials Technology (CAMT),
The University of Sydney,
School of Aerospace, Mechanical and Mechatronic Engineering (J07),
Sydney, NSW 2006, Australia
Fax: +61-2-9351-3760; e-mail: a.dasari@usyd.edu.au

Contributors

Fakirov, S., Prof. Dr.
    Centre for Advanced Composite Materials (CACM),
    Department of Mechanical Engineering,
    The University of Auckland,
    Private Bag 92019, Auckland, New Zealand
    Fax: +64 9 3677181; e-mail: s.fakirov@auckland.ac.nz

Galeski, A., Prof. Dr.
    Centre of Molecular and Macromolecular Studies,
    Polish Academy of Sciences,
    112 Sienkiewicza, 90363 Lodz, Poland
    Fax: +48 42 6847 126; e-mail: andgal@cbmm.lodz.pl

Ginzburg, V. V., Dr.
    Specialty Chemicals R&D,
    The Dow Chemical Company,
    Building 1710, Midland Michigan 48674, USA
    Fax: +1-(989)-638-7003; e-mail: vvginzburg@dow.com

Jancar, J., Prof. Dr.
    Institute of Materials Chemistry,
    Brno University of Technology,
    61200 Brno, Czech Republic,
    Fax: +420 541 141 697; e-mail: jancar@fch.vutbr.cz

Kalfus, J., Prof. Dr.
    Brno University of Technology,
    Institute of Materials Chemistry,
    School of Chemistry, Faculty of Chemistry,
    61200 Brno, Czech Republic,
    Fax: +420 541 141 697; e-mail: kalfus@fch.vutbr.cz

Karger-Kocsis, J., Prof. Dr.-Ing. Dr. h.c.
    Department of Polymer Engineering,
    Faculty of Mechanical Engineering,
    Budapest University of Technology and Economics,
    Müegyetem rkp. 3, H-1111 Budapest, Hungary
    Fax: +36-1-4631527, e-mail: karger@pt.bme.hu
    and
    Department of Polymer Technology,
    Faculty of Engineering and the Built Environment
    Tshwane University of Technology
    Pretoria 0001, Republic of South Africa

Kloczkowski, A., Prof. Dr.
    Iowa State University,
    Department of Biochemistry, Biophysics, and Molecular Biology,
    Ames, IA 50011, USA
    Fax: +1-(515)-294-3841; e-mail: kloczkow@iastate.edu

Ms Lim, S.-H.,
    Centre for Advanced Materials Technology (CAMT),
    The University of Sydney,
    School of Aerospace, Mechanical and Mechatronic Engineering (J07),
    Sydney, NSW 2006, Australia
    Fax: +61-2-9351-3760; e-mail: s.lim@usyd.edu.au

Mai, Y.-W., Prof. Dr.
    University Chair & Personal Chair in Mechanical Engineering,
    Centre for Advanced Materials Technology (CAMT),
    The University of Sydney,
    School of Aerospace, Mechanical and Mechatronic Engineering (J07),
    Sydney, NSW 2006, Australia
    Fax: +61-2-9351-3760; e-mail: y.mai@usyd.edu.au

Mark, J. E., Prof. Dr.
    The University of Cincinnati,
    Department of Chemistry and the Polymer Research Center,
    Martin Luther King Drive, Cincinnati, OH 45221-0172, USA
    Fax: +1-(513)-556-9239; e-mail: markje@email.uc.edu.

Patashinski, A. Z., Prof. Dr.
    Northwestern University,
    Department of Chemistry,
    2145 Sheridan Rd., Evanston, Illinois 60208, USA
    Fax: 1-(847)-491-9982; e-mail: a-patashinski@northwestern.edu

Pegoretti, A., Prof. Dr.
    University of Trento,
    Department of Materials Engineering and Industrial Technologies,
    via Mesiano 77, 38100 Trento, Italy
    Fax:+39 0461 881977; e-mail: alessandro.pegoretti @unitn.it

Regnier, G., Prof.Dr.
    Arts et Métiers ParisTech,
    Laboratoire d'Ingénierie des Matériaux-LIM,
    151 bld de l'Hopital, 75013 Paris, France
    Fax: +33 1 44 24 63 82; e-mail: Gilles.Regnier@paris.ensam.fr

Contributors

Rong, M. Zh., Prof. Dr.
    Zhongshan University,
    Materials Science Institute,
    Key Laboratory for Polymeric Composite and Functional Materials
    of Ministry of Education,
    Guangzhou 510275, P. R. China
    Fax: +86-20-84114008; e-mail: cesrmz@mail.sysu.edu.cn

Ruan, W. H., Assoc. Prof. Dr.
    Zhongshan University,
    Materials Science Institute,
    Key Laboratory for Polymeric Composite and Functional Materials
    of Ministry of Education,
    Guangzhou 510275, P. R. China
    Fax: +86-20-84114008 ; e-mail: cesrwh@mail.sysu.edu.cn

Schrock, A. K., Dr.
    The Dow Chemical Company,
    Building B1608, Freeport, Texas 78155, USA
    Fax: +1-(979)-238-2070; e-mail: akschrock@dow.com

Sen, T. Z., Dr.
    Iowa State University,
    Department of Biochemistry, Biophysics, and Molecular Biology,
    USDA-ARS Corn Insects and Crop Genetics Research Unit,
    Ames, IA 50011, USA
    Fax: +1-(515)-294-8280; e-mail: Taner@iastate.edu

Spencer, P. E., Dr.
    University of Bradford,
    IRC in Polymer Science & Technology,
    School of Engineering Design and Technology,
    Bradford, BD7 1DP, UK
    Fax: +44-1274-234525; e-mail: p.e.spencer@bradford.ac.uk

Stribeck, N., Prof. Dr.
    University of Hamburg,
    Department of Chemistry,
    Institute of Technical and Macromolecular Chemistry,
    Bundesstr. 45, 20146 Hamburg, Germany
    Fax: +49-40-42838-6008; e-mail: norbert@stribeck.de

Sweeney, J., Dr.
    University of Bradford,
    IRC in Polymer Science & Technology,
    School of Engineering Design and Technology,
    Bradford, BD7 1DP, UK
    Fax: +44-1274-234525; e-mail: J.Sweeney@Bradford.ac.uk

Tjong S. C., Prof. Dr.
    City University of Hong Kong,
    Department of Physics & Materials Science,
    Tat Chee Avenue, Kowloon, Hong Kong
    Fax: +852-2788 7830; e-mail: aptjong@cityu.edu.hk

Yu, Zh.-Zh., Prof. Dr.
    Centre for Advanced Materials Technology (CAMT),
    The University of Sydney,
    School of Aerospace, Mechanical and Mechatronic Engineering (J07),
    Sydney, NSW 2006, Australia
    Fax: +61-2-9351-3760; e-mail: z.yu@usyd.edu.au
    and
    Beijing University of Chemical Technology,
    College of Materials Science and Engineering,
    Beijing 100029, China
    Fax: +86-10-6442 8582; e-mail: yuzz@mail.buct.edu.cn

Zhang, M. Q., Prof. Dr.
    Zhongshan University,
    Materials Science Institute,
    Key Laboratory for Polymeric Composite and Functional Materials
    of Ministry of Education,
    Guangzhou 510275, P. R. China,
    Fax: +86-20-84036576; e-mail: ceszmq@mail.sysu.edu.cn

PART I
POLYMERS

Chapter 1

Nano- and Micromechanics of Crystalline Polymers

A. Galeski, G. Regnier

1.1. Introduction

It is currently thought that crystalline polymers consist of lamellar crystals which are separated from each other by a layer of amorphous polymer and are held together by tie molecules through the amorphous phase, see *e.g.* [1]. The lamellae are formed mostly from folded chains. The thickness of the lamellae is determined by parameters such as interfacial energies, glass transition temperature and melting temperature, undercooling, segmental diffusivity, *etc.* The thickness reported lies usually in a narrow range between 3 and 20 nm as obtained from observations in various types of microscopes or calculated from the degree of crystallinity and long period. It has been recognized that chain folding is not as regular as it was believed and molecular packing in lamellae is subject to considerable and irregularly distributed disorder, depending on undercooling regimes of crystallization. It has been demonstrated in various ways that the planar growth front will always break up into fibrous or cellular growth. Also, crystallization of polymers leads to interface instability. More sophisticated treatment of the instabilities involves perturbation analyses of planar interfaces, correlating diffusion, temperature gradients along the interface, and interfacial energy with the size of the growing crystals. All approaches show that all planar interfaces will develop and will yield fibrous or ribbonlike crystals [2–6]. When the crystals get larger, they become stable and originate the growth of spherulites. The arrangement of lamellae within the spherulite can be best described based on the example of polyamide 6 spherulites [7]. The lamellae in bulk polyamide 6 are very long, flat and narrow. One can easily distinguish bundles of lamellae having similar orientation of flat faces and forming separate domains. Three to four lamellae form a small domain having similar orientation. As the spherulite is growing, new lamellae are originated to fill the volume. New lamellae are added either to an existing domain or originate a new domain. The amorphous material is incorporated evenly between lamellae in the amount corresponding to the overall crystallinity. Misfit between domains arising from their different shapes are filled with amorphous material. Occasionally the lamellae may terminate; usually, however, once nucleated they continue to grow until impingement with neighboring spherulites. New lamellae are formed mostly by non-crystallographic branching, by "giant" crystallographic dislocations and/or by crystallographic branching as in the case of isotactic polypropylene. In a number of polymers the stacks of lamellae are twisted to form banded spherulites.

1.2. Tensile deformation of crystalline polymers

The mechanisms of tensile deformation of semicrystalline polymers was a subject of intensive studies in the past [8–20]. It is believed that initially tensile deformation includes straining of molecular chains in the interlamellar amorphous phase which is accompanied by lamellae separation, rotation of lamellar stacks and interlamellar shear. At the yield point, an intensive chain slip in crystals is observed leading to fragmentation but not always to disintegration of lamellae. Fragmentation of lamellae proceeds with deformation and the formation of fibrils is observed for large strains [21–24].

In many papers concerning tensile deformation it was noted that shortly after yielding some cavities are formed in a polymer [11,12,25–27]. Cavitation is not always observed in tensile tests, *i.e.*, the process depends on the material and testing conditions and is never observed in compression of the same material [9]. It was shown in the past that lamellae fragmentation and cavitation during tensile drawing are often detected simultaneously [21,28,29], however, it was never determined whether the lamellae fragmentation causes cavitation or *vice versa*. Since the cavitation appears around yield (at 4–12% of strain), while lamellae fragmentation in tensile drawing occurs at the strain of 30–40% [30], these two phenomena cannot be related directly. It seems more probable that the lamellae fragmentation is triggered by cavities when they reach a certain critical size. The lamellae fragmentation causes a release of 3D-stress, hence, limiting the possibility of further cavitation after fragmentation of lamellae.

1.3. Cavitation in tensile deformation

In the past the cavitation process was treated in the literature as a marginal effect, however, recent studies have shown that it may play an important role in the mechanism of plastic deformation [9,10,17].

Usually the cavitation is detected by small angle X-ray scattering (SAXS), seen as a rapid increase of scattering intensity. Macroscopically the intensive voiding is also observed as whitening of a deformed material.

Liu et al. [13] noticed that the whitening of polypropylene (PP) during tensile deformation was present in specimens drawn more quickly and at lower temperatures, i.e., when the yield stress was higher. Stress-whitening in tensile samples of polyolefins did not occur when the test was conducted under high pressure [14,15]. Yamaguchi and Nitta [16] investigated the intensity of light transmittance while deforming polypropylene. They concluded that the stress-whitening is observed just beyond the yield point, however, a decrease of transmittance begun at a strain of 0.07, i.e., before yielding at a strain of 0.15.

Zhang et al. [12] analyzed mechanisms of deformation for polypropylene differently irradiated, tested at selected temperatures. The authors concentrated mostly on polymer microdeformation but evidenced the presence of voids in most of the samples. They observed that non-irradiated polymer cavitates shortly after yielding, when the temperature is below 60°C. The SAXS scattering patterns indicate that the cavities change their shape during drawing from initially a peanut-shell-like to lemon-like. Zhang et al. [12] proposed two explanations of shape change: a combination of scattering from primary cavities and a scattering from the voids formed during fibrillation, or primary cavities are elongated to form the voids usually associated with fibrillation.

Some authors observed in drawn polypropylene the presence of craze-like features [31–35]. Crazing was identified a long time ago in many glassy polymers, such as polystyrene. Crazes have the shape of planar cracks perpendicular to the local strain direction [36–38]. The edges of a crack are bridged by highly oriented polymer fibrils, usually called tufts. They are parallel to the strain direction and are able to carry the load. The thickness of crazes is in the nanometer range, similar to the thickness of fibrils. Crazes and fibrils can be detected by small angle X-ray scattering due to the nanometer range of their sizes. The characteristic SAXS pattern consists of two components originating from voids and from fibrils, in the form of two perpendicular streaks [36]. One of those streaks, very intense originating from planar crazes, is normal to craze planes while the second, usually less intense, is perpendicular to craze fibrils [36–37].

Crazing is a massive phenomenon involved in plastic deformation of glassy polymers.

The features of craze-like structures observed in tensile drawing of PP at temperatures higher than the glass transition temperature are significantly different. First, their thickness is larger, in the range of 1–5 μm. Fibrils inside crazes, 200–650 nm thick [31], are elongated in the deformation direction, however, a significant fraction of thinner fibrils (20–50 nm) deviate from the principal strain direction [33]. Jang et al. [33] showed that at 23°C crazing is observed for strain rates higher than $0.11~\text{s}^{-2}$. Below this limit the deformation mechanism is the shear yielding [33]. Henning et al. [34] by microscopic observations identified positions of intense crazing in deformed spherulitic structures. The conclusion was that craze-like features are visible for α-polypropylene in polar regions of spherulites. Some large scale, trans-spherulitic crazes were observed by Narisawa et al. [35], passing through equatorial parts of spherulites.

Friedrich [31] and later Kausch [32] reviewed crazing processes in semicrystalline polymers. In their opinion they involve several stages: the initial stage is generally a macroscopically homogeneous deformation involving lamellar tilt and some breakup. At this stage, the strain is accommodated almost entirely by the interlamellar amorphous regions [31]. Disruption of lamellae into 10–30 nm size blocks, voiding (with a length of voids from 10 to 50 nm and their thickness from 2 to 6 nm) and formation of fibrils between voids from partly extended tie molecules and crystal blocks characterize the intermediate stage. The third stage, at higher draw ratios, leads to a perfect stretching of the fibrils. The fibrils, spanning the edges of a craze and possessing the strength of about 1 to 2 orders higher than the yield stress, are able to stabilize the microvoid volume. A lateral coalescence of these voids finally provides a local deformation zone in the shape of a craze. Friedrich limited this description to low temperature crazing. The Kausch description [32] differs from Friedrich's in that in the first stage the interlamellar separation of lamellae oriented perpendicularly to deformation direction is followed by cavitation and/or crystal plastic deformation and not lamellar fragmentation. Lamellar fragmentation occurs at the intermediate stage.

In an earlier paper by Galeski, Argon and Cohen [39] the following scenario of crystalline polymer deformation under uniaxial tension was outlined: the packets of lamellae in the 45° fans of spherulites experience resolved shear stress that promotes chain slip in the lamellae and shear in the interlamellar amorphous regions (see Figure 1.1).

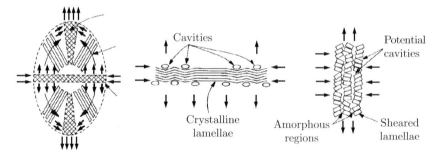

Figure 1.1. Tensile deformation of a spherulite. Lamellae kinking and formation of cavities in equatorial zones of a spherulite. Lamellae fragmentation occurs in polar fans of spherulites, potential cavities may be also expected [39]

Such deformation is accompanied by lattice rotation which generates tensile stresses across the faces in equatorial planes and compressive stresses across the faces in polar fans of spherulites. In addition to these, the overall elongation of spherulites evokes radial pressure on the equatorial planes and additional tensile stresses in the radial direction on the polar fans. Since these accentuated stresses in equatorial lamellae packets have no important shear components on either the interlamellar layers or on the planes of lamellae that promote chain slip, other, more damaging types of local plastic deformation are enforced.

As is well known [40], compression of such composite stacks lamellae and amorphous layers along stiff lamellae gives rise to unstable kinking of the lamellae. This produces periodic undulations in the lamellae ribbons. The accentuated tensile stresses acting across the equatorial disks of a spherulite expand amorphous material within lamellae kinks into pores. Similar and complementary processes are expected to occur in polar fans of spherulites, however, no action is expected from the amorphous phase, but instead, from the lamellae ribbons. In this case, the chain fold planes in the lamellae form unstable stacks in tension. Any inhomogeneous lamellae kinks would be filled with amorphous material under transverse pressure, producing no voids. Thus, the end result should be the array of aligned cavities in the equatorial disks of spherulites.

The above outlined picture of voiding in crystalline polymers above glass transition temperature (T_g) differs significantly from craze formation in glassy amorphous polymers:

1. Formation of aligned voids is triggered by compressive kinking instability along lamellae in equatorial disks of spherulites and not by meniscus instability as in the case of crazes in glassy amorphous polymers. Distances between voids are characteristic for kinking that is dependent on several factors, the most important being the lamellae thickness [41]. The wavelength of kinks in crystalline polymers is usually of the order of 200–500 nm. The voids are narrow and long as the width of kinked fragments of lamellae.

2. Growth of craze-like entities in equatorial disks of spherulites occurs *via* plastic deformation of the amorphous material between laterally aligned voids that is transformed further into fibrils. The thickness and length of these fibrils is related to the amount of the amorphous material between voids and that is connected in the first approximation with the lamellae thickness. These fibrils may mimic tufts in crazes of amorphous glassy polymers.

3. Voids in other parts of a spherulite arise at later stages of deformation and the mechanisms of their formation is different. Since there are no dilatational stresses at 45° fans of spherulites, the interlamellar slip and chain slips in lamellae prevail and no voiding is usually observed.

4. A disruption of lamellae in polar fans of spherulites is required, pores are formed at larger overall strain, when the amorphous material becomes unable to fill significantly increased gaps between lamellae fragments. At even further strain, the neighboring voids coalesce in planes perpendicular to the drawing direction while the amorphous material between these voids is transformed into fibrils. The thickness of those fibrils is related to the thickness of the amorphous layers and depends on the strain at which the voids coalesce and fibrils are formed. These craze-like entities resemble rather thick cracks because fibrils usually break at such high local strain.

The above concept of formation of craze-like objects in crystalline polymers explains the bimodal distribution of fibril thickness and is also supported by recently published atomic force microscopy (AFM) studies on initiation of voids in polybutene spherulites and their transformation into craze-like entities with increasing deformation [42]. The craze-like features in polypropylene are seen by electron microscopy, however, in contrast to crazing in glassy amorphous polymers, the fibrils in PP are usually not detectable in SAXS experiments probably, due to a small number of fibrils and their large thickness.

Some SEM observations (*e.g.*, Jang *et al.* [33]) show fibrils in drawn PP only after etching. However, it is not clear whether these are craze tufts or remnants of initial cross-hatched PP lamellae revealed by etching. The last case seems more reasonable because there are two populations of fibrils, the second one being nearly perpendicular to the first one just as in cross-hatched morphology.

In some other papers (*e.g.* [19]) the term "crazing" is misused for the description of microcracks.

Formation and growth of cavities are the main reasons for volume change during tensile drawing. The concept of volume strain, introduced by Bucknall [43], was recently applied in the studies of semicrystalline polymers [44–48]. The volume strain of cavitating polymers may significantly increase with increasing draw ratio, and reach 0.4 for polyethylene terephthalate (PET) [45] and more than 0.15 for poly(vinylidene fluoride) (PVDF) [46]. Recently, Billon *et al.* [49] observed a rapid increase of volume strain for PP, resulting from nucleation of voids in the core of injection molded polypropylene samples.

It was shown that for most crystalline polymers, including polypropylene and other polyolefins, the tensile drawing proceeds at a much lower stress than kinematically similar channel die compression [10,17]. Lower stress in tension was always associated with cavitation of the material. Usually a cavitating polymer is characterized by larger and more perfect lamellar crystals and cavities are formed in the amorphous phase before plastic yielding of crystals. If the lamellar crystals are thin and defected then the critical shear stress for crystal plastic deformation is resolved at a stress lower than the stress needed for cavitation. Then voiding is not activated. An example of such behavior is low density polyethylene [10].

1.4. Tensile deformation of polyethylene and polypropylene

Recently the cavitation during tensile deformation in high density polyethylene was studied in detail [50]. The samples were prepared by both injection and compression moldings in order to differentiate their internal structure and perfection of lamellae. It was shown that cavitation in high density polyethylene (HDPE) depends on lamellae thickness and their arrangement. For example, when the crystals were oriented perpendicularly to the drawing direction, as in the skin of injection molded specimens, the cavities were formed readily as early as at 1% of elongation (see Figure 1.2).

Cavitation in the core of the same injection molded HDPE sample was observed at yield (above 10% of elongation). The shape of cavities changes with in-

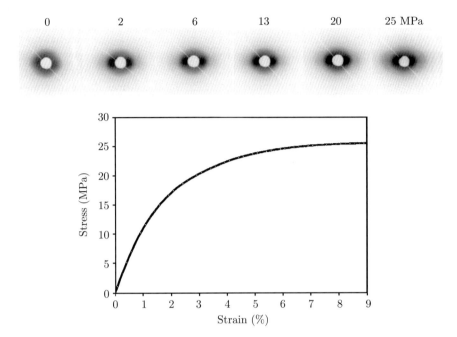

Figure 1.2. The evolution of SAXS scattering pattern with the increase of stress applied to the HDPE injection molded sample. Deformation direction was horizontal [50]

creasing deformation similarly as it was described by Zhang et al. [12]. It was also shown that the same polymer formed by rapid cooling of melt is able to deform without cavitation during tensile drawing. The stress-strain curves for compression molded HDPE cooled in air, in water and in iced water together with respective volume strain measurements are presented in Figures 1.3a and 1.3b.

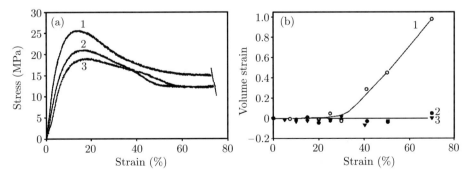

Figure 1.3. (a) Stress-strain curves for compression molded HDPE samples, deformed to 75% of engineering strain. Different cooling procedures were applied during preparation: 1 – sample cooled in air, 2 – sample cooled in water, 3 – sample cooled in water with ice [50]

Polypropylene is a cavitating polymer similar to polyethylene, but the process of cavitation proceeds differently. The role of molecular weight, crystal thickness and testing rate on the formation of cavities and their growth during tensile deformation were the subject of detailed studies [51]. The injected samples of higher molecular weight polypropylene (PPH) and lower molecular weight polypropylene (PPN) deform plastically with necking and significant increase of the volume is observed only after the yield. The increase of volume is accompanied by intensive whitening of samples. The reason for whitening and volume increase is cavitation initiated in the amorphous phase, which was confirmed by SAXS studies. The volume strain is larger for lower molecular weight PP polymer, in which the amorphous phase is less entangled. Apparently, a lower number of entanglements supports easier and more numerous cavitation. The cavities are formed during tensile drawing in the center of injected PP samples shortly before yielding. At small local strains, after the yield, a partial healing of small cavities is possible. The zone of cavitation propagates to the rest of the sample with increasing deformation. In the skin of injected samples, voids are not detected for local strains lower than 2.6 in contrast to the core injection molded bars of PP of lower and also higher molecular weight. That is because in the skin the crystalline structure is less developed which leads to a crystallographic non-cavitational mechanism of deformation. The cavitation process occurs when the crystals are perfect and strong enough, *i.e.*, when the stress for plastic deformation of crystals is higher than the stress needed for cavitation of the amorphous phase. At low local strain the cavities are elongated perpendicularly to the deformation direction. Typical radius of gyration for voids, as obtained from SAXS analysis, in PP of higher molecular weight is 18 nm and 12 nm for voids in PP of lower molecular weight. If we assume an ellipsoidal shape of voids and that their thickness is equal to the amorphous layer thickness (approximately 7 nm) then the lateral size of cavities is around 80 nm in PP of higher molecular weight and around 50 nm in PP of lower molecular weight. The cavities grow with increasing strain and also some new voids are formed.

The volume strain does not increase continuously with deformation. There is a plateau in volume strain for applied local strains between 1.0 and 4.0, where the change of internal crystalline structure – orientation and fragmentation of lamellae – leads to a change in shape of voids from elongated perpendicular to elongated in the deformation direction. A reorientation process sets in for local strains of 0.8–1.0.

The structure of compression molded samples is more uniform than the injection molded bars, so the cavitation occurs across the gauge of the sample. The presence of cavitation in those samples depends on the strain rate. If the deformation is slow, then the crystalline elements are able to deform plastically before reaching the cavitation threshold of the amorphous phase. When the strain rate is higher, the yield stress, related to crystal strength, is higher (38–38.5 MPa) and cavitation occurs first.

The lack of fibrils bridging the walls of cavities results from the mechanism of their formation within spherulitic structure in crystalline polymers above their T_g as explained earlier [39]. Drawing below T_g may produce craze-like features because inhomogeneous lamellae kinks would not be easily filled with glassy amorphous material under transverse pressure, producing voids and drawing crazy tufts in polar fans of spherulites.

The main reasons for cavitation are the misfit between mechanical compliances of heterogeneous elements and the surrounding and the resulting excessive negative pressure. Cavitation appears to be another mechanism of tough response of the material, *e.g.* [52,55]. Although the cavitation does not dissipate very large amount of energy, it enables the surrounding material to undergo further intensive crazing or shear. The possibility of cavitation under external force strongly depends on the character of applied deformation and is larger under tension than under compression. Cavitation in the macro scale is visible as whitening. The voiding was observed in many drawn polymeric materials such as polypropylene [54–57] and high density polyethylene [58], but not in some others such as ethylene-α-olefin copolymer [59,60]. It was established in the past that a low crystallinity level and a fine spherulitic structure are advantageous for the development of crazes in semicrystalline polymers [61]. Unlike in amorphous polymers, crazing in semicrystalline polymers is not restricted to temperatures below T_g but occurs also above the glass transition of the amorphous phase [62]. Below T_g long crazes propagate perpendicularly to the load direction ignoring internal polymer structure, while at a temperature close to T_g crazes are confined to several spherulites [31,62,63]. Radicals formed due to chain scission of macromolecules were observed in oriented polymers during tensile deformation [64,65].

The internal cavitation observed in tension experiments has been referred to as "micro necking" by Peterlin [66]. It was supposed that "micro necking" removes kinematical constraints between lamellae and allows them to untangle. However, from the model as imagined by Peterlin, it follows that the drawing of crystalline polymers inherently involves cavitation as an essential feature. While the above model is reasonable in tensile deformation, it is not correct for modes of deformation not producing cavitation, in which a positive normal stress component prevents the formation of cavities. "Micro necking" is inessential for the development of nearly perfect single crystal textures, as it was shown for several semicrystalline polymers that result from the deformation modes involving positive pressure components [67–72]. One of the possibilities of cavity free deformation is plane strain compression in a channel die [73]. Plane strain compression in a channel die is kinematically very similar to drawing: the sample is extended and its cross-section decreases accordingly. However, no neck is formed and the possibility of voids formation is limited due to a compressive component of stress. It is reasonable to expect that similar elementary plastic deformation mechanisms are initiated under similar true stress in tension and in plane strain compression. It means that any differences in true stress-true strain dependences of drawn and plane strained materials should be attributed to the formation and development of cavities. The number of observations of polymer deformation behavior under positive pressure at the level preventing cavitation which can be found in literature, *e.g.* [14], is limited and no comparison with tensile drawing was made. It is known that the stress in heterogeneous systems is not homogeneously distributed. At sites with a misfit of mechanical properties the stress concentrates; its value may increase locally many folds and trigger cavitation. In the case of crystalline polymers such sites are between differently oriented packets of lamellae and the stress concentration factor can reach 3 or more.

In the paper by Galeski, Argon and Cohen [39] the imprints of cavitation were revealed by osmium tetroxide staining in drawn polyamide 6. The cavitation occurred in the amorphous layers confined by the crystalline lamellae. In fact, the cavities left traces of damaged material having a size of the order of the interlamellar distance, *i.e.*, 4–5 nm. Since cavitational voids are having sizes on the nano-scale level, there is a problem of their stability. On each nanopore the surface tension is exerted from the very beginning of their formation, which tends to close a pore. In order to preserve a pore, an action of negative pressure is required at the level which is reciprocally proportional to the radius of a pore:

$$p = -2\tau_s/r \tag{1.1}$$

where τ_s is the surface tension and r is the size of a pore. It follows that the smallest pores are healed readily while larger pores can be preserved only if the negative pressure is maintained at a sufficiently high level. The lack of negative pressure or its inadequate level will lead to a spontaneous healing of cavitational pores; otherwise they can grow in the course of drawing.

In the papers by Muratoglu *et al.* [74,75] it was shown that when cavitational pores arise, they form initially spherical voids which later grow and become elongated. It is then logical to suppose that at the moment of cavity formation the stress concentrates around a spherical inhomogeneity and, as it was calculated earlier by Goodier [76] for spherical inclusions, is increased by a factor of 2. Also, the negative pressure generated is increased accordingly.

In Table 1.1 the data on surface tension, thickness of the amorphous layers, negative pressure needed to maintain the cavity open (according to Eq. (1.1) assuming the radius of newly formed cavity equal to half of the amorphous layer thickness), and the yield stress in drawing are presented. In the last column of Table 1.1 the negative pressure – generated at yield around newly formed cavities – is listed. These values of pressure result from the stress at yield ($-\sigma_{yield}/3$) increased by a factor of 2 arising from the stress concentration as predicted by Goodier [76].

It is clearly seen that the negative pressure at concentration sites, which is the negative pressure at yield ($-\sigma_{yield}/3$) multiplied by stress concentration factor 3, can cause cavitation in the case of polyamide 6 (PA 6), polymethylene oxide (POM), PP and HDPE because it is higher than the negative pressure for cavitation in polymer melts (HDPE from –3.5 to –10 MPa [80], PP from –13 to –19 MPa [81–83], POM from –10 to –18 MPa [81–83]) but not adequate for cavitation of LDPEs.

Table 1.1. Surface tensions [77–79] and negative pressures involved in cavitation during drawing [159]

Polymer	Surface tension [mJ/m^2]	Amorphous layer thickness* [nm]	Negative pressure** [MPa]	Tensile yield stress [MPa]	Negative pressure*** [MPa]
POM	44.6	4.69	−35.8	63	−42.0
PA6	38.4	5.59	−33.6	42	−28.0
PP	29.4	8.16	−13.7	31	−20.7
HDPE	35.7	9.45	−15.1	23	−15.3
LDPEs	35.3	7.50	−17.6	7	−4.7

* calculated on the basis of SAXS long period and DSC degree of crystallinity
** needed to maintain cavity open
*** generated at yield around newly formed cavities

In POM, PP and HDPE the yield stress generates sufficient negative pressure around cavities to maintain them and to boost their further growth. In fact, the cavitation in those polymers during drawing is evidenced by density decrease and strong small angle X-ray scattering. However, it is also evident that the negative pressure generated around newly formed cavities is not sufficient to keep them open in the case of PA 6; the cavities in PA 6 are unstable and will heal quickly. In fact, transmission electron microscopy (TEM) examination [39] of the same material does not expose any empty voids but only traces of chemically changed material between crystalline lamellae. It must be mentioned here that the growth of cavities proceeds due to the deformation of the surrounding matter, therefore, it is due to the action of dilatational stresses and not due to negative pressure.

1.5. Deformation micromechanisms in crystalline polymers

There are three, currently recognized, principal modes of deformation of the amorphous material in semicrystalline polymers: interlamellar slip, interlamellar separation and lamellae stack rotation [84,85]. Interlamellar slip involves shear of the lamellae parallel to each other with the amorphous phase undergoing shear. It is a relatively easy mechanism of deformation for the material above T_g. The elastic part of the deformation can be almost entirely attributed to the reversible interlamellar slip.

Interlamellar separation is induced by a component of tension or compression perpendicular to the lamellar surface. This type of deformation is difficult since a change in the lamellae separation should be accompanied by a transverse contraction and the deformation must involve a change in volume. Hard elastic fibers are found to deform in such a way. When the lamellae are arranged in the form of stacks embedded in the amorphous matrix, then the stacks are free to rotate under the stress. Any other deformation of the amorphous phase requires a change in the crystalline lamellae; the amorphous material is then carried along with the deforming crystalline material.

Many authors identified two distinct yield points in PE deformed in the tensile mode which are not seen in other deformation modes not producing cavitation. Gaucher-Miri and Seguela [86] tried to clear up the mechanism of these two processes on the micro-structural level, as a function of temperature and strain rate.

The major contribution to toughness comes from plastic deformation of the material which is manifested by ductile behavior. Plastic deformation itself is a complex phenomenon: it concerns the crystalline as well as the amorphous phases. The ductility is expressed by lowering of the stress-strain curve at a certain stress called the yield stress. Yield can be caused either by multiple crazes or by shear yielding. In the first case, the crazes have to be initiated in a relatively large volume of the material in order to contribute significantly to the overall deformation. Shear yielding is the plastic flow without crazing. Crazing is a unique phenomenon occurring in polymers below the glass transition temperature. Crazes themselves are highly localized zones of plastic dilatational deformation, see *e.g.* [87–90]. Edges of crazes are spanned by highly drawn elongated fibrils called tufts, usually having a length of a fraction of 1 mm depending on the molecular weight of a polymer, a diameter of several nanometers and are confined to a small volume of the material. The tufts can carry the load applied to the material and preserve its integrity. In brittle materials crazes are initiated at surfaces. Brittle fracture is usually caused by microcracks originating from breaking crazes initiated from the surface. Crazing occurs mostly in amorphous polymers, although it has been also observed in crystalline polymers in which crazes are propagated between lamellae through spherulite centers as well as through the material between spherulites [91,92]. Localized crazes initiate, propagate and break down to give microcracks at a stress below that necessary to stimulate shear yielding.

Shear yielding can be observed in a wide range of temperature, but only if the critical shear stress for yielding is lower than the stress required to initiate and propagate crazes.

Ductile deformation requires an adequate flexibility of polymer chain segments in order to ensure plastic flow on the molecular level. It has been long known that macromolecular chain mobility is a crucial factor decisive for either brittle or ductile behavior of a polymer [93–95]. An increase in the yield stress of a polymer with a decrease of the temperature is caused by the decrease of macromolecular chain mobility, and vice versa; the yield stress can serve as a qualitative measure of macromolecular chain mobility. It was shown that the temperature and strain rate dependencies of the yield stress are described in terms of relaxation processes, similarly as in linear viscoelasticity. Also, the kinetic elements taking part in yielding and in viscoelastic response of a polymer are similar: segments of chains, part of crystallites, fragments of amorphous phase. However, in crystalline polymers above their glass transition temperature the yield stress is determined by the yield stress required for crystal deformation and not by the amorphous phase. The behavior of crystals differs from that of the amorphous phase because the possibilities of motion of macromolecular chains within the crystals are subjected to severe constraints. Since the mobility of kinetic elements taking part in a plastic deformation is lower at a lower temperature, the energy dissipated increases and produces instability; at those places micronecks are formed because locally the temperature increases. The rate of plastic deformation increases drastically in micronecks and the material may quickly fracture. At higher temperatures the mobility of kinetic elements is higher, so, less energy is dissipated and the local temperature increase is lower. As a result, the neck is stable and tends to occupy the whole gauge length of the sample. The material therefore exhibits a tough behavior.

High plastic deformation can be achieved if the motion of internal kinetic elements is matched with that of the external deformation rate. The relaxation times and the activation energies are the parameters describing the kinetics of the conformation motions of macromolecules and larger elements taking part in the deformation. In crystalline polymers there are essentially three processes of particular importance: the first process which is connected with the presence of crystalline phase, the second corresponding to relaxations related to defects in the crystalline phase, and the third which corresponds to motions of short segments in the amorphous phase related to the glass transition [94].

Both massive crazing and shear yielding dissipate energy; however, shear yielding is often favored over crazing, especially under uniaxial stress, elevated temperature or slow deformation. Shear yielding dissipates energy more efficiently [94]. Switching between crazing and shear yielding is not obvious as it depends also on additional factors such as the shape of an article and the presence of notches or scratches. All other factors being equal, the material will deform according to a most ductile mechanism, which is well explained by the Ludwig-Davidenkov-Orovan hypothesis [96–98] illustrated in Figure 1.4.

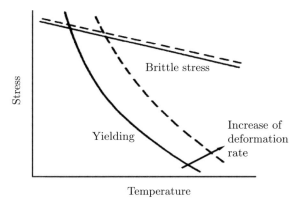

Figure 1.4. Ludwig-Davidenkov-Orovan plot explaining the temperature dependence of ductility and brittleness of a material [99]

The deformation of polymeric materials starts usually at scratches, notches or internal defects because these places are the zones of local stress concentration, sometimes much higher than the applied stress. Toughening of polymeric materials is based on the activation of such plastic deformation mechanisms which are activated at a stress lower than that required for triggering the action of surface and internal defects. Consequently, one of the important means of toughening appears to be a significant lowering of the yield stress of a material [100].

1.6. Molecular mechanisms at a nanometer scale

Taking into consideration the yield behavior of semicrystalline polymers, there are two conflicting approaches concerning the crystals. The first presumes that the process of deformation is composed of a simultaneous melting and recrystallization of polymer under adiabatic conditions [101,102]. The second, developed mainly by Young [103], uses the idea derived from the classical theory of crystal plasticity. The deformation of polymer crystals is considered in terms of dislocation motion within the crystalline lamellae, similarly to slip processes observed in metals, ceramics and low molecular crystals. Crystallographic slips are not processes occurring simultaneously over the whole crystallographic plane. Line and screw dislocations play a great role in the activation and propagation of a slip. Recently the issue of melting-recrystallization *vs.* crystal plasticity was raised again by Seguela [104]. His conclusion is that melting-recrystallization may occur at high strain rates, during lamellae fragmentation and at high strain when collective chain unfolding takes place. However, most of the published data concern the systems in which plastic deformation is accompanied by cavitation – the phenomenon neglected by most of the authors. Also, Seguela in his review [104] does not refer to the relation between lamellae fragmentation and cavitation. Recent papers by Strobl [30], Lv *et al.* [105] and Pawlak *et al.* [51] present attempts to correlate the two coexisting phenomena.

The plastic deformation of polymer crystals, like the plastic deformation of crystals of other materials, is generally expected to be crystallographic in nature and to take place without destroying the crystalline order. The only exception to this is a very large deformation, when cavitation and voiding lead to unraveling the folded chains and break down the crystals completely, new crystals may form with no specific crystallographic relationship with the original structure [106]. Polymer crystals can deform plastically by crystallographic slip, by twinning and by martensitic transformation. The slip mechanism is the most important one since it can produce larger plastic strains than the other two mechanisms. A slip system in a crystal is the combination of a slip direction and a slip plane containing that direction, as shown in Figure 1.5.

The notation for the slip system is (hkl)[$h_1k_1l_1$], where (hkl) is the slip plane while [$h_1k_1l_1$] is the slip direction. A single slip system is only capable of producing a simple shear deformation of a crystal. A general change of the shape of a crystal requires the existence of five independent slip systems [107]. Polymer crystals rarely possess this number of independent slip systems. However, under

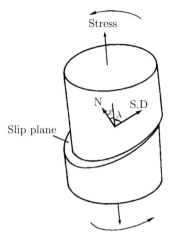

Figure 1.5. Definition of a slip system: slip plane and slip direction

the right conditions the deformation of bulk material can occur without voiding or cracking, perhaps because amorphous regions between lamellae allow for a certain amount of adjustment. In polymer crystals, the slip plane is restricted to planes which contain chain direction. That is because covalent bonds remain unbroken during deformation. In polymers, two types of slip can occur: chain slip, *i.e.*, slip along the chains, and transverse slip, *i.e.*, the slip perpendicular to the chains; both slips occurring in planes containing the chains. The general rule that applies to slip deformation is that the plane of the easiest slip tends to be a close-packed plane in the structure and the slip direction is a close-packed direction. Hence, in crystalline material, it is possible to predict certain mechanical properties associated with crystallographic slip directly from the crystallographic unit cell [108]. In folded chain polymer crystals, the folds at the surface of crystals may in addition impose some restraint on the choice of a slip plane; usually, a slip will be able to occur only parallel to the fold plane. An implication of the geometry of the slip process is that a crystal undergoing a single slip will rotate relative to the stress axis, as is seen in Figure 1.6.

For a single slip the slip direction in the crystal rotates always towards the direction of maximum extension: in uniaxial tension it rotates towards the tensile axis, while in uniaxial compression it rotates away from the compression axis. The angle through which the crystal rotates is a simple function of the applied strain [107]. It must be mentioned that the slip takes place when the resolved shear stress on the slip plane reaches a critical value known as critical resolved shear stress. Currently the critical resolved shear stresses for slips are well known for only few polymers.

They were measured using samples of rather well defined texture. The first measurements performed on polyethylene samples with fiber texture subjected to annealing at high pressure for increasing the crystal thickness yielded a critical shear stress of 11.2 MPa for undisclosed crystal thickness [84]. However,

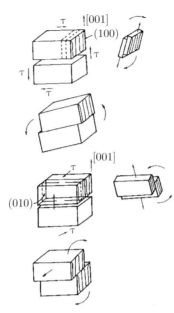

Figure 1.6. The rotation of crystal fragments due to slip: marked by arrows. The resolving of the shear on a plane due to simple tension or compression is also illustrated [67]

with fiber symmetry a combination of easiest slip systems could act simultaneously, disturbing the true value of the critical shear stress. The most exact data concerning critical resolved shear stresses for possible slip systems in polyethylene were obtained by Bartczak et al. [109]. They used single crystal textured polyethylene obtained by plane strain compression. The measurements in uniaxial tension, uniaxial compression and simple shear for samples cut out at various orientations delivered the following: for polyethylene orthorhombic crystals [109] the most active slip system is (100)[001] chain slip at 7.2 MPa, the second is (100)[010] transverse slip at 12.2 MPa, and the third is (010)[001] chain slip at 15.6 MPa. The forth slip system for polyethylene orthorhombic crystals was predicted as (110)[001] with the estimated critical resolved shear stress greater than 13.0 MPa [109]; however, it was never observed separately.

For polyamide 6 crystals the slip systems are: (001)[010] chain slip at 16.24 MPa, (100)[010] chain slip at 23.23 MPa and (001)[100] transverse slip [71]. Relatively little attention was paid to the plastic deformation of other semicrystalline polymers [98,110]. In particular, there are only a few papers [111,112] describing the investigations of the yield behavior and plastic resistance of oriented iPP.

For the determination of critical shear stress of one of the most important deformation mechanisms of iPP crystals, namely the crystallographic slip in the (100) planes along chain direction, i.e., (100)[001] chain slip [113,114], biaxially oriented film was used [115].

According to the theoretical predictions [113] and experimental studies [116], the easiest slip system in iPP crystals is (010)[001] slip, while (100)[001] and (110)[001] systems have higher critical resolved shear stress. The studies of the above mentioned slip system can be made by investigation of the yield behavior of the specimens deformed in tension with the tensile axis in those specimens oriented at various angles to the orientation direction. One can expect that for a certain range of this angle, only the (100)[001] slip system will be activated due to proper orientation of crystallites providing high shear stress on the (100) plane in [001] direction, while other deformation mechanisms will remain inactive due to much smaller resolved shear stresses in appropriate directions [109,116]. The analysis of the yield stress of such samples would give the value of the shear stress necessary to activate the (100)[001] slip [85,109,110, 117]. The critical resolved shear stress for (100)[001] slip for a crystals of polypropylene was determined at a level of 22.6 MPa. Similar studies for oriented iPP in tension and compression were performed by Bartczak et al. [111] but they used samples of uniaxially oriented iPP with a fiber symmetry, so that the critical shear stress they determined, 25 MPa, was an average over slips in a plane oriented around the fiber axis.

Crystal orientation by channel die compression of PET was studied by Bellare et al. [117]. They have found that the macromolecular chain orientation in PET along the flow direction and the texture development are the results of possible crystallographic slips having the following glide planes and directions: (100)[001] chain slip and (100)[010] transverse slip and (010)[001] chain slip. A probable sequence of activities of these slips is the following: the (100)[001] chain slip being the easiest, (100)[010] slip and a sluggish (010)[001] chain slip [118]. Values for critical resolved shear stress for those slips were not determined.

Stress induced martensitic transformation is a transformation from one crystallographic form to another form and associated with a displacement of chains to new positions in the new crystallographic cell in order to accommodate the deformation. An example of martensitic transformation from orthorhombic to monoclinic form was found in oriented polyethylene with well defined texture subjected to uniaxial compression. Martensitic transformation was also found in other polymers: in poly(L-lactic acid) [119] and in nylon 6 with the α-form transforming to the γ-form [120,121].

iPP highly rich in β crystal modification (92%) was deformed by plane-strain compression with constant true strain rate, at room temperature [122]. The results allowed to determine the deformation sequence and the active deformation mechanisms. The most important were interlamellar slip operating in the amorphous layers and resulting in localization of deformation in numerous fine deformation bands and the crystallographic slip systems, including the (110)[001] chain slip and (110)[1̄10] transverse slip. Deformation within deformation bands leads to $\beta\rightarrow$smectic and $\beta\rightarrow\alpha$ solid state phase transformations. At room temperature, the $\beta\rightarrow$smectic transformation appeared to be the primary transformation and $\beta\rightarrow\alpha$ transformation yields only about 4 wt% of a new phase at the same strain. With numerous fine deformation bands, β-lamellae are locally destroyed and fragmented into smaller crystals. Another deformation mechanism is a cooperative kinking of lamellae, leading to their reorientation and formation in a chevron-like lamellar arrangement. At high strains, above true strain equal to 1, an advanced crystallographic slip and high stretch of amorphous material due to interlamellar shear bring further heavy fragmentation of lamellar crystals partially fragmented earlier by deformation bands. This fragmentation is followed by fast rotation of small unconstrained crystallites with chain axis towards flow direction (FD). This process leads to the development of the final texture of the highly deformed β-iPP with molecular axis of both crystalline and smectic phases oriented along the direction of flow. At higher temperatures in the range of 55–100°C, the most important deformation mechanisms found were interlamellar slip operating in the amorphous layers, resulting again in numerous fine deformation bands and the crystallographic slip systems, including the $(110)[001]_\beta$ chain slip and $(110)[1\bar{1}0]_\beta$ transverse slip. Shear within deformation bands leads to $\beta \rightarrow \alpha$ solid state phase transformation in contrast to $\beta \rightarrow$smectic transformation observed at room temperature. Newly formed α crystallites deform with an advancing strain by crystallographic slip mechanism, primarily the $(010)[001]_\alpha$ chain slip. As a result of deformation and phase transformations within deformation bands, lamellae are locally destroyed and fragmented into smaller crystals. Deformation to high strains, above true strain equal to 1, brings further heavy fragmentation of lamellae, followed by fast rotation of crystallites with chain axis towards the direction of flow FD. This process, together with still active crystallographic slip, generates a morphological texture in which the molecular axis of both the α and β phases are oriented along the FD.

Another example of active martensitic transformation is γ modification of polypropylene. Morphology and deformation behavior of iPP homopolymer containing exclusively γ modification with only minor traces of crystals, obtained by isothermal crystallization at high pressure of 200 MPa, were investigated by Lezak et al. [123]. Deformation experiments performed in plane-strain and uniaxial compression demonstrated higher modulus, higher yield stress and flow stress, yet slightly lower ultimate strain of γ-iPP as compared to α-iPP. During plastic deformation, numerous fine shear bands, initiated by the interlamellar shear of the amorphous layers, start to develop already at the yield point. Their propagation across the sample causes a limited destruction of lamellae oriented perpendicularly to the direction of the band. Destroyed fragments of crystallites transform partially into smectic phase. No γ-α phase transformation was detected. With increasing strain the shear bands multiply and tilt towards the flow direction. Lamellae, already fragmented within shear bands, undergo kinking and rotation, resulting in the formation of a chevron-like lamellar morphology. Simultaneously, a relatively weak one-component crystalline texture is developed. This texture is described by the orientation of **c** crystallographic axis along constraint direction (CD), **b** axis 10–30° away from loading direction (LD) towards FD, and **a** axis 10–30° away from FD. Both crystalline texture and lamellae orientation are developed due to the activity of the same deformation mechanism – the interlamellar slip produced by the shear within interlamellar amorphous layers. Activity of any crystallographic deformation mechanism within the crystalline component, including the anticipated (001)[010] transverse slip, was not detected. The interlamellar amorphous shear appears to be the primary deformation mechanism of γ-iPP. The other identified mechanism, γ-smectic phase transformation, plays rather a minor, supplementary role in the deformation sequence. Plastic deformation behavior of iPP homopolymer crystallized exclusively in the γ-modification at higher temperature in the range of 55–100°C was also studied by Lezak et al. [124]. During plastic deformation, numerous fine shear bands, initiated by the interlamellar shear of the amorphous layers, start to develop already at the yield point. Their propagation across the sample causes a limited destruction of γ-lamellae oriented perpendicularly to the direction of the band. Destroyed fragments of crystallites partially reconstruct into either mesophase (smectic) domains or crystals of α phase, depending on the deformation temperature. Mesophase is produced upon deformation at room temperature, while at 55°C and above the crystalline phase is formed instead. With increasing strain, shear bands multiply and tilt toward the flow direction. Fragmented lamellae undergo kinking and rotation, which results in the formation of a chevron-like lamellar morphology. This leads also to a development of a weak crystalline texture. Both crystalline texture and lamellae orientation emerge due to the same deformation mechanism of interlamellar slip, produced by the shear within interlamellar amorphous layers. The activity of any crystallographic deformation mechanism within the crystalline component was not detected at any temperature. Interlamellar amorphous shear appears to be the primary deformation mechanism of γ-iPP. The other identified mechanisms, i.e., γ-smectic and γ-α transformations, play a supplementary role in the deformation sequence.

Twinning may occur in crystals of sufficiently low symmetry: cubic symmetry excludes twinning while orthogonal symmetry allows for twinning. Hexagonal crystal structure allows for twinning of low molecular weight materials, while in polymer crystals of hexagonal symmetry the basic twinning plane would be perpendicular to chains and therefore forbidden. Twinning along other planes in hexagonal crystals is not possible because of their high symmetry. In polyethylene of orthorhombic crystal symmetry the twinning is expected along (110) and (310) planes [125]. Only (110) plane twinning was found in bulk polyethylene. Twinning along the (310) plane is blocked in bulk polyethylene because the fold plane is the (110) plane. In contrast, in rolling in a channel at a high rate and to a high compression ratio, the texture of HDPE sample consists of two components [126]. One of them is the (100)[001] component, while the two others are rotated by ±53° around the rolling direction coinciding with the position of (310) poles clearly indicating the {310} twinning of the basic (100)[001] component. The twinning occurs on unloading, when the sample leaves the deformation zone between the rolls. The partial recovery of the strain produces a tensile stress along the direction of loading. Twinning is activated at high strain rates because the sample does not have sufficient time for stress relaxation, while at high compression ratio the material is highly oriented and contains no more folds in (110) planes. It was estimated earlier that the critical resolved shear stress in the twin plane to activate twinning is around 14 MPa [127]. Therefore, the tensile stress generated along LD on unloading must be at least 28 MPa. Such stress is apparently generated on unloading only when a high deformation rate is applied during deformation.

Besides polyethylene twinning was found only in few polymers including isotactic polypropylene: twinning along (110) plane [114].

Stress induced martensitic transformation and twinning alone are not responsible for large strain deformation.

From the presented review of mechanisms of plastic deformation of amorphous and crystalline phases it follows that the easiest one is the deformation of the amorphous phase since it requires very little stress; the crystallographic mechanisms of plastic deformation need larger stresses. Therefore, it is expected that first the amorphous phase is deformed and then crystallographic mechanisms are activated. An illustrative experimental evidence of this prediction was presented in ref. [128] where the Hermans orientation parameters for the amorphous and crystalline phases of a series of tensile deformed isotactic polypropylene samples are mapped out. The data points are based on the measurements of birefringence, infrared dichroism and wide angle X-ray diffraction of samples fixed in a frame at constant length. At first the amorphous phase becomes oriented, while the crystalline phase remains very slightly oriented or even oriented in the transverse direction (the reason for such behavior of polypropylene is the presence of cross-hatched lamellae). When the draw ratio is further increased, the amorphous phase becomes almost entirely stretched out and then the crystalline phase begins to orient. Finally, at high draw ratio, the macromolecular chain fragments embedded in both phases become highly oriented. Other reports based on a relaxed material after deformation indicate the contrary: amorphous phase orientation is lower than the orientation of crystalline phase, see *e.g.* [129].

It may be concluded that most of the plastic deformation of both crystalline and amorphous phases occurs due to shear stresses and the shear contributes greatly to plastic deformation. It is then obvious that a great amount of plastic deformation is usually found at an acute angle with respect to applied tensile or compressive forces.

There are several ways of achieving plastic deformation of macroscopic samples. Those deformation methods in which some amount of hydrostatic pressure is generated in the material prevent cavitation of the material. Because of the absence of cavitation the material undergoes plastic deformation *via* shear yielding; cavitation is damped. Also, crazing is not preferred under hydrostatic pressure as it involves volume increase and the production of empty spaces between tufts.

In a macroscopic sample of a semicrystalline material subjected to stress, a few or all of the presented mechanisms of plastic deformation are orchestrated. Some of them are preferred, because of low shear stress required, and show up in early stages of deformation; others are activated in later stages under higher stress. The intensity of a particular mechanism may also change, for example, if the possibility of a certain slip is already exhausted or the other slip mechanism rotated the crystals in such a way that the process mentioned is no longer possible. As it was shown, the mechanisms or their intensity may change when changing the temperature, pressure or the deformation rate. Together with the complicated aggregated supermolecular structure the process of deformation in semicrystalline polymers is complicated and not easy to track.

1.7. Dislocations in crystal plasticity

In 1949, Frank [130,131] pointed out the possibility that the growth of crystals could take place because of the formation of dislocations in the crystal so that any real crystal should have a number of dislocations with a screw component, terminating on the face. When growth takes place on these exposed molecular terraces, the edges of these layers develop into spirals centered on the dislocation. The theory of crystal growth based on the dislocation theory as formulated by Frank [131] predicts the presence of growth features in the form of flat, spirally terraced hills on the crystal face which is perpendicular to a screw dislocation line. The height of a terrace should be just one unit of a crystallographic cell. In fact, this was confirmed by multiple-beam interferometry of the growth face of a beryl crystal. Crystallization of any substance is always easier if screw dislocations are engaged, due to a dihedral angle benefit. The observed density of dislocations varies widely with different specimens and different low molecular weight materials, ranging from a few to $\sim 10^4/\text{cm}^2$. It was found that on any crystal they are predominantly of one hand. Crystallization of polymers is also prompted by screw dislocations. Screw dislocation growth mechanism in polymers was reported long ago [132,133] in polymer single crystals. In polymer crystals, the estimate of the number of existing dislocations is a few orders of magnitude larger, 10^5–$10^8/\text{cm}^2$, than in low molecular substances [134]. In terms of the introduction of screw dislocations, Schultz and Kinloch [135,136] presented a model calculation of the twisting correlation of crystallites in a banded spherulite (a spherulite showing periodic extinction bands under polarizing optical microscope) with a row of screw dislocations of the same handedness. Based on their argument, Toda *et al.* proposed the possibility of the twisting correlation due to the introduction of the selective screw dislocations in the chair type PE crystals for the formation of banded spherulites of poly(vinylidene fluoride) [137]. Recently, Ikehara *et al.* [138] showed by 3D electron tomography, apparently searching for the so-called "giant" screw dislocations, that no such screw dislocations are engaged in the formation of banded spherulites of PCL/PVB blends. Nevertheless, the existence and significant role of screw dislocations in melt crystallized polymers is well established, see *e.g.* [139–141].

Screw dislocations observed in single crystals of polymers may have an extraordinarily large Burgers vector, up to 10 nm or more, being equal to the total thickness of lamellar crystals [134]. Such dislocations are easily detectable by microscopes, yet they do not have hollow cores as would be required of such large dislocations in most materials. In these lamellae, the axes of giant screw dislocations are parallel to molecular stems. Such dislocations involve a readjustment in chain folding on a large scale in order to avoid formation of holes. As a consequence, they are immobile. During crystallization they appear sporadically at growth fronts, especially at chain re-entering. The giant screw dislocations play a vital role in lamellar branching and in morphological development of radiating polycrystalline aggregates. Spatial constraints arising during the formation of giant dislocations stimulate adjacent lamellae to develop similar dislocations – the phenomenon often observed in stacks of lamellae. The role of giant dislocations in crystallization by chain folding was already recognized many years ago (see, for example, a review in Geil's book [132], also a discussion on dislocations in polymer crystals [142]).

The amount of mobile dislocations that is present in polymer crystals is sufficient to initiate plastic deformation of polymer crystals. However, during crystallographic slips many more new screw dislocations are generated at crystal edges and propagate through the crystal. This process of emission of dislocations from the edges of the lamellae across the narrow faces that was initially proposed by Peterson [143,144], explored further by Shadrake and Guiu [145], and more rigorously by Young [146], has now been widely accepted [140,141,147–149].

The model of thermal nucleation of screw dislocations by Peterson [143, 144] and Young [150,151] has been shown to account fairly well for the plastic behavior of PE [152,153] and PP [154] and for the yield stress dependency on crystal thickness. Elastic line energy calculations indicate that nucleation of screw dislocations is more favorable than that of edge dislocations [155,156]. Glide is also easier for the former [157]. It has been shown that screw dislocations parallel to the chain stems may be nucleated from the lateral surface of thin polymer crystal platelets upon coupled thermal and stress activation [143,144,158].

The activation volume of the dislocation emission was determined from strain rate jump experiments during uniaxial compression of polyethylene [149]. The mean level of activation volumes was of the order of 350 of crystallographic unit cells and the estimated typical radius of the dislocation was 3.8 nm, and much smaller indeed than the usual lamella thickness of 20 nm [159]. The Burgers vector of such dislocation must be small, of the order of a few lattice units only. Apparently, the dislocations of such characteristics are the easiest to generate. It is then logical to assume similar values for the activity area of screw dislocations arising during melt crystallization.

The detailed knowledge concerns the system of crystallographic slips operating in polymer crystals, critical resolved shear stresses, twinning, martensitic transformation and succession of activation of individual slip mechanisms [111,160,161].

The magnitude of the shear stress needed to move a dislocation along the plane was first determined by Peierls [162] and Nabarro [163]. For the orthorhombic unit cell of a crystal it varies exponentially with the ratio of both unit cell axes perpendicular to the macromolecular chain direction. For a (100) plane being the closed-packed plane (**a** axis larger than **b** axis) the shear stress reaches the minimum. That is because for closed-packed planes the interplanar bonds are weaker, which results in lower activation energy and shear stress. The result is that the dislocations tend to move in the closest packed planes and in the closest packed direction because the Peierls-Nabarro force is smaller for dislocations with short Burgers vectors.

1.8. Generation of dislocations

As this mechanism of dislocation propagation is widely accepted for slip mechanisms in semicrystalline polymers, there are doubts about how the dislocations are generated. The density of dislocation in crystals is estimated at the level from 10^5 to $10^8/\text{cm}^2$ [164]. This number is not enough to give rise to a fine slip most often observed during polymer plastic deformation. There must be a way in which new dislocations are generated within crystals. In metals and other large crystals the identified source of dislocations is a multiplication mechanism known as Frank-Read source [165]. The Frank-Read mechanism involves a dislocation line locked on both ends. When the shear stress is applied above a certain critical value, the dislocation line will move to form first a semicircle and then the line spiral around the two locked ends, finally forming the dislocation ring which will continue to grow outward under the applied stress. At the same time, the original dislocation line has been regenerated and is now free to repeat the whole process. In this way, a series of dislocation rings is generated indefinitely. In polymer crystals, this elegant process is not active because the crystals are usually too thin for the dislocation line to spiral up and to form a dislocation loop.

Due to the peculiar structure of polymer crystals (long molecules which lie across the crystal thickness, fold and re-enter or take part in forming adjacent crystals), there are some restrictions which reduce the number of possible slip systems. Theoretical calculations demonstrated [166] and experiments confirmed [110] that polyethylene lamellae deform easily by slip in the direction of the c-axis. There is a problem of the origin of the dislocations for activation of the slip mechanism. Xu and Argon [166] pointed out that in the case of PE the energy necessary for the creation of a screw dislocation with the Burgers vector parallel to chain direction can be supplied by thermal fluctuations. It was shown that the change in the Gibbs free energy, ΔG, (i.e., the energy which must be supplied by thermal fluctuations) associated with the creation of such dislocation under applied shear stress, τ, is equal to:

$$\Delta G = \frac{Kb^2 l}{2\pi} \ln\left(\frac{r}{r_0}\right) - \tau b l r, \qquad (1.2)$$

where l is the stem length; b is the value of Burgers vector; K is the shear modulus of a crystal; r is the radius of dislocations (the distance from dislocation line to the edge of lamellae); and r_0 is the core radius of dislocations.

There are still doubts if the model can be applied over the whole range of temperatures, i.e., from the temperature of glass transition to the onset of the melting process. The dispute concerns the upper temperature of validity of this approach. Crist [167] suggested that the temperature of γ and α relaxation processes of PE are the limits of applicability of the model. Young [103] and Brooks *et al.* [168] have used that approach to model the yield behavior of bulk crystallized annealed polyethylene at a much higher temperature. Other authors [169] reported the existence of a transition in the range from –60°C to 20°C, depending on the material and strain rate, above which the Young's model cannot be applied. They pointed out that there is a relationship between the transition temperature and β-relaxation and suggested that below the transition temperature the yield process is nucleation controlled, while above it is propagation controlled. However, Galeski *et al.* [161] have shown that in plane strain compression for HDPE at 80°C the beginning of yielding is mainly associated with (100) chain slip within crystalline lamellae. Moreover, it was shown that for linear polyethylene only fine slip occurs below the deformation ratio of 3 associated with chain tilt and thinning of lamellae. Only at higher deformation the widespread fragmentation of those lamellae thinned to one third takes place. This is because further thinning becomes unstable – much like layered heterogeneous liquids respond by capillary waves and breakup of stacks of layers.

Seguela *et al.* [170] proposed that the driving force for the nucleation and propagation of screw dislocations across the crystal width relies on chain twist defects that migrate along the chain stems and allow a step-by-step translation of the stems through the crystal thickness. The motion of such thermally activated defects is responsible for a crystalline relaxation.

When the problems mentioned above were investigated by others, the crystal thickness of PE was controlled by changing the cooling rate, temperature of crystallization, using copolymers of PE characterized by different degrees of branching or crystallizing PE either from solution or from melt. The applied procedures allowed to obtain orthorhombic PE crystals with thicknesses over the range from 3 nm to 35 nm. Crystallization of PE under elevated pressure [171] makes it possible to obtain much thicker crystals. The method exploits the pseudo-hexagonal mobile phase of PE at a certain range of pressure and temperature. The details of how to obtain the samples with crystals of various thicknesses due to crystallization under high pressure are described in [171]. The great advantage of the approach is that the series of samples with various crystal thicknesses is obtained from the same polymer. Most of the previously reported studies of the structural changes caused by deformation were performed in a tensile mode, guided by obvious technological stimuli to explain processes associated with orientation by drawing. However, from the fundamental point of view, the deformation by compression is more important. Uniaxial compression has a great advantage because the deformation is nearly a homogenous process and occurs without any significant deformation instabilities such as necking and cavitation. In the reported study [172], the samples prepared by high pressure crystallization were characterized by crystal thickness covering the range from 20 up to 150 nm. Figure 1.7 presents typical true stress-true strain curves obtained in uniaxial compression for HDPE with

Nano- and Micromechanics of Crystalline Polymers 27

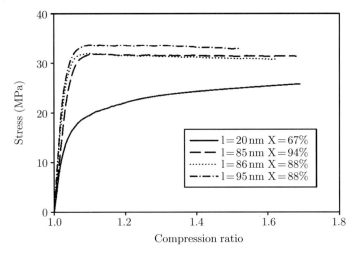

Figure 1.7. Stress *vs.* compression ratio for a series of HDPE samples differing in lamellae thickness and crystallinity. Uniaxial compression rate 0.000055 s^{-1} [149]

various crystal thicknesses and crystallinity uniaxially compressed at room temperature.

After the usual initial elastic response below a compression ratio of 1.05–1.07, there is only a single yield and a region of intense plastic flow which sets in at a compression ratio of 1.12, followed by strain hardening.

The changes in yield stress with increasing crystal thickness are reported in Figure 1.8.

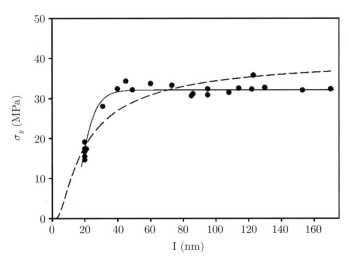

Figure 1.8. Yield stress in uniaxial compression plotted in relation to average lamellae thickness. Compression rate 0.000055s^{-1} [149]. The dashed line is the best fit of Eq. (3) from Xu and Argon [166]

The yield stress increases with crystal thickness up to 40 nm. This part of the data agrees well with the published results by Brooks *et al.* [173], covering the range of lamellae thickness from 9.1 to 28.3 nm. Beyond the region explored in the past, the yield stress still increases in the range of up to 40 nm and then, above 40 nm, the dependence on crystal thickness abruptly saturates at the level of 29.5, 35 and 37 MPa for initial compression rates of 0.000055, 0.0011 and 0.0055 s^{-1}, respectively. The increase of the rate of compression increases the yield stress in the region of crystal thickness below 40 nm and also above 40 nm. However, the saturation of the yield stress occurs at 40 nm independently of the compression rate.

These observations imply that above 40 nm the crystals thickness is no longer the decisive factor for the yield and that some other mechanism takes over the control of the yielding process. Any "coarse slip" and other inhomogeneities in the course of compression which could decrease the yield stress did not occur. Instead, as evidenced in scanning electron microscope (SEM) examination, chain tilt and homogeneous lamellae thinning occur up to a compression ratio above 2 (see Figure 1.9).

It can be concluded that the observed yielding process is a "fine slip" process.

The overwhelming evidence accumulated over several decades is that the crystalline lamellae of PE, much like all crystalline ductile metals, deform plastically by the generation and motion of crystal dislocations and that, while twinning and martensitic shears have also been suspected, these do not make

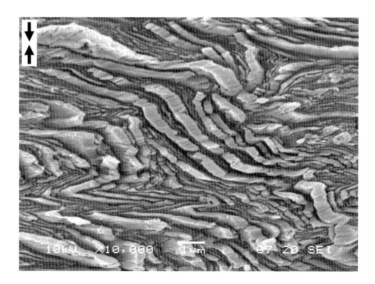

Figure 1.9. SEM micrograph of polyethylene uniaxially compressed to the compression ratio of 3.175 with the rate of 5.5×10^{-5} s^{-1}. Polyethylene sample for compression test was prepared by crystallization under high pressure of 630 MPa and temperature 235°C. Initial average crystal thickness was 156 nm and crystallinity 87% [149]

a substantial contribution. In coarse grained polycrystalline metals the rate mechanism of deformation is governed almost exclusively by either the intrinsic lattice resistance or by the resistance of localized obstacles to dislocation motion.

Dislocation nucleation in crystal plasticity as a rate controlling process is found only in nearly perfect crystals or in polycrystals in the nanoscale range. In PE and similar semicrystalline polymers with chain molecules aligned nearly normal to the wide surfaces of the crystalline lamellae the preferred systems are those of chain slip on planes with the largest interplane separation: (100) in PE; (001) in nylon-6. In such systems, because of the very high stiffness of the chain molecules, it can be expected that the width of the core of the edge dislocations would be very large, resulting in only a negligible lattice resistance, making the screw dislocations with much narrower cores the most likely carriers of plasticity associated with the rate mechanism. This was confirmed indirectly by Peterman and Gleiter [174] who made the first TEM observations of screw dislocations in PE lamellae. The possibility that glide mobility of screw dislocations might be rate controlling was studied by Staniek *et al.* [72] in highly textured nylon-6 with quasi-single crystals, where a generally convincing association was made to chain "cross-overs" in the (001) planes as being the thermally penetrable obstacles to screw dislocation motion. Such chain "cross-overs" which are a topologically unavoidable product of large strain compression flow that was used to produce the quasi-single crystalline samples should, however, be largely absent in melt crystallized lamellae. Considering the fact that the lamellae thickness is in the 10–20 nm range, it was suspected early on that this is the result of a continuously repeating process, whereby dislocations are emitted and nucleated by edges and sides of the lamellae, respectively. This process of emission of dislocations from the edges of the lamellae across the narrow faces was initially proposed by Peterman and Gleiter [175], explored further by Brooks *et al.* [168] and more rigorously by Young [146,151, 176]. The screw dislocation model for yield in polyethylene has now been widely accepted. It was re-examined in 2000 by Brooks and Mukhtar [173]. In the experiments of Kazmierczak *et al.* [149], where PE with lamella thicknesses far exceeding the usual 10–20 nm (up to 170 nm) were studied, it became clear that, when lamella thicknesses exceed 40 nm, the mechanism of emission of screw dislocation lines from lamella edges must be superseded by another, more ubiquitous one that is no longer sensitive to lamella thickness (compare Figure 1.10).

In examining new possibilities for the description of dislocation emission in PE crystals, the recently developed fundamental considerations by Xu *et al.* [177,178], Xu and Zhang [179] and Xu [180] were revisited.

The large strain texturing deformation experiments and computer simulation of the entire process of texture-producing flow (Lee *et al.* [181], van Dommelen *et al.* [182,183]) show that the contribution of the (100)[001] system to the total plastic deformation of PE is dominant and that its rate mechanism also dominates over all others. In the consideration of the rate mechanism of

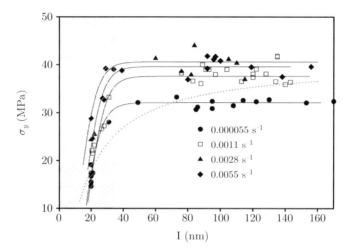

Figure 1.10. Yield stress of HDPE as a function of crystal thickness and compression rate [149]

plastic flow in individual lamellae the attention is limited to deformations resulting from shear on the (100) [001] system as depicted in Figure 1.6. The principal local plastic deformation process is γ_{13} shear, promoted by the resolved t_{13} shear stresses, referred to the PE orthorhombic crystal structure, based on the well established slip on the (100)[001] crystallographic system [181–183].

Figure 1.11 depicts three modes of dislocation nucleation-controlled processes of chain slip on the (100)[001] system. The process identified as A is the previously considered one of nucleation of a fully formed screw dislocation monolithically from the narrow face. Clearly, as the lamella thickness l increases, the ever increasing activation energy of this mode will no longer be kinetically possible. Thus, two other modes were considered: a screw dislocation half loop nucleation, still from the narrow face, and an alternative process of edge dislocation half loop nucleation from the wide face, depicted as processes B and C, respectively, in Figure 1.11.

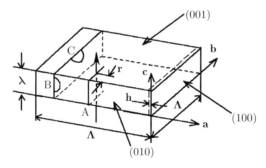

Figure 1.11. Sketch depicting the geometry of a typical lamella showing the principal chain slip system (100)[001] and the three separate modes of dislocation nucleation: *A* monolithic screw, *B* screw loop and *C* edge loop [159]

The energetics of all three processes have been considered rigorously by Xu and Zhang [179] and Xu [180]. In the paper by Argon et al. [159] the previous developments of screw dislocation emission monolithically from edges of thin lamellae were reconsidered and various questionable assumptions were eliminated. It was shown that when this is done, the previously proposed model of screw dislocation emission from lamella edges can provide results as accurate as the more rigorous developments of Xu, based on a variational boundary integral approach. In this case, the Gibbs free energy change, ΔG, due to insertion of a fully formed screw dislocation from the edge, is:

$$\Delta G_A^* = \frac{\mu b^3}{4\pi}\left(\frac{\lambda}{b}\right)\ln\left(\frac{1}{\beta}\right); \quad \beta = \tau/\tau_c \tag{1.3}$$

where μ is the elastic modulus of the crystal, b is the Burgers vector of the lattice unit length, λ is the lamellae thickness as defined in Figure 1.4, τ_c is the ideal shear strength of the (100)[001] chain slip system, and τ is the acting shear stress. This formula is essentially similar to the formula derived by Xu and Argon [166], although the fundamental parameters are expressed here in terms of shear strength of the slip system in exchange to two parameters: the radius of dislocations and their core radius. Both show a linear dependence of ΔG on lamellae thickness.

The two other alternatives, B and C, where the lamella thickness plays no role, were also considered by Argon et al. [159] leading to the following expressions:

Case B:
$$\Delta G^* = \mu b^3 g_B(\beta), \quad g_B(\beta) = (1-\beta^{2/3})/\beta^{1.25} \tag{1.4}$$

Case C:
$$\Delta G_C^* = (\mu b^3/(1-\nu))g_c(\beta), \quad g_c(\beta) = (1-\beta^{1/3})/\beta^{1.15} \tag{1.5}$$

A series of considerations including a correct value of μ modulus and its dependence on temperature, the kinetic crystallographic shear strain rate expression encountered in crystal plasticity, and a Coulomb law where the shear resistance τ is dependent on the normal stress σ_n acting across the glide plane, lead to detailed expressions for yield stresses for all three cases of monolithic, half loop screw and half loop edge dislocations engaged in plastic deformation. The comparison of compressive yield stress of all three mechanisms with experimental data of Kazmierczak et al. [149] is illustrated in Figure 1.12.

The data points for thin lamellae are close to the prediction of the model of monolithic screw dislocations. The plot suggests a sharp departure from the mode of screw dislocation line nucleation to the half loop modes at roughly around 16 nm lamella thickness. The experiments show a more gradual transition at a lamella thickness of roughly 28 nm. For lamellae thicker than 28 nm the data points fall in between the models for nucleation of edge half loops and screw half loops.

The results of experiments by Kazmierczak et al. [149] with PE containing lamellae of much larger thicknesses than those of the usual 10–15 nm have

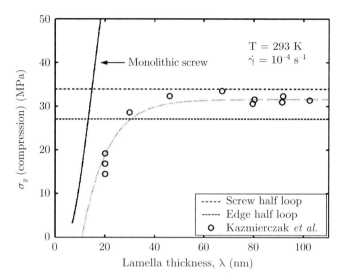

Figure 1.12. Dependence of compressive flow stress of polyethylene at 293 K and $\gamma^{-1} = 10^4$ s^{-1} on lamella thickness: compared with three theoretical models [149,159]

forced a re-examination of not only the model of Young, most recently considered again by Brooks and Mukhtar [173], but also other alternatives that must take over for thicker lamellae. These two new alternatives that are no longer dependent on lamellae thickness have explained well the leveling of the plastic resistance for lamellae thicker than roughly 20 nm. Also, the new alternatives are in far better agreement with the published temperature dependences of the plastic resistance of PE than the previous model by Young (see Figure 1.8) and also predict well the measured levels of activation volume.

The calculated temperature dependent tensile resistances are plotted in Figure 1.13 for the mode of monolithic screw dislocation emission from lamella edges and those of edge and screw half loop emissions from the large and small faces of the lamellae, respectively. The experimental data points of Brooks and Mukhtar [173] lie along a line that parallels the half loop nucleation modes and not the model of the usually considered mode of nucleation of monolithic screw dislocations. However, their positions are considerably below the models of the half loop nucleation. The discrepancy is partly due to the relatively low level of crystallinity of ∼0.538. If the data were converted to account for a crystallinity of 0.9, the experimental points would move up to the dotted curve lying much closer to the model of nucleation of edge dislocation half loops. We note that from the predictions of Figure 1.13, for thinner lamellae the flow stress might be governed by the mechanism of monolithic emissions of screw dislocations; the mechanism can switch over to that of half loop nucleation at lower temperatures.

The kinetic mechanism governing dislocation controlled plasticity can be tested in strain rate jump experiments, where during monotonic deformation,

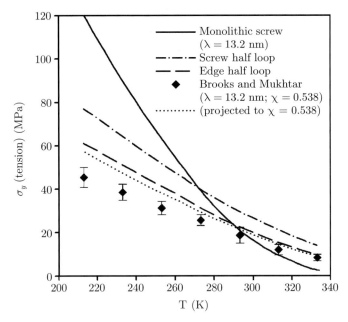

Figure 1.13. Predictions of modes A, B and C models of dislocation nucleations for the temperature dependence of the tensile yield stress of polyethylene compared with experimental results of Brooks and Mukhtar [173] for their PE2 material with moderate level of crystallinity at X = 0.538. Dotted line projects experimental results for X = 0.9 [159]

the strain rate $d\gamma_1/dt$ is abruptly increased or decreased to $d\gamma_2/dt$ and the resulting stress jump $\Delta\sigma$ is recorded. The parameter:

$$\Delta\nu^* = (kT/\Delta\tau)\ln[(d\gamma_1/dt)/(d\gamma_2/dt)] \tag{1.6}$$

is related to the activation volume $\Delta\nu^*$ that characterizes the dependence of the activation free energy on shear stress. In Eq. 1.6, $\Delta\tau$ must be the shear stress change on the glide system. This is related by a factor of 3 to $\Delta\sigma$, which is recorded in a uniaxial experiment. The measurement of $\Delta\nu^*$ allows for classification of material response to strain rate change, and in this case serves to identify the rate controlling mechanisms. In uniaxial stretching the activation volume must be related to the active shear mechanism [72]. The change in the shear resistance of spherulitic PE ($\Delta\tau$) is one third of the change in the uniaxial flow stress ($\Delta\sigma$).

The resulting response to a positive strain rate jump during monotonic deformation is a small yield phenomenon that is developed when a sample with thick crystals are deformed and only very small yield phenomenon for thin crystals. The activation volume increases with the increase of crystal thickness and then stabilizes for thicker crystals, i.e., the changes of activation volume follow the same trend as the yield stress with the increase of crystal thickness which saturates for crystals thicker than 40 nm.

1.9. Competition between crystal plasticity and cavitation

Crystalline polymers are often characterized by two parameters: long period based mostly on SAXS measurements and the crystallinity degree based on WAXS, differential scanning calorimetry (DSC), nuclear magnetic resonanse (NMR) and other combined techniques. There is a large spectrum of long period values from 8.0 nm for PA 6 to 25–30 nm for HDPE. Also, the crystallinities vary substantially: from 5–10% for olefin copolymers, through 20–30% for low density polyethylenes, 35% for PA 6 to 60–70% for HDPE and POM. Comparison of these data with mechanical properties leads to the conclusion that polymers with higher crystallinity and thicker long period, *e.g.*, POM, PP, HDPE are those for which large differences in tensile drawing and plane strain compression responses are visible. The exception from this scheme is PA 6, which is not very crystalline, having low long period, but characterized by high melting temperature. It suggests that its crystals are rather strong, exhibiting high resistance to shear. In fact, polyamide 6 exhibits high critical shear stress for crystallographic slips [72].

The differences in tension/channel die compression tests are the largest in those polymers in which the crystals are more rigid and resistant to plastic deformation, such as POM, PA 6, PP and HDPE. The level of stress necessary to achieve plastic deformation of crystals in the second group of materials is much smaller than in the first group. Negative pressure generated at concentration sites under such stress is lower than the negative pressure needed to cause cavitation (below –20 MPa). In compression, the yield stress reflects the plastic resistance of crystals. The immediate conclusion can be deduced that cavitation during drawing can be observed only in polymers in which the value of stress at yield in compression (the factual yield stress for crystal plasticity) generates a pressure higher than the negative pressure required for cavitation. Otherwise, a crystallographic slip will occur earlier, relaxing the stress, and cavitational pores will not appear. This necessary condition for cavitation can be written in the form:

$$\sigma_{yield}/3 = -p > -p_{cav} \tag{1.7}$$

where σ_{yield} is the yield stress in compression corrected for stress sensitivity factor K as followed from the Coulomb criterion, p is the negative pressure generated in channel die compression at yield, and p_{cav} is the negative pressure required for cavitation. Otherwise, a crystallographic slip will occur earlier, relaxing the stress, and cavitational pores will not appear. This necessary condition for the activation of crystallographic slips can be written in the form:

$$\sigma > 2\tau_0/(1-K) \tag{1.8}$$

where τ_0 is the critical resolved shear stress for the easiest crystallographic slip.

From the above condition it follows that cavitation during drawing can be expected in such polymers as nylons (the easiest slip for PA 6 is 16.24 MPa [72], polypropylene ((010)[001] slip at around 22–25 MPa [184,185], poly(methylene oxide) [186]. Only if thick crystals are present, cavitation can be present in polyethylene, since only thick PE crystals exhibit high enough yield stress in compression [173,187,188]. No cavitation is expected during deformation of low density polyethylene and quenched high density polyethylene, both usually having thin lamellar crystals, but cavitation can be found in HDPE which is slowly cooled when forming thick wall products.

It follows that in drawing there is a competition between cavitation and activation of crystal plasticity: easier phenomena occur first, cavitation in polymers with crystals of higher plastic resistance, and plastic deformation of crystals in polymers with crystals of lower plastic resistance.

1.10. Micromechanics modeling in semicrystalline polymers

1.10.1. *Microstructure and mechanical properties*

At a nanoscale, semicrystalline polymers can be seen as lamellar crystals, which are embedded by amorphous layers. This morphology imparts locally a high mechanical anisotropy, especially when the amorphous phase is in the rubbery state, like in polyolefins.

When crystallizing from the melt without any deformation and any strong thermal gradient, spherulitic textures are formed by the radial growth of lamellae from a nucleation point, with intervening spaces filled by subsidiary lamellae, the polymer chains being more or less perpendicular to the basal plane of lamellae. These spherulites, whose diameter is on the order of magnitude of about 1 to 10 μm, have in fact quite a complex microstructure [189]: they look like a wheat sheaf close to the nucleation point (Figure 1.14). It is generally claimed

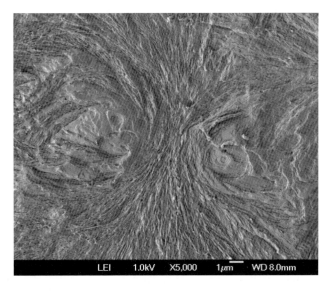

Figure 1.14. Wheat sheaf form of a selective chemical etched PP β-spherulite in an injection molded part

that the spatial distribution of lamellae inside a spherulite is nearly isotropic; the random assembly of spherulites at a macroscopic scale is surely isotropic.

From a mesoscopic point of view, when a semicrystalline polymer is submitted to a large deformation, the spherulitic structure is gradually broken down, while increasing molecular orientation. In a tensile deformation, the spherulitic could be reformed ultimately in a new oriented microstructure: the oriented fibrillar texture [190]. Such cold processed polymer fibers may have a Young's modulus, which could nearly reach 150 GPa for polyaramid fibers (Kevlar®). The strong link between microstructure and mechanical properties can be easily predicted.

When a semicrystalline polymer crystallizes from the melt under a deformation, various crystalline superstructures appear: from deformed or oriented spherulites to very oriented structures such as fibrils or shish-kebabs, which are considered as alternating amorphous and crystalline stacks growing around a fibril. The cooling of a melt is able to create these very oriented crystalline structures in polymer fibers for uniaxial elongation or PET bottles for biaxial elongation, for example. But shearing can also generate very oriented crystalline microstructures [191]. The molecular orientation in the melt, which is being cooled, generates oriented crystallization and therefore, the global orientation in the solid polymer is generally higher than the one that existed in the sheared melt [192]. The crystalline molecular orientation seems to be governed by the strain rate in the melt and the cooling rate, which limits the chain relaxation [193]. The size of the induced crystalline entities depends on the nucleation density, which is directly linked to the melt orientation before crystallization [194]. Therefore, crystalline entities can vary from shish-kebabs of less than 100 nm in the solidified layer during the filling of an injection molded part to isotropic spherulites of several microns in the core of the part.

Unlike amorphous polymers, high variation and anisotropy of mechanical properties can be obtained in polymer processing. In injection molding, for example, these variations of mechanical properties (Figure 1.15) induce anisotropic shrinkage and warpage. The prediction of shrinkage and warpage is necessary to reduce tool developing time and cost in industrial applications. However a thermomechanical calculation is needed for which thermoelastic properties at least are necessary. How do the mechanical properties depend on crystallinity and crystalline orientation [195], amorphous phase orientation or even crystalline macrostructure? Can semicrystalline polymers be considered as composites at a nanoscale and as heterogeneous material at a mesoscopic scale? Could micromechanical modeling give an answer to these questions?

1.10.2. *The micromechanical models*

Considering a heterogeneous microstructure with known component properties, micromechanical modeling is aimed at establishing the behavior law of the equivalent homogeneous medium. Theoretical micromechanical modeling was subject of intensive studies resulting in composite materials and it has been applied with success to heterogeneous materials, such as metal alloys [196,197].

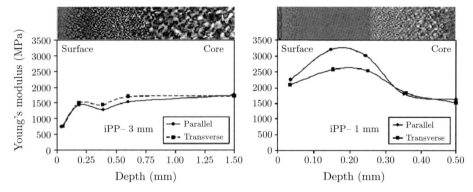

Figure 1.15. Young's moduli vs. depth of 1 mm- and 3 mm-thick injection molded PP plates measured in parallel and transverse directions to the flow: injection time = 1.6 s, mold temperature = 40°C. Superimposition of microscopic view under polarized light of the associated crystalline microstructure

Three main types of heterogeneous materials have been distinguished and treated:

a. Fillers surrounded by a matrix; this first family includes polymers reinforced by mineral or organic fillers such as calcium carbonate particules [199] or glass fibers [200], carbon fibers, and now nanofillers [201]. The fillers are modeled by ellipsoids surrounded by a matrix and the two aspect ratios are the key parameters for such a representation.

b. Laminates. Stacks of unidirectional carbon fiber/epoxy composite sheet used in the aeronautical industry, for example, are part of this family [202].

c. Aggregates of particles, which are generally one phase media, but because of random or partially random orientation of particles or crystals, discontinuities exist across interfaces. The metallic polycrystalline materials are typically members of this family [203].

When only elastic behavior is considered, the constitutive equations for the equivalent homogeneous medium may be written in terms of:

$$\Sigma = \mathbf{C} : \mathbf{E} \quad (1.9)$$

where Σ and $\mathbf{E}$ denote the macro stress and macro strain tensor, respectively, and $\mathbf{C}$ is the stiffness tensor of the semicrystalline polymer. Introducing the representative volume element V of the local properties, which contains a large number of micro-elements but has small dimensions in regard to the structure dimensions, Σ and $\mathbf{E}$ may be given in terms of the volume average stress and strain tensors:

$$\Sigma = \langle \sigma \rangle_V = \frac{1}{V} \int_V \sigma(\mathbf{x}) dV \quad \mathbf{E} = \langle \varepsilon \rangle_V = \frac{1}{V} \int_V \varepsilon(\mathbf{x}) dV \quad (1.10)$$

σ and ε being the microscopic stress and microscopic strain tensors, respectively. Due to Hill [204], strains (resp. stresses) in the heterogeneities are related to the macroscopic strains (resp. stresses) by the concentration-strain (resp. stress) tensor **A**:

$$\varepsilon^c(x) = \mathbf{A} : \mathbf{E} \qquad \sigma^c(x) = \mathbf{B} : \mathbf{\Sigma} \qquad (1.11)$$

The rigidity tensor **C** or the compliance tensor **S** of the homogenized material are simply defined by the relation:

$$\mathbf{C} = \mathbf{C}^{am} + f^c(\mathbf{C}^c - \mathbf{C}^{am})\mathbf{A} \qquad \mathbf{S} = \mathbf{S}^m + f^c(\mathbf{S}^c - \mathbf{S}^m)\mathbf{B} \qquad (1.12)$$

where f^c is the volume fraction of the crystalline phase. The assessment of the concentration-strain tensor **A** (or **B**) depends on the micromechanical model. The simplest models are the Voigt (**A** = **I**, **I** being the identity tensor) and Reuss (**B** = **I**) bounds, which correspond to a homogeneous deformation and homogeneous stress in the material, respectively. Because of the high contrast between amorphous and crystalline phase moduli, both assumptions will give a result very far from the experiments. In another model due to Mori and Tanaka [205], the inclusions experience the influence of other inclusions as equal to the average field in the matrix. This theory only applies to volume fractions up to 30%. Crystalline phase volume fractions encountered in semicrystalline materials are out of the model validity range. The differential scheme is better designed when the volume fraction of the reinforcing phase is high [197]. Then, the reinforcing phase is introduced by infinitesimally small increments using the dilute-distribution assumption of Eshelby [206]. In the self-consistent model, the material surrounding the inclusion is the equivalent homogeneous medium. This non-explicit model was intensively used for all types of materials, but it is better suited for polycrystals or aggregates of inclusions than inclusions embedded in a matrix [207].

1.10.3. *Idealizing the microstructure of semicrystalline polymers*

Semicrystalline materials are often considered as composite materials at a nanoscale and then the use of micromechanical models to predict the macroscopic properties is quite natural. But semicrystalline materials cannot be put so easily in one of the three classical previous modeled families defined for composites and polycrystalline materials because the crystalline microstructure is complex and organized at several scales. Therefore, the first question is: At which scale do we have to consider "this composite material"? At a microscale, that is the scale of the crystalline microstructure (spherulites, shish-kebabs)? Or at the nanoscale of the crystalline lamellae?

The microstructure is composed of spherulites when considering isotropic semicrystalline materials. Most of the works have found that the spherulite size has no significant effect or little effect on mechanical properties [208]. Therefore, the modeling at the nanoscale can be relevant in this case. It could also be true for very oriented polymers such as fibers because the induced crystalline microstructure results mostly in microfibrils with sizes typically in the range of 50–500 nm, even if they are composed of some form of ordered crystalline and amorphous domain [209]. Few researchers applied micromechanics modeling to predict the macroscopic mechanical properties of semicrystalline polymers, and nearly all the research works have only considered the nanoscale. Four general types of microstructure modeling have been performed.

The first attempts in the 1960s and 70s proposed to predict elastic properties by using simple analytical models derived from the analysis of fiber-filled media. Takayanagi [210] described the elastic behavior of unidirectional oriented processed fibers. Halpin and Kardos [211] proposed to use the Halpin-Tsai model [212]. Phillips and Patel [213] also applied this model to PE with an adjustable parameter linked to the crystallite shape ratio. However, these models require the assumption that lamellae be regarded as fibers. Moreover, they are known to fit only the experimental data at low volume fractions of filler and this is not the case for semicrystalline materials, for which the crystallinity can often reach 60 to 70%.

Janzen [214] was one of the first to use the self-consistent model for semicrystalline polymers, although in its most simple form with spherical inclusions and isotropic media. The interaction between a lamella and the others is supposed to be given by the interaction of an inclusion with "an equivalent homogeneous medium" in which it is embedded and whose properties are those of the macroscopic material. A large collection of data on Young's modulus of unfilled and isotropic PE at ambient conditions has been compared with various competing theoretical mixing rules developed for composite micromechanics. It was found that the self-consistent scheme appears to have valid application to this system. The self-consistent scheme with more elaborated constitutive laws for components was used by Parks and Ahzi [215] and Dahoun *et al.* [216] to predict the plastic behavior and the deformation texture of PEEK and iPP [217]. In all models, the behavior of the amorphous phase is accounted for the "softening" of the related polymers on the macroscopic level.

Lee *et al.* [218,219] defined a dedicated polymer micromechanical model in which the material is represented by an aggregate of layered two-phase composite inclusions (Figure 1.16). Large plastic deformation and texture evolution

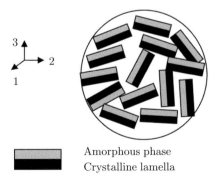

Figure 1.16. Aggregate of layered two-phase composite inclusions [222]

in initially isotropic HDPE was simulated. This aggregate scheme was used later by Nikolov *et al.* [220] to model the viscoelastic behavior of semicrystalline polymers at small deformations and by van Dommelen *et al.* [183] to model the elasto-viscoplastic behavior of semicrystalline polymers under large deformations.

More recently, Bedoui *et al.* [221,222] showed that a classical inclusion/matrix model (Figure 1.17) associated with a differential scheme provides satisfactory results for predicting the elastic behavior of isotropic semicrystalline polymers.

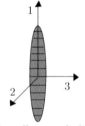

1. Lamella growth direction
2. Lamella width direction
3. Chain direction

Figure 1.17. Schematic illustration of crystal lamellae randomly distributed in the amorphous matrix. Direction 3 goes along the chain direction and direction 1 is the lamella growth direction [221]

1.10.4. *Elastic behavior prediction*

For many applications, however, the small-strain behavior is at least as important as the large deformations. As the prediction of elastic behavior of heterogeneous materials is the most known and well-established, it is worth to know if the current micromechanical models are able to predict the behavior of semicrystalline polymers. Firstly, the elastic properties of the composite components should be determined.

1.10.4.1. *Elastic properties of constitutive phases*

Crystalline phase. The crystalline phase consists of polymer lamellae, which show a highly anisotropic behavior with a very high modulus along the chain axis. Elastic constants have been theoretically calculated and experimentally determined for several polymers and are reviewed by Ward [190].

Polymer chains, which have a helical conformation angle like PP [223] are less stiff in the chain direction. The theoretical rigidity tensor of PP α-monoclinic crystal calculated by Tashiro *et al.* [224] gives a consistent value along the chain direction with experimental determinations of about 40 GPa:

$$\mathbf{C}_{PP}^{c} = \begin{pmatrix} 7.78 & 3.91 & 3.72 & 0 & 0.9 & 0 \\ 3.91 & 11.55 & 3.99 & 0 & -0.36 & 0 \\ 3.72 & 3.99 & 42.44 & 0 & -0.57 & 0 \\ 0 & 0 & 0 & 4.02 & 0 & -0.12 \\ 0.9 & -0.36 & -0.57 & 0 & 3.1 & 0 \\ 0 & 0 & 0 & -0.12 & 0 & 2.99 \end{pmatrix} GPa \qquad (1.13)$$

For orthorhombic forms of PE or α-forms of PA, the theoretical values of the stiffness tensor seem to largely overestimate the measured ones [225]. The polyethylene chain has a planar-zigzag conformation, which justifies a theoretical modulus as high as 300 MPa along the chain direction. However, dynamical calculations have demonstrated the high dependence of the modulus on the chain contraction. When the chain contracts slightly from the planar-zigzag conformation, the modulus along the chain axis drops drastically. Tashiro et al. [223] suggest that, similar to the α-form nylon-6, the PE polymer chain experiences a thermal motion at room temperature and contracts from the planar-zigzag conformation, which induces a drop in modulus. Therefore, the moderate elastic stiffness tensor at ambient temperature could is proposed by Choy [225]:

$$\mathbf{C}_{PE}^{c} = \begin{pmatrix} 7.0 & 3.8 & 4.7 & 0 & 0 & 0 \\ 3.8 & 7.0 & 3.8 & 0 & 0 & 0 \\ 4.7 & 3.8 & 81 & 0 & 0 & 0 \\ 0 & 0 & 0 & 1.6 & 0 & 0 \\ 0 & 0 & 0 & 0 & 1.6 & 0 \\ 0 & 0 & 0 & 0 & 0 & 1.6 \end{pmatrix} GPa \qquad (1.14)$$

PET lamella stiffness was experimentally measured by Hine and Ward [226]; a theoretical estimate was provided by Rutledge [227]. For most components, experimental values of the elastic stiffness tensor are comparable to the theoretical ones. However, a discrepancy is also observed in the chain direction, which may be the result of imperfect crystallites due to chain axes misorientations. A better experimental estimate of this component has been given by Matsuo and Sawatari [228], and the PET crystal stiffness tensor is given by:

$$\mathbf{C}_{PET}^{c} = \begin{pmatrix} 7.70 & 5.46 & 5.07 & 0 & 0 & 0 \\ 5.46 & 7.70 & 5.07 & 0 & 0 & 0 \\ 5.07 & 5.07 & 118. & 0 & 0 & 0 \\ 0 & 0 & 0 & 1.62 & 0 & 0 \\ 0 & 0 & 0 & 0 & 1.62 & 0 \\ 0 & 0 & 0 & 0 & 0 & 1.12 \end{pmatrix} GPa \qquad (1.15)$$

Amorphous phase. In semicrystalline polymers, the amorphous phase is confined between crystalline lamellae and therefore the chain mobility should be reduced

[229]. The confinement can be considered as a phenomenon which reduces the molar mass between entanglements, M_e. This is associated with an increase of the modulus.

When the temperature is above glass transition, some authors [183,214,220] have considered the possibility that the amorphous phase confinement does not significantly affect the rigidity of the amorphous phase and have considered it in a rubbery state. Other authors [230–233] chose a Young's modulus one or two orders of magnitude higher than the rubbery state. Their choice was generally not based on physics modeling or experimental evidence but on micromechanics modeling; the amorphous phase parameters are calculated by an inverse method and highly depend on the model used. Recently, Spitalsky and Bleda [234] have simulated the elastic properties of interlamellar tie molecules in semicrystalline polyethylene by atomistic modeling. In their work, it appears that the Young's modulus of PE tie chains only change drastically for highly extended chains. Considering the lack of experimental and theoretical evidence of a drastic change of Young's modulus of the amorphous phase due to the crystalline environment, assuming the amorphous phase in a rubbery state when the temperature is above glass transition may be a reasonable approach. Then, chain entanglements are the cause of rubber-elastic properties and the kinetic theory of rubber elasticity, which was particularly developed by Flory [235], leads to the following equation:

$$G_N^0 = \frac{\rho RT}{M_e} \qquad (1.16)$$

For example, M_e can be chosen equal to 7 kg/mol for PP and to 1.4 kg/mol for PE [236], then the shear modulus, G_N^0, at ambient temperature is equal to 1.5 MPa for PE and 0.3 MPa for PP.

Since the amorphous phase of the studied polymers is in the rubbery state, its Poisson ratio is very close to 0.5, although slightly lower. To get coherent elastic properties, the Poisson ratio ν is calculated using the classical relation:

$$B = \frac{E}{3(1-2\nu)} \qquad (1.17)$$

where the bulk modulus B can be determined from PVT data thanks to its definition relation:

$$\frac{1}{B} = \frac{1}{V}\frac{\partial V}{\partial p} \qquad (1.18)$$

V with the specific volume [221].

When temperature is lower than the amorphous glass transition, the physical phenomena are quite different and interactions between atoms occur at a much lower scale. Therefore, for small deformations the hypothesis of an elastic behavior of the confined amorphous phase close to the bulk phase may be considered as a valid hypothesis.

1.10.4.2. *Some predictions of elasticity*

Some very sophisticated models have already been performed to simulate the mechanical behavior of semicrystalline polymers at large deformations, but could these models simulate at their limit the simple elastic behavior of isotropic semicrystalline polymers? Bedoui *et al.* [221,222] have considered fully anisotropic crystallites embedded in a matrix and a layered-composite aggregate; several homogenization schemes have been used in those representations and the amorphous phase modulus varied. Finally, only one model gives satisfactory trends and results for both modulus (Figure 1.18) and Poisson ratio (Figure 1.19): the crystalline lamellae embedded in a matrix associated with the differential scheme. The elongation ratio of crystalline lamellae was the only adjustable parameter; it was consistent with microscopic observations. However, the modeling is sensitive to the amorphous modulus and it is difficult to give a physical sense to the differential scheme.

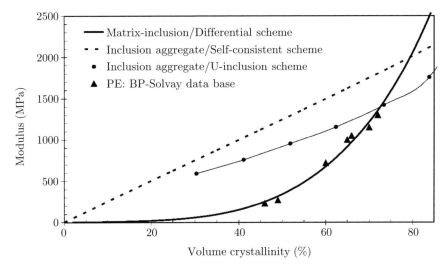

Figure 1.18. Prediction of macroscopic PE Young's modulus *vs.* volume crystallinity. Ellipsoidal shape ratio $a_1/a_3 = 20$ and $a_2/a_3 = 5$ for matrix-inclusion scheme [221]

Considering PET at ambient temperature which shows a low contrast in terms of elastic behavior between amorphous and crystalline phases, several models have proven to provide similar results (Figure 1.20). Here, micromechanics models are only slightly sensitive to the crystallite shape ratio parameters. It explains why models with no crystallite shape ratio, such as the U-inclusion model [183], still apply. Therefore, the obtained results in the case of glassy amorphous phases are contrasted. In terms of absolute values, the Young's modulus is decently approached by the models but in terms of the shape of the curve, the convexity of the experimental data is poorly represented.

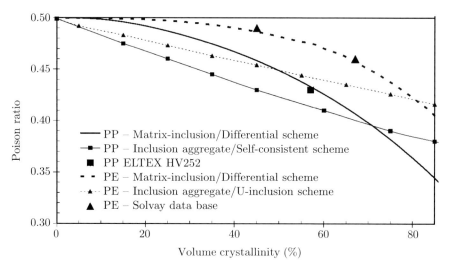

Figure 1.19. Estimates of PE and PP Poisson ratios *vs.* volume crystallinity. Ellipsoidal shape ratios for matrix-inclusion model associated with the differential scheme: $a_1/a_3 = 20$ and $a_2/a_3 = 5$ for PE, $a_1/a_3 = 100$ and $a_2/a_3 = 10$ for PP [221]

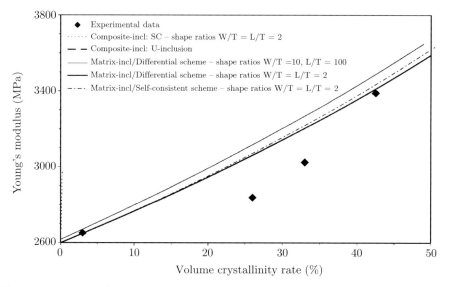

Figure 1.20. Predicted and experimental PET Young's modulus *vs.* crystallinity [222]

As none of the models leads to really satisfactory quantitative results for the most simple case of the elastic predictions of isotropic semicrystalline materials, the spherulitic microstructure should be better modeled. Does an aggregate of randomly oriented crystallites correctly represent this microstructure? Intuitively, it is not surprising that an aggregate of randomly oriented crystallite inclusions associated with a self-consistent scheme is stiffer than it is observed experimentally. But in this case a multiscale modeling is needed.

1.11. Large deformations and bottlenecks

Under large deformations, as described in the previous sections, the initial deformation includes straining of molecular chains in the interlamellar amorphous phase, which is accompanied by lamellae separation, rotation of lamellar stacks and interlamellar shear. At the yield, intensive crystallographic slips are observed leading, subsequently to fragmentation, then to formation of fibrils and generally a disintegration of lamellae. After yielding, some cavities are formed, and the potential appearance of disentanglements in the amorphous phase significantly influences the stress-strain behavior. All these complex mechanisms have been quite well established, but their modeling at two scales (lamella and macrocrystalline structure) is really difficult. The behavior of the crystalline phase can be described as an elastoplastic [216] or elasto-viscoplastic [220] behavior to take the crystallographic slips into account. Is it sufficient to go further than yield? Today, the first bottleneck is certainly to describe the role and the behavior of the confined amorphous phase, especially if it is in the rubbery state. By modeling the dynamic linear viscoelasticity of semicrystalline PET, Diani *et al.* [237] showed that the mechanical properties of the confined amorphous phase are somehow not the ones of the bulk. The second bottleneck is to take into account the initial crystalline macrostructure and its evolution during large deformation. Van Dommelen *et al.* [183] developed a multiscale approach by combining a finite element model describing the deformation of spherulites at a mesoscopic level and a micromechanical model simulating the behavior of the embedded crystalline and amorphous lamellae. The evolution of texture is described in terms of pole figures. This work is actually the most advanced to describe the behavior of semicrystalline polymers in large deformations.

1.12. Phenomenological models of polymer deformation under tensile and compressive stresses

Plane strain compression in a channel die is kinematically very similar to drawing: the sample is extended and its cross-section decreases accordingly. However, the possibility of void formation is limited due to the compressive component of stress. It means that the differences in true stress-true strain dependencies of drawn and plane strained polymers should be attributed to the formation and development of cavities. Slopes of the elastic region of true stress-true strain curves are similar in tension and in plane strain compression. The difference in mechanical properties of polymers sets in at yielding in tension. The scale of difference depends on the particular polymer: the yield in drawing for POM, PA 6, PP and HDPE takes place at a much lower stress than in plane strain compression. For polymers with low crystal plastic resistance, such as LDPEs and ethylene-octene copolymer (EOC), the stresses at selected deformation ratios were very similar for both modes of sample loading. The conclusion is that cavitation during deformation can be observed only in those polymers in which the value of negative pressure generated at yield is higher than the value of negative pressure required for cavitation. Otherwise the crystals will deform earlier, relaxing the stress, and cavitation will not appear.

Beyond the yield in channel die compressed POM, PA 6, PP and HDPE, a short region of plastic flow is followed by a strong strain hardening, while in drawing the plastic flow region is longer and strain hardening is only slightly pronounced before fracture. Nearly no difference in plastic flow and strain hardening is noticed for soft LDPEs and EOC in drawing and plane strain compression.

SAXS, densitometry and TEM [39] show a clear correlation of the formation of cavities in tension with a decrease of the yield stress when compared to the yield stress without cavitation (plane strain compression). It is also evident that cavitation is the reason for low strain hardening and intense chain disentanglement during drawing with cavitation. In turn, no significant disentanglement takes place during plastic deformation of polymers up to a high strain, provided that there is no cavitation.

The mechanisms associated with plastic deformation of crystalline polymers can be explained better on the basis of two examples of isotactic polypropylene subjected to drawing and to plane strain compression in a channel die. The envisioned differences between deformation with and without cavitation are summarized in Figure 1.21.

The initial part of the elastic region of true stress-true strain curves is common for drawing and for plane strain compression. As the strain increases, the material starts to yield in drawing at the stress level which is usually considered as the yield stress of a particular polymer and placed in product information

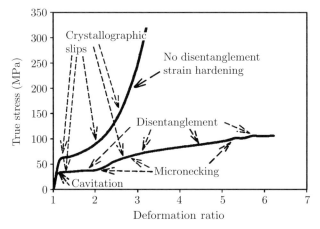

Figure 1.21. Mechanisms associated with plastic deformation of crystalline polymers explained by an example of isotactic polypropylene of molecular weight 3.1×10^5 [10]

sheets. However, the stress level at yield corresponds rather to the onset of cavitation and not necessarily to the onset of plastic deformation of crystals. With the progress of drawing, the cavitational pores extend in the direction of drawing, the polymeric material between them starts to deform plastically, now including crystal plasticity mechanisms, giving rise to the long known "micro-necking" phenomena and for large drawing ratios leading to a "micro-fibrillar" morphology. Along with cavitation and micronecking, chain disentanglement takes place, promoted by the formation of a large fraction of new free surfaces of voids. The chain disentanglement becomes intense; it results in loosening of the entanglement network and limits significantly the strain hardening.

In plane strain compression, the elastic region extends to a larger strain and larger stress because cavitation formation is not in operation, while crystal plasticity needs higher resolved shear stresses to become active. When the stress reaches the level sufficient for triggering the easiest crystallographic slip (for the example of iPP as in Figure 1.21 (100)[001] chain slip system at a critical resolved shear stress of 22.6 MPa [188], also reported at 25 MPa [185]), crystal plasticity takes over the control of further deformation process. The real yield stress for polypropylene from Figure 1.21, as determined by crystal plasticity, is now at the level of 50–55 MPa.

In plane strain compression, the plastic flow ends shortly and strain hardening induced by straightening of the entanglement network sets in. Since no or little chain disentanglement occurs, the stress reaches very high values at a relatively low true strain.

1.13. Conclusions and outlook

The promotion of energy dissipative processes that delay or entirely suppress fracture processes originating from imperfections in the internal structure or scratches and notches is enhanced by cavitation. In almost all cases, cavitation either makes possible further toughening by activating other mechanisms or itself contributes to the plastic response of the polymer. The most energy dissipative processes, crazing and shear yielding, occur at a reduced stress level.

In crystalline polymer systems the tough response, besides cavitation and crazing, is crystallographic in nature. Crystallographic slips are the main plastic deformation mechanisms that require generation and motion of crystallographic dislocations. The concepts of generation of monolithic and half-loop dislocations plausibly explain the observed yield stress dependences on crystal thickness, temperature and strain rate.

It appeared that a successful modeling of mechanical properties of crystalline polymers within the elastic range requires a consideration of lamellae thickness and crystallinity but also the lamellae width and length. Varying elastic properties of the amorphous phase are to be considered when the constraints between lamellar crystals change due to differentiated solidification conditions.

References

1. Keller A (1968) Polymer crystals, *Rep Prog Phys* **31**:623–704.
2. Keith H D and Padden F J Jr (1964) Spherulitic crystallization from the melt. I. Fractionation and impurity segregation and their influence on crystalline morphology, *J Appl Phys* **35**:1270–1285.
3. Keith H D and Padden F J Jr (1964) Spherulitic crystallization from the melt. II. Influence of fractionation and impurity segregation on the kinetics of crystallization, *J Appl Phys* **34**:1286–1296.
4. Coriell S R, McFadden G B, Voorhees P W and Sekerka R F (1987) Stability of a planar interface during solidification of a multicomponent system, *J Cryst Growth* **82**:295–302.
5. Woodruff D P (1973) *The Solid-Liquid Interface*, Cambridge University Press, Cambridge, UK.
6. Calvert P D (1983) Fibrillar structures in spherulites, *J Polym Sci Polym Lett Ed* **21**:467–473.
7. Galeski A, Argon A S and Cohen R E (1987) Morphology of nylon 6 spherulites, *Makromol Chem* **188**:1195–1204.
8. Oleinik E F (2003) Plasticity of semicrystalline flexible-chain polymers at microscopic and mezoscopic levels, *Polym Sci (Russian) Ser C* **45**:17–117.
9. Galeski A (2003) Strength and toughness of crystalline polymersin, *Prog Polym Sci* **28**:1643–1699.
10. Pawlak A and Galeski A (2005) Plastic deformation of crystalline polymers: The role of cavitation and crystal plasticity, *Macromolecules* **38**:9688–9697.
11. Hughes D J, Mahendrasingam A, Oatway W B, Heeley E L, Martin C and Fuller W (1997) A simultaneous SAXS/WAXS and stress-strain study of polyethylene deformation at high strain rates, *Polymer* **38**:6427–6430.
12. Zhang X C, Butler M F and Cameron R E (2000) The ductile-brittle transition of irradiated isotactic polypropylene studied using simultaneous small angle X-ray scattering and tensile deformation, *Polymer* **41**:3797–3807.
13. Liu Y and Truss R W (1994) Study of tensile yielding of isotactic polypropylene, *J Polym Sci B Polym Phys* **32**:2037–2047.
14. Kapur S, Matsushige S, Galeski A and Baer E (1978) The effect of pressure and environment on the fracture and yield of polymers, in: *Advances in Research on Strength and Fracture of Materials* (Ed Taplin D N R) Pergamon Press, New York, pp. 1079–1086.
15. Pae K D, Chiu H C, Lee J K and Kim J H (2000) Healing of stress-whitening in polyethylene and polypropylene at or below room temperature, *Polym Eng Sci* **40**:1783–1795.
16. Yamaguchi M and Nitta K H (1999) Optical and acoustical investigation for plastic deformation of isotactic polypropylene/ethylene-1-hexene copolymer blends, *Polym Eng Sci* **39**:833–840.
17. Galeski A and Bartczak Z (2003) Cavitation and cavity-free deformation of filled crystalline polymer systems, *Macromol Symp* **194**:47–62.
18. Peterlin A and Balta-Caleja F J (1969) Plastic deformation of polypropylene. III. Small-angle X-ray scattering in the neck region, *J Appl Phys* **40**:4238–4242.
19. Dijkstra P T S, Van Dijk D J and Huetnik J (2002) A microscopy study of the transition from yielding to crazing in polypropylene, *Polym Eng Sci* **42**:152–160.
20. Kennedy M A, Peacock A J and Mandelkern L (1994) Tensile properties of crystalline polymers: Linear polyethylene, *Macromolecules*, **27**:5297–5310.

21. Hiss R, Hobeika S, Lynn C and Strobl G (1999) Network stretching, slip processes, and fragmentation of crystallites during uniaxial drawing of polyethylene and related copolymers. A comparative study, *Macromolecules* **32**:4390–4403.
22. Butler M F, Donald A M and Ryan A J (1998) Time resolved simultaneous small- and wide-angle X-ray scattering during polyethylene deformation. II. Cold drawing of linear polyethylene, *Polymer* **39**:39–52.
23. Wang K H, Chung I J, Jang M C, Keum J K and Song H H (2002) Deformation behavior of polyethylene/silicate nanocomposites as studied by real-time wide-angle X-ray scattering, *Macromolecules* **35**:5529–5539.
24. Bartczak Z and Lezak E (2005) Evolution of lamellar orientation and crystalline texture of various polyethylenes and polyethylene copolymers in plane-strain compression, *Polymer,* **46**:6050–6063.
25. Plummer C J G (2004) Microdeformation and Fracture in Bulk Polyolefins, in Long-Term Properties of Polyolefins, *Adv Polym Sci*, **169**:75–119.
26. Aboulfaraj M, G'Sell C, Ulrich B and Dahoun A (1995)*In situ* observation of the plastic deformation of polypropylene spherulites under uniaxial tension and simple shear in the scanning electron microscope, *Polymer* **36**:731–742.
27. Lee J K, Kim J H, Chu H C and Pae K D (2002) Macroscopic observation of healing process in stress-whitened polypropylene under hydrostatic pressure, *Polym Eng Sci* **42**:2351–2360.
28. Plummer C J G, Scaramuzzino P, Kausch H H and Phillipoz J M (2000) High temperature slow crack growth in polyoxymethylene, *Polym Eng Sci* **40**:1306–1317.
29. Plummer C J G, Beguelin P, Grein C, Gensler R, Dupuits L, Gaillard C, Stadelmann P, Kausch H H and Manson J A E (2004) Microdeformation in heterogeneous polymers, revealed by electron microscopy, *Macromol Symp* **214**:97–114.
30. Na B, Zhang Q, Fu Q, Men Y, Hong K and Strobl G (2006) Viscous-force-dominated tensile deformation behavior of oriented polyethylene, *Macromolecules* **39**:2584–2591.
31. Friedrich K (1983) Crazes and shear bands in semi-crystalline thermoplastics, *Adv Polym Sci* **52/53**:225–274.
32. Kausch H H, Gensler R, Grein Ch, Plummer C J G and Scaramuzzino P (1999) Crazing in semicrystalline thermoplastics, *J Macromol Sci Phys* **B38**:803–815.
33. Jang B Z, Uhlmann D R and Van der Sande J B (1985) Crazing in polypropylene, *Polym Eng Sci* **25**:98–104.
34. Henning S, Adhikari R, Michler G H, Balta Calleja F J and Karger-Kocsis J (2004) Micromechanical mechanisms for toughness enhancement in β-modified polypropylene, *Macromol Symp* **214**:157–171.
35. Narisawa I and Yee A F (1993) Crazing and fracture of polymers, in: *Materials Science and Technology. A Comprehensive Treatment. Vol. 12. Structure of Polymers* (Ed. Thomas E L) VCH, Weinheim, pp. 699–766.
36. Brown H R and Kramer E J (1980) Craze microstructure from small-angle X-ray scattering (SAXS), *J Macromol Sci Phys* **B19**:487–522.
37. Mills P J and Kramer E J (1985) Real time small-angle X-ray scattering from polystyrene crazes during fatigue, *J Mater Sci* **20**:4413–4420.
38. Zafeiropoulos N E, Davies R J, Schneider K, Burghammer M, Riekel Ch and Stamm M (2006) The relationship between craze structure and molecular weight in polystyrene as revealed by SAXS experiments, *Macromol Rapid Commun* **27**:1689–1694.
39. Galeski A, Argon A S and Cohen R E (1988) Changes in the morphology of bulk spherulitic nylon 6 due to plastic deformation, *Macromolecules* **21**:2761–2770.
40. Argon A S (1972) Fracture of composites, in: *Treatise on Materials Science and Technology* (Ed. Herman H) Academic Press, New York, Vol. 1, pp. 79–114.
41. Krumova M, Henning S and Michler G H (2006) Chevron morphology in deformed semicrystalline polymers, *Phi Mag* **86**:1689–1712.

42. Thomas C, Ferreiro V, Coulon G and Seguela R (2007) In situ AFM investigation of crazing in polybutene spherulites under tensile drawing, *Polymer* **48**:6041–6048.
43. Bucknall C B (1977) *Toughened Plastics*, Applied Science Publishers Ltd., Barking, Essex, England, pp. 1–359.
44. Quatravaux T, Elkoun S, G'Sell C, Cangemi L and Meimon Y (2002) Experimental characterization of the volume strain of poly(vinylidene fluoride) in the region of homogeneous plastic deformation, *J Polym Sci Part B: Polym Phys* **40**:2516–2522.
45. G'Sell C, Hiver J M and Dahoun A (2002) Experimental characterization of deformation damage in solid polymers under tension, and its interrelation with necking, *Int J Solid Struct* **39**:3857–3872.
46. Castagnet S, Girault S, Gacougnolle J L and Dang P (2000) Cavitation in strained polyvinylidene fluoride: Mechanical and X-ray experimental studies, *Polymer* **41**:7523–7530.
47. Addiego F, Dahoun A, G'Sell Ch and Hivier J M (2006) Characterization of volume strain at large deformation under uniaxial tension in high-density polyethylene, *Polymer* **47**:4387–4399.
48. G'Sell Ch, Bai S L and Hiver J M (2004) Polypropylene/polyamide 6/polyethylene-octene elastomer blends. Part 2: Volume dilatation during plastic deformation under uniaxial tension, *Polymer* **45**:5785–5792.
49. Temimi N, Burr A and Billon N (2006) Damaging processes in polypropylene. Experimental and modelling, in: *Book of Abstracts. Euromech Colloquium 487. Structure Sensitive Mechanics of Polymer Materials: Physical and Mechanical Aspects. Strasbourg 10–13.10.2006* (Eds. Remond Y and Patlazhan S) Strasbourg, pp. 67–68.
50. Pawlak A (2007) Cavitation during tensile deformation of high density polyethylene, *Polymer* **48**:1397–1409.
51. Pawlak A and Galeski A, Cavitation during tensile deformation of polypropylene, *Macromolecules* **41**:2839–2851.
52. Bucknall C B (1978) Fracture and failure of multiphase polymers and polymer composites, *Adv Polym Sci* **27**:121–148.
53. Kramer E J (1983) Microscopic and molecular fundamentals of crazing, *Adv Polym Sci* **52/53**:275–334.
54. Duffo P, Monasse B, Haudin J M, G'Sell C and Dahoun A (1995) Rheology of polypropylene in the solid state, *J Mater Sci* **30**:701–711.
55. Li J X and Cheung W L (1998) On the deformation mechanisms of β-polypropylene. 1. Effect of necking on β-phase PP crystals, *Polymer* **39**:6935–6940.
56. Li J X, Cheung W L and Chan C M (1999) On deformation mechanisms of β-polypropylene. 2. Changes of lamellar structure caused by tensile load, *Polymer* **40**:2089–2102.
57. Li J X, Cheung W L and Chan C M (1999) On deformation mechanisms of β-polypropylene. 3. Lamella structures after necking and cold drawing, *Polymer* **40**:3641–3656.
58. Piorkowska E, Galeski A and Kryszewski M (1982) Heat conduction anisotropy of drawn high density polyethylene samples, *Colloid Polym Sci* **260**:735–741.
59. Butler M F, Donald A M and Ryan A J (1997) Time resolved simultaneous small- and wide-angle X-ray scattering during polyethylene deformation. 1. Cold drawing of ethylene-α-olefin copolymers, *Polymer* **38**:5521–5538.
60. Butler M F and Donald A M (1998) A real-time simultaneous small- and wide-angle X-ray scattering study of *in situ* polyethylene deformation at elevated temperatures *Macromolecules*, **31**:6234–6249.
61. Keith H D and Padden F J Jr (1964) Spherulitic crystallization from the melt. I. Fractionation and impurity segregation and their influence on crystalline morphology, *J Appl Phys* **35**:1270–1285.

62. Monnerie L, Halary J L and Kausch H H (2005) Deformation, Yield and Fracture of Amorphous Polymers: Relation to the Secondary Transitions, *Adv Polym Sci* **187**:215–364.
63. Narisawa I and Ishikawa M (1990) Crazing in semicrystalline thermoplastics, *Adv Polym Sci* **91/92**:353–391.
64. Plummer C J G and Kausch H H (1996) Deformation and entanglement in semicrystalline polymers, *J Macromol Sci Phys* **B35**:637–657.
65. Crist B and Peterlin A (1973) Radical formation during the tensile failure of polymer fibres, *Makromol Chem* **171**:211–227.
66. Peterlin A (1975) Plastic deformation of crystalline polymers, in: *Polymeric Materials* (Ed. Baer E) American Society for Metals, Metals Park, Ohio, pp. 175–195.
67. Galeski A, Bartczak, Argon A S and Cohen R E (1992) Morphological alterations during texture producing plane strain compression of high density polyethylene, *Macromolecules* **25**:5705–5718.
68. Pluta M, Bartczak Z and Galeski A (2000) Changes in the morphology and orientation of bulk spherulitic polypropylene due to plane strain compression, *Polymer* **41**:2271–2288.
69. Bellare A, Cohen R E and Argon A S (1993) Development of texture in poly(ethylene terephthalate) by plane-strain compression, *Polymer* **34**:1393–1403.
70. Boontongkong Y, Cohen R E, Spector M and Bellare A (1998) Orientation of plane strain-compressed ultra-high-molecular-weight polyethylene, *Polymer* **39**:6391–6400.
71. Lin L, Argon A S (1992) Deformation resistance in oriented nylon 6, *Macromolecules* **25**:4011–4024.
72. Staniek E, Seguela R, Escaig B and François P (1998) Plastic behavior of monoclinic polypropylene under hydrostatic pressure in compressive testing, *J Appl Polym Sci*, **72**:1241–1247.
73. Galeski A, Argon A S and Cohen R E (1991) Morphology of bulk nylon 6 subjected to plane strain compression, *Macromolecules* **24**:3953–3961.
74. Muratoglu O K, Argon A S, Cohen R E and Weinberg M (1995) Microstructural processes of fracture of rubber-modified polyamides, *Polymer* **36**:4771–4786.
75. Muratoglu O K, Argon A S, Cohen R E and Weinberg M (1995) Microstructural fracture processes accompanying growing cracks in tough rubber-modified polyamides, *Polymer* **36**:4787–4795.
76. Goodier J (1933) Concentration of stress around spherical and cylindrical inclusions and flaws, *Trans ASME* **55**:39–42.
77. Wu S (1982) *Polymer Interface and Adhesion*, Marcel Dekker, New York, pp. 88–92.
78. Kasemura T, Yamashita N, Suzuki K, Kondo T and Hata T (1978) Studies on surface tension of polymers in relation to their structures. IV. Composition dependence of surface tension of copolymers, *Kobunshi Ronbunshu* **35**:263–268.
79. Brandrup J, Immergut E H and Grulke E A (Eds.) (1999) *Polymer Handbook*, 4th ed., Wiley, New York.
80. Piorkowska E and Galeski A (1993) Crystallization of isotactic polypropylene and high density polyethylene under negative pressure resulting from uncompensated volume change, *J Polym Sci Part B: Polym Phys* **31**:1285–1291.
81. Pawlak A and Piorkowska E (1999) Effect of negative pressure on melting behavior of spherulites in thin films of several crystalline polymers, *J Appl Polym Sci* **74**:1380–1385.
82. Nowacki R, Kolasinska J and Piorkowska E (2001) Cavitation during isothermal crystallization of isotactic polypropylene, *J Appl Polym Sci* **79**:2439–2448.
83. Galeski A, Piorkowska E (2002) Negative pressure development during crystallization of polymers, in: *Liquids Under Negative Pressure* (Eds. Imre A R, Maris H J and Williams P R) NATO Science Series, II. Mathematics, Physics and Chemistry, Vol. 84, pp. 127–136.

84. Bowden P B and Young R J (1971) Critical resolved shear stress for [001] slip in polyethylene, *Nature* **229**:23–25.
85. Haudin J M (1982) Plastic deformation of semicrystalline polymers, in: *Plastic Deformation of Amorphous and Semi-crystalline Materials* (Eds. Escaig B and G'Sell C) Les Editions de Physique, Paris, pp. 291–311.
86. Gaucher-Miri V and Seguela R (1997) Tensile yield of polyethylene and related copolymers: Mechanical and structural evidences of two thermally activated processes, *Macromolecules* **30**:1158–1167.
87. Argon A S and Cohen R E (1990) Crazing and toughness of block copolymers and blends, *Adv Polym Sci* **91,92**:301–351.
88. Kinloch A J and Young R J (1983) *Fracture Behavior of Polymers*, Applied Science Publishers, London, New York, pp. 147–178.
89. Monnerie L, Halary J L and Kausch H H (2005) Deformation, yield and fracture of amorphous polymers: relation to the secondary transitions, *Adv Polym Sci*, **187**:215–364.
90. Kramer E J and Berger L L (1990) Fundamental processes of craze growth and fracture, *Adv Polym Sci* **91,92**:1–68.
91. Olf H G and Peterlin A (1974) Cryogenic crazing of crystalline, isotactic polypropylene, *J Colloid Interface Sci* **47**:621–635.
92. Horst J J and Spoormaker J L (1996) Mechanism of fatigue in short glass fiber reinforced polyamide 6, *Polym Eng Sci* **36**:2718–2726.
93. Boyer R F (1975) Mechanical relaxation spectra of crystalline and amorphous polymers, Midland Macromolecular Institute Symposium on "Molecular Basis of Transitions and Relaxations", Midland, Michigan, Materials distributed by Boyer R F during the Symposium.
94. Ferry J D (1970) *Viscoelastic Properties of Polymers*, 2nd ed., Wiley, New York.
95. Galeski A (2002) Dynamic mechanical properties of crystalline polymer blends. The influence of interfaces and orientation, *e-Polymers* **026**:1–29.
96. Orovan E (1948–49) Fracture and strength of solids, *Rept Prog Phys* **12**:185–232.
97. Vincent P I (1960) The tough-brittle transition in thermoplastics, *Polymer* **1**:425–444.
98. Ward I M and Hadley D W (1993) *An Introduction to the Mechanical Properties of Solid Polymers*, Wiley, Chichester, New York, pp. 271–276.
99. Ward I M and Hadley D W (1993) *An Introduction to the Mechanical Properties of Solid Polymers*, Wiley, New York, pp. 232–245.
100. Piorkowska E, Argon A S and Cohen R E (1990) Size effect of compliant rubbery particles on craze plasticity in polystyrene, *Macromolecules* **23**:3838–3848.
101. Flory P J and Yoon D Y (1978) Molecular morphology in semicrystalline polymers, *Nature* **272**:226–229.
102. Gent A N and Madan S (1989) Plastic yielding of partially crystalline polymers, *J Polym Sci Polym Phys Ed* **27**:1529–1542.
103. Young R J (1976) A dislocation model for yield in polyethylene, *Phil Mag* **30**:86–94.
104. Seguela R (2007) Plasticity of semi-crystalline polymer: Crystal slip versus melting-recrystallization, *e-Polymers* **032**:1–20.
105. Lv R, Xu W, Na B, Zhang Q and Fu Q (2008) Large tensile deformation behavior of oriented high-density polyethylene: A correlation between cavitation and lamellar fragmentation, *J Polym Sci Polym Phys* **46**:1202–1206.
106. Peterlin A (1971) Radical formation and fracture of highly drawn crystalline polymers, *J Macromol Sci Phys* **6**:490–508.
107. Kelly A and Groves G W (1970) *Crystallography of Crystal Defects*, Longman, London, UK, pp. 1–275.
108. Cottrell A H (1953) Dislocations and Plastic Flow, in: *Crystals*, Oxford Univ. Press, London, Ch. 3, pp. 1–223.

109. Bartczak Z, Argon A S and Cohen R E (1992) Deformation mechanisms and plastic resistance in single-crystal-textured high-density polyethylene, *Macromolecules* **25**:5036–5053.
110. Bowden P B and Young R J (1974) Deformation mechanisms in crystalline polymers, *J Mater Sci* **9**:2034–2051.
111. Bartczak Z, Morawiec J and Galeski A (2002) Structure and properties of isotactic polypropylene oriented by rolling with side constraints, *J Appl Polym Sci*, **86**:1413–1425.
112. Caddell R M, Raghava R S and Atkins A G (1973) A yield criterion of anisotropic and pressure dependent solids such as oriented polymers, *J Mater Sci* **8**:1641–1646.
113. Castagnet S and Deburck Y (2006) Relative influence of microstructure and macroscopic triaxiality on cavitation damage in a semi-crystalline polymer, Mater Sci Eng, **A448**:56–66.
114. Bartczak Z and Martuscelli E (1997) Orientation and properties of sequentially drawn films of an isotactic polypropylene; hydrogenated oligocyclopentadiene blends, *Polymer* **38**:4139–4149.
115. Bartczak Z and Galeski A (1999) Yield and plastic resistance of alpha-crystals of isotactic polypropylene, *Polymer* **40**:3677–3684.
116. Lee B J, Argon A S, Parks D M, Ahzi S and Bartczak Z (1993) Simulation of large strain plastic deformation and texture evolution in high-density polyethylene, *Polymer* **34**:3555–3575.
117. Lin L and Argon A S (1994) Review: Structure and plastic deformation of polyethylene, *J Mater Sci* **29**:294–323.
118. Bellare A, Argon A S and Cohen R E (1993) Development of texture in poly(ethylene terephthalate) by plane-strain compression, *Polymer* **34**:1393–1403.
119. Eling B, Gogolewski S and Pennings A J (1982) Biodegradable materials of poly(L-lactic acid). 1. Melt-spun and solution n-spun fibres, *Polymer* **23**:1587–1593.
120. Ferreiro V, Depecker C, Laureyns J and Coulon G (2004) Structures and morphologies of cast and plastically strained polyamide 6 films as evidenced by confocal Raman microspectroscopy and atomic force microscopy, *Polymer* **45**:6013–6026.
121. Chen X, Galeski A and Michler G H (2006) Morphological alteration of polyamide 6 subjected to high plane-strain compression, *Polymer* **4**:3171–3185.
122. Lezak E, Bartczak Z and Galeski A (2006) Plastic deformation behavior of β phase isotactic polypropylene in plane-strain compression at room temperature, *Polymer* **47**:8562–8574.
123. Lezak E, Bartczak Z and Galeski A (2006) Plastic deformation of the γ phase isotactic polypropylene in plane-strain compression, *Macromolecules* **39**:4811–4819.
124. Lezak E and Bartczak Z (2007) Plastic deformation of the γ phase isotactic polypropylene in plane-strain compression at elevated temperatures, *Macromolecules* **40**:4933–4941.
125. Lewis D, Wheeler E J, Maddams W F and Preedy J E (1972) Comparison of twinning produced by rolling and annealing in high- and low-density polyethylene, *J Polym Sci A-2 Notes* **10**:369–373.
126. Bartczak Z (2002) Deformation of high density polyethylene produced by rolling with side constraints. I. Orientation behavior, *J Appl Polym Sci* **86**:1396–1404.
127. Young R J and Bowden P B (1974) Twinning and martensitic transformations in oriented high-density polyethylene, *Phil Mag* **29**:1061–1073.
128. Kryszewski M, Pakula T, Galeski A, Milczarek P and Pluta M (1978) Über einige Ergebnisse der Untersuchung der Korrelation zwischen der Morphologie und den mechanischen Eigenschaften von kristallinen Polymeren, *Faserforsch Textiltech* **29**:76–85 (in German).

129. Mirabella F M Jr (1987) Determination of the crystalline and noncrystalline molecular orientation in oriented polypropylene by infrared spectroscopy, *J Polym Sci Part B: Polym Phys* **25**:591–602.
130. Burton W K, Cabrera N and Frank F C (1949) Role of dislocations in crystal growth, *Nature* **163**:398–399.
131. Frank F C (1949) The influence of dislocations on crystal growth, *Farad Soc Discuss, Crystal Growth* **5**:48–54.
132. Geil P H (1963) *Polymer Single Crystals*, Interscience, New York.
133. Wunderlich B (1973) *Macromolecular Physics, Crystal Structure. Morphology, Defects*, Academic Press, New York, Ch. 4.1.4, pp. 408–414.
134. Keith H D and Chen W Y (2002) On the origins of giant screw dislocations in polymer lamellae, *Polymer* **43**:6263–6272.
135. Schultz J M and Kinloch D R (1969) Transverse screw dislocations: A source of twist in crystalline polymer ribbons, *Polymer* **10**:271–278.
136. Duan Y, Zhang Y, Yan S and Schultz J M (2005) In situ AFM study of the growth of banded hedritic structures in thin films of isotactic polystyrene, *Polymer* **46**:9015–9021.
137. Toda A, Arita T and Hikosaka M (2001) Three-dimensional morphology of PVDF single crystals forming banded spherulites, *Polymer* **42**:2223–2233.
138. Ikehara T, Jinnai H, Kaneko T, Nishioka H and Nishi T (2007) Local lamellar structures in banded spherulites analyzed by three-dimensional electron tomography, *J Polym Sci Polym Phys* **45**:1122–1125.
139. Toda A, Okamura M, Hikosaka M and Nakagawa Y (2005) Three-dimensional shape of polyethylene single crystals grown from dilute solutions and from the melt, *Polymer* **46**:8708–8716.
140. Wilhelm H, Paris A, Schafler E, Bernstorff S, Bonarski J, Ungar T and Zehetbauer M J (2004) Evidence of dislocations in melt-crystallized and plastically deformed polypropylene, *Mater Sci Eng* **A387–389**:1018–1022.
141. Schafler E, Zehetbauer M and Ungar T (2001) Measurement of screw and edge dislocation density by means of X-ray Bragg profile analysis, *Mater Sci Eng* **A319–321**:220–223.
142. Keith H D and Passaglia E (1964) Dislocations in polymer crystals, *J Res Natl Bur Stand* **68A**:513–518.
143. Peterson J M (1966) Thermal initiation of screw dislocations in polymer crystal platelets, *J Appl Phys* **37**:4047–4050.
144. Peterson J M (1968) Peierls stress for screw dislocations in polyethylene, *J Appl Phys* **39**:4920–4928.
145. Shadrake L G and Guiu F (1976) Dislocations in polyethylene crystals: Line energies and deformation modes, *Philos Mag* **34**:565–581.
146. Young R J (1988) Screw dislocation model for yield in polyethylene, *Mater Forum* **11**:210–216.
147. Lin L and Argon A S (1994) Rate mechanism of plasticity in the crystalline component of semicrystalline nylon 6, *Macromolecules* **27**:6903–6914.
148. Butler M F, Donald A M, Bras W, Mant G R, Derbyshire G E and Ryan A J (1995) A real-time simultaneous small- and wide-angle X-ray scattering study of *in-situ* deformation of isotropic polyethylene, *Macromolecules*, **28**:6383–6393.
149. Kazmierczak T, Galeski A and Argon A S (2005) Plastic deformation of polyethylene crystals as a function of crystal thickness and compression rate, *Polymer* **46**:8926–8936.
150. Young R J (1974) Dislocation model for yield in polyethylene, *Philos Mag* **30**:85–94.

151. O'Kane W J, Young R J and Ryan A J (1995) The effect of annealing on the structure and properties of isotactic polypropylene films, *J Macromol Sci, Part B*, **34**:427–458.
152. Crist B, Fisher C J and Howard P (1989) Mechanical properties of model polyethylenes: Tensile elastic modulus and yield stress, *Macromolecules* **22**:1709–1718.
153. Darras O and Séguéla R (1993) Tensile yield of polyethylene in relation to crystal thickness, *J Polym Sci Part B: Polym Phys* **31**:759–766.
154. O'Kane W J and Young R J (1995) The role of dislocations in the yield of polypropylene, *J Mater Sci Lett*, **14**:433–435.
155. Li L (1994) Atomistic simulation of plastic deformation mechanisms in crystalline polyethylene, *MSc thesis*, Massachusets Institute of Technology, Cambridge, MA, pp. 1–58.
156. Shadrake L G and Gui F (1979) Elastic line energies and line tensions of dislocations in polyethylene crystals, *Philos Mag* **39**:785–796.
157. Bacon D J and Tharmalingam K (1983) Computer simulation of polyethylene crystals. Part 3. The core structure of dislocations, *J Mater Sci* **18**:884–893
158. Seguela R (2002) Dislocation Approach to the Plastic Deformation of Semicrystalline Polymers: Kinetic Aspects for Polyethylene and Polypropylene, *J Polym Sci, Part B: Polym Phys* **40**:593–601.
159. Argon A S, Galeski A and Kazmierczak T (2005) Rate mechanisms of plasticity in semi-crystalline polyethylene, *Polymer* **46**:11798–11805.
160. Bartczak Z, Cohen R E and Argon A S (1992) Evolution of crystalline texture of high-density polyethylene during uniaxial compression, *Macromolecules* **25**:4692–4704.
161. Bartczak Z, Morawiec J and Galeski A (2002) Structure and properties of isotactic polypropylene oriented by rolling with side constraints, *J Appl Polym Sci*, **86**:1413–1425.
162. Peierls R (1940) The size of dislocations, *Proc Phys Soc* **289**:34–37.
163. Nabarro F R N (1947) Dislocations in a simple cubic lattice *Proc Phys Soc* **59**:256–272.
164. Honeycomb R W K (1968) *The Plastic Deformation of Metals*, Edward Arnold, London, UK, pp. 41–47.
165. Frank F C and Read W T (1950) Multiplication processes for slow moving dislocations, *Phys Rev* **79**:722–723.
166. Xu G and Argon A S (2001) Energetics of homogeneous nucleation of dislocation loops under a simple shear stress in perfect crystals, Mater Sci Eng **A319-321**:144–147.
167. Crist B (1989) Yielding of semicrystalline polyethylene: A quantitative dislocation model, *Polym Comm* **30**:69–71.
168. Brooks N W, Ghazali M, Duckett R A, Unwin A P and Ward I M (1998) Effects of morphology on the yield stress of polyethylene, *Polymer*, **40**:821–825.
169. Brooks N W J, Ducket R A and Ward I M (1998) Temperature and strain-rate dependence of yield stress of polyethylene, *J Polym Sci Part B: Polym Phys* **36**:2177–2189.
170. Seguela R, Gaucher-Miri V and Elkoun S (1998) Plastic deformation of polyethylene and ethylene copolymers. Part I: homogeneous crystal slip and molecular mobility, *J Mater Sci*, **33**:1273–1279.
171. Kazmierczak T and Galeski A (2002) Transformation of polyethylene crystals by high pressure, *J Appl Polym Sci* **86**:1337–1350.
172. Kazmierczak T and Galeski A (2003) Plastic deformation of polyethylene crystals as a function of crystal thickness and compression rate, in: *Proceedings of 12th International Conference on Deformation, Yield and Fracture of Polymers*, Cambridge, UK, Apr 7–10, pp. 419–422.
173. Brooks N W J and Mukhtar M (2000) Temperature and stem length dependence on the yield stress of polyethylene, *Polymer* **41**:1475–1480.

174. Petermann J and Gleiter H (1972) Direct observation of dislocations in polyethylene crystals, *Philos Mag* **25**:813–816.
175. Peterman J and Gleiter H (1973) Plastic deformationof polyethylene crystals by dislocation motion, *J Mater Sci*, **8**:673–675.
176. O'Kane W J and Young R J (1995) The role of dislocations in the yield of polypropylene, *J Mater Sci Lett* **14**:433–435.
177. Xu G, Argon A S and Ortiz M (1995) Nucleation of dislocations from crack tips under mixed modes of loading: Implications for brittle against ductile behavior of crystals, *Philos Mag* **A72**:415–451.
178. Xu G, Argon AS and Ortiz M (1997) Critical configurations for dislocation nucleation from crack tips, *Philos Mag* **A75**:341–367.
179. Xu G and Zhang C (2003) Analysis of dislocation nucleation from a crystal surface based on the Peierls-Nabarro dislocation model, *J Mech Phys Solids* **51**:1371–1394.
180. Xu G (2002) Energetics of nucleation of half dislocation loops at a surface crack, *Philos Mag* **A82**:3177–3185.
181. Lee B J, Ahzi S and Asaro R J (1995) On the plasticity of low symmetry crystals lacking five independent slip systems, *Mech Mater*, **20**:1–8.
182. van Dommelen J A W, Parks D M, Boyce M C, Brekelmans W A M and Baaijens F P T (2003) Micromechanical modeling of intraspherulitic deformation of semicrystalline polymers, *Polymer* **44**:6089–6101.
183. van Dommelen J A W, Parks D M, Boyce M C, Brekelmans W A M and Baaijens F P T (2003) Micromechanical modeling of the elasto-viscoplastic behavior of semicrystalline polymers, *J Mech Phys Solids* **51**:519–541.
184. Bartczak Z and Galeski A (1997) Plastic resistance of isotactic polypropylene-crystals in bi-oriented polypropylene/hydrogenated oligo(cyclopentadiene)blends, *Fibres and Textiles East Eur* **5**:32–37.
185. Shinozaki D and Groves G W (1973) The plastic deformation of oriented polypropylene: Tensile and compressive yield criteria, *J Mater Sci* **8**:71–78.
186. Kaito A, Nakayama K and Kanetsuna H (1986) Roller drawing of polyoxymethylene, *J Appl Polym Sci* **32**:3499–3513.
187. Kazmierczak T and Galeski A (2003) Plastic deformation of polyethylene crystals as a function of crystal thickness, in: *Proceedings of VIth ESAFORM Conference on Material Forming*, Univ. Salerno, Italy, pp. 431–434.
188. Bartczak Z (2005) Influence of molecular parameters on high-strain deformation of polyethylene in the plane-strain compression. Part II. Strain recovery, *Polymer*, **46**:10339–10354.
189. Basset D C (1981) *Principles of Polymer Morphology*, Cambridge University Press, Cambridge, UK.
190. Ward I M (1997) *Structure and Properties of Oriented Polymers*, Chapman & Hall, London, UK.
191. Fujiyama M and Wakino T (1991) Structures and properties of injection moldings of crystallization nucleator-added polypropylenes. I. Structure-property relationships, *J Appl Polym Sci* **42**:2739–2747.
192. Pople J A, Mitchel G R, Sutton S J, Vaughan A S and Chai C K (1999) The development of organized structures in polyethylene crystallized from a sheared melt, analyzed by WAXS and TEM, *Polymer* **40**:2769–2777.
193. Mendoza R, Régnier G, Seiler W and Lebrun J L (2003) Spatial distribution of molecular orientation in injected molded iPP: Influence of processing conditions, *Polymer* **44**:3363–3373.
194. Koscher E and Fulchiron R (2002) Influence of shear on polypropylene crystallization: Morphology, development and kinetics, *Polymer* **43**:6931–6942.

195. Phillips R, Herbert G, News J and Wolkowicz M (1994) High modulus polypropylene: Effect of polymer and processing variables on morphology and properties, *Polym Eng Sci* **34**:1731–1743.
196. Christensen R M (1991) *Mechanics of Composite Materials*, Krieger Publishing Company, Malabar, Florida.
197. Nemat-Nasser S and Hori M (1999) *Micromechanics: Overall Properties of Heterogeneous Materials*, Second edition, Elsevier Science BV, Amsterdam.
198. Milton G N (2002) *Theory of Composites*, Cambridge University Press, Cambridge, UK.
199. Steenkamer D A and Sullivan J L (1999) The performance of calcium carbonate filled, random fiber composites, *Polym Comp* **20**:392–405.
200. Tucker III C L and Liang E (1999) Stiffness predictions for unidirectional short-fiber composites: Review and evaluation, *Compos Sci Tech* **59**:655–671.
201. Zeng Q H, Yu A B and Lu G Q (2008) Multiscale modeling and simulation of polymer nanocomosites, *Prog Polym Sci* **33**:191–269.
202. Ladeveze P and Lubineau G (2002) An enhanced mesomodel for laminates based on micromechanics, *Compos Sci Tech* **62**:533–541.
203. Liu Y, Gilormini P and Castaneda P P (2003) Variational self-consistent estimates for texture evolution in viscoplastic polycrystals, *Acta Mater* **51**:5425–5437.
204. Hill R (1963) Elastic properties of reinforced solids: Some theoretical principles, *J Mech Phys Solids* **11**:357–372.
205. Mori T, Tanaka K (1973) Average stress in matrix and average elastic energy of materials with misfitting inclusions, *Acta Metall* **21**:571–574.
206. Eshelby J D (1957) The determination of the elastic field of an ellipsoidal inclusion and related problems, *Proc Royal Soc Lond* **A241**:376–396.
207. Gilormini P and Brechet Y (1999) Syntheses: Mechanical properties of heterogeneous media: Which material for which model? Which model for which material?, *Mod Sim Mat Sci Eng* **7**:805–816.
208. Viana J C, Cunha A M and Billon N (2001) The effect of skin thickness and spherulite size on the mechanical properties of injection moldings, *J Mater Sci* **36**:4411–4418.
209. Sawyer L C, Chen R T, Jamieson M G, Musselman I H and Russel P E (1993) The fibrillar hierarchy in liquid crystalline polymers, *J Mater Sci* **28**:225–238.
210. Takayanagi M, Imada K and Kajiyama T (1967) Mechanical properties and fine structure of drawn polymers, *J Polym Sci Part C: Polym Symp* **159**:263–281.
211. Halpin J C and Kardos J L (1972) Moduli of crystalline polymers employing composite theory, *J Appl Phys* **43**:2235–2241.
212. Tsai S W and Pagano N J (1968) Invariant properties of composite materials, in *Proceedings of Composite Materials Workshop* (Eds. Tsai S W, Halpin J C and Pagano N J) Technomic Publishing, Stamford, Connecticut, pp. 233–253.
213. Phillips P J and Patel J (1978) The influence of morphology on the tensile properties of polyethylenes, *Polym Eng Sci* **18**:943–950.
214. Janzen J (1992) Elastic moduli of semicrystalline PE compared with theoretical micromechanical models for composites, *Polym Eng Sci* **32**:1242–1254.
215. Parks D M and Ahzi S (1990) Polycrystalline plastic deformation and texture evolution for crystals lacking five independent slip systems, *J Mech Phys Solids* **38**:701–724.
216. Dahoun A, Aboulfaraj M, G'Sell C, Molinari A and Canova G R (1995) Plastic behavior and deformation textures of poly(etherether ketone) under uniaxial tension and simple shear, *Polym Eng Sci* **35**:317–330.
217. G'Sell C, Dahoun A, Favier V, Hiver J M, Philippe M J and Canova G R (1997) Microstructure transformation and stress-strain behavior of isotactic polypropylene under large plastic deformation, *Polym Eng Sci* **37**:1702–1711.

218. Lee B J, Parks D M and Ahzi S (1993) Micromechanical modeling of large plastic deformation and texture evolution in semi-crystalline polymers, *J Mech Phys Solids* **41**:1651–1687.
219. Lee B J, Argon A S, Parks D M and Ahzi S (1993) Simulation of large strain plastic deformation and texture evolution in high density polyethylene, *Polymer* **34**:3555–3575.
220. Nikolov S, Doghri I, Pierard O, Zealouk L and Goldberg A (2002) Multi-scale constitutive modeling of the small deformations of semi-crystalline polymers, *J Mech Phys Solids* **50**:2275–2302.
221. Bedoui F, Diani J, Regnier G and Seiler W (2004) Micromechanical modeling of elastic properties in polyolefins, *Polymer* **45**:2433–2442.
222. Bedoui F, Diani J and Regnier G (2006) Micromechanical modeling of isotropic elastic behavior of semicrystalline polymers, *Acta Mater* **54**:1513–1523.
223. Tashiro K and Kobayashi M (1996) Molecular theoretical study of the intimate relationships between structure and mechanical properties of polymer crystals, *Polymer* **37**:1775–1786.
224. Tashiro K, Kobayashi M and Tadokoro H (1992) Vibrational spectra and theoretical three-dimensional elastic constants of isotactic polypropylene crystal: An important role of anharmonic vibrations, *Polymer J* **24**:899–916.
225. Choy C L and Leung W P (1985) Elastic moduli of ultradrawn polyethylene, *J Polym Sci Polym Phys Ed* **23**: 1759–1780.
226. Hine P J and Ward I M (1996) Measuring the elastic properties of high-modulus fibres, *J Mater Sci* **31**:371–379.
227. Rutledge G C (1997) Thermomechanical properties of the crystal phase of poly(ethylene terephthalate) by molecular modeling, *Macromolecules* **30**:2785–2791.
228. Matsuo M and Sawatari C (1990) Morphological and mechanical properties of poly(ethylene terephthalate) gel and melt films in terms of the crystal lattice modulus, molecular orientation, and small angle X-ray scattering intensity distribution, *Polym J* **22**:518–538.
229. Gorbunov A A and Skvortsov A M (1995) Statistical properties of confined molecules, *Adv Coll Interf Sci* **62**:31–108.
230. Guan X and Pitchumani R (2004) A micromechanical model for the elastic properties of semicrystalline thermoplastic polymers, *Polym Eng Sci* **44**:433–451.
231. Ahzi S, Parks D M and Argon A S (1995) Estimates of the overall elastic properties in semicrystalline polymers, in *Current Research in the Thermo-Mechanics of Polymers in the Rubbery-Glassy Range* (Ed. Negahban M) ASME, AMD **203**:31–40.
232. Gray R W and McCrum N G (1969) Origin of the γ relaxations in polyethylene and polytetrafluoroethylene, *J Polym Sci Part A-2: Polym Phys* **7**:1329–1355.
233. Boyd R H (1983) The mechanical moduli of lamellar semicrystalline polymers, *J Polym Sci Polym Phys Ed* **21**:493–504.
234. Spitalsky Z and Bleha T (2003) Elastic moduli of highly stretched tie molecules in solid polyethylene, *Polymer* **44**:1603–1611.
235. Flory P J (1953) *Principles of Polymer Chemistry*, Cornell University Press, Ithaca, NY.
236. Wu S (1989) Chain structure and entanglement, *J Polym Sci Polym Phys Ed* **27**:723–741.
237. Diani J, Bedoui F and Regnier G (2008) On the relevance of the micromechanics approach for predicting the linear viscoelastic behavior of semicrystalline PET, *Mat Sci Eng* **A475**:229–234.

Chapter 2

Modeling Mechanical Properties of Segmented Polyurethanes

V. V. Ginzburg, J. Bicerano, C. P. Christenson,
A. K. Schrock, A. Z. Patashinski

2.1. Introduction

Polyurethanes (PU) are widely used in a variety of industries (automotive, biomedical, *etc.*). While the most widespread use of PU in industry is in the form of foams (flexible, rigid, and viscoelastic), an important part is played by PU elastomers [1,2]. The utility of these materials is due to the versatility of their properties, relatively low cost, and well-developed preparation processes such as reaction injection molding (RIM). It is well-established that by varying chemical formulations, one can prepare PU elastomers with Young's modulus ("stiffness") ranging from "soft" (~1–2 MPa) to "hard" (>100 MPa). Understanding the dependence of the properties of PU elastomers on the formulation chemistry is a critical issue in polyurethane science and technology [3–27].

A typical polyurethane, poly(urethaneurea) (PUU) or polyurea elastomer is a copolymer consisting of *hard segment* (HS) *blocks* and *soft segment* (SS) *blocks*. At the use temperature (typically in the range of –30 to +150°C), the hard segment is a glassy (or even semicrystalline) polymer, while the soft segment is already above its glass transition temperature. In many instances, some or most hard segments aggregate into the so-called hard phase domains (glassy), while the rest of the material constitutes the soft phase (rubbery). By varying the relative amounts of the hard and soft phases, one can tailor the physical and mechanical properties of the PU elastomer. Although qualitatively it is clear that increasing the hard segment fraction (HSF) should lead to increasing modulus, strength, and brittleness of the material, a quantitative description of this dependence is rather complicated. One of the main (but certainly not the only) reasons for this complexity is that a change in the chemical formulation typically results in a change in the morphology of the final material.

Development of microphase-separated morphology in polyurethane elastomers has been a topic of theoretical and experimental studies for many years. In the 1960s, Cooper and Tobolsky [3] first proposed that hard segments segregate from the soft phase matrix and form hard phase aggregates that play a significant role in determining the mechanical properties of the material. Later, Koberstein and Stein [4] hypothesized that such a phase separation occurs only if the hard segment length exceeds some threshold value which is dependent on the chemical nature of the hard and soft segments. (Note that the hard segment length is, for a given soft segment length and functionality, proportional to the hard segment volume fraction, so that the hard segment length threshold corresponds to a volume fraction threshold.) By assuming that the hard and soft phases arrange themselves in lamellar domains and applying standard criteria for the order-disorder transition (ODT) in block copolymers, Koberstein *et al.* [4,5] estimated the threshold values for several specific hard and soft segments and showed that such estimates were in good qualitative agreement with experiments.

Subsequent studies of the microphase separated PU and PUU copolymers [6–27] utilized small-angle X-ray scattering (SAXS) and atomic force microscopy (AFM) to characterize the actual morphologies. By the mid-1990s, it was shown theoretically [28–32], that depending on the chemical interaction between the monomers and the relative fractions of the blocks, a block copolymer melt can arrange itself into several possible ordered structures (lamellar, cylindrical, spherical, bicontinuous gyroid, *etc.*). One could, therefore, expect to observe at least some of these structures in polyurethane elastomers. Cylindrical and spherical domains were indeed observed in recent AFM studies by O'Sickey *et al.* [24], and Garrett *et al.* [20,21]. Notably, in polyurethane elastomers, hard phase cylinders do not assemble into a hexagonal lattice (which is typically the case in monodisperse diblock copolymer melts). So far it is not clear whether the lack of hexagonal ordering is due to kinetic limitations and/or to the polydispersity of hard segments (which, for polyurethanes, is usually quite high), or some other factors.

Garrett *et al.* [20,21] also observed that a simple Koberstein-Stein [4] notion of the "threshold HSF" for phase separation breaks down at high HSF due to kinetic limitations. Indeed, as HSF approaches 40–70 wt% (the actual number depending on the chemistry of the hard and soft segments), vitrification often overtakes phase separation, and the system becomes "frozen" in an unstable or metastable configuration. Based on their studies, Garrett *et al.* [20,21] suggested that for any polyurethane elastomer, there should be an optimal HSF range. For lower HSF, thermodynamics prevents the system from microphase separation between hard and soft phases – thus, there is no reinforcement and the material is very soft. For higher HSF, vitrification freezes the system in an unfavorable configuration, and the material is likely to be very brittle.

The above discussion mainly concentrated on the understanding of the polyurethane elastomer morphology. A significant number of studies were also devoted to the relationship between the morphology and mechanical properties. Ryan et al. [12–15] demonstrated that for several model PUU copolymers, shear modulus, G, as a function of HSF can be qualitatively described by Davies' [35–37] equation, $G^{1/5} = \phi_H G_H^{1/5} + \phi_S G_S^{1/5}$ (where G_H, G_S, and ϕ_H, ϕ_S are the shear moduli and volume fractions of the hard and soft phases). Since Davies equation assumes a co-continuous morphology (both hard and soft phases percolate), it is valid primarily in the intermediate-to-high hard segment content and is expected to fail at low hard segment weight fractions.

To understand the dependence of mechanical properties on the hard segment content and chemical structure, researchers at The Dow Chemical Company (Christenson, Turner et al. [6,7]) synthesized several model PU elastomers with very low hard- and soft-segment polydispersity. These studies provided important clues to the ways in which hard segments provide reinforcement of the polyurethane materials. Subsequent work by Sonnenschein et al. [8,9] addressed a similar reinforcement provided by partially crystallizable soft segments. Sonnenschein et al. suggested that the dependence of the Young's modulus on the crystallinity of the polyol could be well described using the concept of the crystalline phase percolation. Other authors have discussed the notion of hard phase percolation (in their case, the hard phase was primarily composed of microphase-separated hard segments). Qualitatively, such a "percolation-type" transition was associated with the Koberstein-Stein concept of *"threshold hard segment length"* for phase separation. However, quantifying this percolation threshold as a function of the system formulation still remains a challenge.

Based on various experimental studies, one can schematically represent the morphology of segmented polyurethane (elastomer or flexible foam polymer) on the nano- and micro-scale as shown in Figure 2.1. For the range of hard segments volume fraction less than 50%, much of the space is occupied by the soft phase matrix. Microphase-separated nano-domains of the hard phase are dispersed in this matrix; they can be individual islands or can form percolated networks. Finally, there could also be some larger (micron-sized) macrophase-separated domains of hard phase, where hard phase domains are ordered at the macro-scale (this is especially true in the case of flexible foams). The relative amounts of all these elements depend on the formulation and processing history.

From the above discussion it follows that in order to understand the properties of segmented PU elastomers, one needs to combine thermodynamic and micromechanical models in a consistent way. In this chapter, we outline our preliminary attempts to develop such a model (see Figure 2.2).

The chapter is structured as follows. In Section 2.2 (see also our recent paper [33]), we discuss recent developments in relating linear elastic properties of polyurethane elastomers to their formulation (hard and soft segment chemistry, soft segment molecular weight, and hard segment weight fraction). In Section 2.3, the extension of the model to describe tensile stress-strain behavior

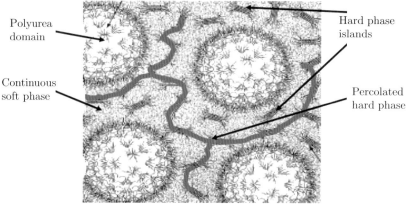

Idealized polymer morphology

Figure 2.1. Complexity in polyurethanes. On a nano- to micro-scale, polyurethanes contain various components, including macrophase separated (micron-size) hard phase domains, dispersed nanometer-sized hard phase "islands", and percolated hard phase "nano-network", all dispersed in the soft phase "matrix"

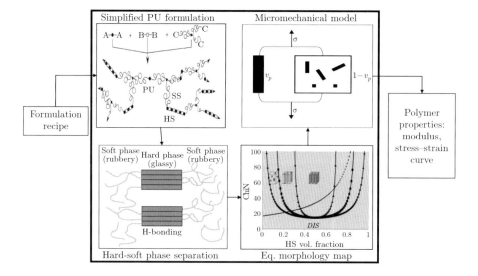

Figure 2.2. Schematic representation of the multi-scale modeling approach outlined in Section 2.2

is reviewed. In Section 2.4, we will briefly touch on other topics such as the role of dynamics and process history in flexible foams, and the role of the soft segment crystallinity in some polyester and polyether elastomers. We conclude by summarizing both successes and challenges in polyurethane theory and modeling.

2.2. Predicting Young's modulus of segmented polyurethanes

2.2.1. *Relationship between Young's modulus and formulation – experimental observations*

We begin our discussion by considering experimental data for various polyurethane elastomers with different hard segment chemistries and weight fractions. Specifically, we consider the dependence of the room temperature Young's modulus on the hard segment weight fraction for three hard segment types shown in Figure 2.3. The data for the MDI-BDO and the MDI-DETDA elastomers are based on the work by Christenson et al. [6,7], while the data for the TDI-PU systems are based on the study by Gier et al. [18] For both MDI-BDO and MDI-DETDA studies, the soft segment was a poly(propylene oxide) (PO) diol with molecular weight of 2000 g/mol; for the TDI-PU study, the soft segment was a PO triol with molecular weight of 3000 g/mol. The mixture composition was varied to obtain different hard segment lengths. Schematic representation of the structure of the resulting segmented polyurethanes and reactions taking place during polymerization is given in Figure 2.4. The hard segments

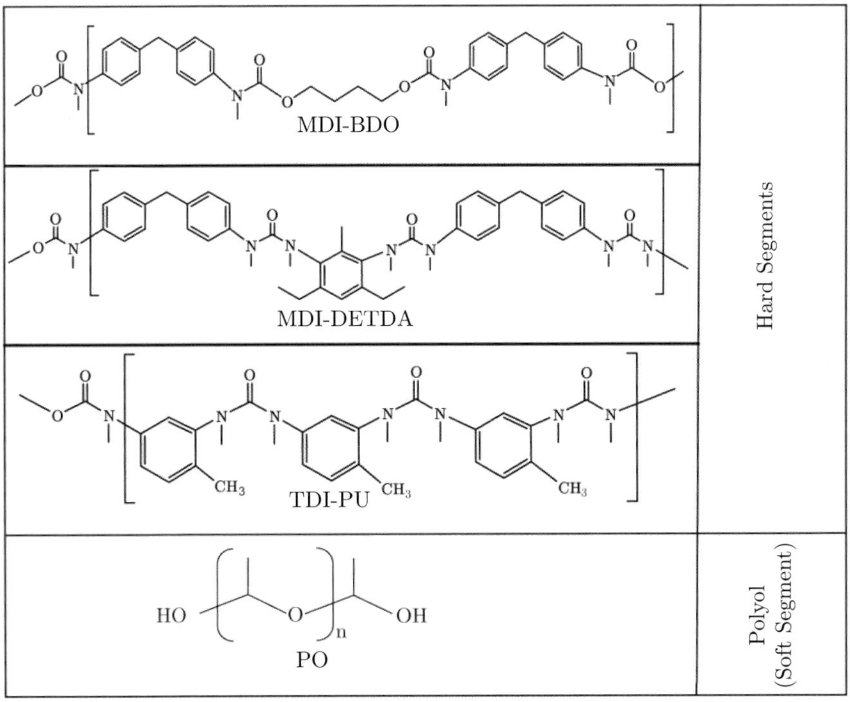

Figure 2.3. Chemical structures of hard and soft segments. Each hard segment contains isocyanate and chain extender. Isocyanates: 4,4-methylene diisocyanate (MDI) and toluene diisocyanate (TDI). Chain extenders: water (for TDI-PU), 3,5-diethyltoluenediamine (for MDI-DETDA), and 1,4-butanediol (for MDI-BDO). Soft segments: poly(propylene oxide) (PO) [33]

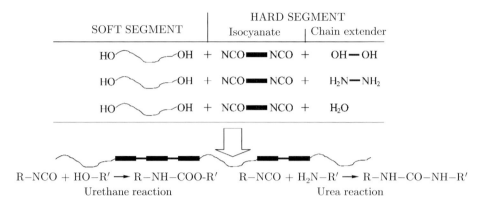

Figure 2.4. Schematic representation of segmented PU formation as a result of polymerization reaction. Soft segment–hard segment linkages occur via urethane reaction (bottom left). Linkages within the hard segment occur *via* urethane reaction in the case of BDO and urea reaction (bottom right) in the case of DETDA and water chain extenders. In the case of water chain extension, urea formation also involves the release of one CO_2 molecule for each urea group formed [33]

contain hydrogen-bonding moieties and therefore could assemble into small glassy domains "fortified" by such hydrogen bonding. The question is how these domains affect the mechanical properties of the polymer.

The dependence of the Young's modulus on the hard segment weight fraction for three different hard segments is shown in Figure 2.5. It can be clearly seen that for any given hard segment weight fraction, the modulus of a TDI-PU polymer is higher than that of an MDI-DETDA, which, in turn, is higher than that of an MDI-BDO polymer. What is the reason for this effect? Qualitatively, it is known that urea groups have higher density of hydrogen bonding than urethane groups, and so one could expect that TDI-PU hard segments would be more "phase separated" than MDI-DETDA hard segments and those, in turn, would be more "phase separated" than MDI-BDO hard segments. Another way of saying the same thing would be to compare cohesive energy densities, E_{coh}, or solubility parameters, $\delta_h = \sqrt{E_{coh}}$, for all of these hard segments. Indeed, calculations using a group contribution approach (see refs [38,39]) show (see Table 2.1) that δ_h (TDI-PU) $> \delta_h$ (MDI-DETDA) $> \delta_h$ (MDI-BDO), which correlates qualitatively with the modulus behavior. However, it would be desirable to develop a quantitative (or at least semi-quantitative) theory that would translate this knowledge into mechanical property prediction. Below, we outline such a theory.

2.2.2. *Theory*

It is known that in segmented polyurethanes, the main reinforcement mechanism comes from the aggregation of the "hard segments" into "hard phase" nanodomains. Thus, our polymer in effect becomes a nanocomposite, with glassy "fillers" (hard phase) in a rubbery matrix (soft phase). However, unlike most nanocomposites, fillers are chemically bonded to the matrix.

Modeling Mechanical Properties of Segmented Polyurethanes

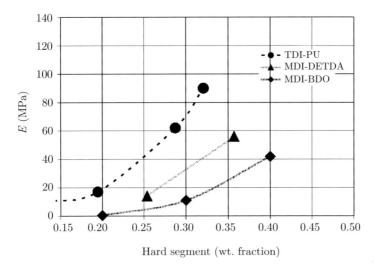

Figure 2.5. Room temperature Young's moduli of various polyurethane elastomers as function of hard segment weight fraction [33]

Table 2.1. Parameters used in model calculations [33]

Name	Hard segment parameters			
	δ_h, $(J/cm^3)^{1/2}$ (van Krevelen est.)	δ_h, $(J/cm^3)^{1/2}$ (best fit)	M_I, (g/mol)	M_C, (g/mol)
MDI-BDO	22.78	23.25	250	90
MDI-DETDA	25.38	24.90	250	178
TDI-PU	28.60	29.15	174	−28[a]
Name	Soft segment parameters			
	δ_h, $(J/cm^3)^{1/2}$ (van Krevelen est.)	δ_h, $(J/cm^3)^{1/2}$ (best fit)	M_S, (g/mol)	T_g, (°C)
PO	17.00	16.50	58	−40[b]

[a] Chain extension with water is associated with the loss of one CO_2 molecule as the byproduct of the urea reaction, thus the effective mass of chain extender, M_C, equals $M(H_2O) - M(CO_2) = 18 - 46 = -28$ g/mol

[b] Actual glass transition temperature for PO is slightly lower, in the range −50 to −60°C. The discrepancy is probably due to the influence of the dispersed hard segments on the soft phase rheology, resulting in the T_g increase (Fox-Flory effect)

As in any composite, the mechanical properties of segmented polyurethanes depend on the volume fraction ("loading") of the fillers and the morphology of the fillers. The problem is now complicated by the fact that both the morphology of the hard phase domains and their effective volume fraction are determined by the thermodynamics ("formulation") and kinetics ("preparation conditions" and process thermal history) of polymerization-induced phase separation. The interplay between thermodynamics and kinetics is especially important in the case of PU foams, both flexible and rigid (see, *e.g.*, Bicerano *et al.* [40] and Gruenbauer *et al.* [41] for detailed discussion). However, when preparation is done under "quasi-equilibrium conditions", for example, using solution polymerization, it is possible to neglect kinetic limitations and concentrate on the thermodynamics of the mixture. If – based on thermodynamic models – one can estimate the volume fraction and morphology of the hard phase nano-domains, it would be then possible to utilize classical composite micromechanical theories to estimate the mechanical (elastic) properties of the polymers.

2.2.2.1. *Thermodynamic model*

Let us consider a hypothetical segmented polyurethane polymer. Its soft segment is made up of a polyol (such as polypropylene oxide (PO)) with hydroxyl functionality f_P and molecular weight M_S. We can also define the equivalent weight of the soft segment as $E_S = M_s/f_p$. In this paper, we neglect the soft segment polydispersity (for many commercial polyols, the polydispersity index (PDI) ranges between 1.03 and 1.20, so this approximation is justified). Thus, for all systems considered here, the soft segment equivalent weight $E_S = 1000$ g/mol. The hydroxyl functionality, f_P, is 2 for MDI-based systems, and 3 for TDI-PU systems.

The hard segment is being formed as a result of a reaction between isocyanate and chain extender (Figure 2.4), so the total hard segment weight fraction, f, is the sum of the weight fractions of the isocyanate and the chain extender. (It is slightly more complicated if the chain extender is water, in which case one would need to subtract the mass of the carbon dioxide produced as a result of a polycondensation-type urea reaction). Since hard segments are created *in situ*, one could expect that there would be a distribution of their lengths (polydispersity). It is easy to estimate the number-average molecular weight of a hard segment:

$$\overline{M}_H = 2E_S \frac{f}{1-f}, \qquad (2.1)$$

where prefactor 2 reflects the fact that each hard segment is attached by urethane bonds to a polyol molecule on both sides. (In deriving Eq. (2.1) we assumed perfect stoichiometry and full conversion – these assumptions would be implicit throughout our further analysis). However, the distribution of the hard segment lengths is a more complicated issue – and its impact on the degree of hard-soft phase separation is even more complex. For the sake of simplicity, we neglect the hard segment polydispersity and assume that all hard segments have the same length. Extending the model to include the hard segment polydispersity will be the subject of future work.

We model the polyurethane polymer as a multiblock copolymer of the $(AB)_n$ type. Phase behavior of multiblock copolymers has been studied theoretically by Benoit and Hadziioannou [31]; it was found that it is very similar to that of the AB diblock copolymers, with all order-disorder transitions being shifted towards lower temperatures. The phase behavior of a given AB-diblock is determined by two main factors: composition parameter, f_A, and the "incompatibility" parameter, $\chi_{AB}(N_A + N_B)$, where χ_{AB} is the Flory-Huggins parameter [42,43], and $N_{A,B}$ are the chain lengths (degrees of polymerization) of A and B blocks. The approximate phase diagram and a graphical description of each "classical" phase (spherical, cylindrical, and lamellar) is given in Figure 2.4; for more details, see references [28–32]. Note that for a given polyol (soft segment) chemistry and molecular weight, increasing the hard segment weight fraction leads to the increase in both f and χN, as is schematically depicted by the dashed curve.

In the case of the hard-soft PU segmented copolymers, we denote the hard block as A and the soft block as B, and set $f_A = f$. To estimate the incompatibility parameter, we make use of the following formulae:

$$\chi_{hs} = v_{ref} \frac{(\delta_h - \delta_s)^2}{RT} = v_{ref} \frac{(\sqrt{(E_{coh})_h} - \sqrt{(E_{coh})_s})^2}{RT} \tag{2.2}$$

$$N_h = \frac{\overline{M_H}}{2(\rho_H v_{ref})} \tag{2.3a}$$

$$N_s = \frac{E_s}{(\rho_S v_{ref})} \tag{2.3b}$$

$$\chi N = \chi_{hs}[N_h + N_s] = \frac{(\delta_h - \delta_S)^2}{RT}\left[\frac{\overline{M_H}}{2\rho_H} + \frac{E_S}{\rho_S}\right] \tag{2.4}$$

Here, $\delta_{h,s}$ is the van Krevelen solubility parameter of the hard or soft segment, and $(E_{coh})_{h,s}$ is the van Krevelen cohesive energy density of the hard or soft segment; $\rho_{H,S}$ denotes the density of the hard or soft segment; v_{ref} is a reference volume (we chose $v_{ref} = 58$ cm^3/mol to roughly correspond to the molar volume of a PO repeat unit).

Equation (2.4) holds the clues to our earlier observation that polyurethanes with higher $(E_{coh})_h$ exhibit higher Young's modulus for a given f, E_S. Indeed, increasing $(E_{coh})_h$ while keeping everything else constant would lead to an increase in hard-soft incompatibility and, therefore, to a higher degree of microphase separation. Similarly, increasing polyol equivalent weight and/or hard segment weight fraction would lead to an increase in microphase separation and, therefore, increase in modulus.

To estimate the phase diagram (including both order-disorder and order-order transitions), we utilize a simple interpolation between the strong-segregation results and the weak-segregation results:

$$(\chi N)_i = (\chi N)_0 \frac{1}{1 - \dfrac{0.25 - f_i(1-f_i)}{0.25 - \alpha_i}}, \qquad (2.5)$$

where $i = DS$ (disordered-spherical), SC (spherical-cylindrical), and CL (cylindrical-lamellar); $\alpha_i = f_i(1-f_i)$; $\alpha_{DS} = 0$, $\alpha_{SC} = 0.098$, $\alpha_{CL} = 0.206$. The prefactor $(\chi N)_0 = 10.5$ within the mean-field approximation for diblock copolymers [28,32]; according to the more accurate fluctuation theory for the diblocks [30], $(\chi N)_0 = 10.5 + C/(N_A + N_B)^{1/3}$. For multiblock copolymers, $(\chi N)_0$ is in the range of 15–20, depending on the total number of blocks per chain. Thus, choosing $(\chi N)_0 = 18$ gives a reasonable approximation to roughly account for the role of both fluctuations and multiblock nature of our polymers. (We postulated here that during the preparation, both blocks are flexible, and the hard block becomes rigid only after much of the phase separation has occurred; otherwise, the phase behavior could be modified slightly, as shown in several studies on the thermodynamics of rod-coil block copolymers [44–47].)

Equation (2.5) is used to predict the degree of phase separation and the morphology of the microphase separated block copolymer. For a given degree of incompatibility, χN, we solve for f_i. Then, if the hard segment weight fraction is less than f_{DS}, all hard segments are uniformly ("molecularly") dispersed in the soft matrix. If $f_{DS} < f < f_{SC}$, hard segments begin to form spheres; for $f_{SC} < f < f_{CL}$, spheres percolate to form cylinders; and for $f_{CL} < f < f_{LC}$, both hard and soft domains are organized into lamellar sheets (or form bicontinuous structures as often is the case in flexible foams). At higher hard segment fractions ($f \to 1$), there are inverse (soft-in-hard) phases which are not considered in this study. (Inverse phases can be important, for example, as one analyzes rigid PU foams, but those systems are known to be extremely non-equilibrium and thus require a completely different type of theoretical treatment.)

Based on the above estimates, we can determine the approximate location of any segmented polyurethane system on the phase diagram (Figure 2.6). Because of their respective solubility parameters (δ_h(TDI-PU) > δ_h(MDI-DETDA) > δ_h(MDI-BDO), see Table 2.1 for numerical values estimated using group contribution approach), the hard segment/soft segment χ-parameter is highest for TDI-PU, intermediate for MDI-DETDA, and lowest for MDI-BDO. Thus, for a given hard segment weight fraction, χN(TDI-PU) > χN(MDI-DETDA) > χN(MDI-BDO). This, in turn, leads to a decrease in the transition hard segment fraction (such as sphere-to-cylinder transitions, marked by black circles in Figure 2.6), as the hard segment is changed from MDI-BDO to MDI-DETDA to TDI-PU. This change has a dramatic impact on the mechanical properties, as we will demonstrate later.

The above discussion assumed that both hard and soft segments were completely monodisperse. As discussed above, this is not always true, and polydis-

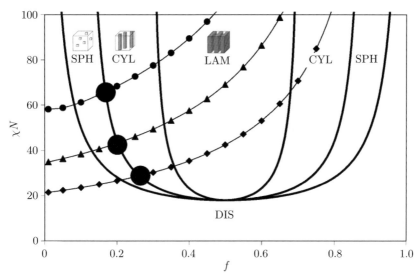

Figure 2.6. Multiblock copolymer phase diagram and its relationship to segmented polyurethanes. Thin lines with symbols show how increasing hard segment weight fraction for a given soft segment changes equilibrium morphology of segmented polyurethanes with TDI-PU (circles), MDI-DETDA (triangles), and MDI-BDO (diamonds) hard segments. Black circles represent the locations of sphere-to-cylinder transition for each system. All estimates are for room temperature and assuming the equivalent weight of PO polyol of 1000 g/mol [33]

persity of hard segments in PU is a well-known phenomenon [34]. Polydispersity of the hard block could, in principle, have a significant effect on the phase diagram of the block copolymer. At present, the influence of polydispersity on the phase behavior is not well understood; there are only a few theoretical studies [48,49].

The thermodynamic model described in this section enables one to estimate the fraction of the hard phase that is microphase-separated and aggregated into nanodomains, and determine whether those nanodomains form spheres, cylinders, or lamellae. The next step is to develop an appropriate micromechanical model to enable prediction of elastic properties (Young's modulus).

2.2.2.2. *Micromechanical model*

We use Kolařik's [50,51] approach to describe the dependence of the modulus of the overall material on the hard-phase nano-domain morphology; a schematic representation of the Kolařik "equivalent box model" (EBM) is given in Figure 2.7. The modulus $E(f,T)$ of a segmented polyurethane as a function of the hard segment weight fraction, f, and temperature, T, is given by the following expression:

$$E(f,T) = v_P E_H + (1 - v_P)E_{FS}, \qquad (2.6)$$

where E_H is the modulus of the hard phase (in this study we set $E_H = 5$ GPa), E_{FS} is the modulus of the "filled soft phase" (to be defined later), and v_P

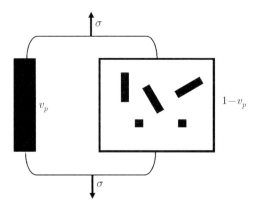

Figure 2.7. Schematic representation of Kolařik micromechanical model used to calculate Young's modulus of segmented polyurethane elastomers. Solid black rectangles represent hard phase domains. Percolated hard phase has effective volume fraction, v_p, and the rest of the hard domains are dispersed in the soft phase matrix [33]

(volume fraction of "elastically active" percolated hard phase) is defined as:

$$v_P = \left[\frac{f - f_{perc}}{1 - f_{perc}}\right]^\delta \qquad (2.7)$$

Here, δ is the percolation exponent (we take $\delta = 2.5$), and f_{perc} is the percolation threshold for the hard phase. The percolation exponent, δ, typically ranges between 1.5 and 2 (see, *e.g.*, ref. [52]), depending on the type of the system and the property described by a percolation model (modulus, conductivity, *etc.*). There are instances, however, when the percolation exponent could be larger than 2 (see, *e.g.*, ref. [53]). Various models (*e.g.*, double percolation – see ref. [54]) have been proposed to explain these high percolation exponents. In our analysis, we refrain from ascribing any specific meaning to exponent $\delta = 2.5$, and treat it simply as an adjustable parameter that is found from the best fit to experimental data.

An important point is that the percolation threshold, f_{perc}, depends on the morphology of the percolating clusters, and thus on the incompatibility between the two phases. We postulate that $f_{perc} = f_{SC}$, so that once hard phase spheres aggregate into infinite cylinders, percolated hard phase is formed. This assumption is a crucial link that relates thermodynamic information about the hard segment (its cohesive energy density or solubility parameter) to mechanical properties of polyurethanes based on that hard segment.

It is important to note that Eq. (2.7) should be used only in the vicinity of the hard phase percolation. Indeed, the meaning of the percolation exponent, δ, is to account for the fact that the percolated pathways are not linear but have some complicated morphology; thus, the efficiency of the reinforcement ("fraction of elastically active hard phase elements") is much less than 100%.

However, at higher hard segment weight fractions, lamellar domains become thicker and longer, leading to more correlated structures. Ultimately, at very high hard segment weight fractions, the dependence of modulus on the hard segment weight fraction should become closer to linear and then saturate. We are not attempting to describe this region of the phase map here, concentrating mainly on the composition range in the vicinity of the percolation threshold that is more typical for polyurethane elastomers and flexible foams.

The above discussion was mainly dealing with the percolated hard phase. The remaining portion of the material is the "filled soft phase" which consists of the soft phase matrix and hard phase spheres. Modulus enhancement provided by the spheres can be estimated by [39]:

$$E_{FS} = E_{SP}(1 + 2.5X + ...), \qquad (2.8)$$

where E_{SP} is the modulus of the unfilled soft phase, and the term in brackets in the right-hand side represents the reinforcement due to fillers. In writing Eq. (2.8), we omitted the higher terms in the Taylor series expansion with respect to the volume fraction of spheres in the soft phase, X. We note that $X = 0$ if $f < f_{DS}$; for $f > f_{DS}$, $X \approx f - v_p - f_{DS}$.

It should be mentioned that in general, hard phase clusters can be nonspherical, as discussed in various earlier papers. In this case, the modulus increase could strongly depend on the aspect ratio; the effect of the aspect ratio can be modeled through the micromechanical models of Halpin and Tsai [55] or Mori and Tanaka [56]. However, as we already commented above, below the spherical-to-cylindrical transition, most of the hard phase nano-domains have an aspect ratio close to 1. Above the spherical-to-cylindrical transition that is in our model associated with percolation threshold, most of the cylinders participate in the formation of the percolated hard phase, while the soft phase primarily contains hard phase "islands" with smaller aspect ratios. Therefore, in our analysis we assume that all the fillers dispersed within the soft phase are spherical (or have aspect ratios close to one).

In order to complete our model, we need to know the soft phase modulus, E_{SP}. It depends on several factors, most notably polyol equivalent weight, E_S, polyol functionality, f_P, and polyol glass transition temperature, T_{gP}. At temperatures substantially above T_{gP} (elastomeric region), the modulus of the soft phase can be described using network elasticity models [57–59]:

$$E_{SP}^r(T) = \frac{3\rho RT}{M_x}\left[1 - \frac{2}{f_P}\right] + \frac{3\rho RT}{M_c}, \qquad (2.9)$$

where $R = 8.31$ J.mol^{-1}.K^{-1} is the gas constant, ρ is the soft phase density, M_x is the molecular weight between crosslinks in the network, M_c is the critical entanglement molecular weight, and T is the absolute temperature. (Note that effective polyol functionality, f_P, could itself be dependent on composition if one accounts for the hydrogen bonding between hard segments in addition to the covalent bonding. We leave the investigation of this effect for the future.)

Equation (2.9) contains two contributions: covalent network (first term) and entanglement network (second term). It is important to note that when polyol functionality $f_P = 2$, polymerization creates only linear chains, so the covalent network contribution to modulus should be zero. Thus, in the systems considered here, effective soft phase modulus is dominated by the entanglement component determined by the critical molecular weight, M_c. For PO, estimated $M_c \approx 2800$ g/mol; we will use this value in subsequent estimates. Note that although the initial polyol chains are usually unentangled due to their relatively low molecular weight (1000–6000 g/mol), the molecular weight increases substantially during polymerization. The final polyurethane molecules might contain of the order of 10–100 soft segments (molecular weights of the order of 10,000–100,000 g/mol), linked by the hard segments. At low-to-intermediate hard segment contents, most of those hard segments are still dissolved in the soft phase, acting as both "chain extenders" and weak crosslinks; only a small part of the hard segments forms hard phase domains. As the hard segment content increases, a larger portion of the hard segments is organized into hard phase domains through hydrogen bonding. To describe the transition from glassy to rubbery behavior for the soft phase, we use the following interpolating function:

$$\ln E_{SP}(T) = 0.5\ln\left(E_{SP}^G E_{SP}^r(T)\right) - 0.5\ln\left(E_{SP}^G / E_{SP}^r(T)\right)\tanh\left[\frac{T - T_{gP}}{\Delta T_g}\right] \quad (2.10)$$

with E_{SP}^G and ΔT_g being adjustable parameters that could be extracted from experimental dynamical mechanical spectroscopy (DMS, also known as DMA) data for lightly-crosslinked soft phase. The meaning of these parameters is reasonably straightforward: E_{SP}^G is the low-temperature ("glassy") modulus of a given PU elastomer, while ΔT_g is the "width" of the glass transition region. For polymers considered in this study, $E_{SP}^G = 2.5$ GPa, which is consistent with accepted values for glassy polymers. Transition width, ΔT_g, was set to 10°C for all systems considered in this study.

We are now in a position to apply our model to describe various test cases and compare its predictions with experiment.

2.2.3. Young's modulus: comparing theory with experiments

In Figure 2.8, we plot model calculations for the room temperature modulus of several TDI-PU, MDI-DETDA, and MDI-BDO systems (see also Figure 2.5). Calculation parameters are given in Table 2.1. The solubility parameters were first calculated using the group contribution technique (second column in Table 2.1) and then slightly adjusted to optimize the agreement between theory and experiment (third column in Table 2.1). It can be seen that the calculated modulus shows a good agreement with experimental data, despite all the simplifications used in the model. It is even more important that the model correctly translates the hard segment cohesive energy density difference into the difference in polyurethane polymer modulus. Indeed, TDI-PU has the high-

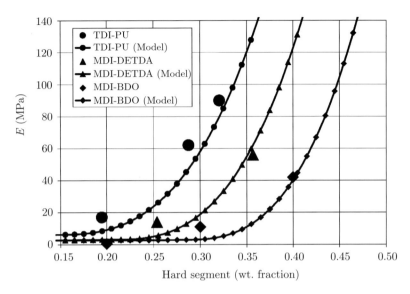

Figure 2.8. Comparison between modeling and experiment for Young's modulus as a function of the hard segment chemistry and weight fraction. All calculations correspond to room temperature ($T = 298$ K) and PO equivalent weight of 1000 g/mol. Functionality $f_P = 2$ for MDI-BDO-PO and MDI-DETDA-PO, and $f_P = 3$ for TDI-PU-PO elastomers [33]

est solubility parameter of the three hard segments; therefore, it also has the highest Flory-Huggins parameter, χ_{HS}, with the PO soft segment. This, in turn, results in the stronger incompatibility and better phase separation for the TDI-PU hard segments. Thus, for the segmented PU with TDI-PU hard segments, spherical-to-cylindrical transition (see Figure 2.6) occurs at lower hard segment fraction than for those with MDI-DETDA hard segments, and for MDI-DETDA systems it is lower than for MDI-BDO elastomers. Hence, percolation threshold for the creation of the percolated hard phase is lowest for TDI-PU, intermediate for MDI-DETDA, and highest for MDI-BDO. For any given hard segment fraction f, the lower the percolation threshold (dictated by the hard segment chemistry and soft segment equivalent weight), the higher the modulus.

In Figures 2.9 and 2.10, we show calculated and measured tensile storage modulus of MDI-BDO-PO and MDI-DETDA-PO polyurethanes as a function of temperature. It can be seen that our model does a fair job in describing temperature dependence of the modulus, although it certainly cannot predict plateau modulus value below the soft phase T_g (low temperatures) and cannot predict the soft phase T_g itself. We note that the value of the soft phase T_g used in our calculations ($-40°C$) is somewhat higher than literature value for poly(propylene oxide) (-50 to $-60°C$). This indicates that dispersed hard segments are likely to elevate the glass transition temperature of the soft phase matrix (see, *e.g.*, Chapter 6 of reference [39]). More detailed investigation of this effect will be the subject of future studies. It is also interesting to observe

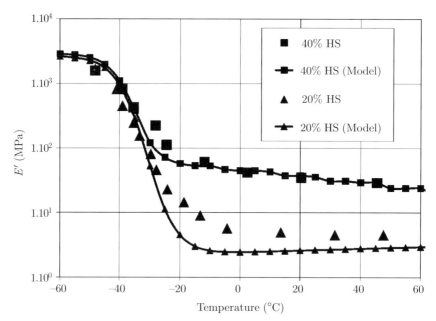

Figure 2.9. Tensile storage modulus, E', and calculated Young's modulus, E, as a function of temperature for several model MDI-BDO-PO segmented polyurethane elastomers [33]

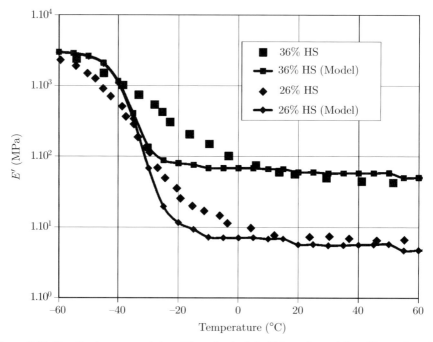

Figure 2.10. Tensile storage modulus, E', and calculated Young's modulus, E, as a function of temperature for several model MDI-DETDA-PO segmented polyurethane elastomers [33]

several other features of experimental DMS curves not captured by the model. For example, the model does not predict the broadening of the glass transition region in some of the systems (*e.g.*, 36% HS MDI-DETDA elastomer). This broadening is probably due to the influence of the dissolved (non-phase separated) hard segments and their impact on the rheology of the soft phase (Fox-Flory effect).

The examples above show the influence of the hard segment structure on mechanical properties. We now concentrate on the role of soft segment length. In their studies of MDI-BDO-PO segmented polyurethanes, Zdrahala and Critchfield [11] varied the length of PO diol and measured DMS curves for three systems with the same hard segment weight fraction yet different diol chain lengths. In their study, the hard segment weight fraction $f = 0.5$, and diol chain lengths were 2000, 3000, and 4000 g/mol. As PO chain length was increased, the room temperature modulus increased as well (Figure 2.11, diamonds). This trend is very reasonable if one recalls that phase separation is dictated by the product χN; increase in the PO molecular weight does not affect χ but increases N. Another (perhaps less formal and more intuitive) way of saying the same is that increasing the soft segment equivalent weight at constant hard segment weight fraction results in the creation of longer (on average) hard segments; longer hard segments are thermodynamically more likely to segregate into strongly reinforcing nano-domains, leading to higher

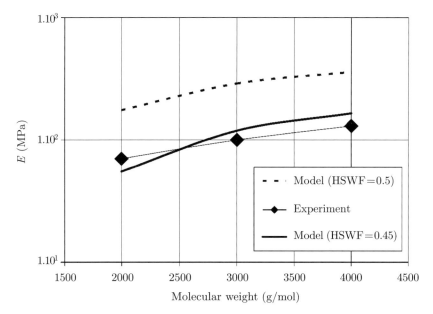

Figure 2.11. Room-temperature Young's modulus, E, as a function of soft segment molecular weight for several model MDI-BDO-PO segmented polyurethane elastomers. Diamonds – experimental data from ref. [11]; dashed line – model calculations assuming $f = 0.5$; solid line – model calculations assuming $f = 0.43$ [33]

modulus and strength. Interestingly, this is contrary to the trend existing in amorphous rubbery networks, where increasing molecular weight between crosslinks leads to the decrease in the network modulus. Our model is able to qualitatively reproduce this trend (Figure 2.11, dashed line), but overestimates the actual numbers for the modulus. This might be due to the incomplete phase separation in the $f = 0.5$ systems or perhaps the change in the "elastically effective" percolated hard phase volume fraction as the system moves away from its percolation threshold. Assuming a slightly lower hard segment weight fraction ($f = 0.43$ instead of $f = 0.5$), we can obtain not just qualitative but also a quantitative agreement with the experiment (Figure 2.11, solid curve). Temperature dependence of modulus (DMS curves) for two systems with different PO molecular weights (2000 and 4000 g/mol) is shown in Figure 2.12. Once again, there is reasonable agreement between theory and experimental data.

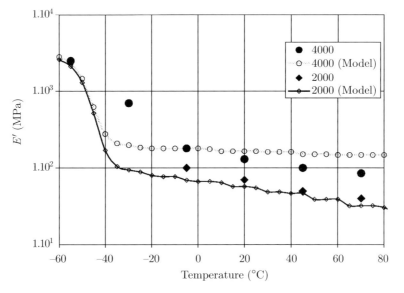

Figure 2.12. Tensile storage modulus, E', and calculated Young's modulus, E, as a function of temperature for several model MDI-BDO-PO segmented polyurethane elastomers. In model calculations, we take $f = 0.43$ [33]

2.3. Modeling tensile stress-strain behavior

While elastic modulus is an important characteristic of a polymeric material, it mainly describes material behavior at small deformations only. This could be important for some applications (note, for example, that scratch resistance or hardness often can be directly linked to the modulus); however, in general, one requires the knowledge of mechanical response of the material over a broad range of deformations. For polyurethane elastomers and TPU's, as well as for polyurethane foams, both tensile and compressive deformations play a significant role in determining their performance.

As discussed by numerous authors [1,40,60–64], stress-strain behavior of polyurethane elastomers and foams is extremely complex and is typically characterized by the following features:

- Strong strain-rate dependence;
- Substantial hysteresis (loading and unloading portions do not coincide);
- Extremely nonlinear behavior (strain softening, often followed by strain hardening).

While many of the above features have been observed even in low-hard-segment polyurethane materials, much of the complexity is certainly added by the hard-phase component. Indeed, as the material is strained, many hard phase domains undergo orientational changes, plastic deformation, break-up and re-forming of numerous hydrogen bonds. The morphology of the material, therefore, is constantly changing. Accordingly, to develop a predictive – or even reasonably descriptive – theoretical model of polyurethane tensile behavior still remains a very challenging problem.

So far, the simplest – and most popular – way of describing stress-strain behavior of polyurethane elastomers is to apply the apparatus of the rubber elasticity theory [58]. Laity *et al.* [60] illustrated the success and limitations of these models by comparing the predictions of the Klüppel-Schramm [65] equations to experimentally observed curves for segmented PU elastomers Pellethane® 2363-55D and 2363-80A (Dow Chemical Company, Midland, MI). The comparison between theory and experiment for the 2363-80A is shown in Figure 2.13. It can be seen that the rubber elasticity approach provides a reasonable description of the observed behavior. However, it is important to note that parameters of the theoretical curve (*e.g.*, "chemical", G_c, and "entanglement", G_e, shear moduli) were not predicted from the first principles, but fitted to obtain the best agreement with experiments. Thus, the value used for G_e (~11.6 MPa) appears to be significantly higher than what should have been expected based on the polyol critical entanglement weight (~1–2 MPa). A similar discrepancy was observed in a study of several PU elastomers done by Špirkova [66], where the effective crosslink density in PU elastomers was always determined to be higher than what should have been expected based on rubber elasticity theory alone. The difference can be ascribed either to hydrogen bonding – or, more likely, to the effect of the hard phase domains (the latter factor certainly becomes dominant when hard segment weight fraction becomes sufficiently large, as discussed in the previous section). Accordingly, it becomes necessary to investigate how the hard segments behave when subject to deformation. This, in turn, would require more complex composite models.

In recent years, mechanical behavior of filled elastomers, block copolymers with glassy blocks, and partially crystalline polymers have been extensively studied by several research groups [61, 62, 67–72]. Drozdov and co-workers [70–72] utilized the concept of "adaptive links" in which the material is

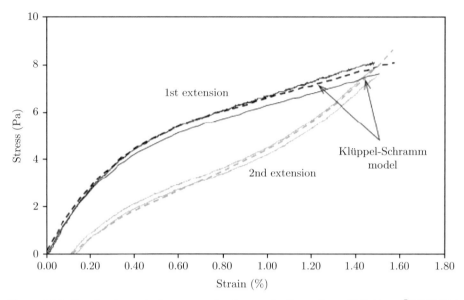

Figure 2.13. Tensile stress-strain curves of a thermoplastic urethane Pellethane® 2363-80A (Dow Chemical Company) under uniaxial extension [60]

represented by a series of parallel elastic springs replacing each other according to some special rule as a function of deformation and/or time. Models based on this concept can successfully reproduce many realistic features of polyurethane stress-strain curves. However, they typically require a number of adjustable parameters, not always predictable from the first principles.

A slightly different approach to tensile behavior of polyurethanes and other multiphase elastomers was adopted by Boyce and co-workers in a series of papers [61,62,68,69]. In a comprehensive 2005 paper, Qi and Boyce [61] investigate the influence of the hard phase (hard fillers in the case of composites) on the large-scale deformation. In particular, they describe the derivation of a constitutive equation for a polyurethane elastomer by splitting the overall stress into the hyperelastic contribution from the filled soft phase and the nonlinear visco-elasto-plastic contribution from the "hard domain" (which is similar to what was referenced as a "percolated hard phase" in the previous section). The schematics of their model and a representative comparison between theory and experiment for a specific thermoplastic urethane (TPU) material with Shore A hardness of 92 are given in Figure 2.14 (note that tensile measurements in this study were compression rather than tension). The model demonstrated a remarkable degree of accuracy in predicting stress *vs.* strain, strain rate dependence, and hysteresis. The number of parameters used in the calculation was fairly limited, and their physical meaning was well-defined.

In calculating the unloading behavior, Boyce and co-workers had to make various assumptions about the evolution of the "effective soft phase volume fraction" as a function of deformation. They postulated that the hysteresis and

Modeling Mechanical Properties of Segmented Polyurethanes

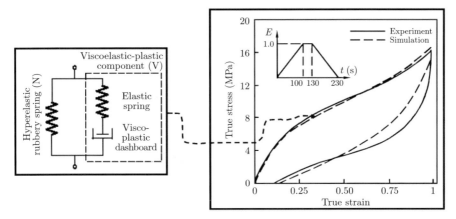

Figure 2.14. Micromechanical model and uniaxial compressive curves (measured and simulated) for a typical thermoplastic urethane [61]

the permanent set of polyurethane elastomers or TPU's come from the fact that soft phase domains were initially "constrained" by the surrounding hard domains and gradually released at large deformations. In addition, Qi and Boyce [61] assumed that the hard domain deformation was always irreversible (this assumption was based on the work of Estes *et al.* [64]). These assumptions were crucial in determining the non-equilibrium, irreversible behavior of the materials. It was important that these assumptions be validated experimentally; however, *in situ* observation of the behavior of nanoscale domains during deformation is a formidable challenge that is only now being addressed. In a recent paper, Christenson *et al.* [27] performed an extensive study of several segmented PU elastomers with varying degrees of phase separation using *in situ* AFM, SAXS, DMA, and tensile measurements. For an intermediate-hard-phase content TPU labeled PEUU (MDI-EDA hard segment, PTMO diol with Mw = 2000 g/mol, hard segment weight fraction 23%), Christenson *et al.* [27] proposed a model of irreversible orientation and re-orientation of hard phase domains during the loading and unloading portions of the conditioning and first cycle (Figure 2.15). AFM and SAXS measurements were also performed for a higher hard-segment content material PEU (40% hard segment) where the hard and soft phases were expected to be co-continuous; yet, the interpretation of experimental results for this system was somewhat more ambiguous. Further analyses would be still needed to better understand the complicated processes taking place in these systems.

To summarize, a successful theory of the stress-strain behavior of polyurethane elastomers has to incorporate the following features:

– Stress should have contributions from a (hyper)elastic soft phase and elasto-plastic (or visco-elasto-plastic) percolated hard phase;

– The hard phase contribution should be "loading history dependent" to reflect the fact that some domains or regions undergo irreversible deformation;

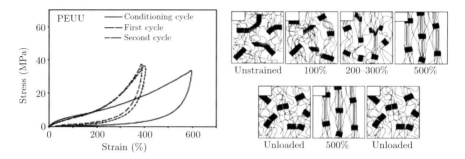

Figure 2.15. Experimental uniaxial elongation curves (conditioning, first, and second cycles) and a qualitative model of deformation-induced morphological changes in a weakly phase-segregated polyurethane elastomer PEUU [27]. Morphological model was proposed by the authors based on the analysis of their AFM and SAXS data

– Relative "weights" or "volume fractions" of the two phases should be functions of the material chemistry – and, perhaps, of its preparation history.

While the existing approaches (such as the model of Qi and Boyce) often provide a good description of polyurethane tensile curves, they typically treat hard and soft phase volume fractions as adjustable (fitting) parameters. In a fully predictive theory, one needs to combine the Qi-Boyce or similar framework with a thermodynamic model to predict hard and soft phase volume fractions, as we discussed in the previous section. Below, we illustrate how one can build such a theory and obtain a qualitative, if not quantitative, agreement with experiment. We start from a micromechanical model of Figure 2.7. The initial value of v_H (volume fraction of the "elastically active regions" of the percolated hard phase) is determined on the basis of thermodynamic considerations and the percolation model, as described in the previous section. We assume that each "elastically active region" of the hard phase can be described as an elasto-plastic material:

$$\sigma(\varepsilon) = E_H \left(1 - \exp\left[-\frac{\varepsilon}{\varepsilon_Y}\right]\right) \qquad (2.11)$$

The yield strain, ε_Y, is likely to be a function of the temperature and the deformation rate, but for a moment we will assume it to be constant (which is the case if all tensile experiments are done at a given temperature and strain rate). Hard phase modulus, E_H, has the same meaning as in the previous section, and is again taken to be equal to 5000 MPa.

We now have to determine how the number of the elastically active hard phase elements evolves with time (or, assuming the constant strain rate, s, as a function of strain, ε). It is important to note that initially, the fraction of elastically active elements, $v_H = [(f-f_{SC})/(1-f_{SC})]^\delta$, is much smaller than the total volume fraction of the percolated hard phase, $\Xi = f - f_{SC}$. It is then natural to assume (in a spirit similar to the Qi-Boyce model) that many "inactive" hard segments would become "activated" during deformation, primarily due to their interactions with other hard segments nearby. To describe this process, we utilize an approach similar to the one proposed by Patashinski et al. [73] in their study of the pre-yield behavior of glassy polymers. Following that formalism, we write the hard phase contribution to the total stress as follows:

$$\frac{d\sigma_H}{d\varepsilon} = E_H\left[v_H e^{-\varepsilon/\varepsilon_Y} + \int_0^\varepsilon d\xi \frac{d\eta}{d\xi} e^{-(\varepsilon-\xi)/\varepsilon_Y}\right] \quad (2.12)$$

The function $\eta(\varepsilon)$ describes the evolution of the hard phase elements as some of them become "activated" via orientation in the deformation direction and/or interaction with other hard phase regions. We assume the following function for $\eta(\varepsilon)$:

$$\eta(\varepsilon) = v_H + (\Xi - v_H)g(\varepsilon) \quad (2.13a)$$

$$g(\varepsilon(t)) = 1 - \exp\left[-\frac{r^2(t)}{d_0^2}\right] = 1 - \exp\left[-\frac{Dt}{d_0^2}\right] = 1 - \exp\left[-\frac{D\varepsilon}{sd_0^2}\right] = 1 - \exp\left[-\frac{\varepsilon}{\varepsilon_H}\right] \quad (2.13b)$$

Here, ε_H is the effective parameter depending on the effective "diffusion constant" for the displacement of the hard phase domains in the transverse direction, D; the strain rate, s, and the average inter-domain distance, d_0. Assuming, once again, that s and D are the same in all experiments (constant temperature and strain rate), we can postulate that $\varepsilon_H \cong b\Xi^{-1/2}$. For the comparison with experiments described below, we set $b = 1.5$.

To describe the soft phase contribution, one needs to develop a hyperelastic model taking into account: (i) rubber elasticity behavior; (ii) strain amplification due to the trapped hard phase inclusions; (iii) strain hardening as the chains approach their maximum extensibility. Typically, one could approximate these effects using an inverse Langevin function or its Pade approximation (see ref. [39], Chapter 11), and using a strain multiplication factor. Here, we use a somewhat simplified expression that retains most of the required features:

$$\frac{d\sigma_S}{d\varepsilon} = (1-\Xi)G_S X \frac{1+2/[1+\varepsilon]^3}{(1+\varepsilon)(1-\varepsilon/\varepsilon_{\max})}, \quad (2.14)$$

where $\varepsilon_{\max} = N_S^{1/2} - 1$, $G_S = (\rho RT/Ew)(1-2/f_P)$ is the soft phase shear modulus, and X is the strain amplification factor ($X \approx 1 + 2.5(f_{SC} - f_{DS})$).

This proposed description of the tensile behavior captures many of the important features of the Qi-Boyce and Drozdov models, while maintaining the direct relationship between the model parameters and the polyurethane formulation (through the formalism described in the previous section). To show how the model compares with experiment, in Figure 2.16 we plot tensile stress-strain curves for three MDI-DETDA/PO elastomers with different hard segment contents (31, 33, and 35 wt% hard segment). The hard segment is MDI-DETDA (see Figure 2.3), and the soft segment is PO 6000 triol (data courtesy of Dr. H. Lakrout, Dow Chemical Company). It can be seen that the

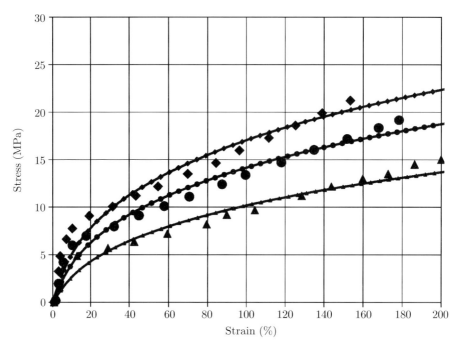

Figure 2.16. Experimental and theoretical uniaxial elongation curves for several MDI-DETDA/PO elastomers (see text for more details). Hard segment weight fractions: 35% (diamonds), 33% (circles), and 31% (triangles)

model gives a good semi-quantitative description of the tensile behavior as a function of the hard segment content.

To conclude this section, we note that the prediction of tensile behavior of segmented polyurethanes is still in its infancy. While there are several qualitative theories describing uniaxial tensile or compressive loading, the description of unloading and permanent ("set") residual deformation is still a challenge and very much depends on numerous assumptions about the hard domain relaxation. There is also very limited understanding of the ultimate strength and ultimate elongation (corresponding to the material failure). It is likely that for different compositions and/or temperatures, failure mechanisms can change from brittle to ductile, thus requiring very different theoretical analyses.

2.4. Linear viscoelasticity

The "composite" nature of polyurethane elastomers strongly affects their linear viscoelastic properties. It is known that for most polymers, linear viscoelastic moduli (storage modulus, $E'(\omega, T)$, and loss modulus, $E''(\omega, T)$) are characterized by the so-called time-temperature superposition (TTS) (see, *e.g.*, Ferry [74]). Such behavior can be understood if one assumes that E' (and E'') is always a function of the product $\omega\tau(T)$, where $\tau(T)$ is effective relaxation time.

The dependence of the relaxation time on temperature could be described by various theories such as the Doolittle free volume theory [75], Arrhenius activation theory, or the polymer-specific Williams-Landell-Ferry (WLF) [76] theory. The TTS was demonstrated to be applicable to a variety of polymers, including many low-hard-segment polyurethanes (see, *e.g.*, Dušek [63]). However, its applicability to higher-hard-segment content PU elastomers is not clear. One expects that in a composite arrangement such as the one schematically depicted in Figure 2.7, the hard and soft phases would be characterized by different relaxation times, thereby rendering TTS impossible (except, perhaps, in a short temperature range). In that case, one could still analyze the temperature dependencies of E' and E'' taken at a given frequency (*e.g.*, 1 Hz), but it would be difficult to translate this information to what happens at much faster – or much slower – frequencies.

For typical PU elastomers, DMA curves as a function of temperature (at a given frequency) look similar to those shown in Figure 2.17. At lower temperatures, there is a dramatic drop in the storage modulus, E', and a strong maximum in the loss tangent, $\tan \delta = E''/E'$, corresponding to the glass transition of the soft phase. (Note that the presence of crosslinks, dissolved hard segments, and hydrogen bonds can increase the glass transition temperature of the soft phase compared to that of a pure polyol polymer.) As the temperature increases, the material exhibits elastomeric behavior, and E' is gently decreasing with temperature ("elastomeric plateau region"). Finally, at higher temperatures, the system undergoes softening due to the hard phase glass transition, melting, or chemical degradation (whichever comes first).

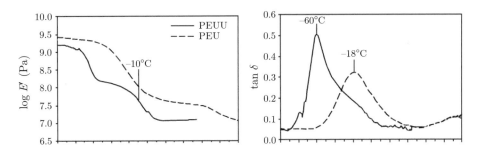

Figure 2.17. Typical DMA curves of polyurethane elastomers: tensile storage modulus, E', (left) and loss tangent, $\tan \delta = E''/E'$, (right). Measurements were performed at a frequency of 1 s^{-1}. The PEUU curves are typical for weakly phase-segregated elastomer, while PEU curves are typical for strongly phase-separated elastomer with percolated hard phase [27]

To our knowledge, there is no comprehensive and predictive theory of polyurethane elastomer linear viscoelasticity. We believe that it should be possible to extend our earlier model (Section 2.2) to describe not just "quasi-static" modulus, but also the time (or frequency) dependent moduli and overall effects of linear viscoelasticity. This is a subject of ongoing research.

2.5. Non-equilibrium factors and their influence on mechanical properties

So far, our discussion of the mechanical properties of polyurethane elastomers was predicated on the implicit assumption that the morphology of the microphase-separated material is at equilibrium or close to equilibrium. It is, therefore, important to understand the limitations of this assumption and to see how the preparation history might influence the properties.

It has been known for a long time that the preparation history has a significant impact on the polyurethane polymer morphology, especially on the degree of the phase separation between hard and soft phases. Ryan and co-workers investigated the dynamics of morphology evolution in a number of flexible and rigid foam systems using *in situ* small-angle X-ray scattering (SAXS) and other experimental techniques [77–81]. They found that in a typical flexible foam, microphase separation between hard phase and soft phase can advance fairly substantially before being "arrested" by the hard phase vitrification (Berghmans' point). In rigid foam, on the other hand, vitrification occurs much more rapidly, and the degree of phase separation is extremely low, much lower than expected from purely thermodynamic considerations [41]. For elastomeric systems, the preparation processes differ very substantially, and the degree of the hard-soft phase separation depends on the processing temperature, processing time, and the formulation. As discussed by Garrett *et al.* [20], for elastomers with intermediate (20–40 wt%) hard segment loadings, the degree of phase separation is often very close to equilibrium predictions, while for those with higher (above 45 wt%) hard segment loadings, the degree of phase separation is somewhat lower than expected. These kinetic considerations should certainly be taken into account in both theoretical model development and practical material design of segmented polyurethanes.

2.6. Conclusions and Outlook

The promise of polyurethanes today remains as broad as ever, and new polyurethane components, materials, and applications are constantly being introduced in the market. While the design and synthesis of new materials is still more art than science, the developments of the past two decades enable a much better understanding of the structure–property relationships in PU than was available in the early days of the industry.

In this chapter, we reviewed some of the recent developments in modeling and predicting the mechanical properties of polyurethane elastomers as a function of their formulation. Based on the knowledge of the formulation and processing history, one can roughly predict the degree of the microphase separation between the hard and soft phases. That, in turn, enables the construction of constitutive "micromechanical" models for the calculation of elastic, viscoelastic, and nonlinear tensile and compressive properties.

The development of physical and mechanical models for polyurethanes, of course, is still at the very beginning. As we already discussed, today's models are based on equilibrium (thermodynamic) predictions, and take little account of sample history. Even in dealing with equilibrium systems, one would need to account for the lack of flexibility in the hard segments; hence, the phase behavior of segmented polyurethanes would likely be closer to that of rod-coil block copolymers [44–47] than to that of conventional flexible chains. Moreover, the use of micromechanical models (as opposed to, say, Finite Element Analysis (FEA) [82,83] or the similar Lattice Springs Model (LSM) [84]) speeds up the calculations but potentially limits their accuracy. Other complexities such as polydispersity of the soft segment, partial crystallinity of the soft phase, and polydispersity of the hard segments must be also taken into account. This is especially important as new molecules (such as natural oil-based polyols [85,86]) are being developed and commercialized. As the new models are being introduced to deal with all these complexities, the researchers will have new tools to help in designing new molecules to fit various applications and bring new products to the market.

Acknowledgment

We thank The Dow Chemical Company for supporting this research. We are indebted to Drs. D. Latham, B. Obi, A. Shafi, M. Sonnenschein, A. Birch, H. Gruenbauer, H. Lakrout, C. Leibig (Dow), M. Elwell (Huntsman) and D. J. Read (Leeds) for helpful discussions.

References

1. Macosko C W (1989) *RIM Fundamentals*, Hanser, Munich.
2. Sweeney F M (1987) *Reaction Injection Molding Machinery and Process*, Marcel Dekker, New York.
3. Cooper S L and Tobolsky A V (1966) Properties of Linear Elastomeric Polyurethanes, *J Appl Polym Sci* **10**:1837–1844.
4. Koberstein J T and Stein R S (1983) Small-angle X-ray scattering studies of microdomain structure in segmented polyurethane elastomers, *J Polym Sci Polym Phys* **21**:1439–1472.
5. Leung L M and Koberstein J T (1985) Small-angle scattering analysis of hard-microdomain structure and microphase mixing in polyurethane elastomers, *J Polym Sci Polym Phys* **23**:1883–1913.
6. Christenson C P, Harthcock M A, Meadows M D, Spell H L, Howard W L, Creswick M W, Guerra R E and Turner R B (1986) Model MDI/Butanediol polyurethanes: Molecular structure, morphology, physical and mechanical properties, *J Polym Sci Part B: Polym Phys* **24**:1401–1439.
7. Christenson C P and Turner R B (unpublished data).
8. Sonnenschein M F, Rondan N, Wendt B and Cox J M (2004) Synthesis of transparent thermoplastic polyurethane elastomers, *J Polym Sci Part A: Polym Chem* **42**:271–278.
9. Sonnenschein M F, Lysenko Z, Brune D A, Wendt B L and Schrock A K (2005) Enhancing polyurethane properties via soft segment crystallization, *Polymer* **46**:10158–10166.
10. Turner R B, Spell H L and Vanderhider J A (1982) The Effect of Hard Segment Content on a Crosslinked Polyurethane RIM System, in *Reaction Injection Molding and Fast Polymerization Reactions* (Ed. Kresta J E) Plenum Press, New York, pp. 63–70.

11. Zdrahala R J and Critchfield F E (1982) RIM Urethanes Structure/Property Relationships for Linear Polymers, in *Reaction Injection Molding and Fast Polymerization Reactions* (Ed. Kresta J E) Plenum Press, New York, pp. 55–62.
12. Ryan A J, Willkomm W R, Bergstrom T B, Macosko C W, Koberstein J T, Yu C C and Russell T P (1991) Dynamics of (Micro)phase Separation during Fast, Bulk Copolymerization: Some Synchrotron SAXS Experiments, *Macromolecules* **24**:2883–2889.
13. Ryan A J (1990) Spinodal decomposition during bulk copolymerization: reaction injection moulding, *Polymer* **31**:707–712.
14. Birch A J, Stanford J L and Ryan A J (1989) The effects of hard segment content on microphase separation and physical properties of non-linear, segmented copolyureas formed by RIM, *Polym Bull* **22**:629–635.
15. Ryan A J, Bergstrom T B, Willkomm W R and Macosko C W (1991) Thermal, mechanical, and fracture properties of copolyureas formed by reaction injection molding: Effects of hard segment structure, *J Appl Polym Sci* **42**:1023–1039.
16. Nakamae K, Nishino T, Asaoka S and Sudaryanto S (1996) Microphase Separation and Surface Properties of Segmented Polyurethane – Effect of Hard Segment Content, *Int J Adhesion and Adhesives* **16**:233–239.
17. Rosthauser J W, Haider K W, Steinlein C and Eisenbach C D (1997) Mechanical and Dynamic Mechanical Properties of Polyurethane and Polyurethane/Polyurea Elastomers Based on 4,4*-Diisocyanatodicyclohexyl Methane, *J Appl Polym Sci* **64**:957–970.
18. Gier D R, O'Neill R E, Adams M R, Priester, Jr. R D, Lidy W A, Barnes C G, Rightor E G and Davis B L (1998) Fillers, Hard Phases and Copolymer Polyols – Their Different Mechanism to Reinforce Flexible Polyurethane Foams, in: *Polyurethanes Expo '98*, SPI Polyurethanes Division, Dallas, pp. 227–229.
19. Sanchez-Adsuar M S, Papon E and Villenave J J (2000) Rheological characterization of thermoplastic polyurethane elastomers, *Polym Inter* **49**:591–598.
20. Garrett J T, Runt J and Lin J S (2000) Microphase Separation of Segmented Poly(urethane urea) Block Copolymers, *Macromolecules* **33**:6353–6359.
21. Garrett J T, Runt J and Siedlecki C A (2001) Microdomain Morphology of Poly(urethane urea) Multiblock Copolymers, *Macromolecules* **34**:7066–7070.
22. Kim H D, Huh J H, Kim E W and Park C C (1998) Comparison of Properties of Thermoplastic Polyurethane Elastomers with Two Different Soft Segments, *J Appl Polym Sci* **69**:1349–1355.
23. Kim H D, Lee T J, Huh J H and Lee D J (1999) Preparation and Properties of Segmented Thermoplastic Polyurethane Elastomers with Two Different Soft Segments, *J Appl Polym Sci* **73**:345–352.
24. O'Sickey M J, Lawrey B D and Wilkes G L (2002) Structure-property relationships of poly(urethane urea)s with ultra-low monol content poly(propylene glycol) soft segments. I. Influence of soft segment molecular weight and hard segment content, *J Appl Polym Sci* **84**:229–243.
25. Yang J H, Chun B C, Chung Y C and Cho J H (2003) Comparison of thermal/mechanical properties and shape memory effect of polyurethane block-copolymers with planar or bent shape of hard segment, *Polymer* **44**:3251–3258.
26. Sheth J P, Aneja A, Wilkes G L, Yilgor E, Attila G E, Yilgor I and Beyer F L (2004) Influence of system variables on the morphological and dynamic mechanical behavior of polydimethylsiloxane based segmented polyurethane and polyurea copolymers: a comparative perspective, *Polymer* **45**:6919–6932.
27. Christenson E M, Anderson J M, Hiltner A and Baer E (2005) Relationship between nanoscale deformation processes and elastic behavior of polyurethane elastomers, *Polymer* **46**:11744–11754.

28. Leibler L (1980) Theory of Microphase Separation in Block Copolymers, *Macromolecules* **13**:1602–1617.
29. Semenov A N (1985) Contribution to the theory of microphase layering in block-copolymer melts, *Sov Phys JETP* **61**:733–742.
30. Fredrickson G H and Helfand E (1987) Fluctuation effects in the theory of microphase separation in block copolymers, *J Chem Phys* **87**:697–705.
31. Benoit H and Hadziioannou G (1988) Scattering theory and properties of block copolymers with various architectures in the homogeneous bulk state, *Macromolecules* **21**:1449–1464.
32. Matsen M and Bates F S (1996) Unifying Weak- and Strong-Segregation Block Copolymer Theories, *Macromolecules* **29**:1091–1098, and references therein.
33. Ginzburg V V, Bicerano J, Christenson C P, Schrock A K and Patashinski A Z (2007) Theoretical Modeling of the Relationship Between Young's Modulus and Formulation Variables for Segmented Polyurethanes, *J Polym Sci Part B: Polym Phys* **45**:2123–2135.
34. Yontz D J and Hsu S L (2000) A Mass Spectrometry Analysis of Hard Segment Length Distribution in Polyurethanes, *Macromolecules* **33**:8415–8420.
35. Davies W E (1971) The theory of composite dielectrics, *J Phys D Appl Phys* **161**:318–328.
36. Davies W E (1971) The elastic constants of a two-phase composite material, *J Phys D Appl Phys* **161**:1176–1181.
37. Davies W E (1971) The theory of elastic composite materials, *J Phys D Appl Phys* **161**:1325–1339.
38. van Krevelen D (1990) *Properties of Polymers*, 3rd edition, Elsevier, Amsterdam.
39. Bicerano J (2002) *Prediction of Polymer Properties*, 3rd edition, Marcel Dekker, New York.
40. Bicerano J, Daussin R D, Elwell M J A, van der Wal H R, Berthevas P, Brown M, Casati F, Farrisey W, Fosnaugh J, de Genova R, Herrington R, Hicks J, Hinze K, Hock K, Hunter D, Jeng L, Laycock D, Lidy W, Mispreuve, H, Moore R, Nafziger L, Norton M, Parish D, Priester R, Skaggs K, Stahler L, Sweet F, Thomas R, Turner R, Wiltz G, Woods T, Christenson C P and Schrock A K (2004) Flexible Polyurethane Foams: A Review of the State-of-the-Art, in *Polymeric Foams: Mechanisms and Materials* (Eds. Lee S T and Ramesh N S) CRC Press, Boca Raton, pp. 173–252.
41. Gruenbauer H J M, Bicerano J, Clavel P, Daussin R D, de Vos H A, Elwell M J A, Kawabata H, Kramer H, Latham D D, Martin C A, Moore S E, Obi B C, Parenti V, Schrock A K and van der Bosch R (2004) Rigid Polyurethane Foams: A Review of the State-of-the-Art, in *Polymeric Foams: Mechanisms and Materials* (Eds. Lee S T and Ramesh N S) CRC Press, Boca Raton, pp. 253–310.
42. Huggins M J (1941) Solutions of Long Chain Compounds, *J Chem Phys* **9**:440–440.
43. Flory P J (1941) Thermodynamics of High Polymer Solutions, *J Chem Phys* **9**:660–661.
44. Semenov A N and Vasilenko S V (1986) Theory of the nematic–smectic-A transition in a melt of macromolecules consisting of a rigid and a flexible block, *Sov Phys JETP* **63**:70–78.
45. Matsen M W and Barrett C J (1998) Liquid-crystalline behavior of rod-coil diblock copolymers, *J Chem Phys* **109**:4108–4118.
46. Li W and Gersappe D (2001) Self-Assembly of Rod-Coil Diblock Copolymers, *Macromolecules* **34**:6783–6789.
47. Olsen B D and Segalman R A (2005) Structure and Thermodynamics of Weakly Segregated Rod-Coil Block Copolymers, *Macromolecules* **38**:10127–10137.

48. Sides S W and Fredrickson G H (2003) Theory of Polydisperse Inhomogeneous Polymers, *Macromolecules* **36**:5415–5423.
49. Jiang Y, Yan X, Liang H and Shi A C (2005) Effect of Polydispersity on the Phase Diagrams of Linear ABC Triblock Copolymers in Two Dimensions, *J Phys Chem B* **109**:21047–21055.
50. Kolařik J (1998) Simultaneous prediction of the modulus, tensile strength and gas permeability of binary polymer blends, *Eur Polym J* **34**:585–1590.
51. Kolařik J (1996) Simultaneous prediction of the modulus and yield strength of binary polymer blends, *Polym Eng Sci* **36**:2518–2524.
52. Stauffer D (1985) *Introduction to Percolation Theory*, Taylor and Francis, London.
53. Fizazi A, Moulton J, Pakbaz K, Rughooputh S, Smith P and Heeger A J (1990) Percolation on a self-assembled network: Decoration of polyethylene gels with conducting polymer, *Phys Rev Lett* **64**:2180–2183.
54. Levon K, Margolina A and Patashinsky A Z (1993) Multiple percolation in conducting polymer blends, *Macromolecules* **26**:4061–4063.
55. Halpin J C and Kardos J L (1976) The Halpin-Tsai equations: A review, *Polym Eng Sci* **16**:344–352.
56. Mori T and Tanaka K (1973) Average Stress in Matrix and Average Elastic Energy of Materials With Misfitting Inclusions, *Acta Metall Mater* **21**:571–574.
57. Rubinstein M and Colby R (2003) *Polymer Physics*, Oxford University Press, Oxford.
58. Erman B and Mark J E (1997) *Structure and Properties of Rubberlike Networks*, Oxford University Press, New York – Oxford.
59. Flory P J (1953) *Principles of Polymer Chemistry*, Cornell University Press, Ithaca, New York.
60. Laity P R, Taylor J E, Wong S S, Khunkamchoo P, Cable M, Andrews G T, Johnson A F and Cameron R E (2006) Morphological Behaviour of Thermoplastic Polyurethanes During Repeated Deformation, *Macromol Mater Eng* **291**:301–324.
61. Qi H and Boyce M C (2005) Stress-strain behavior of thermoplastic polyurethanes, *Mech Mater* **37**:817–839.
62. Yi J, Boyce M C, Lee G F and Balizer E (2006) Large deformation rate-dependent stress–strain behavior of polyurea and polyurethanes, *Polymer* **47**:319–329.
63. Ilavsky M, Šomvársky J, Bouchal K and Dušek K (1993) Structure, Equilibrium and Viscoelastic Mechanical Behaviour of Polyurethane Networks Based on Triisocyanate and Poly(oxypropylene)Diols, *Polymer Gels and Networks* **1**:159–184.
64. Estes G M, Seymour R W and Cooper S L (1971), Infrared studies of segmented polyurethane elastomers. II, *Macromolecules* **4**:452–457.
65. Klüppel M and Schramm J (2000) A generalized tube model of rubber elasticity and stress softening of filler reinforced elastomer systems, *Macromol Theory Simul* **9**:742–754.
66. Špirkova M (2002) Polyurethane Elastomers Made from Linear Polybutadiene Diols, *J Appl Polym Sci* **85**:84–91.
67. Tobushi H, Okumura K, Hayashi S and Ito N (2001) Thermomechanical constitutive model of shape memory polymer, *Mech Mater* **33**:545–554.
68. Arruda E M and Boyce M C (1993) A three-dimensional constitutive model for the large stretch behavior of elastomers. *J Mech Phys Solids* **41**:389–412.
69. Qi J H and Boyce M C (2004) Constitutive model for stretch-induced softening of the stress-stretch behavior of elastomeric materials, *J Mech Phys Solids* **52**:2187–2205.
70. Drozdov A A and Dorfmann A (2003) A micro-mechanical model for the response of filled elastomers at finite strains, *Int J Plasticity* **19**:1037–1067.
71. Drozdov A A and Dorfmann A (2001) The stress-strain response and ultimate strength of filled elastomers, *Comp Mater Sci* **21**:395–417.

72. Drozdov A A (1999) Mechanically induced crystallization of polymers, *Int J Non-Linear Mech* **34**:807–821.
73. Patashinski A, Moore J, Bicerano J, Mudrich S, Mazor M and Ratner M (2006) Stress-Biased Rearrangements and Preyield Behavior in Glasses, *J Phys Chem B* **110**:14452–14457.
74. Ferry J D (1961) *Viscoelastic Properties of Polymers*, John Wiley and Sons, New York – London.
75. Doolittle A K (1951) Studies in Newtonian Flow. II. The Dependence of the Viscosity of Liquids on Free-Space, *J Appl Phys* **22**:1471–1475.
76. Williams M L, Landel R F and Ferry J D (1955) The Temperature Dependence of Relaxation Mechanisms in Amorphous Polymers and Other Glass-forming Liquids, *J Amer Chem Soc* **77**:3701–3707.
77. Elwell M J, Ryan A J, Grünbauer H J M and van Lieshout H C (1996) FTIR Study of Reaction Kinetics and Structure Development in Model Flexible Polyurethane Foam Systems, *Polymer* **37**:1353–1361.
78. Elwell M J, Ryan A J, Grünbauer H and van Lieshout H C (1996) *In-Situ* Studies of Structure Development during the Reactive Processing of Model Flexible Polyurethane Foam Systems Using FT-IR Spectroscopy, Synchrotron SAXS, and Rheology, *Macromolecules* **29**:2960–2968.
79. Elwell M J, Mortimer S and Ryan A J (1994) A Synchrotron SAXS Study of Structure Development Kinetics during the Reactive Processing of Flexible Polyurethane Foam, *Macromolecules* **27**:5428–5439.
80. Li W, Ryan A J and Meier I (2002) Morphology Development via Reaction-Induced Phase Separation in Flexible Polyurethane Foam, *Macromolecules* **35**:5434–5442.
81. Li W, Ryan A J and Meier I (2002) Effect of Chain Extenders on the Morphology Development in Flexible Polyurethane Foam, *Macromolecules* **35**:6306–6312.
82. Gusev A A (2001) Numerical Identification of the Potential of Whisker- and Platelet-Filled Polymers, *Macromolecules* **34**: 3081–3093, and references therein.
83. Read D J, Teixeira P I C, Duckett R A, Sweeney J and McLeish T C B (2002) Theoretical and finite-element investigation of the mechanical response of spinodal structures, *Eur Phys J E* **8**:15–31.
84. Buxton G and Balazs A C (2002) Lattice spring model of filled polymers and nanocomposites, *J Chem Phys* **117**:7649–7658.
85. Babb D, Larre A, Schrock A K, Bhattacharjee D and Sonnenschein M F (2007) Triglycerides as feedstocks for polyurethanes, *Polymer Preprints* **48**:855–856.
86. Sanders A, Babb D, Prange R, Sonnenschein M, Delk V, Derstine C and Olson K (2006) Producing Polyurethane Foam from Natural Oil, in *Catalysis of Organic Reactions* (Ed. Schmidt S R) CRC Press, Boca Raton, pp. 377–384.

PART II
NANOCOMPOSITES: INFLUENCE OF PREPARATION

Chapter 3

Nanoparticles/Polymer Composites: Fabrication and Mechanical Properties

M. Q. Zhang, M. Z. Rong, W. H. Ruan

3.1. Introduction

Polymeric nanocomposites have been an area of intense industrial and academic research for the past twenty years. No matter the measure – articles, patents, or R&D funding – efforts in this respect have been exponentially growing worldwide over the last ten years. It is believed that the tremendous interfacial area helps to greatly influence the composite's properties. As compared to neat polymers or micro-particulate filled polymer composites, polymer nanocomposites exhibit markedly improved properties, including modulus, strength, impact performance, and heat resistance at low concentration of the inorganic components (1~10 wt%) [1]. In this context, the nanocomposites are much lighter in weight and easier to be processed.

With respect to manufacturing of polymeric nanocomposites, direct incorporation of inorganic nano-scale building blocks into polymers represents a typical way. Because of the strong tendency of nanoparticles to agglomerate, however, nano-size fillers are hard to be uniformly dispersed in polymers by conventional techniques. The most important issue for producing nanocomposites lies in surface modification of the nano-fillers. An ideal measure should be able to increase the hydrophobicity of the fillers, enhance the interfacial adhesion *via* physical interaction or chemical bonding, and eliminate the loose structure of filler agglomerates.

The present chapter reviews the recent achievements in the fabrication of nanoparticles/polymer composites acquired in the authors' laboratory, including the specific surface pre-treatment approaches and their applications. In addition, effects of the treated nanoparticles on mechanical properties of the composites are highlighted.

In general, the strategies of polymer nanocomposites manufacturing developed by the authors fall into three categories. That is, (i) dispersion-oriented, (ii) dispersion and filler/matrix interaction-oriented, and (iii) dispersion, filler/filler interaction and filler/matrix interaction-oriented. The corresponding technical routes are based on one common nanoparticles surface modification method – graft polymerization. Grafting macromolecules onto inorganic nanoparticles has some advantages over the modification by low molecular surfactants or coupling agents. The polymer-grafted particles can possess desired properties through a proper selection of the species of graft monomers and the graft conditions, so that the interfacial characteristics between the treated nanoparticles and the matrix can be tailor-made when manufacturing nanocomposites. Besides, the fragile nanoparticle agglomerates become stronger because they turn into a nanocomposite microstructure comprising the nanoparticles, the grafted and the ungrafted (homopolymerized) polymer (Figure 3.1). A series of works has been done in this field of growing interest for the purposes of improving the dispersibility of the nanoparticles in solvents and their compatibility in polymers [2]. Mostly, the graft polymerization is conducted *via* two routes: (i) monomers are polymerized from active compounds (initiators or comonomers)

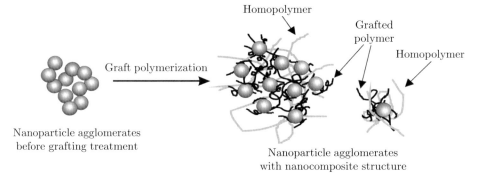

Figure 3.1. Schematic drawing of the structural change of nanoparticle agglomerates before and after graft polymerization treatment

covalently attached to the inorganic surface (called "grafting from"), and (ii) ready-made polymers with reactive end-groups react with the functional groups on the particles (called "grafting to"). The latter benefits the control of the molecular weight of the grafting polymer. But, generally, the former approach has advantage over the latter one as it is more difficult for polymers to penetrate into the particulate agglomeration than monomers. Besides, the "grafting to" technique would lead to a polymer coating on the particles, which might hinder the attachment of other polymers.

Practically, surface grafting onto nanoparticles and compounding of different ingredients to make nanocomposites can be completed by a one-step or a two-step process. In the following text, details of the fabrication and mechanical properties of nanocomposites are discussed with nano-SiO_2/polypropylene (PP) as the model material.

3.2. Dispersion-oriented manufacturing of nanocomposites

3.2.1. *Conventional two-step manufacturing*

3.2.1.1. *High energy irradiation induced graft coupled with melt compounding*

To well disperse the nanoparticles in the polymer matrix, γ-ray irradiation graft polymerization was applied to modify SiO_2 nanoparticles first, and then the treated particles were mechanically mixed with PP as usual [3–5]. Under the high energy radiation, the surfaces of the nanoparticles outside and inside the agglomerates are equally activated. Therefore, the low molecular weight monomers are allowed to react with the activated sites of the nanoparticles throughout the agglomerates [6].

Typical tensile stress-strain curves of neat PP and its filled version are shown in Figure 3.2, indicating that both a reinforcing and a toughening effect of the nanoparticles on the polymeric matrix was brought into play. That is, a structural weakness, which would have resulted from the agglomerating behavior of the nanoparticles, has been eliminated by the grafting macromolecular chains.

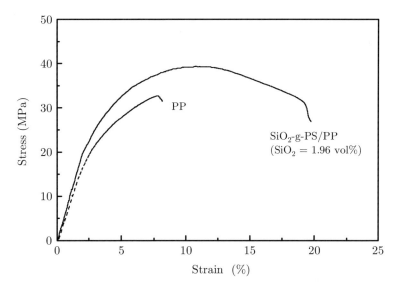

Figure 3.2. Typical tensile stress-strain curves of the neat PP matrix resin (MFI = 6.7 g/10 min), and the one filled with SiO_2-g-PS [3]. (SiO_2-g-PS means polystyrene grafted nano-SiO_2)

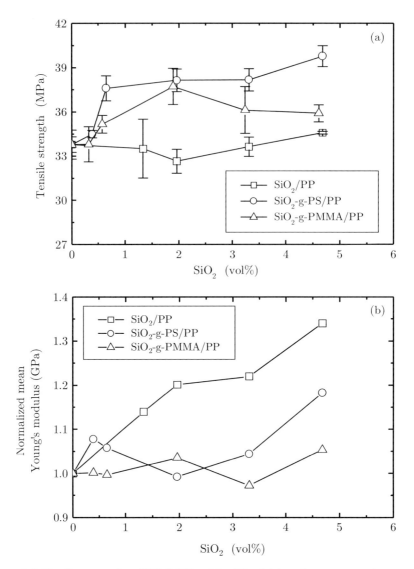

Figure 3.3. Tensile properties of PP (MFI = 6.7 g/10 min) based nanocomposites as a function of SiO$_2$ content: (a) tensile strength; (b) Young's modulus; (c) elongation to break; and (d) area under stress-strain curve [3]. (SiO$_2$-g-PMMA means poly(methyl methacrylate) grafted nano-SiO$_2$)

By further examining the composition-dependent tensile properties of the materials (Figure 3.3), it can be seen that the incorporation of untreated nano-SiO$_2$ lowers the tensile strength of PP in the lower loading region, but then leads to a slight increase in strength when the particle fraction reaches 4.68 vol% (Figure 3.3a). When the nanocomposites are filled with polystyrene (PS) or poly(methyl methacrylate) (PMMA) grafted SiO$_2$, however, the situation is

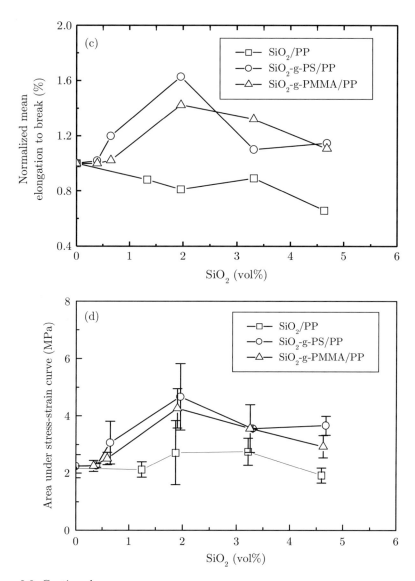

Figure 3.3. Continued

quite different. There is a considerable rise in the strength of SiO$_2$-g-PS/PP composites at a SiO$_2$ content as low as 0.65 vol%, then the strength remains almost unchanged with further addition of the filler. A similar behavior can be observed in the case of SiO$_2$-g-PMMA/PP, except that a slight drop in strength occurs when the filler content exceeds 1.96 vol%. Nevertheless, compared to the untreated case, it can be stated here that mechanical loading seems to be more effectively transferred from the matrix to the filler particles owing to the interfacial bonding effect of the grafting polymers, which might be rooted in an interdiffusion and entanglement between the molecules of the grafting polymers and the matrix. The decrease in tensile strength of SiO_2-g-PMMA/PP above 1.96 vol% can be interpreted by a change in the dispersion status of the fillers. It is believed that a higher filler loading is detrimental to its uniform dispersion in the polymer matrix. In fact, transmission electron microscopy (TEM) observation showed that the size of the dispersed phases increases with increasing filler content in the nanocomposites (Figure 3.4). It should be noted that the dispersed phases in the nanocomposites illustrated in Figure 3.4b–e are clearly smaller than the untreated SiO_2 (Figure 3.4a), but they are still much larger than the size of the primary nano-SiO_2 particles (7 nm). Therefore, these dispersed phases are actually microcomposite agglomerates consisting of primary particles, grafting polymer, homopolymer, and a certain amount of matrix.

It is well-known that the interface adhesion markedly influences the mechanical behavior of particulate filled polymer composites [7–11]. The present irradiation grafting pre-treatment provides possibilities for interfacial design,

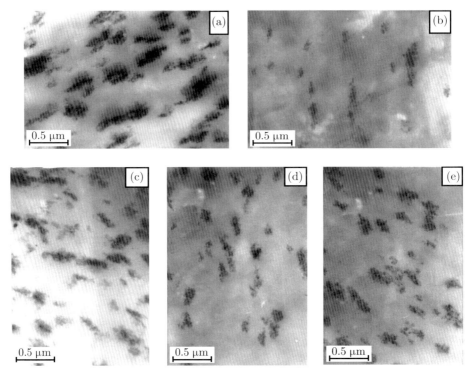

Figure 3.4. TEM micrographs of PP (MFI = 6.7 g/10 min) based nanocomposites filled with (a) SiO_2 as-received (content of SiO_2 = 1.96 vol%), (b) SiO_2-g-PS (content of SiO_2 = 1.96 vol%), (c) SiO_2-g-PS (content of SiO_2 = 6.38 vol%), (d) SiO_2-g-PMMA (content of SiO_2 = 1.96 vol%), and (e) SiO_2-g-PMMA (content of SiO_2 = 6.38 vol%). Magnification = 2×10^4 [3]

i.e., both surface characteristics of the nanoparticles and the particle agglomerates/matrix interface can be changed by using different grafting monomers. As a result, the mechanical properties of the nanocomposites can be tailored accordingly. Table 3.1 gives the mechanical properties of nanocomposites filled with SiO_2 particles, grafted with various polymers, at a fixed SiO_2 fraction. Although the monomers of the grafting polymers should have different miscibilities with PP, all the grafting polymers except poly(ethyl acrylate) (PEA) exhibit a reinforcement effect on the tensile strength of the nanocomposites.

Table 3.1. Mechanical properties of PP (MFI = 8.5 g/10 min) based nanocomposites[a] filled with different polymers grafted SiO_2 [3]

Grafting polymers	Nanocomposites						Neat PP
Property	PS	PBA[b]	PVAc[c]	PEA[d]	PMMA	PMA[e]	
Tensile strength (MPa)	34.1	33.3	33.0	26.8	35.2	33.9	32.0
Young's modulus (GPa)	0.92	0.86	0.81	0.88	0.89	0.85	0.75
Elongation to break (%)	9.3	12.6	10.0	4.6	12.0	11.9	11.7
Area under tensile stress-strain curve (MPa)	2.4	3.3	2.3	0.8	3.2	2.9	2.2
Unnotched Charpy impact strength (kJ/m²)	19.8	19.4	22.9	14.6	20.5	4.7	8.0

[a] Content of SiO_2 = 3.31 vol%. [b] PBA = polybutyl acrylate. [c] PVAc = polyvinyl acetate. [d] PEA = polyethyl acrylate. [e] PMA = polymethyl acrylate

These results contribute to a further understanding of the modified nanoparticles and their role in the composites. That is, interdiffusion and entanglement of the grafting polymer segments with the polypropylene molecules, instead of miscibility between the grafting polymer and the matrix, dominate the interfacial interaction in the nanocomposites. This leads to the conclusion that a PP matrix with a higher molecular weight should even be entangled more effectively with the nanoparticle agglomerates, thus leading to a higher tensile strength increment (comparing Figure 3.2 with Table 3.1). In fact, Kendall and Sherliker [12] studied the effect of the polymer molecular weight on colloidal silica filled thermoplastics and observed a similar phenomenon.

For the moment, it can be concluded that the reinforcing effect of nanoparticles on the polymeric matrices can be realized as long as the particles are grafted and a proper dispersion of the modified particles can be formed. Besides, the tensile properties of the nanocomposites can be purposely adjusted according to the interfacial viscoelastic properties provided by different grafting monomers.

As a parameter closely related to the static stress transfer at the interface, Young's moduli of the composites show another aspect of the role played by the grafting polymers, or rather by the interphase related to them (Table 3.1).

The increase in stiffness of the nanocomposites is obviously a result of the high modulus of the particulate fillers (modulus$_{\text{silica}}$ = 70 GPa). However, considering that the tensile modulus was determined within only a small strain range (where Hooke's law is still valid), the formation of a relatively compliant layer at the interface (PBA, PVAc and PEA, for example) tends to hinder a complete stress transfer under such a low stress level and thus masks the stiffness of the filler particles. Figure 3.3b gives evidence for this mask effect of the grafting polymers, showing that the modulus of the composites filled with untreated SiO_2 increases almost linearly with the addition of SiO_2, while grafting PS and PMMA greatly decreases the stiffening effect of SiO_2. Walter and co-workers obtained similar results in differently treated Kaolin/high density polyethylene systems [13].

The elongation to break of the nanocomposites exhibits a more complicated relationship to the interfacial characteristics in the case of a lower molecular weight PP matrix (Table 3.1). Relative to the neat matrix, the values of the grafting polymer treated particle reinforced nanocomposites remain on the same level, except for the PEA treated system. For the matrix with the higher molecular weight (Figure 3.3c), however, the filler content dependence of the elongation to break seems to be clearly evident. The reduction in elongation to break of the composite by the addition of untreated SiO_2 implies that the fillers cause a reduction in matrix deformation due to an introduction of mechanical restraints. When the nanoparticles are grafted with PS and PMMA, on the other hand, an increase in elongation to break can be found because both interfacial viscoelastic deformation and matrix yielding can contribute to this value. The further tendency of decrease in elongation to break above a filler content of about 1.96 vol% suggests that matrix deformation is not only related to the interface feature but also to the dispersion state of the fillers.

Compared with elongation to break, the area under the tensile stress-strain curve is able to characterize more reasonably the toughness potential of the nanocomposites under static tensile loading circumstances. From the data in Table 3.1, it can be seen that the addition of modified nano-SiO_2 helps to improve the ductility of the PP, except in the case of SiO_2-g-PEA. It is believed that localized plastic deformation or drawing of the matrix polymer, being the main energy absorption process in particulate filled polymer systems, can be induced more efficiently by SiO_2-g-PS and SiO_2-g-PMMA than by untreated SiO_2 (Figure 3.3d). The deteriorated effect of grafting PEA on the tensile behavior of the nanocomposites cannot be explained by the present results, *i.e.* further detailed studies have to be carried out.

3.2.1.2. *Chemical graft coupled with melt compounding*

Besides the aforesaid irradiation induced graft polymerization, chemical graft is also able to introduce different polymers onto the surfaces of silica nanoparticles. In this way, finely controlled graft products can be obtained. Comparatively, this approach is more suitable for fundamental research that leads to understanding of the underlying parameters in correlation with structure–property relationship of nanocomposites. A typical procedure is described as follows.

To initiate polymerization from the particles, double bonds (*i.e.* reactive groups) had to be firstly attached to the surfaces of SiO_2 nanoparticles through the pre-treatment of silane (γ-methacryloxypropyl trimethoxy silane) [14,15]. Then, the silane pre-treated nanoparticles were mixed with toluene under sonication. Afterwards the initiator, isobutyronitrile (AIBN), was added with stirring into the reactor that had been kept at a certain temperature and protected with N_2. The monomers were incorporated into the system a few minutes later. The reaction went on for several hours, and then the graft product could be received from the filtration of the resultant suspension. Eventually, the grafted nano-SiO_2 was mixed with PP powder and extruded.

Morphologies of the nanoparticles before and after grafting polymerization are illustrated in Figure 3.5. Large agglomerates of particles (~400 nm) can be observed for the silica as-received, while grafting treatment helps to decrease the size of the agglomerates to about 150 nm. A very thin layer of grafting polymer covering the silica aggregates can be identified, which should be responsible for the interface altering when the particles are used to reinforce PP. Meanwhile, the grafting polymer and particles build up a nanocomposite structure, which eliminates the loose structure of the agglomerated nanoparticles and should be beneficial to the mechanical performance of silica/PP composites.

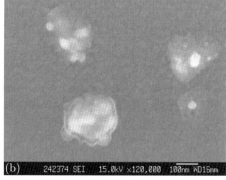

Figure 3.5. Scanning electron microscopy(SEM) images of (a) untreated SiO_2, and (b) SiO_2-g-PMMA [14]

Intuitively, it is believed that the feature of filler/matrix interfacial layer is responsible to a great extent for the impact property of PP composites. When ungrafted nano-silica is added to PP, a mild increase in impact strength is measured up to 0.8 vol% (Figure 3.6). Above this filler content, there is a decreasing trend in impact strength with increasing the content of silica. It can be attributed to the worse dispersion of nanoparticles in PP matrix at higher

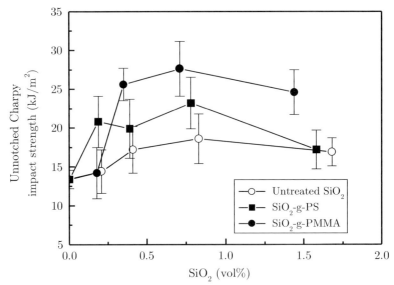

Figure 3.6. Unnotched Charpy impact strength of PP composites filled with untreated SiO$_2$, SiO$_2$-g-PS (γ_g = 5.6%) and SiO$_2$-g-PMMA (γ_g = 14.1%) as a function of SiO$_2$ content. γ_g: percentage grafting

filler content. The loosened clusters of the nanoparticles are surely detrimental to the impact toughness of the composites.

The effects of different grafted nano-silica on the impact strength of PP composites are plotted as a function of filler content in Figures 3.6 and 3.7.

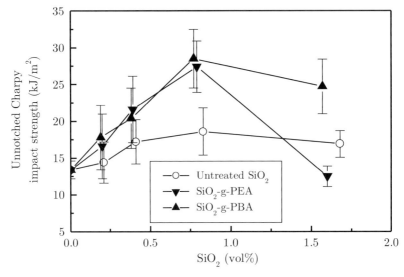

Figure 3.7. Unnotched Charpy impact strength of PP composites filled with untreated SiO$_2$, SiO$_2$-g-PEA (γ_g = 4.4%) and SiO$_2$-g-PBA (γ_g = 6.5%) as a function of SiO$_2$ content

The grafting polymers, which encapsulate silica particles, result in an evident increase in impact strength as compared with the untreated silica counterpart. It means that the formation of an interphase consisting of the grafting polymers, especially those with higher molecular mobility like poly(ethyl acrylate) (PES) and poly(butyl acrylate) (PBA), might act as a "bumper" interlayer around the fillers. It absorbs the impact energy and prevents the initiation of cracks. Besides, the nanoparticles/matrix adhesion created by the entangled interphase between the grafting polymer and the matrix can also obstruct crack propagation at the interface. It is worth noting that there exists an optimum value in the dependence of impact strength on filler content (~0.8 vol%). This can still be explained by the poorer dispersion of the particles at higher filler content like in the case of untreated silica filled PP.

The above discussion is focused mainly on the effects of species of grafting polymer and contents of grafted nano-SiO_2 on the impact strength of nano-silica filled PP. In fact, structure and viscoelastic characteristics of the interphase also play an important role. At a constant filler loading, the influence of the amount of grafting polymers on nano-SiO_2 is given in Figure 3.8. It can be seen that high percentage grafting is generally disadvantageous to the improvement of impact strength for the nanocomposites. If the interphase thickness can be assumed to increase with the grafting polymer fraction, the impact strength should increase due to the possible increase in interface entanglement. However, that is not the case. Therefore, the evident toughening effect perceived at such low percentage grafting in the present nanocomposites implies that other factors should account for the mechanism involved besides interfacial

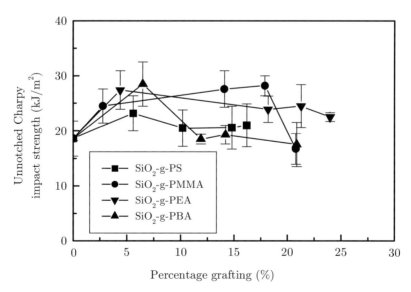

Figure 3.8. Unnotched Charpy impact strength of PP composites filled with SiO_2-g-PS (γ_g = 5.6%), SiO_2-g-PMMA (γ_g = 14.1%), SiO_2-g-PEA (γ_g = 4.4%) and SiO_2-g-PBA (γ_g = 6.5%) as a function of percentage grafting at a constant filler loading of 0.8 vol% [14]

adhesion that restricts the propagation of cracks. Hasegawa *et al.* [16] investigated the distribution of polymer-grafted particles in a polymer matrix and found that there is an optimum grafting density for well dispersing the particles. Tada and co-workers [17] also pointed out the importance of optimum molecular weight of the grafting polymer in the dispersibility of TiO_2 particles. It can be imagined that severe entanglement of grafting polymers at high percentage grafting might hinder the separation of nanoparticle agglomerates in the matrix polymer. Consequently, the greatest contribution of grafted SiO_2 nanoparticles to the improvement of impact properties of PP composites is observed at low percentage grafting. For the moment, the optimum percentage grafting that facilitates nanoparticle dispersion for the current composites is hard to be determined on the basis of the impact performance data, because most composites filled with grafted nanoparticles having different percentage grafting possess higher impact strength than that of untreated nano-SiO_2/PP composites (Figure 3.8). Much further work is needed to have a deeper understanding of this matter.

3.2.1.3. *Mechanochemical graft coupled with melt compounding*

Considering that both irradiation induced graft and chemical graft prior to blending with a polymer matrix would produce some compact agglomerates that are hard to be split by the limited shear forces offered by conventional mixers [18], the grafting polymerization can be conducted *via* mechanochemical methods to solve the problem. In more detail, a planetary ball mill is designed to deal with the surface pre-treatment of nanoparticles [19,20]. The intentions of ball milling are to make grafting monomer penetrate into the nanoparticle agglomerates covering the surfaces of nanoparticles uniformly, and to initiate slight graft polymerization onto the particles by mechanochemical effect.

Ball milling is an industrial method for preparing ultrafine materials through pulverization, or for mixing materials, with or without liquid, in a rotating cylinder or conical mill partially filled with grinding media such as balls or pebbles. In planetary action, centrifugal forces alternately add and subtract. The grinding balls roll halfway around the bowls and then are thrown across the bowls, impacting on the opposite walls at high speed. Grinding is further intensified by interaction of the balls and sample. Planetary action gives up to 20 g acceleration and reduces the grinding time to about 2/3 of a simple centrifugal mill (the one that simply spins around). With the aid of the concentrated high energy, the strong mechanochemical effect will damage the long-range ordering structure of the materials, activate them, and result in unsaturated valence bonds. It is thus possible to be used for carrying out graft polymerization onto inorganic nanoparticles, with the advantages of mass production, low-cost, simple processing and being solvent-free.

The typical ball milling induced graft polymerization proceeded as follows. Before being mixed with the monomers, the SiO_2 nanoparticles were preheated at 140°C for 5 h to eliminate possible absorbed water on the surface of the particles and the initiator benzoyl peroxideide (BPO) or azo-bis-iso-butyronitrile (AIBN) was dissolved in the monomer. The monomer used for grafting, butyl acrylate (BA), was purified by reduced pressure distillation. Then the mixture of SiO_2 nanoparticles, monomer and initiator was incorporated into the planetary ball mill and the reaction took place under preset conditions.

In fact, a series of parameters are closely related to the graft reaction in ball milling. Besides the characteristics of the nanoparticles, monomer concentration, species and dosage of initiator, milling time, rotation speed, raw materials-to-balls volume ratio, *etc.*, are all important, and moreover, the influencing factors interact with one another [21]. Therefore, orthogonal design had to be applied to get reasonable arrangement of the experiments and to optimize the graft conditions. It can be concluded that rotation speed, monomer concentration and milling time are the main influencing factors for the ball milling induced graft polymerization on silica nanoparticles. The optimum reaction conditions are rotation speed = 500 rpm, weight ratio of BA to nano-silica = 2:1, milling time = 4 h, initiator = AIBN, dosage of the initiator = 0.5 wt% of the amount of BA.

For performance study, the mechanochemically grafted SiO_2 nanoparticles were melt mixed with PP to produce bulk composites. Besides, two other types of nanocomposites were also manufactured. One consisted of PP and nano-silica as-received (*i.e.* SiO_2/PP), and the other was prepared by melt mixing PP, nano-silica and butyl acrylate without the pre-treatment in the ball mill.

The basic tensile properties of the composites, including tensile strength and Young's modulus, are shown in Figure 3.9. It can be seen that the addition of untreated nano-silica results in a decrease in the composites' tensile strength, while the particles modified by *in situ* grafting are able to generate a reinforcing effect (Figure 3.9a). Clearly the composites with untreated nanoparticles follow the general law of micro-particulate composites, *i.e.*, the incorporation of the fillers always weakens the strength of the unfilled matrix [22]. This implies that the nanoparticles appear in the composites in the form of large agglomerates and cannot provide load-bearing capacity. In contrast, the *in situ* grafted nanoparticles exhibit reinforcing ability even at rather low filler content. The applied stress can be effectively transferred to most of the particles due to the appearance of the grafting macromolecular chains as expected. The results demonstrate that (i) the grafting treatment has brought the positive effect of the nanoparticles into play, and (ii) the pre-treatment using ball milling is necessary because of the even mixing of the particles and the grafting monomers. Compared to the case of strength property, the grafted nanoparticles are not as effective as their untreated version in stiffening the matrix polymer (Figure3. 9b). In spite of the fact the Young's moduli of all the composites increase with a rise in the filler content, the lower Young's moduli of the composites with the grafted nanoparticles result from the shielding effect of the grafting polymers, which build up a compliant interlayer and lower the stress transfer efficiency in the elastic deformation region. The significantly lower stiffening ability of the SiO_2-

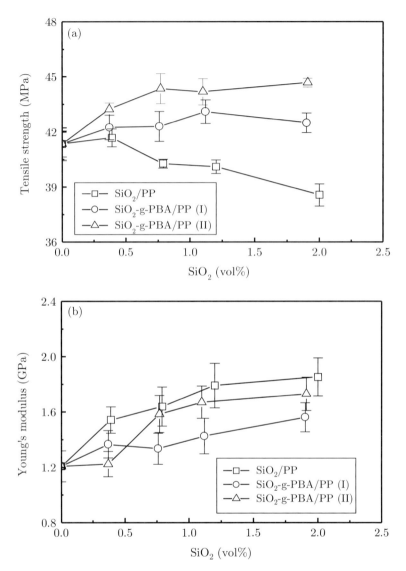

Figure 3.9. Tensile properties of PP composites as a function of nano-SiO_2 volume fraction: (a) tensile strength; (b) Young's modulus. SiO_2/PP denotes untreated nano-silica filled PP composites. SiO_2-g-PBA/PP (I) symbolizes the composites prepared by adding all the ingredients into the mixer of the torque rheometer, and SiO_2-g-PBA/PP (II) the composites obtained by using the ball milling pre-grafted nano-silica [19]

g-PBA/PP composites prepared without ball milling pre-treatment of the nanoparticles is again indicative of the uneven distribution of the grafted nanoparticles.

Besides the increase in strength and modulus, it is seen from Figure 3.10 that the impact strengths of the nanocomposites can also be improved by low

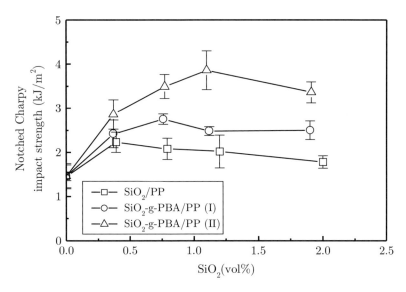

Figure 3.10. Notched Charpy impact strength of PP composites as a function of nano-SiO_2 volume fraction. The symbols of the composites have the same meanings as those described in the caption of Figure 3.9 [19]

loading nano-silica, especially when the particles have been treated by grafting. In comparison to the PP composites filled with nano-silica pre-treated by irradiation graft polymerization [3,4], which possess high notch sensitivity, the current versions prove that the methodology based on mechanochemical grafting treatment is a feasible way to improve the notch ductility of PP.

3.2.2. *Specific two-step manufacturing*

3.2.2.1. *Pre-drawing induced dispersion of nanoparticles*

Our early investigation revealed that when tension is applied, the nanoparticle agglomerates in composites can be extended and broken along the tensile direction [23]. The degree of deformation of the grafted nanoparticle agglomerates is much higher than that of the untreated ones, because the grafted nanoparticles are interconnected with the matrix by the grafted polymer chains. Accordingly, the applied load can be transferred to numbers of nanoparticles that are in intimate contact with the matrix, inducing large-scale plastic yielding of the matrix polymer nearby. For the composites filled with untreated nanoparticles characterized by low extensionality, voiding and disintegration of the nanoparticle agglomerates play the main role. Agglomeration of the nanoparticles stays unchanged in nature. On the basis of this finding, it is suggested that when a grafted nanoparticles/polymer composite is pre-stretched to induce separation of the agglomerated nanoparticles and then re-molded, the improved nanoparticles' dispersion would not completely recover in the ultimate composite due to obstruction of the entanglement between the grafted polymer and

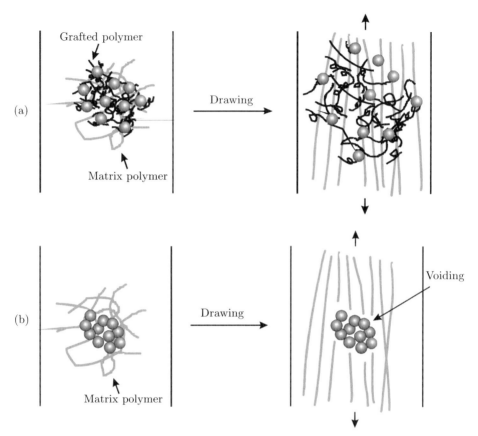

Figure 3.11. Cross-sections of (a) grafted nanoparticles/polymer composites before and after stretching, and (b) untreated nanoparticles/polymer composites before and after stretching [25]

the matrix (Figure 3.11a). If this is the case, the nanocomposites' resistance to crack propagation ought to be greatly raised owing to the increased fraction of interface. Besides, because the nanoparticles are well distributed in the matrix and correlated with one another throughout the entire composite, an overall improvement of mechanical properties might be obtained. With respect to untreated nanoparticles/polymer system, as the agglomerated fillers are nearly undeformed during stretching as a result of the poor interfacial interaction (Figure 3.11b), no obvious increase in properties of the ultimate composites can be expected.

On the basis of the above considerations, a pre-drawing technique was developed in our laboratory (Figure 3.12) [24–26]. That is, inorganic nanoparticles were first treated by graft polymerization to convert their hydrophilic surfaces into hydrophobic while breaking apart the particles with grafted polymer chains. Then, the treated nanoparticles were melt-compounded with matrix polymer and extruded producing composite filaments *via* drawing. Taking advantage

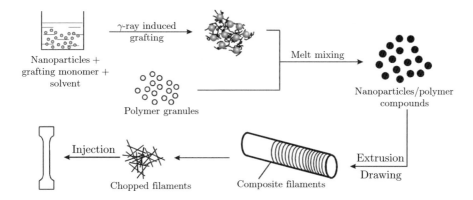

Figure 3.12. Flow diagram of the pre-drawing technique [25]

of the grafted polymer that was covalently connected to the nanoparticles and entangled with the surrounding matrix polymer [3], the interstitial intervals among the nanoparticles had to be remarkably expanded along with the stretching process. Eventually, the composite filaments were injection molded into bulk materials. The improved dispersion status of the nanoparticles had to be somewhat "frozen" in the ultimate composites as a result of the strong interfacial interaction. Accordingly, the nanoparticles were well distributed throughout the entire composites leading to intimate filler/matrix contacts.

The notch impact fracture behavior of the PP nanocomposite that experienced pre-drawing is compared with that of conventional PP in Figure 3.13.

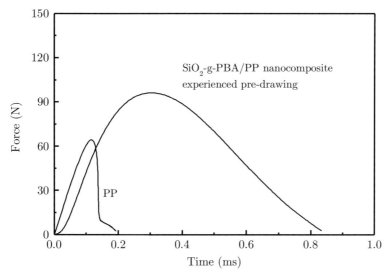

Figure 3.13. Force-time curves of PP and its nanocomposites recorded during notch impact tests [24]

Table 3.2. Mechanical properties of PP and its nanocomposites[a] [24]

Materials	Young's modulus (GPa)	Tensile strength (MPa)	Elongation at yield (%)	Elongation at break (%)	Area under tensile stress-strain curve (MPa)	Notch impact strength (kJ/m^2)
PP	1.20±0.05	36.10±0.68	9.32±0.39	228±14.3	79.4±4.92	1.40±0.05
PP (experienced pre-drawing)	1.62±0.05(35.0%)	35.75±0.77(-1.0%)	10.00±0.29(7.3%)	297.5±22.0(30.5%)	93.80±6.78(18.1%)	1.34±0.02(-4.3%)
SiO$_2$/PP	2.65±0.03(120.8%)	36.68±0.36(1.6%)	7.01±0.52(-24.8%)	89.3±3.59(-60.8%)	15.64±0.89(-80.3%)	1.20±0.02(-14.3%)
SiO$_2$/PP (experienced pre-drawing)	2.32±0.02(93.3%)	36.80±0.64(1.9%)	7.87±0.33(-15.6%)	102.1±6.25(-55.2%)	21.96±1.01(-72.3%)	1.72±0.02(22.9%)
SiO$_2$-g-PBA/PP	1.31±0.04(9.2%)	38.64±0.36(7.0%)	7.88±0.10(-15.5%)	112.9±10.32(-50.5%)	35.08±1.76(-55.8%)	2.78±0.08(98.6%)
SiO$_2$-g-PBA/PP (experienced pre-drawing)	1.96±0.02(63.3%)	38.10±0.57(5.5%)	11.5±0.16(23.4%)	399.0±9.2(75.0%)	114.83±2.85(44.6%)	4.09±0.05(192.1%)
PBA/PP[b] (experienced pre-drawing)	0.98±0.03(-18.3%)	33.90±0.65(-6.1%)	8.98±0.28(-3.6%)	375.6±15.3(64.7%)	91.64±4.69(15.4%)	2.33±0.03(66.4%)

[a]The percentage numerals inside the parentheses represent the changes in the mechanical performance of the materials relative to PP.
[b]The PBA for preparing PBA/PP blend was isolated from SiO$_2$-g-PBA by extraction and etching, respectively. The content of PBA in PBA/PP is 3 vol%, approximately equivalent to the amount of PBA in SiO$_2$-g-PBA/PP.

Evidently, they are quite different. The latter exhibits typical brittle failure and presents unstable crack propagation after the maximum force. In contrast, in the nanocomposite crack formation took much longer time and crack propagation was somewhat stabilized before final fracture [27]. The results clearly manifest that the above-proposed methodology is feasible to improve notch ductility of PP. What is more, the area under the tensile stress-strain curve, a measure of static toughness of the nanocomposite, is also prominently increased (Table 3.2).

Table 3.2 summarizes mechanical properties of a group of related materials. The unfilled PP specimen made from PP filaments has similar tensile strength as the original PP, but its modulus, elongation to break and area under stress-strain curve is slightly higher. Although the orientation structure obtained from drawing has been disordered by injection molding, the differences in tensile properties between the PP specimens with and without pre-stretching imply that the drawing treatment still affects the ability of spherulites slipping and orientation of PP during tensile tests [28]. Addition of untreated nano-silica increases the modulus of PP composites due to the rigidity of the particles. However, tensile strength, elongation to break and area under stress-strain curve of untreated SiO_2/PP become worse as compared with PP. Drawing treatment has little effect on performance improvement in this case. This should be attributed to the fact that the untreated nanoparticles appear in the composites in the form of large agglomerates and cannot provide load-bearing capacity as a result of poor interfacial interaction. When PP is filled with grafted nano-silica, the situation is changed. Tensile strength and modulus of the composites increase owing to the improved stress transfer efficiency at the interface. For the SiO_2-g-PBA/PP composite that experienced pre-drawing, interdiffusion and entanglement between the molecules of the grafted polymer and the matrix help to keep the separation status of the nanoparticles during injection molding. Consequently, distribution of the nanoparticles is greatly improved, and thus much more surrounding matrix polymer is involved in large-scale plastic deformation. Either elongation to break and area under stress-strain curve or notch impact strength of the composite are much higher than those of SiO_2-g-PBA/PP composite without pre-drawing treatment. Reduction of failure strain is common for nanoparticles/polymer composites [29,30] except that nanoparticle mobility is guaranteed [28]. Although the grafted PBA itself is able to toughen PP, deterioration of modulus and strength of PP is detected. From the above results we can conclude that only when the composite of grafted nanoparticles and PP was drawn and then injection molded, stiffness, strength, and especially toughness of the composites can be improved without the side-effect of embrittlement.

Transmission electron microscopy images along the injection direction of the impact specimen bars provide direct inspection of the nanoparticles in the PP matrix (Figure 3.14). In the case of untreated nano-silica/PP composite, large agglomerates (> 500 nm) appear and there is no trace of elongation of

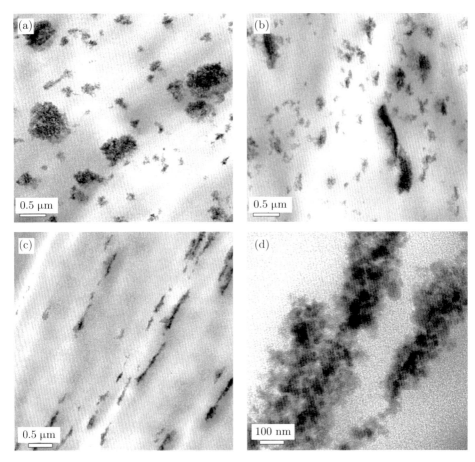

Figure 3.14. TEM micrographs of PP based nanocomposites: (a) SiO_2/PP (experienced pre-drawing); (b) SiO_2-g-PBA/PP; (c) and (d) SiO_2-g-PBA/PP (experienced pre-drawing) [24]

agglomerates resulting from pre-drawing (Figure 3.14a). Having been grafted with PBA, the nanoparticles can be more homogeneously dispersed but large agglomeration is still perceivable (Figure 3.14b). This means that the kinetic energy supplied by the compounding machine is not high enough to break apart the nanoparticle agglomerates in the polymer melt despite the fact that the filler/matrix compatibility is improved. When pre-drawing is applied, distinct elongation and alignment of the nanoparticle agglomerates in the SiO_2-g-PBA/PP composite can be seen, while the severely agglomerated nanoparticles (that might act as weak sites causing stress concentration) disappear (Figure 3.14c). The deformed nanoparticle agglomerates look like fibers and their dispersion becomes much more uniform. The magnified photo (Figure 3.14d) shows that the agglomerates have network microstructure, in which the nanoparticles have indeed been pulled apart, while the grafted polymer appears inside and outside the agglomerates. In this context, it can be expected that the tiny reinforcements can give full play and lead to overall improvement of mechanical performance of the nanocomposites.

Devaux and Chabert suggested that drawing of semicrystalline polymers might lead to the formation of a phase that is not accessible under normal conditions [31]. For PP, the result lies in the appearance of β-crystals, in addition to the commonly occurring α-phase [32]. Because of the metastable nature of β-crystal PP, however, the β-phase is difficult to be retained when the drawn samples are molten again and cooled without further drawing [33]. Figure 3.15 shows the *wide angle X-ray diffraction* (WAXD) spectra of PP and its nanocomposites.

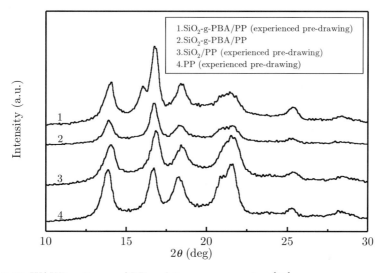

Figure 3.15. WAXD patterns of PP and its nanocomposites [24]

Besides the peaks at $2\theta = 14.2°$, $17.0°$, $18.8°$, $21.4°$ and $22.0°$, which represent the existence of α-crystal of PP, only on the spectrum of SiO_2-g-PBA/PP that had experienced drawing pre-treatment, there is a small peak at $2\theta = 16.2°$ corresponding to the (300) reflection of β-crystal of PP. It implies that some β-phase remains in this composite. Since TEM observation confirms that the extended structure of nanoparticle agglomerates can be aligned in SiO_2-g-PBA/PP (Figure 3.14c), we might infer that with the help of the anchoring effect of the aligned nanoparticles, the crystal structure of the PP matrix developed in the drawing process can be retained in the re-molded nanocomposites. In consideration of the high ductility of β-crystals, it is believed that the specific crystalline morphology of the matrix in SiO_2-g-PBA/PP with drawing pre-treatment should also contribute to the toughening of the composites.

3.2.2.2. *In situ bubble-stretching*

In consideration of the fact that stretching is superior to shearing in separating aggregated granules, Wu and co-workers proposed an *in situ bubble-stretching*

(ISBS) *model* [34,35]. It was believed that the rapid inflation of a polymer bubble could result in the polymer melt surrounding the bubble being subjected to high stretching rates that may lead to effective stretching dispersion of the dispersed phase in the polymer melt. Theoretically, when the bubble expands, the order of magnitude of stretching rate of the surrounding polymer can reach 10^5–10^6 s^{-1} [34,36], which is about two orders of magnitude higher than the shearing rate (10^3–10^4 s^{-1}) obtained with screw extrusion. The minimum grain size of the dispersed granules with this method should therefore be two orders of magnitude less than that observed when the particles are dispersed by screw extrusion.

With the enlightenment of Wu's work, we planned to combine graft polymerization and *in situ* bubble-stretching into a new approach for pre-treating nanoparticles [37–39]. By graft polymerization of monomers containing specific side groups that can be decomposed to gas under certain temperature onto nanoparticles, the nanoparticles are thus connected with macromolecular foaming agent. When the nanoparticles of this kind are incorporated into a polymer melt, the foaming agent would be triggered producing *in situ* bubble-stretching effects to remarkably separate the agglomerated nanoparticles. In addition, the remaining backbones of the grafted polymers might still help to increase interfacial compatibility, which would also contribute to the improvement of stretching-induced extensionality and hence dispersion of the nanoparticle agglomerates [23]. Compared to the case of Wu *et al.* [34], in which nano-$CaCO_3$ and foaming agent were individually mixed with high density polyethylene, the method suggested by the authors ensures that the foaming agent is intimately attached to the nanoparticles. Consequently, the efficiency of bubble-stretching might be much higher, or in other words, much less foaming agent would be sufficient to achieve the same goal. This is particularly important when mechanical properties of the ultimate composites are considered. Because the amount of the side groups on the grafted polymer chains serving as foaming agent is so little that not many small-scale bubbles can be generated, defoaming would be no longer required in the subsequent injection molding for manufacturing bulk composites using these micro-foamed compounds.

In accordance with this idea, *p*-vinylphenylsulfonylhydrazide was synthesized and grafted onto silane coupling agent pre-treated nano-silica *via* aqueous radical polymerization [37,40]. Then, the grafted nanoparticles were melt compounded with PP as usual. The monomer was selected because it possesses (i) a double bond that can be used to react with the functional groups on the nanoparticles and to grow into macromolecules, and (ii) a chemical structure similar to the foaming agents with sulfonyl hydrazide groups that can be decomposed to gas at around 120~160ºC (matching the processing temperature of PP) [41]. As a result, melt blending of poly(p-vinylphenylsulfonylhydrazide) grafted nano-silica with PP should cause a bubble-stretching effect to improve the nanoparticles' dispersion in the PP matrix (Figure 3.16).

Nanoparticles/Polymer Composites

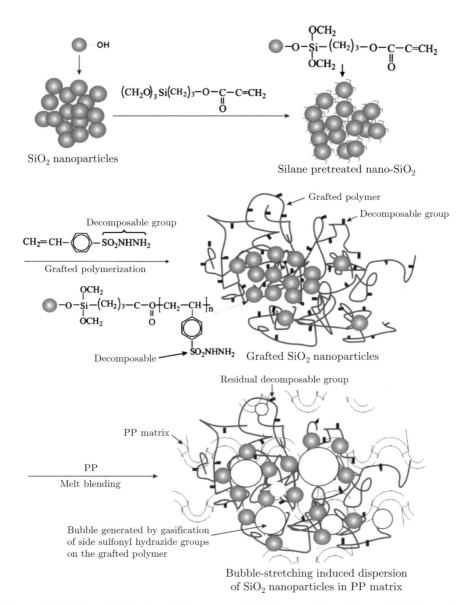

Figure 3.16. Schematic drawing of the proposed bubble-stretching effect realized *via* graft macromolecular foaming agent onto silica nanoparticles [37]

Dispersion status of the grafted nanoparticles in the PP matrix was examined by SEM in comparison with the untreated nanoparticles (Figure 3.17). The micrographs were taken from the result of melt compounding of the ingredients. Although no visible foam-like structure can be found from the appearance of the composite with grafted nano-silica, some porous portions are observed under high magnification, while the composite with silica as-received is characterized

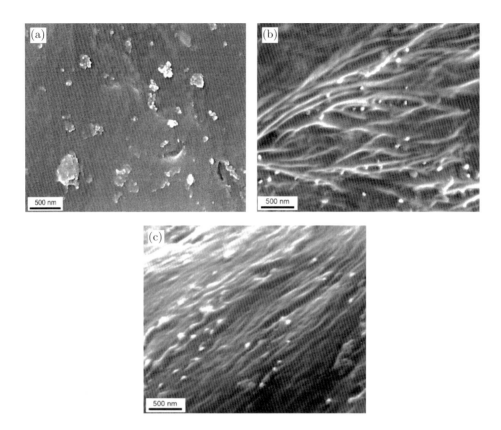

Figure 3.17. SEM images of freeze-fractured surfaces of (a) untreated nano-silica/PP (nano-silica content = 0.95 vol%), (b) poly(p-vinylphenylsulfonylhydrazide) grafted nano-silica/PP composites (nano-silica content = 1.43 vol%), and (c) poly(p-vinylphenylsulfonylhydrazide) grafted nano-silica/PP composites (nano-silica content = 2.41 vol%)

by a compact morphology (Figure 3.17). The striped traces of the matrix and the separated nanoparticles along specific flow directions are indicative of the bubble-stretching effect (Figure 3.17b and c). It implies that the decomposable groups on the grafted polymer must have taken effect and the foamed parts are limited under the present manufacturing conditions. To confirm the estimation in a quantitative way, porosity of the composites, η, was determined using density measurement (listed in Table 3.3). The results (refer to η_1 in Table 3.3) demonstrate that (i) there are detectable voids in the composites, and (ii) the amount of the voids increases with increasing the grafted nanoparticles content, *i.e.*, the content of the blowing agent.

In the last step of this work, the compounds of PP and nano-silica were broken and injection molded to make standard specimens for mechanical testing. Prior to the tests, porosity of the composites filled with the grafted nanoparticles was determined again. As seen in Table 3.3 (refer to η_2), the internal cavities

Table 3.3. Porosity of poly(p-vinylphenylsulfonylhydrazide) grafted nano-silica/PP composites [37]

Content of nano-silica (vol%)	0.47	0.95	1.43	2.41
$\eta_1{}^a$ (%)	3.60	7.53	11.87	15.02
$\eta_2{}^b$ (%)	0.07	0.13	0.08	0.19

^{a} η_1: Porosity of the compounds produced by the mixer of the torque rheometer
^{b} η_2: Porosity of the injection molded specimens

have been effectively eliminated after injection molding, and this should be favorable for the mechanical properties of the materials.

Figure 3.18 shows notched Charpy impact strength of PP composites as a function of nano-silica content. It is seen that the contribution of the untreated fillers to the improvement of toughness under high speed deformation and to the reduction of notch sensitivity is nearly negligible as compared with the grafted ones. The impact strength of untreated nano-silica/PP composites slightly increases with the filler content and reaches its maximum at about 0.95 vol% of silica. For poly(p-vinylphenylsulfonylhydrazide) grafted nano-silica/PP systems, however, the highest impact strength that appears at 1.43 vol% of silica is about 1.26 times higher than that of the composite with untreated nano-silica. So far, the crack front bowing mechanism has been widely used for explaining the toughening effect in micro-sized particulate filled polymers [42]. Nevertheless, Chan et al. [43] inferred that in the case where the size of

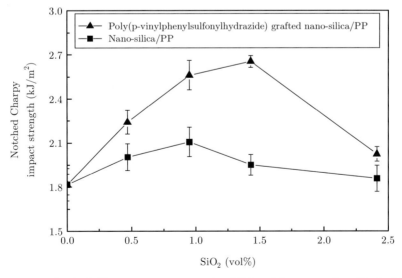

Figure 3.18. Notched Charpy impact strength of PP and its composites filled with nano-silica as a function of silica content [37]

the rigid particles is of the order of 50 nm or less, the applicability of the bowing mechanism is questionable, because such small size rigid particles may not be able to resist the propagation of the crack. From J-integral tests, they suggested that nanoparticles would trigger large-scale plastic deformation of the matrix, which consumes tremendous fracture energy. Since the above morphological investigation reveals that the dispersion scale of the nanoparticles is less than 50 nm in the composites with the grafted nano-silica, it is believed that the mechanism proposed by Chan and co-workers should also be valid in the current study.

3.2.3. *One-step manufacturing*

The above sub-sections 3.2.1 and 3.2.2 discussed the nanocomposites prepared by two-step techniques. Nano-silica is grafted in advance, and then melt compounded with matrix PP. In fact, these procedures can be simplified into a one-step approach as follows [44]. The nanoparticles are pre-treated by a silane coupling agent to introduce reactive C=C double bonds onto the surface. Afterwards, a mixture of the silane pre-treated nanoparticles, graft monomer, initiator and matrix polymer is melt compounded. The graft monomer is expected to react with the silane coupling agent on the nanoparticles and then to form grafted macromolecules *in situ* in the course of compounding. As a result of side reactions, slight graft polymerization on the matrix polymer might take place. Nevertheless, it should be an effective way to eliminate agglomeration of nanoparticles and improve intimate interaction between the nano-fillers and matrix.

The silane pre-treatment of nano-silica is based on the consideration of the fact that graft polymerization onto nanoparticulate surfaces in molten PP is more difficult than in solution of the polymer. To further increase collision between nano-silica and the monomer under melt mixing, concentrated batches should be made by incorporating nano-silica and graft monomer into small amounts of matrix polymer. Then, the concentrated batches can be diluted into desired concentrations to prepare nano-silica/PP composites.

Dilution of concentrated batch into desired concentrations was accomplished by Haake torque rheometer mixer in our lab. Processing conditions of the mixer must greatly influence the performance of the ultimate nanocomposites. As shown in Figure 3.19, notched impact strength and tensile strength of the nanocomposites decrease with increasing mixing temperature and have an optimum value with a rise in rotation speed and mixing time, and fluctuate with rotation speed more severely than with mixing temperature and time. In fact, PP is more sensitive to shear rate so that an acceleration of rotation speed would effectively decrease its melt viscosity, but elevated temperature tends to cause degradation of PP. Decrease of melt viscosity is desirable to improve nanoparticle dispersion in the matrix. Nevertheless, an increase in rotation speed and mixing time might also lead to matrix degradation. As a result, mechanical properties of the nanocomposites change as a function of rotation

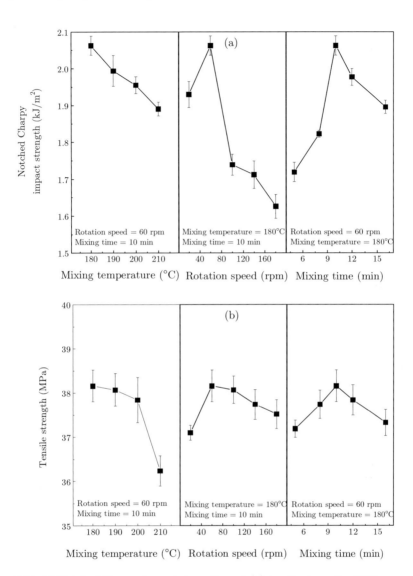

Figure 3.19. (a) Notched Charpy impact strength, and (b) tensile strength of PP composites as a function of mixing temperature, rotation speed and mixing time of the mixer. Content of nano-SiO$_2$ = 1.36 vol% [44]

speed and mixing time, having an optimum value. To balance mechanical properties and energy consumption, the reasonable melt mixing conditions can be set as 180°C, 10 min and 60 rpm.

Table 3.4 collects the differential scanning calorimetry (DSC) studies of non-isothermal crystallization and melting behavior of PP and its nanocomposites produced by the one-step technique. The data show that the melting point of PP is nearly not affected by the nano-silica. It means that no transition of

Table 3.4. Non-isothermal crystallization and melting data of PP and its nanocomposites [44]

Samples	T_m^a (°C)	T_c^b (°C)	ΔT^c (°C)	X_c^d (%)
Neat PP	162.96	112.89	50.07	46.7
SiO$_2$/PP	162.92	118.39	44.53	44.9
SiO$_2$-g-PBA/PP (percent grafting = 11.76%)	162.90	122.12	40.78	46.62
SiO$_2$-g-PBA/PP (percent grafting = 15.91%)	161.76	122.21	39.55	45.53

[a] T_m: peak melting temperature; [b] T_c: peak crystallization temperature; [c] $\Delta T = T_m - T_c$, denoting supercooled temperature; [d] X_c: matrix crystallinity
Content of nano-SiO$_2$: 1.36 vol%

crystal forms occurs, which coincides with the result of wide angle X-ray diffraction study. On the other hand, supercooling temperature, DT, of the polymer decreases with the addition of untreated SiO$_2$ and SiO$_2$-g-PBA, indicating that crystallization of the nanocomposites becomes easier owing to the nucleation effect of the nanoparticles. The SiO$_2$-g-PBA/PP composites have lower DT but higher crystallinity than SiO$_2$/PP. Some researchers [45,46] suggested that particulate fillers influence crystallization behavior of polymer matrices not only because the fillers act as nucleation agents, but also because the filler/matrix interfacial stress has an effect. Uniformly dispersed fillers would produce higher interfacial stress accumulation and stronger interfacial bonding that ensures better interfacial stress transfer. Consequently, the rate of crystallization and crystallinity of the composites with well-dispersed fillers would be higher than those of the systems containing poorly dispersed fillers. According to their consideration, the data in Table 3.4 might evidence that SiO$_2$-g-PBA/PP composites have more uniform particle dispersion and enhanced interfacial adhesion than the untreated silica system.

3.3. Dispersion and filler/matrix interaction-oriented manufacturing of nanocomposites

3.3.1. *Two-step manufacturing in terms of in situ reactive compatibilization*

In the above text, the advantages of graft polymerization onto SiO$_2$ nanoparticles followed by melt compounding with PP for making PP based nanocomposites have been shown. It is worth noting that the interface in these grafted nanoparticles/polypropylene composites is believed to be enhanced by the chain entanglement between the grafted polymers and the matrix. That is, physical interaction predominates the interfacial interaction.

Actually, the graft treatment of nanoparticles leads to specific structures that can be tailored by changing graft monomers and graft conditions. If the grafting polymers attached to the nanoparticles possess reactive groups, chemical bonding at the filler/matrix interface might thus be built up during melt compounding with the help of certain compatibilizers. As a result, a fine adjustment of interfacial interaction in the subsequent composites can be completed by controlling the amount and species of the reactive groups. The technical route belongs to reactive compatibilization, which means the blending process is accompanied by chemical reaction and is capable of generating strong polymer/polymer or filler/polymer interactions at the interface [47–49].

The concept of reactive compatibilization has been used for preparing compatible polymer blends so long as the components have functional groups capable of generating covalent, ionic, donor-acceptor, or hydrogen bonds between them. Accordingly, the addition of a reactive polymer, miscible with one blend component and reactive towards functional groups attached to the second blend component, would result in the "*in situ*" formation of block or grafted copolymers [47]. It appears to produce the best blend compatibilization [48]. Nevertheless, few works focusing on manufacturing of polymer composites filled with inorganic particles in this way have been published.

To further bring the positive effects of nanoparticles into play, we introduced reactive compatibilizing into nanoparticles filled PP composites as follows [50–53]. Firstly, the monomer glycidyl methacrylate (GMA) with reactive epoxide groups was introduced onto SiO_2 nanoparticles by chemical graft polymerization [54]. Then, the grafted nanoparticles (denoted by SiO_2-g-PGMA) were melt-blended together with PP and functionalized PP (aminated PP (PP-g-NH_2) [55] or maleic anhydride grafted PP (PP-g-MA)) at different compositions. The latter was applied as a reactive compatibilizer as it is able to react across the melt phase boundary with the pendant oxirane ring on the grafting chains on the nanoparticles.

For confirming the reaction between the components in the course of melt compounding, a model mixture of SiO_2-g-PGMA/PP-g-NH_2 (20/80) was manufactured using the same processing conditions as those adopted for preparing the composites. Then it was extracted by toluene for 8 days to remove the unreacted PP-g-NH_2. Figure 3.20 compares the FTIR spectra of the residue of the compounds after extraction, SiO_2-g-PGMA and PP-g-NH_2. On the spectrum of SiO_2-g-PGMA, the absorption band at 910 cm^{-1} represents epoxide functional groups. It proves that the nano-silica particles have acquired the desired reactivity. From the FTIR spectrum of the residue of the compounds after extraction, it is found that the peak of epoxide groups on SiO_2-g-PGMA becomes smaller, while the characteristic absorptions of PP (*i.e.*, the multiple peaks at 2820~2962 cm^{-1} corresponding to the stretching modes of C–H in –CH_2–, –CH– and –CH_3, and the peaks at 1377~1457 cm^{-1} for the bending modes of –CH_2– and –CH_3) and SiO_2 (*i.e.*, the anti-symmetric stretching mode of Si-O-Si at 1100 cm^{-1}) are perceivable. Evidently, some of the epoxide groups on SiO_2-g-PGMA have been consumed through the reaction with the amine groups on PP-g-NH_2 during melt blending.

Figure 3.21 illustrates the pyrolytic behaviors of PP, PP-g-NH_2, SiO_2-g-PGMA and the residue of their composites (SiO_2-g-PGMA/PP-g-NH_2/PP =

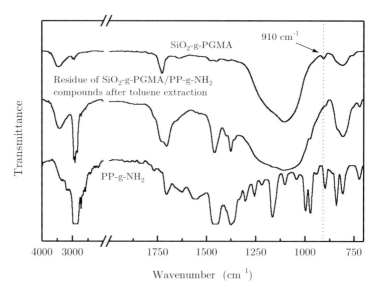

Figure 3.20. FTIR spectra of PP-g-NH_2, SiO_2-g-PGMA and the residue of SiO_2-g-PGMA/PP-g-NH_2 (80/20) compounds after toluene extraction [53]

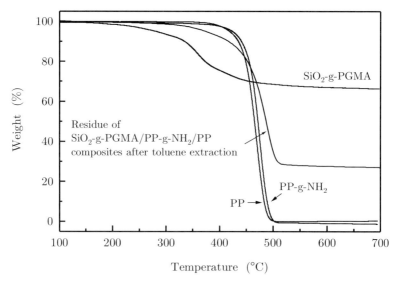

Figure 3.21. Thermal decomposition curves of PP, PP-g-NH_2, SiO_2-g-PGMA and the residue of SiO_2-g-PGMA/PP-g-NH_2/PP (87/3/10) composites after toluene extraction [53]

3/10/87) after toluene extraction, from which the amount of epoxide groups consumed during the reaction of SiO_2-g-PGMA and PP-g-NH_2 can be estimated. Since weight loss of the residue of SiO_2-g-PGMA/PP-g-NH_2/PP after extraction (*i.e.* the reaction product of SiO_2-g-PGMA/PP-g-NH_2) is 73%, and that of SiO_2-g-PGMA is 33%, it can be calculated that the weight ratio of reacted SiO_2-g-PGMA/PP-g-NH_2 is 40/60. Accordingly, it is known that about 8% of epoxide groups on the grafted PGMA were consumed in the reaction with PP-g-NH_2.

On the other hand, Figure 3.21 shows that the onset pyrolytic temperature of the residue of the composites after toluene extraction is lower than those of PP and PP-g-NH_2, but higher than that of SiO_2-g-PGMA. The result again evidences the generation of copolymer of PGMA and PP-g-NH_2.

It is known that the mechanical properties of particulate filled polymer composites depend to a great extent on interfacial adhesion and filler dispersion. As exhibited in Figure 3.22a, the untreated nano-silica particles are agglomerated in the matrix, and a rise in the particles concentration results in a higher degree of agglomeration (Figure 3.22b). Having been grafted with PGMA, the nanoparticles are characterized by much smaller size and uniform distribution (Figure 3.22c). When PP-g-NH_2 was compounded, the grafted nanoparticle agglomerates were further broken up into stretched bands (Figure 3.22d). The phenomenon agrees with the common law of polymer blends containing compatibilizer.

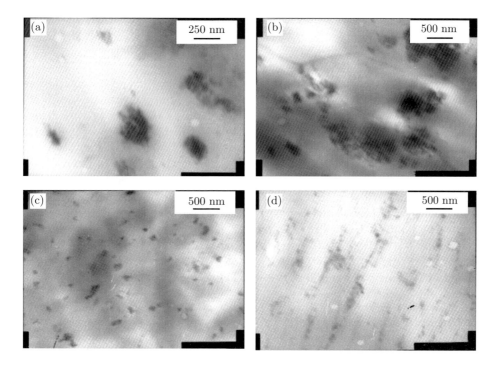

Figure 3.22. TEM micrographs of (a, b) untreated SiO_2/PP, (c) SiO_2-g-PGMA/PP, and (d) SiO_2-g-PGMA/PP-g-NH_2/PP composites. Contents of nano-SiO_2: (a) 3 wt%; (b) 5 wt%; (c) 3 wt%; (d) 3 wt% [53]

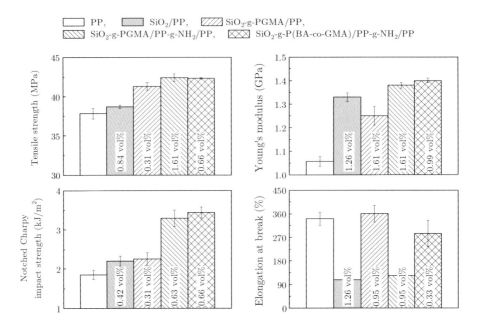

Figure 3.23. Comparison of tensile strength, Young's modulus, notched Charpy impact strength and elongation at break of PP and its composites. The volume percentage of nano-SiO_2 is marked in each column, while the content of PP-g-NH_2 is 10 wt% [53]

Figure 3.23 summarizes mechanical properties of PP and its composites filled with untreated and grafted nano-SiO_2, in the presence or absence of functionalized PP. Each composite shown in this figure was chosen because it has the highest performance in its family. With respect to tensile strength, for example, the highest value of SiO_2-g-PGMA/PP-g-NH_2/PP appears at a nano-silica content of 1.6 vol% within the entire filler loading range of interest and hence the composite with the specified filler content was selected. Evidently, the composites with reactive compatibilization have an advantage over the others. Although the reactive compatibilization approach seems to be detrimental to the failure strain of the composites, this weakness can be overcome by increasing the flexibility of the grafted polymers. That is, so long as SiO_2-g-PGMA is replaced by SiO_2-g-P(BA-co-GMA), the interfacial viscoelastic deformation can be largely induced and hence the elongation at break of the nanocomposites is enhanced.

3.3.2. *One-step manufacturing in terms of in situ graft and crosslinking*

The treatment based on graft polymerization onto nanoparticles provides polymeric composites with a tailorable interphase. Filler/matrix interfacial adhesion can be improved by the grafting polymers adhered to the nanoparticles, which create entanglements or chemical bonding with the matrix molecules. The characteristics of grafting polymers on nano-silica, including molecular stiffness, molecular solubility with the matrix, and grafting density, exert great influence on the reinforcement effectiveness [15]. The interphase thickness in the grafted nanoparticle composites increases with the percentage of grafting. A thicker interphase is not bound to improve the nanocomposites' performance. Apart from the interaction between the grafting polymers and the surrounding matrix polymer, however, there is no interaction between the treated nanoparticle agglomerates themselves.

For more effective strengthening of the agglomerates and improving the interaction among nanoparticles and matrix, an *in situ* crosslinking method was proposed accordingly [56,57]. That is, reactive monomer (BA), initiator (AIBN), crosslinking agent (trimethylolpropane triacrylate, TMPTA), silica nanoparticles and PP were melt mixed together. In the course of compounding, the reactive monomers were polymerized and grafted onto nano-silica, and then the grafted polymers were crosslinked. Since the linear PP chains penetrated through the crosslinking networks, the nanocomposites microstructure can be regarded as a semi-interpenetrating polymer network (semi-IPN) (Figure 3.24). In this way all the nanoparticles were interconnected, and the weak physical interaction or "no" interaction among the grafted nanoparticle agglomerates in the composites would be switched to a strong interaction. Such a semi-IPN structure should greatly enhance the interfacial interaction and facilitate stress transfer inside as well as outside the nanoparticle agglomerates, hindering motion of matrix polymer chains. It is believed that the nanocomposites might thus gain more significant performance improvement.

Since reactive monomer BA and crosslinking agent TMPTA were melt compounded together with nano-silica and PP, the effect of *in situ* crosslinking polymerization on the microstructure of nano-silica/PP composites should be

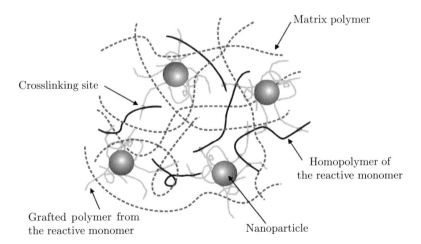

Figure 3.24. Schematic drawing of the proposed semi-IPN structure in nanoparticles filled polymer composites [57]

identified at the very beginning. Here poly(butyl acrylate) crosslinked nano-SiO_2 filled PP compounds are denoted as SiO_2-c-PBA/PP, while the composites filled with pre-treated silica/BA/AIBN without the crosslinking agent TMPTA are denoted as SiO_2-g-PBA/PP. The torque-time curves provided by a Haake torque rheometer are capable of illustrating the occurrence of the crosslinking reaction (Figure 3.25). For SiO_2-g-PBA/PP composites without TMPTA, only one peak (*i.e.*, the feeding peak) appears. When TMPTA is added, however, an additional peak is perceived shortly after the feeding peak. It means that the *in situ* crosslinking took place as soon as the feeding process was completed and the reaction was so intense that the torque increased drastically. Within

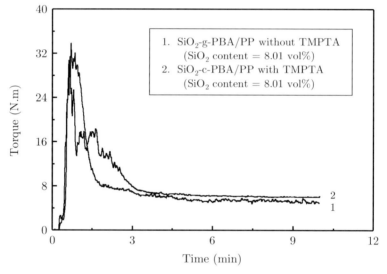

Figure 3.25. Torque-time curves of PP based nanocomposites recorded during melt compounding (content of nano-silica = 1.36 vol%)

about four minutes, the *in situ* crosslinking polymerization came to an end and the torque gradually descended. The slightly higher equilibrium torque of the composites with TMPTA indicates that the melt viscosity is increased owing to the crosslinking reaction of PBA. The subsequent thermogravimetric (TG) analysis shows that the weight gain of the nano-silica extracted from SiO_2-c-PBA/PP is 37.9 wt%, while that of the nano-silica extracted from SiO_2-g-PBA/PP is only 15.9 wt%. The difference should originate from the crosslinked portions in SiO_2-c-PBA/PP.

For examining the microstructure of the materials, untreated nano-silica and the extracted resultants of the composites were observed by SEM (Figure 3.26). It is seen that the nanoparticles as-received are loosely aggregated together (Figure 3.26a). For SiO_2-g-PBA/PP prepared without the crosslinking agent TMPTA (Figure 3.26b), the nanoparticle agglomerates become much smaller and uniformly dispersed, but no polymer networks around the nanoparticles

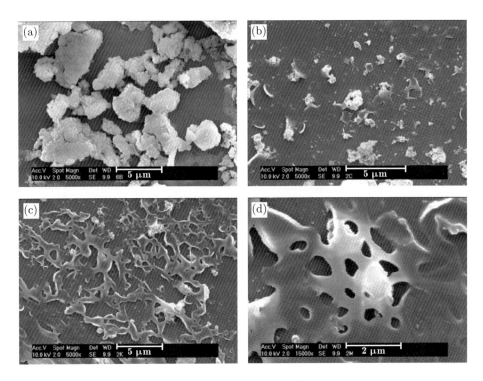

Figure 3.26. SEM micrographs of (a) SiO_2 as-received, (b) extracted resultant of SiO_2-g-PBA/PP, (c) and (d) extracted resultant of SiO_2-c-PBA/PP [57]

remain after the extraction. In contrast, the image of extracted SiO_2-c-PBA/PP shows the interconnected polymer networks that wrap the nanoparticles (Figure 3.26c). The magnified photo (Figure 3.26d) further reveals that the extractable components have left many cavities in the networks. The result suggests that the desired semi-IPN structure in which linear PP molecules penetrate into the networks of crosslinked nanoparticle micro-composites has indeed been established. The estimation receives support from TG measurements. It is found that the weight gain of nano-silica in the extracted solid of SiO_2-c-PBA/PP is 37.9 wt%, while that in the extracted solid of SiO_2-g-PBA/PP is only 15.9 wt%. The difference should originate from the crosslinked (unextractable) portions in SiO_2-c-PBA/PP.

The creep behavior of the PP nanocomposites is compared with that of unfilled PP in Figure 3.27. In general, a creep strain *vs.* time curve can be classified as four stages: (i) initial rapid elongation, (ii) primary creep, (iii) secondary creep, and (iv) tertiary creep [58]. The first stage is independent of time and the rapid elongation is attributed to the elastic and plastic deformation of the polymer specimen under the applied constant load. In the primary creep stage, the creep rate starts at a relatively high value, and then decreases rapidly with time owing to the slippage and reorientation of polymer chains under

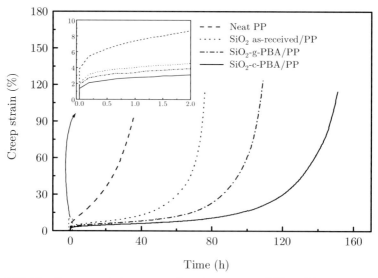

Figure 3.27. Tensile creep strain *vs.* test duration-curves of PP and its composites under 80% of the static ultimate tensile strength (UTS) at room temperature [57]

persistent stress. After a certain period, the creep rate reaches a steady-state value and the secondary creep stage starts. Usually, the duration of this stage is relatively long. Finally, the material falls into the tertiary creep stage, in which the creep rate increases rapidly and final creep fracture occurs.

As shown in Figure 3.27, neat PP exhibits a particularly short creep life (less than 30 h), and higher creep strain as compared with the other materials at the same time level. For the composites of untreated nano-SiO_2/PP and SiO_2-g-PBA/PP without TMPTA, their creep lives are longer (~70 and 90 h, respectively) and the creep strains are significantly reduced over the former three creep stages due to the restriction effect of the nanoparticles. It is noteworthy that the introduction of *in situ* crosslinked nanoparticle networks greatly improves the creep life (>130h) and further lowers the creep strain of PP (refer to the curve of SiO_2-c-PBA/PP in Figure 3.27). A careful survey of the inset in Figure 3.27 demonstrates that SiO_2-c-PBA/PP system also exhibits the lowest initial creep rate at the transition between the initial and the primary creep stage. As for the secondary creep stage, the creep rates of the materials are ranked in the following order: PP > SiO_2 as-received/PP > SiO_2-g-PBA/PP > SiO_2-c-PBA/PP. In practice, reduction of creep strain and creep rate as well as extension of creep life is very important for structural materials. The above experimental results prove that a proper pre-treatment of the nanoparticles in combination with the *in situ* reactive processing technique is more effective in increasing durability and service life of PP than the addition of untreated nanoparticles.

Figure 3.28 plots the time dependences of creep compliance of the materials, which give more information about their creep behaviors. Under 80% of the

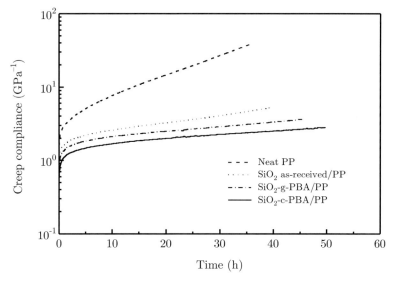

Figure 3.28. Tensile creep compliance *vs.* test duration curves of PP and its composites under 80% of the static UTS at room temperature [57]

static UTS, neat PP possesses the fastest rate of increase in creep compliance and the highest creep compliance over the entire range of testing time. For the loading time of 35 h, for instance, its creep compliance is 37.1 GPa^{-1}, nearly losing its value of use. Addition of untreated nano-silica does not help much: the creep compliance of SiO$_2$ as-received/PP is 4.67 GPa^{-1} at the same loading time. When the nanoparticles were grafted by PBA, and especially crosslinked with TMPTA, the increase in creep compliance is slowed down. The creep compliance after 35 h of loading amounts to 3.09 GPa^{-1} (for SiO$_2$-g-PBA/PP) and 2.39 GPa^{-1} (for SiO$_2$-c-PBA/PP), respectively. The results manifest that the load-bearing capability and dimensional stability of PP under constant load are remarkably enhanced by the grafted nanoparticles and the semi-IPN structure.

3.4. Dispersion, filler/filler interaction and filler/matrix interaction-oriented manufacturing of nanocomposites

Recently, Gersappe has made molecular dynamics simulations and suggested that the mobility of the nanofillers in a polymer controls their ability to dissipate energy, which would increase toughness of the polymer nanocomposites in the case of proper thermodynamic state of the matrix [59]. By examining elongation to break and area under stress-strain curve of treated nanoclay filled poly(vinylidene fluoride) (PVDF) composites, Giannelis and his co-workers evidenced the above hypothesis [28]. They showed that the incorporation of the nanoclay brought about a significant toughening effect when the tensile test was carried out above the glass transition temperature (T_g) of the matrix polymer, whereas embrittlement was detected at a temperature below the matrix' T_g. It was believed that mobility of the polymer matrix is a precondition for this mechanism, which dictates the mobility of the nanoparticles. However, verification with the literature data reveals that for the polymer composites consisting of nanoparticles without layered structures, toughness increase is not bound to be perceived even if the matrix possesses higher mobility. The room temperature ductility of natural rubber reinforced by nano-Fe, nano-Ni, nano-SiC particles and single-walled carbon nanotubes (SWNT), for example, is lower than that of the matrix [60,61].

Accordingly, we prepared nano-silica/thermoplastics composites *via* a specific route and demonstrated that the concept of nanoparticles mobility is still valid for toughening non-layered nanoparticles/polymer composites if both reduced interparticulates interaction and enhanced nano-filler/matrix interaction are guaranteed besides sufficient mobility of the polymer matrix [62]. These guidelines are applicable to the formulation of a technical route for manufacturing nanoparticles/polymer composites with increased toughness, or for increasing the degree of improvement in toughness of the nanocomposites.

The key issue of preparation of the nanocomposites lies in the application of fluoropolymers (poly(dodecafluoroheptyl acrylate) (PDFHA) or poly(hexafluorobutyl acrylate) (PHFBA)) as the grafting polymers. Figure 3.29 exhibits tensile behaviors of the nanocomposites. Evidently, the addition of untreated nano-silica deteriorates the toughness of PP as viewed from the failure strain (Figure 3.29a), no matter whether the environmental temperature is above or below the T_g of PP (~18°C). This contradicts the expectation that rubbery matrices benefit the most from the addition of nanofillers when the mechanism of mobile nanoparticles induced energy-dissipation is concerned [59]. To look into the reason, the specimens before and after the tensile tests at 25°C were cut by microtome in the direction of the applied stress and observed under TEM (Figure 3.30). It is seen that the untreated nano-silica particles are severely agglomerated in the PP matrix (Figure 3.30a) and the agglomerates appear not to be alleviated and oriented under tension (Figure 3.30b), implying that the mobility of the polymer matrix cannot be imparted to the nanoparticles. These large nanoparticle agglomerates have to simply act as sites of stress concentration, making no contribution to the toughening of the matrix polymer.

In fact, the non-layered nanoparticles are easy to adhere to each other due to their high surface areas. Unlike organo-clay that can be intercalated or even exfoliated by matrix polymer during compounding [28,63], the non-layered nanoparticle agglomerates are hard to be broken up by conventional mechanical mixing. In addition, the compatibility between the hydrophilic nanoparticles and hydrophobic polymer matrix is poor in nature. Therefore, weakening the interaction between the nanoparticles while enhancing nanofiller/matrix interaction should be critical for providing individual nanoparticles with certain mobility through the traction of the neighbor mobile matrix. If merely establishment of bonding at filler/matrix interface is emphasized, nanoparticle

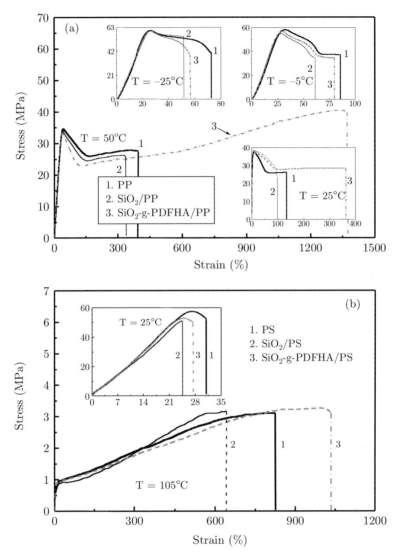

Figure 3.29. Typical tensile stress-strain curves of nano-SiO$_2$/polymer composites measured below and above T_g's of the matrix polymers. Matrix polymer: (a) PP (T_g ~18°C); (b) PS (T_g ~100°C) [62]

agglomerates in the composites would neither be fragmented nor contribute to energy consumption through rearrangement under applied force. As a result, toughening is hard to be realized, as revealed by the existing experimental results [61,64].

In this context, grafting of poly(dodecafluoroheptyl acrylate) onto the nanoparticles might be a solution. On the one hand, the grafting pre-treatment is analogous to that of the nanoparticles put on a "slippery coating". The fluoride

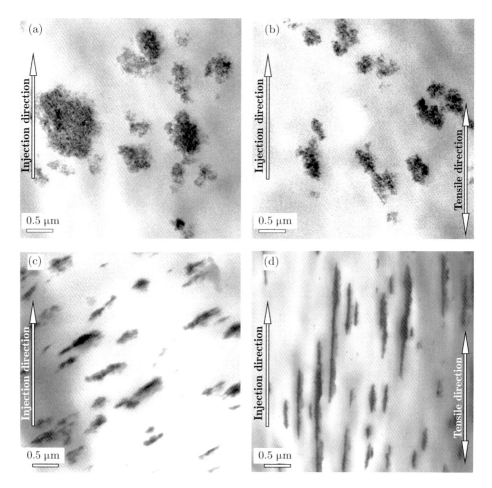

Figure 3.30. TEM micrographs of nano-SiO$_2$/PP composites before and after tensile tests: (a) untreated nano-SiO$_2$/PP as-injection molded; (b) untreated nano-SiO$_2$/PP experienced tensile test at 25°C; (c) SiO$_2$-g-PDFHA/PP as-injection molded; (d) SiO$_2$-g-PDFHA/PP experienced tensile test at 25°C [62]

facilitates relative sliding of the grafted nanoparticles under applied stress. It is particularly important for the particles inside the agglomerates. On the other hand, the grafted polymer chains would get entangled with the matrix polymer in the course of melt mixing, leading to improved nanoparticles/matrix interaction.

The results in Figure 3.29a prove that our idea works. So long as the matrix PP stays in the rubber-like state, the grafted nano-silica is capable of bringing about a toughening effect in the composites, as evidenced by the remarkably higher areas under the stress-strain curves than those of neat PP determined at 25 and 50°C. Besides, the higher the testing temperature than T_g of the matrix PP, the tougher the nanocomposites. It means that the additional dissipative mechanism resulting from the mobility of the nanoparticles must have taken effect. TEM micrographs in Figure 3.30c and 3.30d clearly display that the grafted nanoparticle agglomerates in the composites were stretched and realigned by the tensile force at a temperature above the T_g of the PP matrix. They present a striking contrast to the case of untreated nano-silica (Figure 3.30a and 3.30b) and provide morphological evidence for the mobility of the nanoparticles. The thermodynamic state of the grafted polymer seems to be less important in this case (T_g of PDFHA is –15.6°C and it starts to flow above 15°C), as the composites' toughness is always lower than that of unfilled PP at either –5°C or –25°C. In the glass state of PP, the movement of silica nanoparticles was greatly frozen. During deformation of the composites, the nanoparticles could not form the temporary bonds between the surrounding polymer chains that allow them to dissipate energy [59], and had to behave like conventional fillers initiating cracks and embrittling the matrix.

To further confirm the preconditions for providing non-layered nanoparticles with mobility, a series of specific experiments were carried out and discussed in the following. Figure 3.31 records mechanical response of the mineral oil incorporated with nanoparticles during dynamic rheological analysis. Because mineral oil is a non-polar liquid free of hydrogen-bonding capacity and hence has extremely weak interaction with the particles, the measured dynamic elastic modulus is a function of particle-particle interaction excluding the influence of the dispersion medium [65]. Accordingly, it is shown that the interparticulate interaction in PDFHA grafted nano-silica is indeed much lower than that in the untreated nano-silica, as the latter has higher dynamic elastic modulus (Figure 3.31). On the other hand, the degree of interaction between the

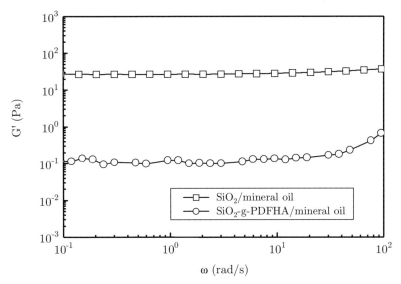

Figure 3.31. Dynamic elastic modulus, G′, versus frequency, ω, of untreated and grafted silica nanoparticles in mineral oil at 30°C [62]

nanoparticles and matrix polymer can be estimated from dynamic mechanical analysis by the method proposed by Sumita *et al.* [66] It turns out to be 6.7 for untreated nano-SiO$_2$/PP and 11.3 for SiO$_2$-g-PDFHA/PP composites at 25°C (measured at 1 Hz), respectively, meaning that the filler/matrix interaction in the latter has been strengthened due to chain entanglement between the grafted polymer and matrix polymer chains [15].

Since the interactions in the interior of the nanoparticle agglomerates and at the nanoparticles-matrix interface can be purposely modified, the variation in microstructure of the nanocomposites with untreated and treated nanoparticles under applied force must be different. Table 3.5 lists the relative intensity $I(110)/I(040)$ of the materials measured by wide angle X-ray diffraction (WAXD) to evaluate the degree of crystal orientation of PP [67] before and after tensile tests at 25°C. It can be seen that the crystallites in both unfilled PP and SiO$_2$/PP prior to tensile tests have similar orientations as a result of injection molding. Tensile tests also lead to similar increment of crystalline orientation of PP. The appearance of untreated nano-silica has nearly no effect on increasing PP orientation during composites manufacturing and tensile testing. In contrast, the nanocomposites containing PDFHA grafted nano-silica exhibit significantly increased orientation of the polymer crystallites after being stretched. The results coincide with the TEM observations in Figure 3.30. For a rubbery polymer matrix, the improved interaction between nanofillers and matrix would impulse the nanoparticles to move upon the elongation. Because the transferred stress might not afford to drive oversize nanoparticle agglomerates, weaker attraction among the nanoparticles helps more particles to be involved in the relative movement. Meanwhile, deformation and alignment of nanoparticle agglomerates in turn leave space for arrangement of molecular chains of the matrix polymer. Consequently, the mobile nanoparticles might act as "plasticizer", improving orientation of the matrix [68]. This might induce the formation of a phase that is not accessible under normal conditions. For PP, the result lies in the appearance of β-crystals, in addition to the commonly occurring α phase [32].

Table 3.5. Crystal orientation of PP and its composites before and after tensile tests at 25°C characterized by relative WAXD intensity $I(110)/I(040)$ [62]

Samples	$I(110)/I(040)$ (before tensile test)	$I(110)/I(040)$ (after tensile test)
PP	1.44	2.36
SiO$_2$/PP	1.14	2.63
SiO$_2$-g-PDFHA/PP	1.47	5.38

The above results demonstrate that the mobility of the non-layered nanoparticles can be acquired in case the particles were properly pre-treated. When the nanocomposite is subjected to tensile stress above T_g of the matrix polymer, sufficiently high interfacial adhesion between nanoparticles and matrix ensures that the mobility of polymer chains could be transferred to the nanoparticles. In the meantime, the weakened interactions among the nanoparticles encourage breakage of the agglomerates so that a great number of nanoparticles is allowed to take part in the deformation of the composite and energy consumption as well. For untreated nanoparticles, the poor filler-matrix interaction and strong interparticulate adhesion keep the structural integrity of the agglomerates. No additional energy dissipation mechanism is available even in the rubbery state of the matrix.

To evaluate the effectiveness and applicability of our findings, we reduced the number of the fluorine on the grafted polymer attached to the nano-silica by replacing PDFHA with PHFBA. Such change would certainly lower the self-lubricating effects among the grafted nanoparticles and hence the nanoparticles mobility. As a result, the nanocomposites would be less effectively toughened. The experimental results prove this is the case. The area under the tensile stress-strain curve of SiO_2-g-PHFBA/PP determined at 25°C is 65.2 MPa, which is higher than that of PP (~37.9 MPa) but lower than that of SiO_2-g-PDFHA/PP (~107.5 MPa).

Also, we added untreated and PDFHA grafted nano-silica into polystyrene, a typical amorphous thermoplastic polymer, and conducted tensile tests of the nanocomposites (Figure 3.29b). Similarly, the nano-SiO_2 as-received always reduced the toughness of the matrix polymer above or below the T_g of PS (~100°C), but the grafted nano-silica results in an increase in the toughness of the nanocomposites at 105°C. It agrees with our consideration dealing with validation of nanoparticles mobility.

3.5. Conclusions

The researches carried out so far have proved that incorporation of inorganic nanoparticles into polymers can effectively improve the latter's properties. It opens a new road for polymer modification that is superior to the addition of microparticles. Because of the specific surface feature of the nanoparticles, a variety of possibilities is offered by developing novel manufacturing techniques, as shown here.

Compared to the substantial experimental attempts, however, theoretical consideration of the mechanisms involved in adequate predictability is not well reported. Design and production of polymer nanocomposites have to be mostly conducted on a trial and error basis. Empirical extrapolation of the parameters related to components selection and processing technique is not very successful.

We raised several questions a couple of years ago with regard to the development of polymer nanocomposites [1]: (i) What are the theoretical upper limits of the mechanical properties of a polymer nanocomposite? (ii) How does the interphase influence the mechanical performance of a polymer nanocomposite? (iii) What kind of surface treatment is most effective for the purposes of separating nanoparticle agglomerates and improving filler/matrix interfacial adhesion? (iv) What is the optimum phase morphology of nanocomposites, and how can this morphology be obtained by choosing proper processing conditions (including surface treatments of nanoparticles and mixing variables)? (v) Are there any relationships between the species of the nanoparticles and the improvement in mechanical performance of the nanocomposites? (vi) What are the applications for which only the nanocomposites are competent? By surveying the results reported in recent years, it is known that most of the above questions are still open.

Before the appearance of revolutionized manufacturing techniques, there is still large room for inventing new tricks on the basis of existing industrial measures and equipments. Interdisciplinary studies through tight collaboration among scientists are a prerequisite for overcoming the difficulties. The key point to attain achievements lies in a deep understanding of the nature of nanoparticles, polymers, processing, surface chemistry and physics, nanocomposites performance, *etc.*

Acknowledgements

The authors are grateful for the support of the National Natural Science Foundation of China (Grant: 50473019, 50773095), and the Natural Science Foundation of Guangdong, China (Grants: 2004A10702001, 2005A10702001 and 5003267).

References

1. Zhang M Q, Rong M Z and Friedrich K (2003) Processing and properties of non-layered nanoparticle reinforced thermoplastic composites, in *Handbook of Organic-Inorganic Hybrid Materials and Nanocomposites, Vol. 2: Nanocomposites* (Ed. Nalwa H S) American Scientific Publishers, California, Chapter 3, pp. 113–150.
2. Rong M Z, Zhang M Q and Ruan W H (2006) Surface modification of nanoscale fillers for improving properties of polymer, *Mater Sci Technol* **22**:787–796.
3. Rong M Z, Zhang M Q, Zheng Y X, Zeng H M, Walter R and Friedrich K (2001) Structure-property relationships of irradiation grafted nano-inorganic particle filled polypropylene composites, *Polymer* **42**:167–183.
4. Wu C L, Zhang M Q, Rong M Z and Friedrich K (2002) Tensile performance improvement of low nanoparticles filled–polypropylene composites, *Compos Sci Technol* **62**:1327–1340.
5. Wu C L, Zhang M Q, Rong M Z and Friedrich K (2005) Silica nanoparticles filled polypropylene: effects of particle surface treatment, matrix ductility and particle species on mechanical performance of the composites, *Compos Sci Technol* **65**:635–645.
6. Zhang M Q, Rong M Z, Zeng H M, Schmitt S, Wetzel B and Friedrich K (2001) An atomic force microscopy study on structure and properties of irradiation grafted silica particles in polypropylene based nanocomposites, *J Appl Polym Sci* **80**:2218–2227.

7. Sahnoune F, Lopez-Cuesta J M and Crespy A (1999) Effect of elastomer interfacial agents on tensile and impact properties of $CaCO_3$ filled HDPE, *J Mater Sci* **34**:535–544.
8. Long Y and Shanks R A (1996) PP-elastomer-filler hybrids. I. Processing, microstructure, and mechanical properties, *J Appl Polym Sci* **61**:1877–1885.
9. Pukanszky B (1990) Influence of interface interaction on the ultimate tensile properties of polymer composites, *Composites* **21**:255–262.
10. Quazi R T, Bhattacharya S N and Kosior E (1999) The effect of dispersed paint particles on the mechanical properties of rubber toughened polypropylene composites, *J Mater Sci* **34**:607–614.
11. Demjen Z and Pukanszky B (1998) Evaluation of interfacial interaction in polypropylene/surface treated $CaCO_3$ composites, *Compos Part A – Appl S* **29**:323–329.
12. Kendall K and Sherliker F R (1980) Effect of polymer molecular weight on colloidal reinforcement, *Brit Polym J* **12**:111–113.
13. Walter R, Friedrich K, Privalko V and Savadori A (1997) On modulus and fracture toughness of rigid particulate filled high density polyethylene, *J Adhesion* **64**:87–109.
14. Rong M Z, Zhang M Q, Pan S L and Friedrich K (2004) Interfacial effects in polypropylene-silica nanocomposites, *J Appl Polym Sci* **92**:1771–1781.
15. Rong M Z, Zhang M Q, Pan S L, Lehmann B and Friedrich K (2004) Analysis of the interfacial interactions in polypropylene/silica nanocomposites, *Polym Int* **53**:176–183.
16. Hasegawa R, Aoki Y and Doi M (1996) Optimum graft density for dispersing particles in polymer melts, *Macromolecules* **29**:6656–6662.
17. Tada H, Saito Y and Hyodo M (1993) Dependence of molecular weight of grafting polymer on dispersibility of TiO2 particles in poly(dimethylsiloxane), *J Colloid Interf Sci* **159**:249–252.
18. Rong M Z, Zhang M Q, Zheng Y X, Zeng H M and Friedrich K (2001) Improvement of tensile properties of nano-SiO_2/PP composites in relation to percolation mechanism, *Polymer* **42**:3301–3304.
19. Ruan W H, Zhang M Q, Rong M Z and Friedrich K (2004) Polypropylene composites filled with in situ grafting polymerization modified nano-silica particles, *J Mater Sci* **39**:3475–3478.
20. Ruan W H, Zhang M Q, Rong M Z and Friedrich K (2004) Mechanical properties of nanocomposites from ball milling grafted nano-silica/polypropylene block copolymer, *Polym Polym Compos* **12**:257–267.
21. Venkataraman K S and Narayanan K S (1998) Energetics of collision between grinding media in ball mills and mechanochemical effects, *Powder Technol* **96**:190–201.
22. Nicolais L and Narkis M (1971) Stress-strain behavior of styrene-acrylonitrile/glass bead composites in the glass region, *Polym Eng Sci* **11**:194–199.
23. Wu C L, Zhang M Q, Rong M Z, Lehmann B and Friedrich K (2003) Deformation characteristics of nano-SiO_2 filled polypropylene composites, *Polym Polym Compos* **11**:559–562.
24. Ruan W H, Huang X B, Wang X H, Rong M Z and Zhang M Q (2006) Effect of drawing induced dispersion of nano-silica on performance improvement of polypropylene based nanocomposites, *Macromol Rapid Commun* **27**:581–585.
25. Ruan W H, Mai Y L, Wang X H, Rong M Z and Zhang M Q (2007) Effects of processing conditions on properties of nano-SiO_2/polypropylene composites fabricated by pre-drawing technique, *Compos Sci Technol* **67**:2747–2756.

26. Ruan W H, Rong M Z and Zhang M Q (2007) Drawing-induced dispersion of nanoparticles and its effect on structure and properties of thermoplastic nanocomposites, *Key Eng Mater* **334–335**:717–720.
27. McCrum N G, Buckley C P and Bucknall C B (1997) *Principles of Polymer Engineering*, Oxford University Press, New York, Chapter 5, pp. 242–245.
28. Shah D, Maiti P, Jiang D D, Batt C A and Giannelis E P (2005) Effect of nanoparticle mobility on toughness of polymer nanocomposites, *Adv Mater* **17**:525–528.
29. Liu T, Tjiu W C, He C, Na S S and Chung T S (2004) A processing-induced clay dispersion and its effect on the structure and properties of polyamide, *Polym Int* **53**:392–399.
30. Usuki A, Kato M, Okada A and Kurauchi T (1997) Synthesis of polypropylene-clay hybrid, *J Appl Polym Sci* **63**:137–138.
31. Devaux E and Chabert B (1991) Nature and origin of the transcrystalline interphase of polypropylene/glass fibre composites after a shear stress, *Polym Commun* **32**:464–468.
32. Varga J and Karger-Kocsis J (1993) The occurence of transcrystallization or row-nucleated cylindritic crystallization as a result of shearing in a glass-fiber-reinforced polypropylene, *Compos Sci Technol* **48**:191–198.
33. Varga J and Karger-Kocsis J (1996) Rules of supermolecular structure formation in sheared isotactic polypropylene melts, *J Polym Sci Polym Phys* **34**:657–670.
34. Wu D M, Meng Q Y, Liu Y, Ding Y M, Chen W H, Xu H and Ren D Y (2003) In situ bubble-stretching dispersion mechanism for additives in polymers, *J Polym Sci Polym Phys* **41**:1051–1058.
35. Meng Q Y and Wu D M (2004) A study of bubble inflation in polymers and its applications, *Phys Lett A* **327**:61–66.
36. Yoo H J and Han C D (1982) Oscillatory behavior of a gas bubble growing (or collapsing) in viscoelastic liquids, *AIChE J* **28**:1002–1009.
37. Cai L F, Huang X B, Rong M Z, Ruan W H and Zhang M Q (2006) Effect of grafted polymeric foaming agent on the structure and properties of nano-silica/polypropylene composites, *Polymer* **47**:7043–7050.
38. Cai L F, Huang X B, Rong M Z, Ruan W H and Zhang M Q (2006) Fabrication of nanoparticle/polymer composites by in situ bubble-stretching and reactive compatibilization, *Macromol Chem Phys* **207**:2093–2102.
39. Cai L F, Mai Y L, Rong M Z, Ruan W H and Zhang M Q (2007) Interfacial effects in nano-silica/polypropylene composites fabricated by in situ chemical blowing, *eXPRESS Polym Lett* **1**:2–7.
40. Cai L F, Rong M Z, Zhang M Q and Ruan W H (2007) Graft polymerization of *p*-vinylphenylsulfonylhydrazide onto nano-silica and its effect on dispersion of the nanoparticles in polymer matrix, *Key Eng Mater* **334–335**:729–732.
41. Hunter B A and Schoene D L (1952) Sulfonyl hydrazide blowing agents for rubber and plastics, *Ind Eng Chem* **44**:119–122.
42. Lange F F (1970) Interaction of a crack front with a second-phase dispersion, *Philos Mag* **22**:983–992.
43. Chan C M, Wu J S, Li J X and Cheung Y K (2002) Polypropylene/calcium carbonate nanocomposites, *Polymer* **43**:2981–2992.
44. Zhou T H, Ruan W H, Wang Y L, Mai Y L, Rong M Z and Zhang M Q (2007) Performance improvement of nano-silica/polypropylene composites through in situ graft modification of nanoparticles during melt compounding, *e-Polymers* No.058.

45. Long Y and Shanks R A (1996) PP/elastomer/filler hybrids. II. Morphologies and fracture, *J Appl Polym Sci* **62**:639–646.
46. Rybnikar F (1989) Orientation in composite of polypropylene and talc, *J Appl Polym Sci* **38**:1479–1490.
47. Koning C, van Duin M, Pagnoulle C and Jerome R (1998) Strategies for compatibilization of polymer blends, *Prog Polym Sci* **23**:707–757.
48. Lu Q W and Macosko C W (2004) Comparing the compatibility of various functionalized polypropylenes with thermoplastic polyurethane, *Polymer* **45**:1981–1991.
49. Sadhu V B, Pionteck J, Potschke P, Jakisch L and Janke A (2004) Creation of crosslinkable interphases in polymer blends, *Macromol Symp* **210**:165–174.
50. Friedrich K, Rong M Z, Zhang M Q and Ruan W H (2006) Preparation of nano-silica/polypropylene composites using reactive compatibilization, *Key Eng Mater* **312**:229–232.
51. Friedrich K, Rong M Z and Zhang M Q (2007) Reactive compatibilization in nano-silica filled polypropylene composites, *Solid State Phenom* **121–123**:1433–1436.
52. Zhou H J, Rong M Z, Zhang M Q and Friedrich K (2006) Effects of reactive compatibilization on the performance of nano-silica filled polypropylene composites, *J Mater Sci* **41**:5767–5770.
53. Zhou H J, Rong M Z, Zhang M Q, Ruan W H and Friedrich K (2007) Role of reactive compatibilization in preparation of nanosilica/polypropylene composites, *Polym Eng Sci* **47**:499–509.
54. Zhou H J, Rong M Z, Zhang M Q, Lehmann B and Friedrich K (2005) Grafting of poly(glycidyl methacrylate) onto nano-SiO_2 and its reactivity in polymers, *Polym J* **37**:677–685.
55. Wu C L, Zhang M Q, Rong M Z, Lehmann B and Friedrich K (2004) Functionalisation of polypropylene by solid phase graft polymerisation and its effect on mechanical properties of silica nanocomposites, *Plast Rubber Compos* **33**:71–76.
56. Zhou T H, Ruan W H, Rong M Z and Zhang M Q (2007) *In situ* crosslinking induced structure development and mechanical properties of nano-silica/polypropylene composites, *Key Eng Mater* **334–335**:733–736.
57. Zhou T H, Ruan W H, Yang J L, Rong M Z, Zhang M Q and Zhang Z (2007) A novel route for improving creep resistance of polymers using nanoparticles, *Compos Sci Technol* **67**:2297–2302.
58. Standard Test Methods for Tensile, Compressive, and Flexural Creep and Creep Rupture of Plastics [ASTM D2990–01] West Conshohocken, PA, ASTM International, 2001.
59. Gersappe D (2002) Molecular mechanisms of failure in polymer nanocomposites, *Phys Rev Lett* **89**:058301.
60. El-Nashar D E, Mansour S H and Girgis E (2006) Nickel and iron nano-particles in natural rubber composites, *J Mater Sci* **41**:5359–5364.
61. Kueseng K and Jacob K I (2006) Natural rubber nanocomposites with SiC nanoparticles and carbon nanotubes, *Eur Polym J* **42**:220–227.
62. Zhou T H, Ruan W H, Rong M Z, Zhang M Q and Mai Y L (2007) Keys to toughening of non-layered nanoparticles/polymer composites, *Adv Mater* **19**:2667–2671.
63. Alexandre M and Dubois P (2000) Polymer-layered silicate nanocomposites: preparation, properties and uses of a new class of materials, *Mater Sci Eng R* **28**:1–63.
64. Bikiaris D N, Vassiliou A, Pavlidou E and Karayannidis G P (2005) Compatibilisation effect of PP-g-MA copolymer on iPP/SiO_2 nanocomposites prepared by melt mixing, *Eur Polym J* **41**:1965–1978.

65. Khan S A and Zoeller N J (1993) Dynamic rheological behavior of flocculated fumed silica suspensions, *J Rheol* **37**:1225–1235.
66. Sumita M, Tsukihi H, Miyasaka K and Ishikawa K (1984) Dynamic mechanical properties of polypropylene composites filled with ultrafine particles, *J Appl Polym Sci* **29**:1523–1530.
67. Obata Y, Sumitomo T, Ijitsu T, Matsuda M and Nomura T (2001) The effect of talc on the crystal orientation in polypropylene/ethylene-propylene rubber/talc polymer blends in injection molding, *Polym Eng Sci* **41**:408–416.
68. Yalcin B and Cakmak M (2005) Molecular orientation behavior of poly(vinyl chloride) as influenced by the nanoparticles and plasticizer during uniaxial film stretching in the rubbery stage, *J Polym Sci Polym Phys* **43**:724–742.

Chapter 4

Rubber Nanocomposites: New Developments, New Opportunities

L. Bokobza

4.1. Introduction

Polymer composites have attracted a great deal of interest in recent years. In most cases, fillers are used as additives for improving the mechanical behavior of the host polymeric matrix. The reinforcement of elastomers by mineral fillers is essential to the rubber industry, because it yields an improvement in the service life of rubber compounds. The state of filler dispersion and orientation in the matrix, the size and aspect ratio of the particles as well as the interfacial interactions between the organic and inorganic phases have been shown to be crucial parameters in the extent of property improvement [1,2].

Carbon black and silica have been widely used as reinforcing agents for elastomeric compounds. These conventional fillers must be used at high loading levels to impart the desired properties to the material [3]. On the other hand, the filler particles are usually blended into the polymers before the cross-linking reaction. The particles tend to agglomerate and the resulting materials are rather inhomogeneous. The use of silane coupling agents, in combination with silica, in nonpolar polymers is commonly recommended to help dispersion essentially in silica-filled hydrocarbon polymers, where the interfacial adhesion between the two phases is poor.

The last few years have seen an extensive use of nanoparticles because of the small size of the filler and the corresponding increase in the surface area allowing to achieve the required mechanical properties at low filler loadings. Nanometer-scale particles including spherical particles such as silica or titanium dioxide generated *in situ* by the sol-gel process [4–8], layered silicates [9–12], carbon [13] or clay fibers [14], and single-wall or multi-wall carbon nanotubes

[15,16] have been shown to significantly enhance the physical and mechanical properties of rubber matrices.

4.2. General considerations on elastomeric composites

A filled elastomeric network may be regarded as a two-component-system of rigid particles surrounded by an elastomeric network formed by flexible chains permanently linked together by chemical junctions during the cross-linking process. In the basic terminology, primary filler particles are fused together to build up *stable aggregates* which form a persistent and irreversible structure. These aggregates stick together to form loosely bonded agglomerates which are fully reversible and are known as *transient structures*. As a typical example, Figure 4.1 shows atomic force microscopy (AFM) images of a poly(dimethylsiloxane) network (PDMS) filled with 40 phr (parts per hundred parts of rubber, by weight) of a pyrogenic silica which has undergone a surface treatment in order

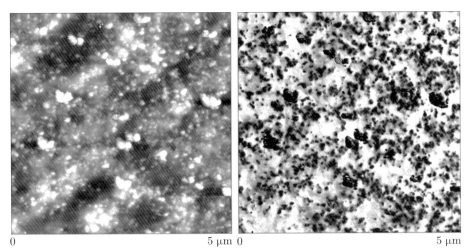

Figure 4.1. Height (left) and phase (right) images of a PDMS vulcanizate filled with 40 phr of treated silica

to reduce agglomeration. The "height" (or topological image) on the left gives a topographic mapping of the surface with bright areas corresponding to silica aggregates or agglomerates protruding above the surface. The "phase image" on the right provides compositional mapping of the same area and allows a reliable discrimination between the elastomeric matrix and the filler particles, and a better resolution than the height image [17].

As seen in Figure 4.2a, the addition of silica in PDMS leads to an increase in the elastic modulus and to a significant improvement in the ultimate properties such as the stress at rupture and the maximum extensibility. The increase in stiffness results from the inclusion of rigid particles in the soft matrix and from additional cross-linking sites at the particle-matrix interface. In silica-

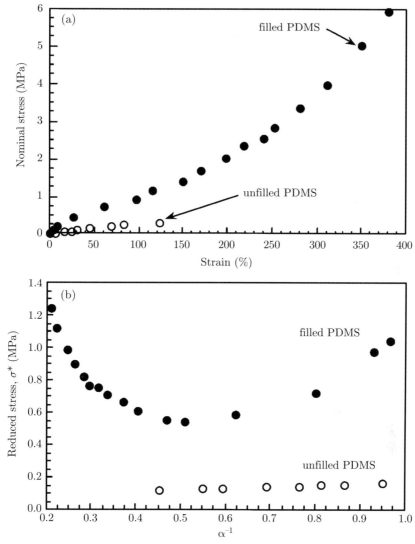

Figure 4.2. Stress-strain curve for an unfilled PDMS and for a PDMS filled with 40 phr of fumed treated silica (a), Mooney-Rivlin plots (b)

filled silicone rubbers, the interaction between the network chains and the filler is ensured by hydrogen bonds between the silanol groups, present on the silica surface, and the oxygen atoms of the PDMS chains. The density of polymer-filler attachments has been evaluated by the analysis of chain orientation carried out by infrared dichroism or by equilibrium swelling measurements. The tensile data can be plotted in a Mooney-Rivlin representation (Figure 4.2b), based on the so-called Mooney-Rivlin equation:

$$[\sigma^*] = \sigma/(\alpha - \alpha^{-2}) = 2C_1 + 2C_2 \alpha^{-1} \tag{4.1}$$

where σ is the nominal stress (force divided by the undeformed area of the sample), α is the extension ratio (ratio of the final length of the sample in the direction of stretch to the initial length before deformation) and $2C_1$ and $2C_2$ are constants independent of α.

Figure 4.2b compares the Mooney-Rivlin plots for neat PDMS and for the composite. While the unfilled network exhibits an almost constant value of the reduced stress, the filled sample displays an upturn in the modulus at high deformations which is typical, here of limited chain extensibility. A large increase in stress appearing at large strains in unfilled networks has been attributed to strain-induced crystallization in the case of strain crystallizing rubbers such as natural rubber. In the case of amorphous rubbers, which do not crystallize, the upturns in stress-strain curves are attributed to non-Gaussian effects arising from the limited extensibilities of the network chains [18,19]. The values of σ^* as well as the elongation at the upturn have been related to the cross-linking density. In bimodal networks consisting of mixtures of very short chains and relatively long chains, an increase in the short-chain concentration leads to the occurrence of the upturn in the modulus at a lower extension ratio and to a large increase in the reduced stress [20,21]. This improvement in properties is similar to that provided by reinforcing fillers where the increase in the cross-linking density is created by polymer-filler interactions [1,22].

Taking into account the limiting stretch of the chains forming the polymeric network provides a basis for including other aspects of rubber elastic behavior such as the stress-softening phenomenon widely known as the *Mullins effect* [23]. The Mullins effect is characterized by a significant reduction in the stress at a given level of strain on unloading during the first and successive cycles (Figure 4.3a). Mooney-Rivlin representation of Figure 4.3a is shown in Figure 4.3b. The curves in Figure 4.3b show that the limited chain extensibility of the network chains occurs at increasing strain values for second extensions performed at increasing strain. This suggests a distribution of chain lengths within the sample reaching their limit of extensibility at different macroscopic deformations.

The Mullins effect, which can be considered as a hysteretic mechanism related to energy dissipated by the material during deformation, corresponds to a decrease in the number of elastically effective network chains. It results from chains that reach their limit of extensibility by strain amplification effects caused by the inclusion of undeformable filler particles [24,25]. Stress-softening in filled rubbers has been associated with the rupture properties and a quantitative relationship between total hysteresis (area between the first extension and the first release curves in the first extension cycle) and the energy required for rupture has been derived [26,27].

4.3. Spherical *in situ* generated reinforcing particles

In addition to the conventional technique of blending filler particles into the elastomer prior to cross-linking, reinforcement of rubbers has also been obtained by *in situ* generated filler particles. This technique uses the sol-gel method,

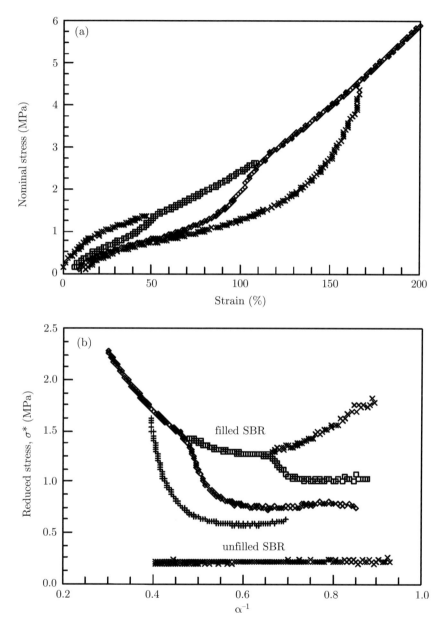

Figure 4.3. Cycles at various deformations for a styrene-butadiene rubber (SBR) filled with 58 phr of silica used in combination with a silane coupling agent (a) and corresponding Mooney-Rivlin plots (b)

based on the polymerization of inorganic precursors, the most common being tetraethoxysilane (TEOS). Pioneered by Mark for essentially silica filling of silicone rubbers [4,28–33], the sol-gel process has been successfully extended to various other polymers [5,6–34–41]. The *in situ* organic-inorganic systems are often called nanocomposites on account of the small size of the generated structures.

The work previously reported on the reinforcement of PDMS by *in situ* generated silica is essentially related to silicone elastomers formed by the use of functionally-terminated PDMS fluids that undergo catalyzed cross-linking reactions. Typically, silica precipitation can be carried out before or after or alternatively during curing [42]. In the third procedure, hydroxyl-terminated PDMS chains are blended with enough TEOS that will simultaneously end-link the precursor chains and provide the silica particles.

Two reactions are generally used to describe the sol-gel synthesis of silica from TEOS:

$$Si(OC_2H_5)_4 \xrightarrow{hydrolysis} Si(OH)_4 + C_2H_5OH$$

$$Si(OH)_4 + Si(OH)_4 \xrightarrow{polycondensation} (OH)_3Si-O-Si(OH)_3 + H_2O$$

This process, which can be carried out at relatively low temperatures in host polymeric matrices, leads to samples exhibiting excellent mechanical properties and optical transparency.

The generated inorganic structures highly depend on the hydrolysis and condensation conditions and essentially on the nature of the catalyst used to accelerate the gelation process. The role played by the catalyst in the sol-gel processing of silica has been thoroughly examined by Pope and Mackenzie [43]. On the other hand, the mechanisms and kinetics of hydrolysis-condensation reactions have been widely discussed on the basis of the water content and on the pH dependence of the gelation process [44,45].

Large spherical particles are expected in the case of a base-catalyzed reaction while linear chain growth is suggested *via* acid catalysis. Silica polymerization in the solution precursor is generally described by models of nucleation, growth and aggregation. Kinetic models based on fractal geometry concepts were developed and applied for the analysis of small-angle X-ray scattering profiles in order to gain information on the morphology of the inorganic entities.

A wide range of complex structures were obtained in *in situ* silica-filled poly(dimethylsiloxane) networks prepared by various synthetic protocols [4,33]. Nevertheless, the typical fractal patterns and morphologies described in the case of polymerization of silica in solution are not exactly those observed when the polymerization is carried out in PDMS [4]. This is most probably due to the constraints provided by the polymer environment.

Generating silica structures within a preformed network is expected to prevent the formation of large silica aggregates.

The experimental procedure for the polymerization of silica in already-formed networks consists of swelling polymer films in TEOS in the presence of a tin catalyst: dibutyltin diacetate (DBTDA) or dibutyltin dilaurate (DBTDL). The swelling time determines the degree of TEOS absorption and thus the filler loading. The TEOS-swollen film and a beaker containing water have to be placed for 24 hours into a desiccator maintained at a constant temperature (30°C), thus exposing the swollen film to saturated water vapor. The film is then vacuum-dried at 80°C for several days to constant weight in order to remove any alcohol generated from the reaction and also the remaining TEOS that has not been hydrolyzed. Samples can also be prepared according to the procedure already reported [4,33], which consists of immersing the TEOS-swollen network into an aqueous solution containing 3 wt% of diethylamine (DEA) for one day. The films are then removed from the solution and vacuum-dried as described above. The amount of filler incorporated into the network was calculated from the weights of the films before and after the generation of the filler.

Figure 4.4 shows transmission electron microscopy (TEM) images of a DEA- and a DBTDL-catalyzed PDMS filled with 10 phr of *in situ* silicas. While silica particles are uniformly dispersed in the polymer phase in both samples, quite different inorganic structures are revealed. The particles generated under the DEA catalysis are seen to be spherical (around 20 nm in diameter) and larger than those generated using the tin catalyst, which in turn gives rise to a highly porous material.

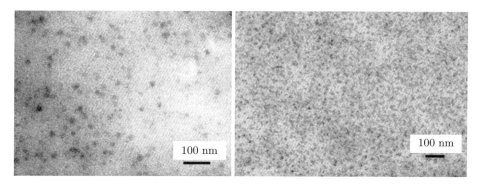

Figure 4.4. TEM images of composites filled with 10 phr of SiO_2 precipitated into a PDMS network catalyzed by 3 wt% DEA (left) and by 3 wt% (DBTDL) (right)

In the case of poly(dimethysiloxane) filled with *in situ* precipitated silica under dibutyltin diacetate catalysis (Figure 4.5), the strain dependence of the stress shows large improvement in modulus and rupture properties.

Compared to a classical composite filled with fumed silica, the sol-gel process provides inorganic particles with greatly improved reinforcing properties on account of the small size of the particles and of the hydrophilic character of the filler surface, thus allowing strong interactions with the polymer chains. At high filler loadings and above the percolation threshold, a change in the shape of the curve is observed: the mechanical behavior is close to that of a thermoplastic polymer with a well-defined yield point followed by a smaller strain dependence of the stress. This plastics-like stress-strain curve is probably related

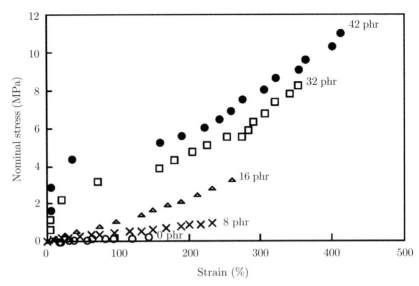

Figure 4.5. Stress-strain curves at room temperature for PDMS filled with *in situ* generated silica in the presence of dibutyltin diacetate as catalyst

to the existence of an interfacial layer where the network chains are immobilized by interaction with the glassy silica domains. In addition to the α relaxation associated with the glass transition of the polymer matrix, an additional slower α relaxation, assigned to polymer chains close to the polymer/filler interface and with restricted mobility, is observed [46]. The confinement of polymer chains in a quasi-glassy state may be responsible for the plastics-like behavior. On the other hand, the results of differential scanning calorimetry (DSC) show that the temperature of crystallization of PDMS as well as the degree of crystallinity decrease with the filler content. This effect may be regarded as the result of an increase in the cross-linking density, the ability of the polymer to crystallize being reduced when the apparent molecular weight between cross-links decreases [47].

Another interesting point of comparison between conventional and *in situ* generated silicas concerns the Mullins effect. Figure 4.6 compares first and second stretchings performed at different strain values of a PDMS composite filled with fumed silica with those of a composite prepared by an *in situ* sol-gel process. After stretching and release, the filled elastomers show a permanent deformation, which was contributed by Kilian *et al.* [48] to an irreversible deformation of the filler network. After stretching at 200%, it is about 10% for the silica-filled PDMS sample obtained by the usual mechanical mixing (Figure 4.6a) and much more pronounced for the samples submitted to the *in situ* filling process (Figures 4.6b and 4.6c). The composite filled with *in situ* precipitated particles is presumed to possess a very fine and interpenetrated polymer-silica morphology. The significant residual deformation obtained in that case may be explained by the breaking up of both the filler network and the rubber-to-filler bonds. Moreover, the amount of stress-softening for the

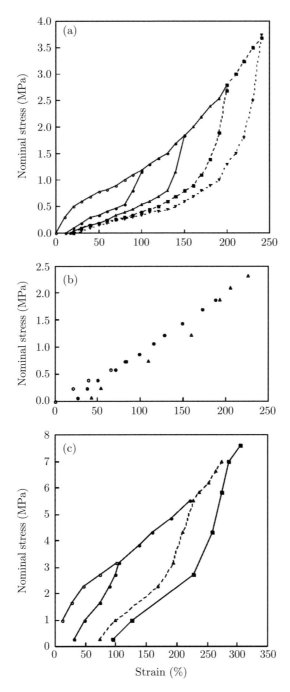

Figure 4.6. Comparison of the amount of stress-softening displayed by a PDMS filled with 40 phr of fumed silica (a) with that obtained for a composite filled with 16 phr (b) and 27 phr of *in situ* generated particles (c)

sample filled with 16 phr of silica (Figure 4.6b) is much smaller than that exhibited by the high loaded polymer. This may be due to the good dispersion of silica and consequently to a narrow distribution of chain lengths between fillers.

The incorporation of fillers to elastomeric compounds strongly modifies the viscoelastic behavior of the material at small strains and leads to the occurrence of a non-linear behavior known as *"Payne effect"* [49] characterized by a decrease in the storage modulus with an increase in the amplitude of small-strain oscillations in dynamic mechanical tests. This phenomenon has attracted considerable attention in the past decade on account of its importance for industrial applications [50–54]. The amplitude $\Delta G' = (G'_0 - G'_\infty)$ of the Payne effect, where G'_0 and G'_∞ are the maximum and minimum values of the storage modulus respectively, increases with the volume fraction of filler as shown in silica-filled PDMS networks (Figure 4.7a). At a same silica loading, the PDMS network filled with untreated silica displays a much higher G' value than the treated one and is much more resistant to the applied deformation (Figure 4.7b).

In silica-filled styrene-butadiene rubbers, due to weak polymer-filler interactions, the silica network is highly developed (Figure 4.7c). The use of a coupling agent improves the dispersion and consequently reduces substantially the amplitude of the Payne effect and also tg δ, which is an important parameter in the rolling resistance of tires.

The origin of the Payne effect is still controversial but the most commonly accepted picture is the destruction of filler networks upon application of the oscillatory shear. Note that filler networks are due to filler-filler interactions which lead to filler agglomerates and clusters. Another analysis of the Payne effect based on interactions between network chains and the filler surface was proposed by Maier and Goritz [55]. This model explains the breakdown of the dynamic modulus of filled elastomers to a desorption of the chains with increasing amplitude. In a recent work of Ramier *et al.* [54], filler-matrix interactions *via* chemical grafting were tailored in order to evaluate the influence of the filler surface modification on the Payne effect in silica-filled SBR vulcanizates. The main idea was to discriminate between the two different pictures commonly used to explain the Payne effect and involving filler-filler interaction or filler-matrix interaction. Unfortunately, the authors were not able to distinguish the two scenarios because filler-filler and filler-matrix interactions are modified in the same manner by the grafting covering agent. Micro-mechanical concepts of non-linear viscoelasticity based on fractal approaches of filler networking were also developed [52]. They consider the arrangement of filler particles in clusters with well-defined fractal structure and the elasticity or fracture of such clusters under external strain. Under the approximation that the elastic modulus of the clusters is much higher than that of the rubber, above the gel point, a simple power law relation: $G'_0 \sim \Phi^{3.5}$ was derived for the dependency of the small strain modulus, G'_0, of the composite on filler concentration, Φ. According to the above concepts, the temperature dependence of G'_0 is determined by the bending-twisting energy of the filler-filler bonds controlled by the bound rubber phase

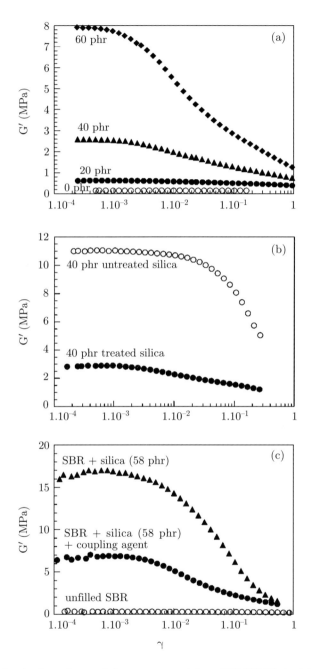

Figure 4.7. Strain dependence of the storage modulus of PDMS filled with various amounts of treated pyrogenic silica (a); comparison of the Payne effect in PDMS filled with 40 phr of treated and untreated silica (b); silica-filled SBR vulcanizates with and without a coupling agent (c)

around the filler clusters and should be that of a glassy polymer. For highly filled rubbers above the glass transition temperature, an Arrhenius temperature behavior typical for polymers in the glassy state is expected, which is in agreement with the experimental findings.

For the unfilled network, the G' modulus increases proportionally with temperature due to its entropic nature (Figure 4.8a). For the filled systems, the

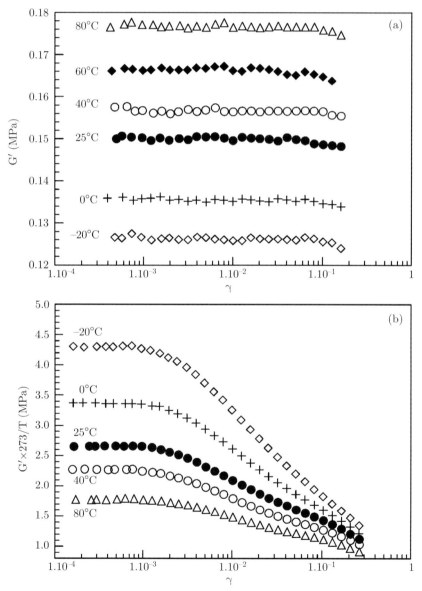

Figure 4.8. Influence of temperature on the storage modulus of an unfilled PDMS and of a network filled with 40 phr of treated silica

modulus values should then be corrected by the entropic factor 273/T in order to get rid of the modulus dependence of the rubber part. In the case of the systems filled with conventional fillers such as fumed silica, G'_0 is seen to decrease with increasing temperature showing an enthalpic variation opposite to the entropic one characteristic of the unfilled network.

For PDMS filled with *in situ* generated silicas, no Payne effect is observed: G' remains constant in the whole range of deformation investigated and its level increases very sharply above 30 phr (Figures 4.9a and 4.9b). This could be due to the excellent dispersion of the filler particles in the elastomeric matrix. In the description of the Payne effect based on agglomeration/desagglomeration of the filler structures, if no aggregates or agglomerates are formed, no destruction of the filler structures could occur by applying the oscillatory shear. At higher filler loadings, a continuous filler network is formed that is not broken at least within the strain range. This would show that the silica network of generated particles is much more resistant to the applied mechanical work than that formed by conventional silicas. These quite different behaviors may be regarded as different nanocomposite morphologies: a common particle-matrix morphology where silica particles tend to form aggregates in the continuous matrix is expected for a classical material, while an open mass-fractal silica structure believed to be bicontinuous with the polymer phase, at a molecular level, is expected in the case where the mineral part is synthesized *in situ*. Moreover and surprisingly with regard to what is usually obtained in elastomeric networks filled with classical fillers, the storage modulus increases proportionally with temperature (Figure 4.9), displaying, as for the unfilled sample, an entropic nature. The slope of the curve representing G' *versus* temperature, is related to the cross-linking density (Figure 4.9c). It increases with the filler content thus reflecting the addition of physical cross-links arising from polymer-filler interactions.

4.4. Carbon nanotube-filled rubber composites

Considerable interest has been devoted recently to carbon nanotubes for their potential use as reinforcing fillers on account of their high aspect ratio, their exceptionally high Young's modulus and their excellent electrical and thermal properties [56,57]. Nevertheless, poor dispersion and poor interfacial bonding limit the full utilization of carbon nanotubes for reinforcing polymeric media.

Carbon nanotubes can be visualized as graphene layers rolled into cylinders consisting of a planar hexagonal arrangement of carbon-carbon bonds. Their outstanding properties are a consequence of this unique bonding arrangement combined with topological defects required for rolling up the sheets of graphite into cylinders. During growth, depending on the synthesis methods, they can assemble either as concentric tubes (multi-wall nanotubes, MWNTs) or as individual cylinders (single-wall nanotubes, SWNTs). Their diameters range from about a nanometer to tens of nanometers with lengths ranging from several micrometers to millimeters or even centimeters.

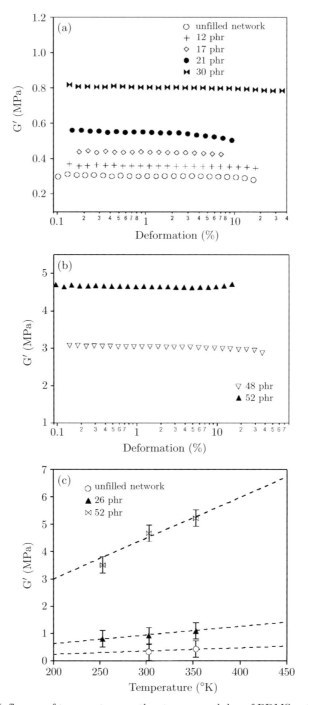

Figure 4.9. Influence of temperature on the storage modulus of PDMS networks filled with different amounts of *in situ* generated silicas

Carbon nanotubes can be synthesized by different techniques including arc-discharge [58,59], laser ablation [60–63] and various catalytic chemical vapor depositions (CCVD) [64–67].

Figure 4.10 shows typical images of a toluene suspension of multi-wall carbon nanotubes previously sonicated before being put onto copper grids for observation. Figures 4.10a and 4.10b reveal a broad distribution in lengths and diameters in the range of 0.1–5 µm and 10–50 nm, respectively. Figures 4.10c and 4.10d show entanglements and curled structures of individual tubes.

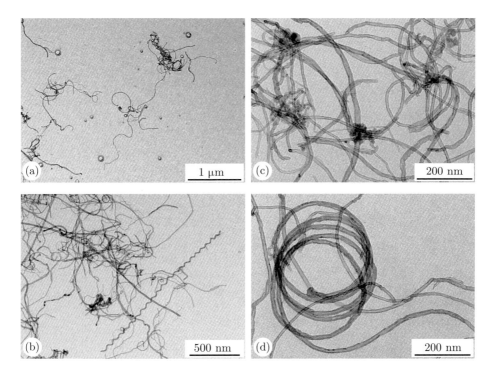

Figure 4.10. TEM images of multi-wall carbon nanotubes (scale bars are respectively 1 µm (a), 500 nm (b), 200 nm (c and d))

As mentioned above, the incorporation of conventional fillers into elastomeric systems usually causes an increase in the modulus ascribed to the inclusion of rigid particles and to the creation of additional cross-links by polymer-filler interactions. The anisometry of particle aggregates or agglomerates is also expected to increase the modulus as well as the occluded rubber, considered as a typical mechanical interaction and consisting of elastomer chains trapped inside the filler aggregates. This occluded rubber, which is assumed to be partially shielded from deformation, increases the effective filler concentration.

On account of their extremely high moduli (~1 TPa for SWNTS [68]), carbon nanotubes are expected to provide much higher reinforcement effects than conventional fillers. Significant improvements in mechanical properties of elastomeric matrices by CNTs have been observed [15,16,69,70].

As a typical example, stress-strain curves for a pure styrene-butadiene rubber and for MWNTs/SBR composites are displayed in Figure 4.11. The elastic moduli, determined from the slope of the true stress against (a^2-a^{-1}) as well as other mechanical characteristics, are listed in Table 4.1. For the sake of comparison, are given, results obtained for a sample filled with 10 phr of carbon black (CB N330 from Cabot) and for a sample containing a double filling (5 phr of MWNTs + 10 phr of CB).

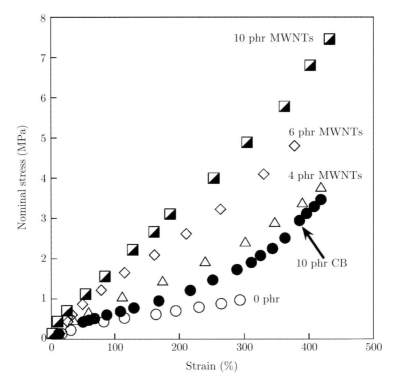

Figure 4.11. Stress-strain curves for an unfilled SBR, MWNTs/SBR composites and a 10 phr carbon black (CB)/SBR composite

It can be seen that, with regard to the pure matrix, the elastic modulus, tensile strength and strain at break increase with nanotube loading. On the other hand, the addition of 10 MWNTs impart to the matrix a higher level of reinforcement than 10 phr of carbon black. Many factors could potentially explain the superior reinforcing efficiency of carbon nanotubes and among them, the filler aspect ratio is expected to play an important role in the mechanical response of the composite. Experimental data can be usefully compared to theoretical model predictions, especially those of Guth [71] and Halpin-Tsai [72].

Table 4.1. Mechanical properties of SBR and SBR composites

Sample	Elastic modulus (MPa)	Stress at 200% (MPa)	Tensile strength (MPa)	Strain at break (%)
Neat SBR	0.24	0.71	0.97	293
SBR + 1 phr MWNTS	0.40	0.91	2.36	480
SBR + 2 phr MWNTS	0.48	1.13	2.75	465
SBR + 4 phr MWNTS	0.71	1.60	3.78	419
SBR + 5 phr MWNTS	0.80	1.85	4.20	414
SBR + 6 phr MWNTS	1.00	2.44	4.80	378
SBR + 10 phr MWNTS	1.37	3.33	7.45	432
SBR + 10 phr CB (N330)	0.55	1.13	3.46	418
SBR + 10 phr CB (N330) +5 phr MWNTs	1.39	2.93	9.34	486

The Guth model [71] only based on the aspect ratio, f, and volume fraction, ϕ, of the filler has been widely used to account for the change in modulus in filled elastomers:

$$E/E_0 = (1 + 0.67 f\phi + 1.62 f^2 \phi^2), \qquad (4.2)$$

(E and E_0 are the moduli of the composite and the unfilled elastomer, respectively).

The Halpin-Tsai model [72] also predicts the stiffness of the composite as a function of the aspect ratio. The longitudinal modulus measured parallel to perfectly oriented fibers, is expressed in the general form:

$$E/E_0 = (1 + 2f\phi\eta)/(1-\phi\eta) \qquad (4.3)$$

where η is given by:

$$\eta = [(E_f/E_0 - 1)/(E_f/E_0 + 2f)] \qquad (4.4)$$

(E_f being the modulus of the filler).

In elastomeric composites, $E_f \gg E_0$, so Eq. 4.2 reduces to:

$$E/E_0 = (1 + 2f\phi)/(1-\phi) \qquad (4.5)$$

In Figure 4.12, the experimental values of E/E_0 are compared with the Guth and Halpin-Tsai predictions using the respective aspect ratios of 40 and 45 to fit the data. While the Guth model departs from the experimental results at the highest filler loadings, the Halpin-Tsai model, for perfectly aligned fibers, satisfactorily captures the experimental elastic modulus behavior of the SBR/MWNTs composites. It is interesting to mention that both models yield almost similar aspect ratios. The values required to fit the experimental data

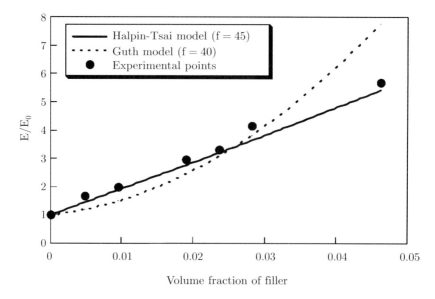

Figure 4.12. Dependence of the experimental moduli ratio of the composite and the matrix E/E_0 on the filler volume fraction and comparisons with theoretical predictions

are lower than those calculated from the average dimensions of the MWNTs. It is most probably due to aggregation of the nanotubes which reduces the aspect ratio of the reinforcement.

As seen in Table 4.1, incorporating simultaneously carbon black and carbon nanotubes into the polymer matrix leads to better mechanical characteristics with regard to single-filler materials on account of synergistic effects arising between the two different fillers.

Figure 4.13 compares the Mooney-Rivlin plots for pure SBR and for composites loaded with single and double fillers. For filled samples, an upturn in the reduced stress is observed at high deformations and is seen to occur sooner for the matrix containing the double filling (carbon black + MWNTs). As already discussed above, the large increase in stress appearing at large strains in filled systems has been attributed to an increase in the cross-linking density created by polymer-filler interactions. In view of the results presented in Figure 4.13, one may conclude that the presence of carbon black leads to a better interaction between the nanotubes and the polymer chains.

Incorporating carbon black and carbon nanotubes into the polymer matrix leads to conductive materials. A lot of work has been devoted to investigations of the electrical properties of these filled materials. The electrical conduction process depends on several parameters such as processing techniques used to mix fillers with rubber, filler content and filler characteristics (particle size and structure) as well as polymer-filler interactions.

Rubber Nanocomposites: New Developments, New Opportunities

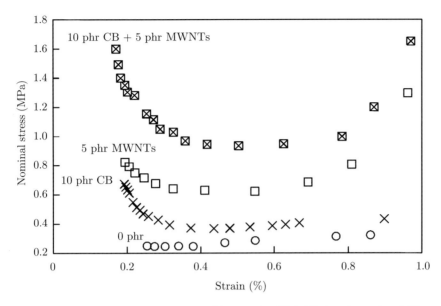

Figure 4.13. Mooney-Rivlin plots for pure SBR and for SBR filled with 10 phr of CB, 5 phr (MWNTs), and a mixture of fillers: 10 phr (CB) + 5 phr (MWNTs)

At a certain filler loading, a conducting path is created throughout the sample reaching the so-called *percolation threshold* and the *insulator-conductor transition*. At this critical filler concentration, the resistivity decreases sharply by several orders of magnitude.

The results shown in Figure 4.14a are related to the effect of carbon black concentration on the volume resistivity of SBR/CB compounds. Carbon nanotubes form the interconnecting filler network between 2 and 3 phr and adding nanotubes in composites containing a constant CB content of 10 phr reduces the percolation threshold (Figure 4.14b). These results suggest that the incorporation of nanotubes improves the formation of connected structures by bridging the uncontacted CB particles.

Since the resistivity is very sensitive to any change in filler distribution, electrical measurements were carried out under an uniaxial deformation. Typical strain dependences of volume resistivity are shown in Figure 4.15. A gradual increase in resistivity is obtained upon applying increased deformations. This increase may be attributed to the breakdown of the conductive network and the orientation of carbon black aggregates and nanotubes. After total unloading of the sample, the resistivity is much higher than that obtained for the undeformed material thus showing that the filler network is not re-formed after removal of the stress. A second stretching leads to a decrease in resistivity as a result of formation of new conductive pathways.

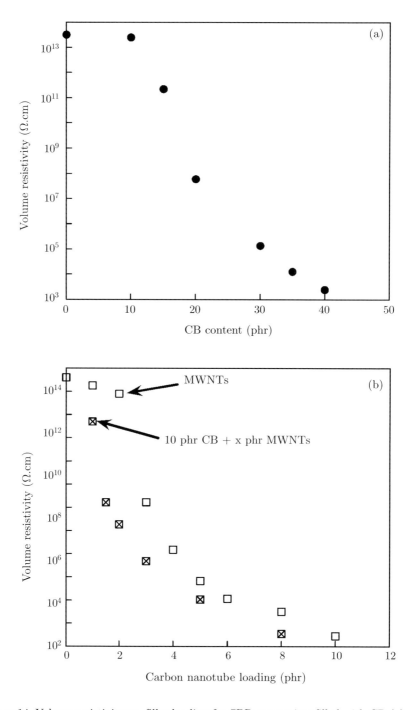

Figure 14. Volume resistivity vs. filler loading for SBR composites filled with CB (a), with MWNTs and mixtures (10 phr CB + x phr MWNTs) (b)

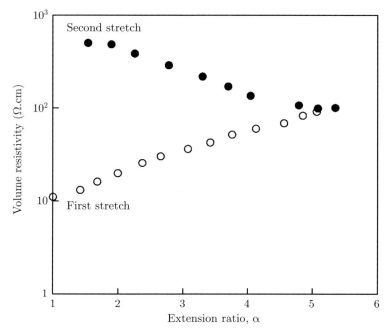

Figure 4.15. Strain dependence of the electrical resistivity for the SBR filled with 10 phr of MWNTs

4.5. Conclusions

This chapter gives examples of the current state of the art in studying reinforcement of rubber-like networks filled with different types of fillers. After recalling the specific features of filled elastomers and the main parameters which play a role in the mechanisms of reinforcement, we focus on materials filled with particles of different morphologies: spherical particles generated *in situ* by the sol-gel process and carbon nanotubes. In both cases, significant improvement in the mechanical properties of the composites are observed with regard to conventional fillers. In nanocomposites synthesized by the sol-gel technique, the small size of the particles, their state of dispersion as well as the hydrophilic character of the filler surface allowing strong interactions with the polymer chains, can account for the greatly improved reinforcing properties. The exceptional mechanical and physical properties of carbon nanotubes make them ideal candidates as reinforcing fillers for elastomeric matrices. Nevertheless, the poor dispersion, the lack of interfacial interactions between the tubes and the polymer chains, reduce the expected property improvements of the resulting composite. The results demonstrate the need for optimizing the processing conditions to achieve good dispersion and good interactions with the matrix. Appropriate functionalization of the tube surface or the use of a double filling (carbon black + carbon nanotubes) may provide convenient routes to improve dispersion and modify interfacial properties.

References

1. Bokobza L (2004) The reinforcement of elastomeric networks by fillers, *Macromol Mater Eng* **289**:607–621.
2. Bokobza L (2004) Elastomeric Composites. I. Silicone Composites, *J Appl Polym Sci* **93**:2095–2104.
3. Bokobza L and Rapoport O (2002) Reinforcement of natural rubber, *J Appl Polym Sci* **85**:2301–2316.
4. McCarthy D W, Mark J E and Schaffer D W (1998) Synthesis, structure, and properties of hybrid organic-inorganic composites based on polysiloxanes I Poly(dimethylsiloxane) elastomers containing silica, *J Polym Sci Part B Polym Phys* **36**:1167–1189.
5. Kohjiya S, Murakami K, Iio S, Tanahashi T and Ikeda Y (2001) In situ filling of silica onto "green" natural rubber by the sol-gel process, *Rubber Chem Tech* **74**:16–27.
6. Yoshikai K, Ohsaki T, Furukawa M (2002) Silica reinforcement of synthetic diene rubbers by sol-gel process in the latex, *J Appl Polym Sci* **85**:2053–2063.
7. Bandyopadhyay A, De Sarkar M and Bhowmick A K (2005) Polymer-filler interactions in sol-gel derived polymer/silica hybrid nanocomposites, *J Polym Sci Part B Polym Phys* **43**:2399–2412.
8. Dewimille L, Bresson B and Bokobza L (2005) Synthesis, structure and morphology of poly(dimethylsiloxane) networks filled with in situ generated silica particles, *Polymer* **46**:4135–4143.
9. Arroyo M, López-Manchado M A and Herrero B (2003) Organo-montmorillonite as substitute of carbon black in natural rubber compounds, *Polymer* **44**:2447–2453.
10. Varghese S and Karger-Kocsis J (2003) Natural rubber-based nanocomposites by latex compounding with layered silicates, *Polymer* **44**:4921–4927.
11. Sadhu S and Bhowmick A K (2004) Preparation and properties of styrene-butadiene rubber based nanocomposites: the influence of the structural and processing parameters, *J Appl Polym Sci* **92**:698–709.
12. Liao M, Zhu J, Xu H, Li Y and Shan W (2004) Preparation and structure and mechanical properties of poly(styrene-b-butadiene)/clay nanocomposites, *J Appl Polym Sci* **92**:3430–3434.
13. Gauthier C, Chazeau L, Prasse T and Cavaillé J Y (2005) Reinforcement effects of vapour grown carbon nanofibres as fillers in rubbery matrices, *Compos Sci Technol* **65**:335–343.
14. Bokobza L and Chauvin J-P (2005) Reinforcement of natural rubber: use of in situ generated silicas and nanofibres of sepiolite, *Polymer* **46**:4144–4151.
15. López-Manchado M A, Biagiotti J, Valentini L and Kenny J M (2004) Dynamic mechanical and Raman spectroscopy studies on interaction between single-walled carbon nanotubes and natural rubber, *J Appl Polym Sci* **92**:3394–3400.
16. Bokobza L and Kolodziej M (2006) On the use of carbon nanotubes as reinforcing fillers for elastomeric materials, *Polym Intern* **55**:1090–1098.
17. Clément F, Lapra A, Bokobza L, Monnerie L and Ménez P (2001) Atomic force microscopy investigation of filled elastomers and comparison with transmission electron microscopy – application to silica-filled elastomers, *Polymer* **42**:6259–6270.
18. Arruda E M and Boyce M C (1993) A three-dimensional constitutive model for the large stretch behavior of rubber elastic materials, *J Mech Phys Solids* **41**:389–412.
19. Yeoh O H and Fleming P D (1997) A new attempt to reconcile the statistical and phenomenological theories of rubber elasticity, *J Polym Sci Part B Polym Phys* **35**:1919–1931.

20. Andrady A L, Llorente M A and Mark J E (1980) Model networks of end-linked poly-dimethylsiloxane chains IX Gaussian, non-Gaussian, and ultimate properties of the trifunctional networks, *J Chem Phys* **73**:1439–1445.
21. Curro J G and Mark J E (1984) A non-Gaussian theory of rubberlike elasticity based on rotational isomeric state simulations of network chain configurations II Bimodal poly(dimethylsiloxane) networks, *J Chem Phys* **80**:4521–4525.
22. Mark J E (1985) Bimodal networks and networks reinforced by the in situ precipitation of silica, *British Polym J* **17**:144–148.
23. Mullins L (1969) Softening of rubber by deformation, *Rubber Chem Tech* **42**:339–362.
24. Bueche F (1960) Molecular basis for the Mullins effect, *J Appl Polym Sci* **4**:107–114.
25. Clément F, Bokobza L and Monnerie L (2001) On the Mullins effect in silica-filled polydimethylsiloxane networks, *Rubber Chem Tech* **74**:847–870.
26. Grosch K A, Harwood J A C, Payne A R (1968) Hysteresis in polymers and its relation to strength, *Rubber Chem Tech* **41**:1157–1167.
27. Harwood J A C, Payne A R (1968) Hysteresis and strength of rubbers, *J Appl Polym Sci* **12**:889–901.
28. Wang S, Xu P and Mark J E (1991) Shear and biaxial extension measurements of reinforcement from in-situ precipitated silica, *Rubber Chem Tech* **64**:746–759.
29. Mark J E (1992) Novel reinforcement techniques for elastomers, *J Appl Polym Sci: Applied Polymer Symposium* **50**:273–282.
30. Wen J and Mark J E (1994) Precipitation of silica-titania mixed-oxide fillers into poly(dimethylsiloxane) networks, *Rubber Chem Tech* **67**:806–819.
31. McCarthy D W, Mark J E, Clarson S J and Schaeffer D W (1998) Synthesis, structure, and properties of hybrid organic-inorganic composites based on polysiloxanes II Comparisons between poly(methylphenylsiloxane) and poly(dimethylsiloxane), and between titania and silica, *J Polym Sci Part B Polym Phys* **36**:1191–1200.
32. Breiner J M, Mark J E and Beaucage G (1999) Dependence of silica particle sizes on network chain lengths, silica contents, and catalyst concentrations in in situ-reinforced polysiloxane elastomers, *J Polym Sci Part B Polym Phys* **37**:1421–1427.
33. Yuan Q W, Mark J E (1999) Reinforcement of poly(dimethylsiloxane) networks by blended and *in-situ* generated silica fillers having various sizes, size distributions, and modified surfaces, *Macromol Chem Phys* **200**:206–220.
34. Tanahashi H, Osanai S, Shigekuni M, Murakami K, Ikeda Y and Kohjiya S (1998) Reinforcement of acrylonitrile-butadiene rubber by silica generated in situ, *Rubber Chem Tech* **71**:38–52.
35. Ikeda Y and Kohjiya S (1997) In situ formed silica particles in rubber vulcanizate by the sol-gel method, *Polymer* **38**:4417–4423.
36. Hashim A S, Azahari B, Ikeda Y and Kohjiya S (1998) The effect of bis(3-triethoxysilylpropyl) tetrasulfide on silica reinforcement of styrene-butadiene rubber, *Rubber Chem Tech* **71**:289–299.
37. Kohjiya S and Ikeda Y (2000) Reinforcement of general purpose grade rubbers by silica generated in situ, *Rubber Chem Tech* **73**:534–550.
38. Murakami K, Iio S, Tanahashi T, Kohjiya S, Kajiwara K and Ikeda Y (2001) Reinforcement of NR by silica generated in situ: comparison with carbon black stock, *Kautschuk Gummi Kunststoffe* **54**:668–672.
39. Matejka L, Dukh O and Kolarik J (2000) Reinforcement of crosslinked rubbery epoxies by *in-situ* formed silica, *Polymer* **41**:1449–1459.
40. Matejka L and Dukh O (2001) Organic-inorganic hybrid networks, *Macromol Symp* **171**:181–188.

41. Jang J and Park H (2002) In situ sol-gel process of polystyrene/silica hybrid materials: effect of silane-coupling agent, *J Appl Polym Sci* **85**:2074.
42. Mark J E and Erman B (1988) *Rubberlike Elasticity: A Molecular Primer*, Wiley-Interscience, New York.
43. Pope E J A and Mackenzie J D (1986) Sol-gel processing of silica, *J Non-Cryst Solids* **87**:185–198.
44. Pouxviel J C, Boilot J P, Beloeil J C and Lallemand J Y (1987) NMR study of the sol/gel polymerization, *J Non-Cryst Solids* **89**:345–360.
45. Brinker C J (1988) Hydrolysis and condensation of silicates: effects on structure, *J Non-Cryst Solids* **100**:31–50.
46. Fragiadakis D, Pissis P, Bokobza L (2005) Glass transition and molecular dynamics in poly(dimethylsiloxane)/silica nanocomposites, *Polymer* **46**:6001–6008.
47. Patel M, Morrell P R and Skinner A R (2002) Physical & thermal properties of model siloxane rubbers: impact of crosslink density and tin concentration, *Macromol Symp* **180**:109–123.
48. Kilian H G, Strauss M, Hamm W (1994) Universal properties in filler-loaded rubbers, *Rubber Chem Tech* **67**:1–16.
49. Payne A R (1962) The dynamic properties of carbon black-loaded natural rubber vulcanizates, *J Appl Polym Sci* **VI**:57–63.
50. Wang M J (1998) Effect of polymer-filler and filler-filler interactions on dynamic properties of filled vulcanizates, *Rubber Chem Tech* **71**:520–589.
51. Drozdov A D and Dorfmann A (2002) The Payne effect for particle-reinforced elastomers, *Polym Eng Sci* **42**:591–604.
52. Klüppel M (2003) The role of disorder in filler reinforcement of elastomers on various length scales, *Adv Polymer Sci* **164**:1–86.
53. Clément F, Bokobza L and Monnerie L (2005) Investigation of the Payne effect and its temperature dependence on silica-filled polydimethylsiloxane networks Part I: Experimental results, *Rubber Chem Tech* **78**:211–231.
53bis Clément F, Bokobza L and Monnerie L (2005) Investigation of the Payne effect and its temperature dependence on silica-filled polydimethylsiloxane networks Part II, *Rubber Chem Tech* **78**:232–244.
54. Ramier J, Gauthier C, Chazeau L, Stelandre L and Guy L (2007) Payne effect in silica-filled styrene-butadiene rubber: influence of surface treatment, *J Polym Sci Part B Polym Phys* **45**:286–298.
55. Maier P G and Göritz D (1996) Molecular interpretation of the Payne effect, *Kautschuk Gummi Kunststoffe* **49**:18–21.
56. Thostenson E T, Ren Z, Chou T S (2001) Advances in the science and technology of carbon nanotubes and their composites: a review, *Compos Sci Technol* **61**:1899–1912.
57. Moniruzzaman M, Winey K I (2006) Polymer nanocomposites containing carbon nanotubes, *Macromolecules* **39**:5194–5205.
58. Hutchison J L, Kiselev N A, Krinichnaya E P, Krestinin A V, Loufty R O, Morawsky A P, Muradyan V E, Obraztsova E D, Sloan J, Terekhov S V, Zakharov D N (2001) Double-walled carbon nanotubes fabricated by a hydrogen arc discharge method, *Carbon* **39**:761–770.
59. Saito Y, Nakahira T and Uemura S (2003) Growth conditions of double-walled carbon nanotubes in arc discharge, *J Phys Chem B* **107**:931–934.
60. Zhang Y and Iijima S (1999) Formation of single-wall carbon nanotubes by laser ablation of fullerenes at low temperature, *Appl Phys Lett* **75**:3087–3089.
61. Scott C D, Arepalli S, Nikolaev P and Smalley R E (2001) Growth mechanisms for single-wall carbon nanotubes in a laser-ablation process, *Appl Phys A* **72**:573-580.

62. Arepalli S (2004) Laser ablation process for single-walled carbon nanotube production, *J Nanosci Nanotechnol* **4**:317–325.
63. Jiang W, Molian P and Ferkel H (2005) Rapid production of carbon nanotubes by high-power laser ablation, *J Manuf Sci Eng* **127**:703–707.
64. Hiraoka T, Kawakubo T, Kimura J, Taniguchi R, Okamoto A, Okazaki T, Sugai T, Ozeki Y, Yoshikawa M, Shinohara H (2003) Selective synthesis of double-wall carbon nanotubes by CCVD of acetylene using zeolite supports, *Chem Phys Lett* **382**:679–685.
65. Darabont A, Nemes-Incze P, Kertész K, Tapaszó L, Koós A A, Osváth Z, Sárközi Zs, Vértesy Z, Horváth Z E and Biró L P (2005) Synthesis of carbon nanotubes by spray pyrolysis and their investigation by electron microscopy, *J Optoelectr Adv Mater* **7**:631–636.
66. Aghababazadeh R, Mirhabibi A R, Ghanbari H, Chizari K, Brydson R M and Brown A P (2006) Synthesis of carbon nanotubes on alumina-based supports with different gas flow rates by CCVD method, *J Phys Conf Ser* **26**:135–138.
67. Endo M, Hayashi T, Kim Y A and Muramatsu H (2006) Development and application of carbon nanotubes, *Jap J Appl Phys* **45**:4883–4892.
68. Salvetat J P, Briggs G A D, Bonard J M, Bacsa R R, Kulik A, Stöckli T, Burnham N A and Forró L (1999) Elastic and shear moduli of single-walled carbon nanotube ropes, *Phys Rev Lett* **82**:944–947.
69. Atieh M A, Girun N, Ahmadun F R, Guan C T, Mahdi A S and Baik D R (2005) Multi-wall carbon nanotubes/natural rubber nanocomposite, *AZojono J Nanotechnol Online* **1**:1–11.
70. Frogley M D, Ravich D and Wagner H D (2003) Mechanical properties of carbon nanoparticle-reinforced elastomers, *Compos Sci Technol* **63**:1647–1654.
71. Guth E (1945) Theory of filler reinforcement, *J Appl Phys* **16**:20–25.
72. Halpin J C (1969) Stiffness and expansion estimates for oriented short fiber composites, *J Compos Mater* **3**:732–734.

Chapter 5

Organoclay, Particulate and Nanofibril Reinforced Polymer-Polymer Composites: Manufacturing, Modeling and Applications

D. Bhattacharyya, S. Fakirov

5.1. Introduction

In recent times, polymeric composite materials have replaced traditional metals in a variety of applications. Light weight, coupled with enhanced properties, are the main reasons for their market acceptance and growth, and the optimization of their performance is a challenge worldwide. The high aspect ratio of the reinforcing particle and its adhesion to the matrix are of great importance, because they control the final properties of the composites. In the last decades, a variety of nanomaterials, *i.e.*, materials with at least one dimension below 100 nm, appeared, which can be successfully used as reinforcements of polymer composites. The term *nanocomposites* generally refers to composite materials which contain fillers whose the dimension, in at least one direction, is less than 100 nm.

It has been shown that, when using fiber-reinforced polymers, the maximum amount of fibers in the matrix is about 70 vol%. In practice, the fiber volume fractions in these materials vary between 20 and 60%. Let a volume element of 1 cm^3 be reinforced with discontinuous fibers with 10 µm diameter, particles (*e.g.*, talcum) with 1 µm diameter and carbon nanotubes (CNT) with 10 nm diameter. The aspect ratios are generally expected to be 20, 100, and 1000, respectively. If it is further assumed that a volume content of 30% exists for both the fibers and the particles, and only 3% for the nanotubes, the results are quite interesting – the filler element numbers become ~10^6 fibers, ~10^{10} particles and ~10^{15} nanotubes. The surface areas amount to ~0.1 m^2 for the fibers, ~1 m² for the particles and ~100 m² for the nanotubes. There is another important aspect to be considered and that is the distance between the filler elements. For the fibers one can calculate a distance of ~10 μm, for the particles ~1 μm, and for the nanotubes ~100 nm [1]. These values highlight some problems to be overcome: nanocomposites have to deal with huge surface areas, a large number of nanofiller particles and a small distance between the two adjoining reinforcement phases.

Nanoparticle reinforcement in certain polymeric matrix materials can lead to significant property improvements, whereas in others it results only in marginal gains or in some cases even in worsening of the properties [2–4]. This is due to the strong tendency of nanoparticles to agglomerate in polymeric matrices and can hardly be controlled by conventional processing techniques owing to the limited shear force generated during the mixing procedure. Consequently, clusters and aggregates of nanoparticles lead to properties that are worse than conventional particle/polymer systems [5].

The best performance of polymeric nanocomposites is achieved when the nanofiller is dispersed in the polymer matrix without any agglomeration. Currently, numerous procedures for the preparation of polymeric nanocomposites have been proposed using the following approaches: (i) direct incorporation of nanoscale building blocks into a polymer melt or solution [6–9], (ii) *in-situ* generation of nanoscale building blocks in a polymer matrix (*e.g.*, vacuum evaporation of metals, thermal decomposition of precursors, reduction of metal ions through an electrochemical procedure) [10–12], (iii) polymerization of monomers in the presence of nanoscale building blocks [13,14], (iv) a combination of polymerization and formation of nanoscale building blocks (*e.g.*, sol-gel method, intercalation of monomers into layered structures followed by polymerization) [15,16]. The key issue in these techniques is that the geometry, spatial distribution, and volume content of the nanofillers must be effectively controlled by adjusting the preparation conditions so as to ensure the structural requirements for nanocomposites stated earlier.

From the engineering point of view, large-scale and low-cost production routes, as well as a broad applicability of thermoplastic nanocomposites as structural materials should be considered. This may be achieved by the employment of commercially available non-layered nanoparticles, and by the use of blending techniques that are already widely applied in the plastics industry [17].

It is worth noting again that the problem with dispersive mixing is that the commercially available nanoparticles usually exist in the form of agglomerates, which are difficult to disconnect by the limited shear force during mixing. This is true even when a coupling agent is used [18] it can only react with the nanoparticles on the surface of the agglomerates. The latter maintain their friable structure in the composites and can hardly provide any property improvement. Even more complex is the case when the fillers represent natural clay minerals because additional specific steps, such as *exfoliation* and *intercalation*, are required.

It is quite evident that the huge potential for a drastic improvement of the performance of a polymer nanocomposite is not easy to realize, mostly because of the problems related to the dispersion process.

5.2. Polypropylene/organoclay nanocomposites: experimental characterisation and modeling

5.2.1. *Peculiarities of polymer/clay nanocomposites*

Recent research and development in polymer nanocomposites have shed some light on the use of inorganic layered silicates or smectite clays, well dispersed into an organic polymer for increasing its mechanical, thermal, barrier and fire retardant properties [19–25]. Layered silicates are made up of several hundred thin platelets stacked in orderly particles or tactoids of the size of 8–10 µm. Each disk-shaped platelet has a very large aspect ratio of about 100–1000 and is easily agglomerated due to the interlayer van der Waals force. Accordingly, clay particles should be homogeneously dispersed and exfoliated as individual platelets within the polymer matrices in order to achieve good property enhancement, schematically demonstrated in Figure 5.1.

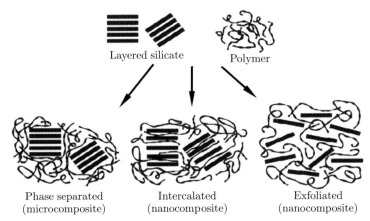

Figure 5.1. Schematic representation of the different composite types arising from the interaction of layered silicates and polymers [26]

In addition, lower clay content (≤ 5 wt%) is also found essential to obtain reasonably good dispersion by alleviating the clay aggregation and to attain large contact surface area between the fillers and the polymer matrix [24]. Melt blending is normally regarded as one of the most promising approaches today to prepare nanocomposites due to its cost-effectiveness in using conventional polymer compounding techniques. In particular, a high shear mixer or twin screw extruder has been proven to facilitate clay delamination in the molten polymers under their respective physical mechanisms [27–31].

Polyolefin/clay nanocomposites have attracted the largest commercial interest on the part of material innovators and researchers. In particular, polypropylene (PP) has been investigated as a widely used commodity polymer due to its good processability, low cost and enormous applications in the automotive and packaging industries. However, the incompatibility between the non-polar hydrophobic PP and hydrophilic clays impedes a good clay dispersion. Consequently, great efforts have been made to use the functional polyolefin oligomers as compatibilizers, modified with either maleic anhydride (MA) or hydroxyl groups (OH). Such modifiers include a certain number of polar groups (*e.g.*, –OH, –COOH) to intercalate between silicate layers through hydrogen or other chemical bonding to the oxygen group or negative charges of the silicate layers [28]. Evidently, clay aggregates are easily broken up and uniform dispersion of clay particles can be achieved in the presence of maleic anhydride modified polypropylene (MAPP) [27,29,31]. However, the addition of MAPP mostly leads to the improvement of dominant intercalated structures for melt blended PP/organoclay nanocomposites rather than the dispersion of clay in single platelets (*i.e.*, complete exfoliation, Figure 5.1) [32]. Their mechanical properties can be directly affected by the intercalation/exfoliation levels in nanocomposite morphology, which depends on the input material characteristics such as clay type and content, compatibilizer content and matrix viscosities. Therefore, it is crucial to effectively investigate the effects of these material parameters in order to achieve superior mechanical properties.

At the same time, due to the excellent property improvements and the versatile applications in automotive and packaging industries, polymer/clay nanocomposites are intensively researched as an innovative replacement of the conventional composite materials [20–24]. Such nanocomposites can be ideally classified to be *intercalated* or *exfoliated/delaminated* according to their different morphological structures (Figure 5.1). Nevertheless, in reality, partially intercalated and exfoliated nanocomposites are normally formed, which creates a very complex heterogeneous material system. The irregular shapes and sizes of the dispersed clay platelets with random orientation states are inevitably challenging to understand the relationship between the processing, the microstructure and the resulting properties. Numerical modeling techniques such as *molecular dynamics* (MD) *simulation* and *finite element method* (FEM) have been found for decades to be more effective approaches compared to the analytical models. Since most nanocomposites contain nanoscale fillers such as carbon nanotubes, nanoclays and nanofibers, molecular dynamic simulation [33,34] has been employed to accurately predict the properties of nanocomposites with the tedious computational efforts to scale up the time and the integrated length. Alternatively, *representative volume elements* (RVE) have been implemented with either well-aligned [35,36] or Monte Carlo type randomly distributed [37–39] patterns, which might not represent the actual complex and highly heterogeneous nanocomposite structures. Consequently, effective modeling and prediction of the mechanical properties of nanocomposites are still at an early stage for the design and development of such high-performance materials to be used by composites manufacturers and engineers.

Recently [40], a novel numerical approach, the *object-oriented finite element* (OOF) [41,42] analysis has been utilized by mapping the real micro/nano morphological images of PP/clay nanocomposites with varied clay contents between 1 and 10 wt%. Such morphological images are captured by two different microscopic techniques, the scanning electron microscopic (SEM) and transmission electron microscopic (TEM) analyses. The tensile moduli of nanocomposites are numerically predicted and subsequently compared with the tensile test data. Furthermore, the available composites models are used to validate the numerical approach developed in the same [40] study. Finally, the effect of particle distribution on the deformation behavior is also evaluated through the tensile stress and elastic strain contours of such nanocomposites [40].

5.2.2. *Parametric study and associated properties of PP/organoclay nanocomposites*

As said earlier, in order to achieve superior mechanical properties, it is essential to know the effects of processing parameters individually and/or collectively as a guide for the material design and formulation.

In the following study, three commercial polypropylenes with different melt viscosities were twin screw extruded with three types of organoclay in the presence of MAPP to prepare PP/organoclay nanocomposites. A new approach to use both Taguchi design of experiment (DoE) method and ANOVA was implemented to statistically determine the significant factors influencing the enhancement of mechanical properties of such nanocomposites. The ultimate goal was to obtain anocomposites tailored to their specific usages with desirable mechanical properties in view of technical and economic considerations.

5.2.2.1. *Materials, fabrication process and characterization*

Three grades of commercially available polypropylenes, denoted as PP-Co M710, PP-Hom Y130 and PP-Hom H380F with various melt flow indices (MFI), along with MAPP Exxelor PO 1020 (MA content of 0.5–1 wt%) as compatibilizer were used. In particular, PP-Co M710 was a high impact PP copolymer with an ethylene content of 6 wt%.

The layered silicates used were three types of organomodified montmorilonite (MMT) NANOLIN™ clay denoted as DK1N, DK2 and DK4. They are normally characterized by an average fully dispersed thickness of clay platelets of about 25 nm (aspect ratio 100–1000) in 95–98% purified smectite content. The three types of organoclay were modified by different levels of cation exchange reaction with octadecylammonium salt to broaden their interlayer spacing according to the aforementioned method [43]. The organoclays DK1N, DK2 and DK4 were distinguished by an interlayer spacing of 2.29, 2.25 and 3.56 nm, respectively, and the same density of 1.8 g/cm^3 for all of them [44].

PP/organoclay nanocomposites were prepared by melt blending in a co-rotating intermeshing twin screw extruder Brabender® DSE 20 (screw diameter

$D = 20$ mm, $L\!:\!D = 40\!:\!1$). The temperature profile of four zones was controlled at 185, 195, 200 and 210°C, respectively, from the feeding zone to the heating zones and die temperature was over 203°C.

A twice direct compounding method was employed. The first step was to feed the PP and MAPP pellets into the high speed extruder and then feed the organoclay powders downstream into the fully melted polymeric mixture (PP and MAPP) to prepare initial nanocomposite batches. In the second step, such nanocomposite pellets were recompounded at a relatively low screw speed to make the final nanocomposites. Subsequently, organoclay powders were fed downstream into melted PP and MAPP. The first direct compounding was made at the screw speed of 200 rpm and the feed rate for all PP grades about 3.0 kg/hr. Initial nanocomposite batches were then recompounded at 100 rpm with the same feed rate.

Injection molded test samples were prepared on a BOY® 50A injection molding machine with a temperature profile of 195, 200, 210 and 190°C from the first, second, third feeding zones to the nozzle. The die temperature was kept at 25°C and the injection pressure was about 60–80 bars. In order to produce the reference materials, neat PP underwent only the first direct compounding and injection molding in the same processing conditions as the corresponding nanocomposites to reduce the possibility of thermal degradation of PP. Mechanical testing and characterization of the samples was performed as described for the previous case of PP/organoclay nanocomposites.

5.2.2.2. Taguchi method

The *Taguchi method* is a well known approach in the modern robust design process, mainly based upon statistical design of experiments. This special engineering method is used to optimize the process conditions with minimal sensitivity to the various causes of variation and also to produce high-quality products with low development and manufacturing costs [45]. Normally, there are two core tools employed in the Taguchi method, namely *signal-to-noise ratio* to measure variation induced quality as well as *orthogonal arrays* to accommodate many design factors simultaneously.

The primary goal of the study [44] was to detect the most significant factors for achieving the maximum enhancement of mechanical properties of prepared nanocomposites and work out the preferable conditions in a limited number of trials. There are two major categories of factors to be considered, namely processing parameters such as temperature profile, screw speed, feed rate, die pressure in extrusion compounding and injection molding processes as well as the raw material grades and contents. Nevertheless, such a large number of influencing factors inevitably lead to the complexity of the experimental work. As a result, the effects of using different combinations of PP and organoclays with varied types and contents in the presence of MAPP were evaluated. Thus, all the process conditions remained the same during twin screw extrusion and injection molding processes.

Four factors of clay type and content, MAPP content and PP type with three different levels of low, medium and high settings were selected in the DoE work, Table 5.1. This setup resulted in a typical three-level four factors L_9 Taguchi DoE layout compared with a traditional full-factorial 81 trials to complete the entire experimental work. To keep the analysis simple, the factorial interactions were not considered. An identical L_9 DoE layout for both first direct compounding and recompounding processes has been elaborated in the sequence of random order trial numbers (R – first direct compounding process, and RR – recompounding process). The ultimate responses were set to maximize tensile/flexural moduli and strengths, and impact strengths of prepared nanocomposites. Hence, a "larger-the-better" characteristics [45,46] formula was used to identify the combination of factors with the maximum response shown as

$$S/N = -10\log\left(\frac{1}{n}\sum_{i=1}^{n}\frac{1}{y_i^2}\right) \tag{5.1}$$

where S/N is the signal-to-noise ratio based on the improvement of quality and measurement *via* variability reduction [45], n is the number of samples in each trial and y is the measured response value, namely the normalized moduli or strengths over those of corresponding neat PP.

Table 5.1. Four factors and three levels used in L_9 DoE layout [55]

Factor	Level		
	1	2	3
A: Clay type	DK1N	DK2	DK4
B: Clay content (wt%)	3	5	10
C: MAPP content (wt%)	5	10	20
D: PP type	PP-Co M710	PP-Hom Y130	PP-Hom H380F

5.2.2.3. Pareto ANOVA analysis

Pareto analysis of variance [45,47] is a simplified ANOVA method using the 80/20 Pareto principle which means that in anything a few (20 percent) are vital and many (80 percent) are trivial. This is a quick and easy method to analyze the results of parameter design without the requirement of ANOVA table and F-tests. Moreover, significant factors and interactions as well as the relevant optimum factor levels can be easily detected by this special Pareto-type analysis. The general criterion to determine the significant factors in this study was based upon the derived cumulative contribution percentage of about 90%. Economic and technical issues were considered for the other non-significant factors.

5.2.3. Evaluation of the experimental data by means of Taguchi and Pareto ANOVA methods

5.2.3.1. Recompounding effects on the enhancement of mechanical properties

Mechanical testing was performed on the injection molded samples prepared in both first direct compounding and recompounding processes in twin screw extrusion. Figure 5.2 displays as an example the relevant tensile modulus and tensile strength of the neat PP types used and the corresponding nanocomposites in both processes of the L_9 DoE layout. Comparisons of all the mechanical properties of nanocomposites were based on those of corresponding neat PP in each trial.

Basically, the improvement of tensile modulus was found to be quite marginal for PP-Co M710- PP-Hom Y130 based nanocomposites (around 5%), while recompounding played a favorable role in the enhancements of tensile moduli – RR1 demonstrated an increase up to 24%. Higher tensile moduli were observed for R8 and R9 with the respective enhancement of 22 and 27%. Meanwhile, after recompounding, the tensile moduli for RR8 and RR9 were significantly enhanced by 45 and 39%, respectively (Figure 5.2a).

As far as the tensile strengths were concerned, for PP-Co M710 and PP-Hom Y130 based nanocomposites they remained almost the same as those of the corresponding neat PP in the first compounding and were relatively good after recompounding. It appeared that the major increases of tensile strength were most likely to take place in PP-Hom H380F based nanocomposites (R5, R8–R9 and RR5, RR8–RR9). Furthermore, the increasing trend was also well maintained after recompounding with the maximum increases of tensile strength up to 12% for R9 and RR9.

As for the flexural properties of PP-Co M710 based nanocomposites, both R4 and RR4 showed decreased flexural moduli in contrast to R7 and RR7, which demonstrated improvements in flexural moduli by about 25%, very likely due to the addition of 10 wt% DK4 organoclays. For PP-Hom H380F based nanocomposites, R5 and RR5 showed increases over 12 and 19% but more remarkable enhancements happened in R8 (49%) and R9 (~46%) as well as RR8 (~60%) and RR9 (45%) compared to the flexural modulus of PP-Co H380F. It seemed that recompounding had more significant effects on the improvement of flexural modulus for PP-Hom H380F based nanocomposites with a higher clay content (~10 wt%).

With respect to the flexural strengths, PP-Co M710 based nanocomposites showed poor results with reductions of about 5% for both R4 and R7 and maintained a similar trend after recompounding in RR4 and RR7. The situation was marginally better with the other compositions and only for PP-Hom Y130 based nanocomposites, an increase of up to 10% was found for R5 and RR5 and more remarkable strength improvements over 20% took place for R8–R9 and RR8–RR9 in both first direct compounding and recompounding processes.

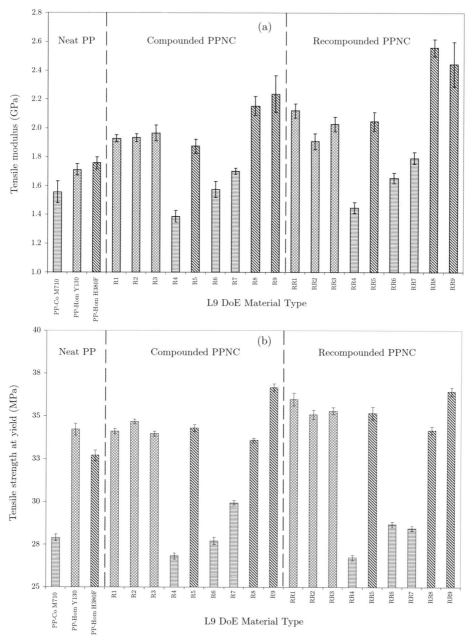

Figure 5.2. Tensile properties of injection molded samples of PP/organoclay nanocomposites in L_9 DoE: (a) tensile modulus, and (b) tensile strength [44]

Regarding impact properties, PP-Co M710 based nanocomposites demonstrated dramatic decreases of impact strength (over 60%). Further recompounding helped in alleviating this declining trend and especially the strength decrease for RR4 became 20% compared to 33% for R4. The addition of organoclays did not greatly affect the impact strengths of PP-Hom Y130 based nanocomposites (R1–R3) in the first direct compounding. After recompounding, a strength increase of about 20% was found in RR3 compared to 1% increase in R3.

In general, recompounding processes could enhance the mechanical properties of nanocomposites, depending on the type of polymer. It appeared that reprocessing of nanocomposites in a twin screw extruder at a low screw speed (*e.g.*, 100 rpm) without altering the other processing conditions might contribute to a more uniform clay dispersion by extending the compounding residence time. Consequently, the determination of preferred combinations of factors in this L_9 DoE was solely based on the evaluation of mechanical testing data obtained from recompounded nanocomposites RR1–RR9.

5.2.3.2. *Evaluation of significant factors*

The Taguchi method suggests the analysis of the *S/N* ratios with the conceptual approach, which involves graphing the effects and visually identifying the significant factors. Moreover, preferred combinations of factors can be further determined based on Pareto ANOVA. *S/N* ratios are regarded as measures of effects of noise factors on the performance characteristics, which consider both the amount of variability in response data and the closeness of average response to target [46]. As said earlier, the calculated *S/N* ratios were evaluated with the "larger-the-better" characteristics for maximizing tensile, flexural and impact properties. Each of the normalized response values was derived from five varied sample testing data for recompounded nanocomposites.

The Pareto ANOVA technique was performed for each of the mechanical properties and the related Pareto ANOVA diagrams were constructed [44]. As an example, such histograms for the tensile modulus are depicted in Figure 5.3. As can be seen, PP type (factor D) and clay content (factor B) have the most significant effects with the contribution percentages of 60.4 and 33.2%, respectively, whereas the effects of clay type (factor A) and MAPP content (factor C) are trivial with the sum of the contribution percentages being less than 10%. The non-significant effect of MAPP content is possibly attributed to the MAPP grade used in this study. Since it has a low molecular weight, short length chains of MAPP are mainly penetrating in the clay platelets leading to a limited increase of the interlayer spacing. In addition, the excessive amount of MAPP causes less further enhancement of the tensile and flexural properties of nanocomposites, which could be attributed to the negative matrix plasticization effect of MAPP to soften the prepared nanocomposites in competition with the good clay dispersion. Therefore, it is suggested that PP type and clay content greatly influence the enhancement of tensile modulus with the cumulative contribution percentage beyond 90%.

From a similar Pareto ANOVA diagram [44] the same trend was found for the tensile strength as for the tensile modulus (Figure 5.3). Further, for flexural properties, the enhancements of both flexural modulus and strength again come

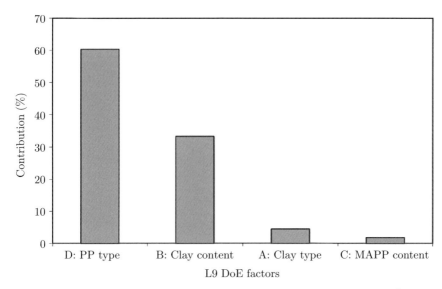

Figure 5.3. Pareto ANOVA analysis for enhancing the tensile modulus of PP/organoclay nanocomposites [44]

from PP type and clay content. As observed in Pareto ANOVA diagrams [44], PP type contributes 53.2 and 77.1%, while the clay content contributes 42.6 and 21.1%, respectively.

In the case of impact strength, its improvement was influenced by PP type, MAPP content and clay type with the contribution percentages of 66.7, 20.2 and 7.92%, respectively [44]. It is suggested that the MAPP content, as the second significant factor, plays a more important role in the impact properties compared to tensile and flexural properties.

5.2.3.3. *Determination of preferred combinations of factors*

It is well understood for the "larger-the-better" characteristics in the Taguchi DoE method that mathematically the higher the sum of S/N ratio is, the better is the response for the factorial effects. It is worth noting that the presented study [44] was focused solely on the maximization of individual mechanical properties since the global property maximization of nanocomposites would be very difficult to achieve due to some contradictory effects. Consequently in practice, a good balance has to be maintained while increasing the tensile/flexural properties without greatly sacrificing the impact strength.

As seen from Table 5.2, the best combination of the significant factors to get the highest value for the sum of S/N ratio is at level 3 of clay content (*i.e.*, 10 wt% organoclays) and level 3 of PP type (*i.e.*, PP-Hom H380F). Thus, the preferred condition for all factors becomes $A_3B_3C_3D_3$ with level 3 for both clay type (DK4 organoclays) and MAPP content (20 wt% MAPP) as the two non-significant factors. To maximize the tensile strength, the corresponding best

Table 5.2. Summary of preferred combinations of factors for improving the mechanical properties of PP/organoclay nanocomposites [44]

Larger-the-better L_9 DoE Response (normalized)		Preferred factorial level combination	Composition (wt%)	Estimate of error variance
Tensile properties	Tensile modulus	$A_3B_3C_3D_3$	DK4/MAPP/H380F (10/20/70)	0.347
	Tensile strength	$A_3B_2C_2D_3$	DK4/MAPP/H380F (5/10/85)	0.036
Flexural properties	Flexural modulus	$A_3B_3C_3D_3$	DK4/MAPP/H380F (10/20/70)	0.313
	Flexural strength	$A_2B_3C_1D_3$	DK2/MAPP/H380F (10/5/85)	0.052
Impact properties	Impact strength	$A_1B_1C_1D_3$	DK1N/MAPP/H380F (3/5/92)	9.702

combination yields $A_3B_2C_2D_3$ (*i.e.*, 5 wt% DK4 organoclays, PP-Hom H380F and 10 wt% MAPP), Table 5.2. Similarly, in terms of enhancing the flexural modulus and strength, the best combinations can be given by $A_3B_3C_3D_3$ (*i.e.*, 10 wt% DK4 organoclays, PP-Hom H380F and 20 wt% MAPP) and $A_2B_3C_1D_3$ (*i.e.*, 10 wt% DK2 organoclays, PP-Hom H380F and 5 wt% MAPP), respectively, as shown in Table 5.2. In contrast, for the impact strength, the majority of calculated sums of S/N ratios at each factor level appear to be negative [44]. This implies that the addition of organoclays in the presence of compatibiliser in PP matrices tends to deteriorate the impact properties, particularly for PP-Co M710 based nanocomposites. This is not surprising, considering the fact that MAPP can induce some material brittleness to nanocomposites [48]. Hence, the best combination to get the highest impact strength is $A_1B_1C_1D_3$ (*i.e.*, 3 wt% DK1N organoclays, PP-Hom H380F and 5 wt% MAPP).

5.2.4. *Materials, manufacturing and characterization of nano composites*

NANOLIN DK2 organoclays were provided by Zhejiang Feng Hong Clay Chemicals Co., Ltd, China. The *organoclays* were prepared *via* modification by cation exchange reaction with octadecylammonium salt to broaden their interlayer spacing according to a method described in the literature [49]. The interlayer spacing corresponded to 2.25 nm as found from raw organoclay powders using X-ray diffraction (XRD) analysis [50]. Polypropylene copolymer Hi-rene M710 (density 0.9 g/cm^3 and melt flow index 0.6 g/10 min at 230°C) was obtained from Clariant New Zealand Ltd. Maleated polypropylene (MAPP) Exxelor PO 1020 (density 0.9 g/cm^3 and melt flow index ~430 g/10 min at 230°C) was used as a compatibilizer, which was supplied by ExxonMobil Chemical, Germany.

PP/organoclay nanocomposites were prepared in pellet form using a co-rotating intermeshing twin screw extruder Brabender DSE 20 with a screw length-diameter ratio of 40:1 (diameter D = 20 mm) at 185–210°C and 200 rpm. Two-step masterbatch compounding was employed in this study where 1:1 weight ratio of premixed organoclays and MAPP powders were first melt compounded to make the masterbatch. It was then further sized into PP pellets to prepare nanocomposites with clay contents of 1, 3, 5 and 10 wt% (denoted as NC1, NC3, NC5 and NC10, respectively). The dried nanocomposite pellets were then injection molded using a BOY 50A injection molder to make the tensile specimens at a temperature profile of 190–210°C and injection pressure of 70–80 bars. Neat PP was also processed in the same conditions for comparison.

Morphological images were based on SEM and TEM analyses for the geometric generation of nanocomposite structures. SEM surfaces were generated by cryofracturing tensile specimens along the longitudinal direction for further OOF modeling work. Similarly, TEM ultra thin film samples (nominal thickness of 70 nm) were also cryosectioned in the longitudinal direction at –80°C. The original SEM and TEM images were processed using Photoshop 7.0 software to adjust their contrast and the brightness for the best image quality. Subsequently, the region of interest for nanocomposite morphology was selected based on the microstructural features as well as the availability of the computational capacity in OOF modeling. A region of interest was taken in the size of approximately 6.15 μm × 4.10 μm from the SEM image captured at 20 000× magnification and 6.62 μm × 6.62 μm from the TEM image at 15 000× magnification.

The tensile properties of the specimens were measured according to ASTM D638. The chord tensile modulus (0.05–0.25% strain) and tensile strength at yield were determined at the crosshead speed of 5 mm/min and 50 mm/min. More than ten specimens were tested with the calculated mean values and standard deviations used in the final results.

5.2.5. *Analytical models for composites*

5.2.5.1. *Halpin-Tsai model*

The *Halpin-Tsai model* [49] is most widely used in the fiber composites industry to predict the tensile modulus of unidirectional composites as a function of aspect ratio. The Halpin-Tsai model can deal with a variety of reinforcement geometries, including discontinuous filler reinforcement such as fiber-like or flake-like fillers. The Young's modulus of a composite material in the Halpin-Tsai model is proposed as

$$\frac{E_c}{E_m} = \frac{1 + \zeta \eta \phi_f}{1 - \eta \phi_f} \tag{5.2}$$

$$\eta = \left(\frac{E_f}{E_m} - 1\right) / \left(\frac{E_f}{E_m} + \zeta\right) \tag{5.3}$$

where E_c, E_f, E_m are Young's modulus of composites, inclusions and polymer matrix, respectively. ϕ_f is the filler volume fraction and ζ is a shape parameter dependent on filler geometry and loading direction. In particular, $\zeta = 2(l/d)$ for fibers or $2(l/t)$ for disk-like platelets. l, d, t are the length, diameter and thickness of the dispersed fillers.

As a matter of fact, 2-D disk-like clay platelets can make less contribution to modulus than 1-D fiber-like inclusion. A *modulus reduction factor* (MRF) [51] is introduced to modify the Halpin-Tsai model as

$$\frac{E}{E_m} = \frac{1 + \zeta(MRF)\eta\phi_f}{1 - \eta\phi_f} \tag{5.4}$$

5.2.5.2. Hui-Shia model

The *Hui-Shia model* [52,53] is developed to predict the tensile modulus of composites including unidirectional aligned platelets with the simple assumption of perfect interfacial bonding between the polymer matrix and platelets, which is given by

$$\frac{E_c}{E_m} = \frac{1}{1 - \dfrac{\phi_f}{4}\left[\dfrac{1}{\xi} + \dfrac{3}{\xi + \Lambda}\right]} \tag{5.5}$$

$$\xi = \phi_f + \frac{E_m}{E_f - E_m} + 3(1 - \phi_f)\frac{\left[(1-g)\alpha^2 - \dfrac{g}{2}\right]}{\alpha^2 - 1} \tag{5.6}$$

$$g = \frac{\pi}{2}\alpha \tag{5.7}$$

$$\Lambda = (1 - \phi_f)\left[\frac{3(\alpha^2 + 0.25)g - 2\alpha^2}{\alpha^2 - 1}\right] \tag{5.8}$$

where α is the inverse aspect ratio of dispersed fillers and $\alpha = t/l$ for disk-like platelets ($\alpha \leq 0.1$).

5.2.5.3. Voigt upper bound and Reuss lower bound

When $\xi \to \infty$, the Halpin-Tsai model equations reach the upper bound, which is normally called *Voigt rule of mixtures* (ROM) [54] where fiber and matrix have the same uniform strain (*i.e.*, isostrain approach):

$$E_c = \phi_f E_f + (1 - \phi_f)E_m \tag{5.9}$$

Conversely, when $\xi \to 0$, the Halpin-Tsai model equations reach the lower bound under equal stress (*i.e.*, isostress approach), the so-called *Reuss inverse rule of mixtures* (IROM) [55]:

$$\frac{1}{E_c} = \frac{\phi_f}{E_f} + \frac{(1 - \phi_f)}{E_m} \tag{5.10}$$

Both Voigt and Reuss models provide initial estimates of the upper and lower bounds of elasticity of multiphase composites with the only consideration of the inclusion volume fraction but irrespective of inclusion shape/geometry, orientation and spatial arrangement.

5.2.5.4. *OOF modeling*

A *two-dimensional object-oriented finite element code* (OOF2) version 2.0.3 [41], developed by the National Institute of Standards and Technology (NIST), USA, was installed in a Fedora Red Hat Core 4 Linux system. All finite element analyses were conducted on a Dell INSPIRON 8600 laptop with the dual boot system of Windows XP and Linux using 1.6 GHz processors and 2 GB of RAM.

OOF2 can correlate the heterogeneous material structure based on the real images with the bulk physical properties. Furthermore, OOF2 normally deals with a complex small-scaled and disordered system and includes details of microstructures such as particle size, shape and real orientation [42]. In general, the OOF modeling procedure consists of six major steps: (i) capture the real image, (ii) assemble the material properties for each constituent, (iii) assign the material properties to the microstructure represented by groups of pixels, (iv) generate the skeleton, the only geometry of FE mesh, (v) create FE mesh with boundary conditions and mathematical equations, (vi) solve the models.

For the purposes of this study, the computational tool OOF2 was used to predict the tensile moduli of PP/clay nanocomposites at various clay contents from 1 to 10 wt%. Since nanocomposites stretched by small strains (less than 1%) are still subject to elastic deformations, both clay particles and PP matrix were assumed to be linear elastic materials with perfect interfacial bonding between the two constituents. The material properties of the constituents are listed in Table 5.3. Typical OOF models of 5 wt% filled nanocomposites based on respective SEM and TEM morphological images are demonstrated in Figure 5.2. In terms of boundary conditions, as depicted in Figures 5.4a and 5.4b, both X and Y displacements U_x, U_y on the left boundary were set to zero (*i.e.*, fully constrained) and U_x on the right boundary was equivalent to 0.05 and 0.25% tensile strain, respectively. The tensile modulus was calculated from the tensile stress resulting from the sum of forces applied on the right boundary. The entire work assumed the unit thickness and plane stress for all OOF models.

Table 5.3. Material properties of constituents in PP/clay nanocomposites [40]

Property	PP matrix	Clay particles	References
Young's modulus (GPa)	1.66[a]	178[b]	[a][40]; [b][56]
Poisson's Ratio	0.35	0.20	[56]
Density (g/cm^3)	0.9	1.8	material data sheet

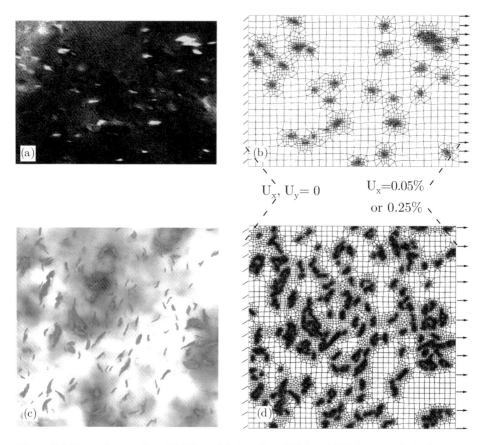

Figure 5.4. Typical examples of OOF models based on SEM and TEM morphological images of 5 wt% filled nanocomposites: (a) and (c) are selected pixels containing clay particles (highlighted areas); (b) and (d) show FE meshes [40]

5.2.6. *Comparisons of experimental results with the calculated values*

5.2.6.1. *Tensile modulus*

OOF modeling results based on SEM and TEM morphological images are demonstrated along with the experimental data and the theoretical composite models in Figure 5.5. Both modeling results and experimental data have been satisfied within the Voigt upper bound and Reuss lower bound, the general limits for the elasticity of the multi-phase composites. Modeling results obtained from both SEM and TEM images have a fairly good agreement with the experimental data despite the slight overestimate of tensile moduli, particularly for 1 and 10 wt% filled nanocomposites. The possible reason might be attributed to the fact that this study ignored MAPP as the compatibilizer and the interphases between clay particles and PP matrix, both of which could lead to the

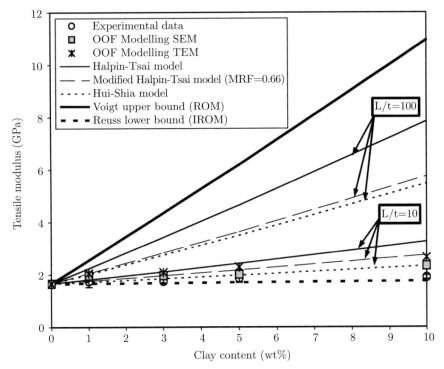

Figure 5.5. Prediction of tensile moduli of PP/clay nanocomposites at various clay contents [40]

different mechanical properties. In addition, very low contrast and poor phase distinction, more obvious in TEM images, are observed between clay particles and PP matrix. Hence, artifacts in image segmentation could also result in the inaccuracy of meshing and assignment of material properties based on the size and distribution of selected pixels. On the other hand, the microstructural region of interest in OOF modeling becomes more subjective for clay contents of 1 and 10 wt%, due to the non-uniform clay dispersion by initial pulse feeding and clay agglomeration, respectively [57].

The OOF modeling results are also compared with theoretical composite models at two fixed aspect ratios ($L/t = 10$ and $L/t = 100$), the thresholds of intercalated and exfoliated structures [58], respectively. It is suggested that a higher level of intercalation is more prevalent due to the much closer relationship of experimental and numerical results with the theoretical predictions at $L/t = 10$. Furthermore, the Hui-Shia model curve with $L/t = 10$ gives the best agreement with the experimental data and OOF modeling (SEM) results, in comparison to original and modified Halpin-Tsai models. It is most likely due to the assumption of perfect interfacial bonding in the Hui-Shia model. As expected, the modified Halpin-Tsai model curve agrees slightly better than that of the Halpin-Tsai model, fitting OOF modeling (TEM) results very well. Therefore, introducing an MRF of 0.66 [51] due to 2-D disk-like platelet geometry could be effective for the tensile modulus prediction of PP/clay nanocomposites.

5.2.6.2. *Effect of particle distribution*

In order to understand the deformation mechanism of PP/clay nanocomposites, stress contours (σ_{xx}) under a uniaxial applied strain of 0.25% in OOF modeling (SEM) work are demonstrated in Figure 5.6, which contains different particle distributions resulting from varied clay contents between 1 and 10 wt%. More evidently, higher stress has been observed within the clay particles or in the PP matrix immediately adjacent to the particles, which reveals the classical shear lag mechanism of the load transfer from the matrix to the particles. Furthermore, less variation of the stress distribution takes place in Figures 5.6b and 5.6c due to the uniform clay dispersion into the PP matrix in 3 and 5 wt% filled nanocomposites. On the other hand, it is expected that the stress concentrations appear at the sharp corners of clay particles along the axial loading direction. In addition, concentration of stress can also be seen in the clustered regions where the clay particles are close to each other. Typical region A in Figure 5.6d shows locally an intensified maximum stress of over 0.29 GPa for the matrix between two large clay clusters. As a result, the matrix in the clustered regions tends to yield at much lower levels of the macroscopic applied stress, which can inevitably cause the onset of clay particle cracking, initiating the damage of the nanocomposites [59,60].

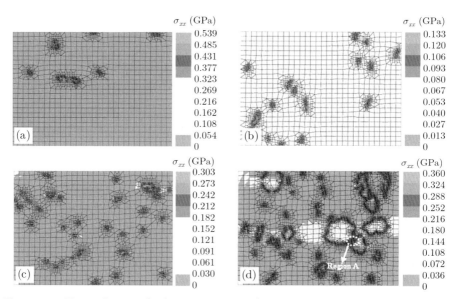

Figure 5.6. Uniaxial stress (σ_{xx}) contours of PP/clay nanocomposites with varied clay contents ($Ux = 0.25\%$ strain): (a) 1 wt%, (b) 3 wt%, (c) 5 wt%, and (d) 10 wt% [40]

The elastic strain contours are also clearly shown in Figure 5.7. Since the clay particles as platelet inclusions are over 100 times more rigid than the PP

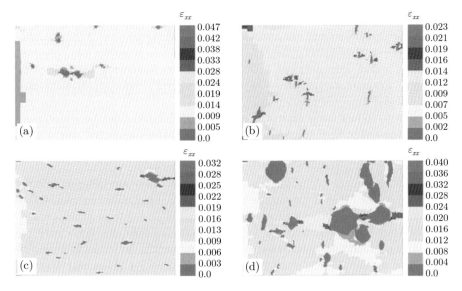

Figure 5.7. Elastic tensile strain (ε_{xx}) contours of PP/clay nanocomposites with varied clay contents ($Ux = 0.25\%$ strain): (a) 1 wt%, (b) 3 wt%, (c) 5 wt%, and (d) 10 wt% [40]

matrix, the elastic strain becomes much lower inside the clay particles indicative of the most constraints of the matrix by the surrounding particles [60]. Similarly, higher levels of elastic strain are detected around the sharp edges of the clay particles again parallel to the loading direction. The distribution of elastic strain in the homogeneous microstructures of 3 and 5 wt% filled nanocomposites is much more uniform than that of the clustered microstructure of 10 wt% filled nanocomposites. This finding, in view of the deformation mechanism, could explain the importance of clay dispersion in PP matrix as well as the deterioration effects of the clay clusters in nanocomposites.

5.3. The dispersion problem in the case of polymer-polymer nanocomposites

Let us now go back to the basic problem in manufacturing of polymer nanocomposites, namely, the dispersion of the reinforcing component to single nanoparticles. In this respect, it seems useful to cite here the opinion of Wendorff and Greiner [61] expressed in their excellent review of electrospinning as a method for preparation of ultrathin fibers and their application as reinforcing material for nanocomposites. After listing the advantages of the electrospun nanofibers as reinforcement over macroscopic fibers as owing to their extremely high aspect ratio, little refraction of light (due to the small diameters) resulting in transparent reinforced matrices, the authors [61] conclude: *"Given the advantages of nanofibers for reinforcement, the number of investigations on this topic is rather small. The main problems, to which there are no convincing solutions yet, are the dispersion of the electrospun nanofiber webs and the control of the nanofiber orientation in the polymer matrix. The felt-mat structure of the nonwoven is for the most part maintained upon the incorporation of the nanofibers into the matrix. Attempts to disperse single nanofibers from the nonwoven mats using ultrasound, kneaders, or high-speed stirrers have only been marginally successful. Very similar problems occurred in matrix reinforcement with carbon nanotubes and nanofilaments [62]".*

Recently, it has been shown that these problems could be overcome by applying the microfibril reinforced composites (MFC) concept, starting from polymer blends [63–74]. The MFC use *thermodynamically immiscible polymers* with different (at least 30°C) melting temperatures (T_m).

In contrast to the common macrocomposites where the reinforcement represents glass- or other fibers, in the case of MFC, the reinforcing role is played by microfibrils. Since these fibrils are not available separately as a form of material, they have to be created. The preparation of MFC includes three basic steps: (i) melt blending *via* extrusion of two immiscible polymers having different melting temperatures (*mixing step*); (ii) cold drawing of the extrudate with good molecular orientation of the two components (*fibrillation step*); (iii) thermal treatment at a temperature between the two melting temperatures of the two blend partners (*isotropization step*).

While during the second step (fibrillation) the two polymers are converted into a highly oriented state (*i.e.*, one deals with a highly oriented blend), the third step results in melting of the lower-melting component and its transformation into an isotropic matrix, reinforced with the microfibrils of the component with the higher T_m. Technologically, this transition to an MFC structure can take place during processing of the drawn blend *via* injection- or compression molding. The essential requirement is that the processing window is not too close to T_m of the microfibrils, otherwise they will melt and return to their original spherical shape.

The MFC concept is relatively new and seems to be a useful approach for the processing of polymer blends and manufacturing of polymeric materials and articles with environmentally friendly properties as no mineral reinforcement or additives are needed. Another characteristic feature of these materials is the extremely homogeneous distribution of the reinforcement in the matrix polymer, which might be of significant importance, depending on the end usage. The problems discussed earlier regarding the homogeneous dispersion of the nano-/micro-reinforcing elements do not exist in the case of the MFC approach because of the inherent nature of this method. The fine and homogeneous dispersion of the reinforcing fibrillar component in the matrix starts from the very beginning as spheres, and later as micro- and/or nanofibrils. In addition, the fibrils add the benefit of having extremely high aspect ratios (often between 100 and 1000). Last but not the least, in contrast to other polymer-polymer composites, MFC comprise only *flexible-chain* thermoplastics.

In order to obtain fibrils with diameters in the nano range, the idea was to use a more intensive melt blending technique and obtain as fine as possible spheres of the dispersed component, applied in conjunction with high draw ratios and the smallest possible extrusion die. These targets could be achieved by using commercial equipment for spinning of synthetic textile fibers.

It is evident that the suggested concept does not use starting nanomaterial which has to be blended with the matrix; instead, the reinforcing nanofibrils are created from the blending of two polymers, one of them later playing the role of reinforcement. In this way, the most common problem of dispersion, encountered in nanocomposite technology, is overcome.

It should be mentioned here that quite a similar approach for preparation of thermoplastic nanocomposites has been used by Wu *et al.* [75]. They compounded 75 wt% of high-density polyethylene (HDPE) and 25 wt% of polypropylene melt blended in an extruder. The resulting compound was fibrillated by drawing it through a pair of steel rollers at 138°C. The fibrillated tape was cut into small pieces and then re-molded at temperatures above the melting point of HDPE. Transmission electron microscopy results indicated that PP fibrils of about 30–150 nm in diameter were created in the HDPE matrix. The mechanical properties of HDPE were greatly improved with a yield stress above 60 MPa and a Young's modulus of 3.5 GPa [75]. While in this publication [75] a drastic improvement in the mechanical performance of the composite material was demonstrated, the morphological characteristics of the reinforcement were not illustrated in the same convincing way.

In the present chapter, in addition to the description of these polymer-polymer composites, it is demonstrated that the reinforcing nanofibrils can be isolated as a separate material and used for completely different purposes. This nanofibrillar composites (NFC) concept [76] allows one to transform any polymer into a fibrillar nanomaterial suitable for further processing and apply it for technical or biomedical purposes.

5.3.1. *Manufacturing of nanofibrillar polymer-polymer composites*

Commercial fiber grade PET and commercial grade PP (type Novolen with MFI 5, as well as Novolen 1100L, provided by Basell, Germany) were used as the reinforcing component and matrix material, respectively. The dried PET pellets were dry-mixed with PP to give a weight ratio of PP:PET as 80:20. The melt blending was performed in a twin-screw Brabender extruder. The temperature profile, starting from the feeding zone to the die, was 260, 270, 260 and 245°C. The extruded bristles (2 mm in diameter) were cooled down to room temperature and pelletized. To this PP/PET blend (80/20 by wt) were added 20 wt% chips of PP type Novolen 1100L so that the effective amount of PET in the blend was reduced to 16.7 wt%.

The new blend was subjected to spinning on commercial equipment used for the manufacturing of synthetic fibers, at the Thuringian Institute for Textile and Plastics Research, Rudolstadt, Germany. In this way filaments of 58.5 dtex f12 (30.3 μm in diameter) were prepared *via* drawing (1:3.47) at 130°C from 300.0 dtex f12. Subsequently yarns comprising many single filaments were prepared and used for knitting, which was performed on a 10 gauge Dubied v-bed knitting machine, producing a piece of balanced fabric with 1×1 rib structure. Samples of the knitted nanofibrillar fabric were then compression molded in a hot press at 180°C and 650 kPa pressure, in order to make the PP matrix isotropic and form sheets of NFC.

Scanning electron microscopic observations were performed on cryofractures as well as on samples with PP removed with boiling xylene using the recently modified Soxhlet apparatus [77].

Thermal analysis was carried out using differential scanning calorimetry (DSC) from –50 to 280°C at a scan rate of 10°C/min and the degree of crystallinity, w_c (DSC), was calculated by means of the following equation:

$$w_c \text{ (DSC)} = \Delta H_{exp}/\Delta H_{id} \qquad (5.11)$$

where ΔH_{exp} and ΔH_{id} are the measured and the ideal (for 100% crystalline sample) values of the fusion heat, respectively. For PP and PET, the values of ΔH_{id} were adopted as 209 and 140 J/g, respectively [78].

5.3.2. *Nanofibrillar vs. microfibrillar polymer-polymer composites and their peculiarities*

Figure 5.8 shows the photographs (Figures 5.8a, d and e) as well as the SEM images (Figures 5.8b, c and f) of the various stages of preparation of MFC and NFC. Figure 5.8a illustrates the cross-plied bristles after drawing being the precursor for MFC and Figure 5.8d shows the textile fibers after spinning (used later for NFC preparation), while Figure 5.8e depicts the knitted material thereof. The last three images (Figures 5.8b, c and f) are taken from the surfaces of the cryofractured compression molded samples at various magnifications, from the samples with MFC (Figures 5.8b and c) and with NFC structure (Figure 5.8f).

The fact that the blend PP/PET (83.3/16.7 by wt) is extruded in the shape of fine commercial textile fibers (with diameter of 30 µm, Figure 5.8d) allows the preparation of a prepreg by means of various commonly used textile techniques such as knitting (Figure 5.8e), weaving, non-woven textile, *etc.*, which can be further processed by applying a typical manufacturing method, such as compression molding (CM).

It is important to note here that the materials displayed in Figures 5.8a, d and e consist of *highly drawn blends* but by no means do they represent a composite material. The isotropic state (matrix) is obtained only after thermal treatment at temperatures between the two melting temperatures of two blend constituents (in the present case PP and PET with $T_m = 170$ and 250°C, respectively). As a result of such treatment, the lower-melting component (PP) is converted into an isotropic matrix reinforced with PET fibrils. Such a composite structure is demonstrated in Figures 5.8b, c and f at different magnifications.

From the same micrographs it can be seen that the PP matrix is reinforced with a large amount of fibrils with diameters of a couple of microns (in the case

Organoclay, Particulate, and Nanofibril Reinforced Composites

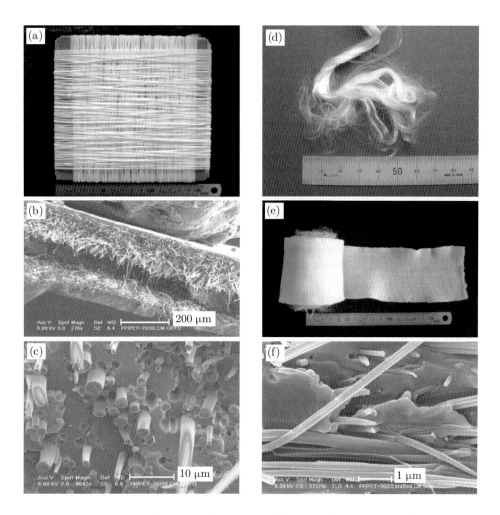

Figure 5.8. Photographic (a, d and e) and SEM (b, c and f) images of PP/PET (80/20 by wt) after: (a) and (d) blending and spinning, (e) knitting, (b), (c) and (f) cryofracture of compression molded knitted composite

of MFC, Figures 5.8b and c) or of around 100 nm (Figure 5.8f), *i.e.*, one is really dealing with a nanofibrils reinforced polymer composite in the second case. Furthermore, it is clearly illustrated in the last image (Figure 5.8f) that the nanofibrils are placed in a rather regular pattern, being perpendicular to each other and thus reflecting the knitted character of the reinforcement (Figure 5.8e). What is of particular importance is the fact that bundle-like aggregates of microfibrils (Figures 5.8b and c) or nanofibrils, (Figure 5.8f) are not observed, *i.e.*, each fibril is individually surrounded by matrix material. In other words, *a very good distribution of even the nanosized reinforcing material* has been achieved, which is in contrast to the cases with many other nanomaterials used

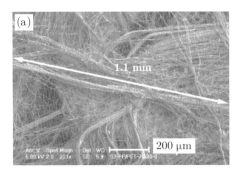 

Figure 5.9. SEM micrographs of PET reinforcing nanofibrils after selective removal of the PP matrix demonstrating the sizes of nanofibrils: (a) the length, (b) the diameter

for manufacturing polymer nanocomposites. There is hardly any need to stress the importance of this novelty of the developed polymer-polymer nanocomposites.

Another important characteristic of the nanofibrils is their *aspect ratios*, *i.e.*, length to diameter ratios. From the images displayed in Figure 5.9 it may be concluded that the reinforcing nanofibrils are distinguished by an extremely high aspect ratio – above 7 000 in the present case (Figure 5.9a). Such an aspect ratio surpasses even that of carbon nanotubes, reported to be in the vicinity of 1000 [1].

The micrographs in Figure 5.9 allows another important conclusion to be made supporting the statement regarding the desirable distribution of nanofibrils in the matrix. These images have been taken after the selective dissolution of the matrix (PP) with boiling xylene [77]. It is clear that the remaining PET nanofibrils are completely loose, separated from each other (Figure 5.9b), which verifies that they are actually individually surrounded by the matrix material, *i.e.*, no bundles or aggregates of them are present.

5.4. Directional, thermal and mechanical characterization of polymer-polymer nanofibrillar composites

5.4.1. *Directional state of NFC as revealed by wide-angle X-ray scattering*

Figure 5.10 summarizes the patterns of wide-angle X-ray scattering (WAXS) images of non-drawn and drawn samples of neat PP and its blend with PET after spinning and compression molding of the knitted sample. It is evident that after extrusion, PP is in a completely isotropic state, the well defined crystalline reflections represent isointensity circles, *i.e.*, there is no indication of any preferred orientation. This situation changes completely after cold drawing of the extruded PP fibers when rather perfect uniaxial orientation takes place – the circular reflections (Figure 5.10a) dissipate into multiple distinct spots instead (Figure 5.10b).

With the PP/PET (83.3/16.7 by wt) blends similar pictures are observed but in the opposite sequence. The spun filaments of the blend (Figure 5.10c) show

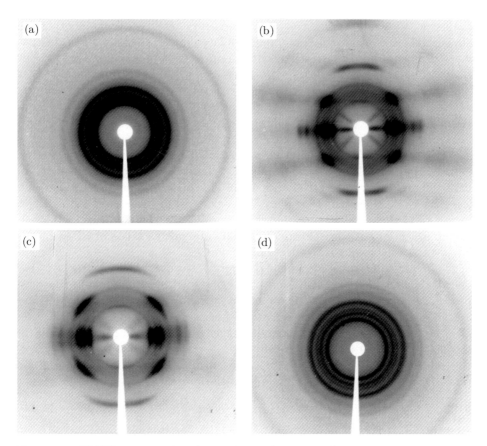

Figure 5.10. WAXS patterns taken from undrawn (as extruded) and cold drawn samples of neat PP and its blend with PET: (a) PP as extruded, (b) PP drawn, (c) PP/PET (73.3/16.7 by wt) spun (textile filament), (d) knitted and compression molded PP/PET (73.3/16.7 by wt) filament followed by annealing at 145°C

the same very high orientation as the drawn homopolymer PP (Figure 5.10b), which disappears after compression molding of the knitted sample (Figure 5.10d) and the scattering pattern seems to be identical to that of the non-drawn PP (Figure 5.10a). Obviously, the melting of PP during the CM results in an isotropization of PP.

With respect to the observation of any lack of orientation after CM of the knitted sample (Figure 5.10d) it seems important to note here that the knitting process also contributes to this type of WAXS pattern. WAXS inspection of the knitted PP/PET yarn (before compression molding) shows the same pattern as that displayed in Figure 5.10d, *i.e.,* an absence of any orientation effect regardless of the fact that any thermal treatment was applied and the two yarn components are highly uniaxially oriented.

While this observation seems strange at first glance, the reason for it is related to the fact that the WAXS reflects the orientation state of the molecular chains, which while having been stretched and aligned with one another are at the same time essentially randomly displaced due to the knitting of the filaments. In other words, one still deals with orientation on a molecular level but also with isotropization on a macro level, *i.e.*, a pseudo isotropization takes place. The same phenomenon has been observed for injection molded samples with MFC structure when the highly uniaxially oriented starting blend is converted into an isotropic matrix reinforced with randomly distributed microfibrils being responsible for the isotropic WAXS after injection molding [71].

The WAXS patterns displayed in Figure 5.10 possess another peculiarity, namely, there does not appear to be any well defined reflections arising from PET. There are at least three reasons for this observation: (i) there is much less PET than PP in the blend (only 16.6 wt%), (ii) PET crystallizes to a lesser extent than PP, *i.e.*, it dominates the amorphous phase, and (iii) PET crystalline reflections are in the same angular intervals as those of PP, *i.e.*, the latter mask the weak PET reflections.

Summarizing the WAXS results, it is worth to note that in the case of NFC one deals with the situations (and changes during the manufacturing and processing steps) that are already well documented for MFC [63–74]. These are: (i) a lack of orientation immediately after extrusion, (ii) perfect orientation after cold drawing (necking), and (iii) regeneration of the isotropic state of the matrix after processing, as well as (iv) eventual pseudo isotropization if the reinforcing microfibrils loose their parallel configuration during the processing stages, as the knitting does in the present case.

5.4.2. *Thermal characterization of NFC*

In addition to the orientational characterization *via* WAXS, some information regarding the phase state and the degree of crystallinity was obtained from the DSC analysis. The measurements were performed on the starting components and the intermediate and final products, as well as on some of them after annealing. The evaluated data are summarized in Table 5.4.

In the first approximation, DSC curves indicated that the two polymers are in a semi-crystalline state regardless of whether they are in a neat or blended

Table 5.4. Melting temperature, T_m, and degree of crystallinity, w_c(DSC), of the starting, intermediate and final NFC products as evaluated from the DSC curves

Sample	Melting temperature (°C)		Crystallinity (%)	
	PP	PET	PP	PET
PP	168	–	70.4	–
PET-pellets	–	231/246	–	24.5
Undrawn blend fiber	164.9	245	50.0	38.1
Drawn blend fiber	167.1	245	71.5	39.9
Hot pressed knitted plate (NFC)	164.7	246.1	72.7	27.5

state. While the neat PET demonstrates some additional crystallization during heating in the DSC instrument, it shows a very poor crystallinity, similarly to the situation after compression molding, contrasting to the drawn and non-drawn blends (Table 5.4). This observation suggests that PET does not reach a substantial level of crystallinity during melt cooling after compression molding (the isotropization step of the NFC preparation), Table 5.4. It is well known that PET crystallizes very slowly and simple cooling of the melt does not result in a well developed crystal structure, as already reported for MFC based on LDPE/PET [71].

It should be noted that the presence of an amorphous phase was not resolved *via* its glass transition temperature, T_g, even for the PET component. The reason for this observation can be related to the relatively small amount of PET in the blends under investigation (20 wt%) and the available crystallinity (around 25%, Table 5.4).

Summarizing the results of thermal characterization by means of DSC samples with NFC structure and their precursors, it should be noted that the situation is quite similar to the case of MFC samples and their precursors based on the same blend partners, PP and PET [79].

5.4.3. *Mechanical properties of NFC*

Concerning the mechanical performance of PP/PET nanofibrillar composites, it should be stressed that dealing with a perfect distribution of the reinforcing nanofibrils in the matrix (Figure 5.8), as characterized by a superior aspect ratio (Figure 5.9a), one could expect to see a very high reinforcing effect in these composites. Figure 5.11 shows the results of tensile testing of neat PP, knitted PP/PET (83.3/16.7 by wt) and knitted PP/glass fibers (GF) (80/20 by wt) samples, all of them after compression molding of plates at the same processing conditions that were described above. The more precise data of the same measurements are summarized in Table 5.5.

Some improvement of the tensile strength and the tensile modulus can be observed for the NFC as compared to the neat PP samples (Table 5.5). The values of 50% and 22% (for E and σ, respectively) are much smaller than that for the PP/GF composites. Before discussing the possible reasons for not achieving the expected large improvements as reported for the LDPE/PP nanocomposites [75], let us note a peculiarity within the results displayed in Figure 5.11 and Table 5.5.

Comparing the results of the neat PP and PP/PET nanocomposites with the PP/GF composites, a striking conclusion can be drawn. In the latter case, one deals with an extremely large error range (Figure 5.11), indicating that some of the PP/GF samples have the same mechanical performance as that of the PP/PET nanocomposites where the error range is negligibly small. This difference in the mechanical behavior of the two types of composites can be explained by the different quality of the reinforcement distribution and also the wetting quality – in the present nanocomposite both of them are perfect and therefore there is no difference from specimen to specimen during testing.

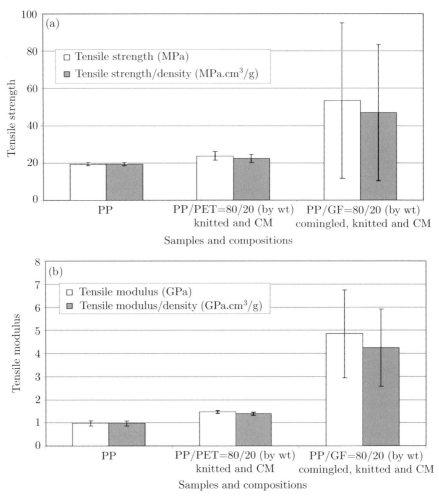

Figure 5.11. Nominal and specific (shadowed columns) values of the tensile strength (a) and tensile modulus (b) of compression molded plates of neat PP, blend of PP/PET (80/20 by wt) with NFC structure, and PP/glass fiber (GF) (80/20 by wt)

Table 5.5. Tensile strength, σ, and tensile modulus, E, (and their specific values, σ/ρ and E/ρ) of knitted and compression molded neat PP, PP/PET nanofibrillar composite and PP/glass fiber (GF) composite.

Sample	Composition PP/PET/GF (by wt)	σ (MPa)	E (GPa)	σ/ρ (MPa.cm³/g)	E/ρ (MPa.cm³/g)
PP	100/0/0	19.4±0.8	0.97±0.10	19.4±0.8	0.97±0.10
PP/PET	80/20/0	23.7±2.3	1.47±0.06	22.3±2.2	1.39±0.06
PP/GF	80/0/20	53.5±41.7	4.85±1.91	46.9±36.6	4.25±1.67

Regarding the difference in the mechanical performance of the two types of nanocomposites, the LDPE/PP [75] and the PP/PET, one can assume at least three reasons for this difference. These are related to the nature of the matrix and reinforcement and their amounts. In the case of LDPE/PP the reinforcing component is some 50% more, *i.e.*, 25 wt% against 16.7 wt% of the PP/PET composite.

The second reason for this could be the adhesion quality on the interphase boundary. One can expect much better interaction between LDPE and PP as compared to the PP and PET. For example, the authors [75] demonstrated an intensive presence of transcrystalline layers of LDPE on the PP nanofibrils, which considerably contribute to the improvement of the matrix-reinforcement adhesion and thus enhance the final mechanical properties. Using TEM, the same phenomenon was observed in an earlier study [80,81] on stained thin slices of injection molded samples of LDPE/PET with MFC structure. What is more, the transcrystallization phenomenon is less typical for MFC based on PP/PET [82]. This observation indicates that LDPE is much more inclined to form transcrystalline layers on the same type of fibrils (PET) than PP is.

Finally, the mechanical performance of a composite material or more specifically, the reinforcing effect as compared to the neat matrix depends on the type of matrix used. In the cases under discussion, PP is characterized by superior mechanical properties when compared to LDPE. The importance of the particular matrix applied was already demonstrated using the same reinforcement (PET) for preparation of MFC with LDPE or PP matrix. In the case of LDPE/PET (70/30 by wt) the injection molded samples with MFC structure showed tensile properties comparable with those of short glass fiber reinforced (30 wt%) LDPE [71]. Further, the Young's modulus was up to 5 times larger and the improvement of the tensile strength was up to 10 times greater compared to the neat LDPE [73]. At the same time, the MFC based on PP/PET (70/30 by wt) showed only about 50% improvement in the mechanical properties as compared to the neat PP [72]. The conclusion is that the poorer the mechanical performance of the matrix used, the stronger the reinforcing effect.

Analysing the performance of the NFC with the glass fiber reinforced samples (Figure 5.11 and Table 5.5), the following observation should be stressed upon. Comparing the nominal values of the tensile strength and tensile modulus with their specific values one finds that σ for the glass fiber reinforced material drops by 15%, while for the sample with NFC structure it practically does not change (Figure 5.11, Table 5.5). The same trend is observed for E, although the absolute change is slightly smaller. In this way, the difference in the mechanical performance between the polymer-polymer nanofibrillar composites and glass fiber reinforced composites is getting smaller in favor of NFC (for σ the difference between the nominal values is 29.8 MPa and between the specific values is 24.6 MPa, *i.e.*, a drop of some 20%, as also in the case of E values is observed). Interestingly, this fact represents one of the main advantages of polymer-polymer composites – in addition to the fact that they are more environmentally friendly, they do not show a drop in their specific properties as compared to the nominal ones.

Concluding this discussion on the mechanical behavior of NFC, it should be stressed that an improvement of 22% (for σ) and 50% (for E) is quite typical for all polymer nanocomposites [83]. According to Bousmina [84], only in exceptional circumstances can one observe an improvement greater than 30% in the mechanical performance of nanocomposites, including their barrier properties [84,85,88]. More specifically, Zhang et al. [17] studied thoroughly the case of PP/SiO_2 nanocomposites in which nanoparticles have been coated by various polymers to improve the interfacial adhesion. They reported mechanical properties (Young's modulus and tensile strength) only 20–25% higher than those of the neat PP [17].

The above conclusions are supported by the statements of Schaefer and Justice in their recent review "How Nano Are Nanocomposites?" [86]: "*Composite materials loaded with nanometer-sized reinforcing fillers are widely believed to have the potential to push polymer mechanical properties to extreme values. Realization of anticipated properties, however, has proven elusive.*

Such systems have attracted enormous interest from the materials community because they theoretically promise substantial improvement of mechanical properties at very low filler loadings. In addition, nanocomposites are compatible with conventional polymer processing, thus avoiding costly layup required for the fabrication of conventional fiber-reinforced composites. The appeal of nanocomposites is illustrated by considering single-walled carbon nanotubes (SWCNTs). With tensile moduli in the terapascal range and lengths exceeding 10 µm, simple composite models predict order-of-magnitude enhancement in modulus at loadings of less than 1%.

Introductory paragraphs similar to the above can be found in hundreds of nanocomposite papers. With the exception of reinforced elastomers, nanocomposites have not lived up to expectations. Although claims of modulus enhancement by factors of 10 exist, these claims are offset by measurements that show little or no improvement...The lackluster performance of nanocomposites [87] has been attributed to a number of factors including poor dispersion, poor interfacial load transfer, process-related deficiencies, poor alignment, poor load transfer to the interior of filler bundles, and the fractal nature of filler clusters" [86].

5.5. Potentials for application of nanofibrillar composites and the materials developed from neat nanofibrils

The approach described for the preparation of NFC offers the opportunity to isolate neat nanofibrils of any immiscible polymer blend in a variety of material shapes. To do so one needs to treat the drawn blend before or after compression molding for extraction of the matrix component by means of a selective solvent.

The poor solubility of polymers, particularly at room temperature, is a very well known fact. The most frequently used instrument for such extractions, the Soxhlet apparatus, was recently modified [77] in such a way that the poly-

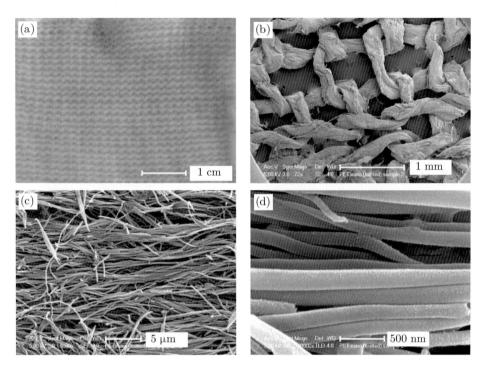

Figure 5.12. Photograph of knitted PP/PET (73.3/16.7 by wt) filaments (a) and SEM micrographs (b–d) of textile from PET nanofibrils at various magnifications of the same spot

mer extraction takes place with boiling solvent. In this way, the extraction can be completed in minutes instead of hours or days.

Figure 5.12 shows the photographic picture of the knitted sample (Figure 5.12a) and the SEM images (Figures 5.12b–d) of the fabrics from PET nanofibrils. All the SEM micrographs are taken from the same spot using increasing magnifications, and are from knitted fabrics of PP/PET (83.3/16.7 by wt) filaments (Figure 5.8d) after extraction by means of the modified Soxhlet apparatus [77] and using xylene as the solvent. The extraction time in this particular case was 30–40 min.

The knitted character of the fabrics is quite evident (Figures 5.12a and b) as well as the nanosizes of the constituent fibrils (Figures 5.12c and d). Note that in the vast amount of nanofibrils there are no bundles of interconnected (stuck together) nanofibrils. This observation indicates once again the fact that the nanofibrils are distributed extremely uniformly throughout the matrix, *i.e.*, each individual fibril is surrounded by matrix material.

Regarding the potential applications of these nanofibrillar materials it should be noted that being polymer-polymer composites, *i.e.*, free of mineral fillers, NFCs seem to be a challenge for the car manufacturing industry, particularly in Europe, due to the initiation of environmental regulations and legislation such as the European Union End-of-Life Vehicles regulation [89].

Starting from 2015, the regulation allows for a incineration quota of only 5% of disused cars [89].

Further, as mentioned above, the suggested approach for the preparation of NFC offers the possibility of isolating of nanofibers as a separate material, which has many opportunities for biomedical and technical applications. In the latter case, the preparation of gas and/or liquid nanofilters as non-woven textiles is a possibility.

Much more challenging seems to be their biomedical application, for instance, in the field of regenerative medicine. The need for scaffold materials suitable for tissue engineering is growing. Taking into account the fact that the scaffold performance depends on the surface of the scaffolding materials, nanofibrils are an excellent candidate for this job. In addition, the NFC approach can be applied to biodegradable and biocompatible polymers. Such experiments started some years ago [74].

The experience gained with micro- and nanofibrillar scaffolds based on PET [90] was further used to develop poly(lactic acid) (PLA)-based microfibrillar scaffolds. Following the basic concept of the MFC approach, an appropriate second blend component was selected. It should have a melting temperature close to that of PLA and be thermodynamically non-miscible with PLA, exhibiting good drawing properties. Poly(vinyl alcohol) (PVA) was found to be a suitable candidate. After melt blending, drawing was performed in a continuous way and the obtained fine filaments were submerged in water for a couple of days on a filter paper [91]. The residue looked like a thin film, whose SEM inspection showed unexpected results regarding the intimate structure. The SEM micrographs at lower magnifications indicated that the "film" has a thickness of some tenths of microns. The observation at higher magnification revealed that the structure of the "film" as comprising microfibrils of relatively small length (around 20 µm). At the same time, the microfibrils had an approximate diameter of about 1 µm for the thickest, while the finest ones approach the typical nanosizes [90].

The most surprising finding was made at an even higher magnification, when it was established that these microfibrils are interconnected, thus forming a three-dimensional network (Figure 5.13a). Such a spatial arrangement of microfibrils results in the formation of an extremely porous material, where the pores have sizes in the nano range (Figure 5.13a). In other words, in the present case one is dealing with biodegradable (PLA-based) nanoporous scaffolds distinguished by an extremely high specific surface. What is of particular importance in this case is the fact that the scaffolds were prepared in a way excluding any contact with organic solvents or other toxic components since the extraction of PVA was carried out by means of water only.

The results of the biomedical testing with living cells are rather promising. The living cells attach well to the scaffold surface (Figure 5.13b), proliferate and grow further forming a continuous tissue [90].

Another possible application in the same area is the use of nanofibrils as carriers for controlled drug delivery. For this purpose, the fibrils should have

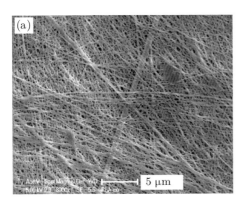 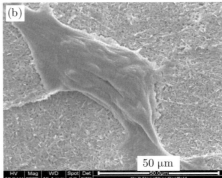

Figure 5.13. SEM images of nanofibrillar nanoporous PLA-based scaffold (a) and a stage of the cells growth on the same scaffold (b)

a rough surface (or even better, be porous) in order to enhance the drug loading capabilities. Again, the fine sizes and huge surface area relative to volume that is available on a nanofibril for drug loading make it an attractive option.

5.6. Conclusions and outlook

Object-oriented finite element analysis, OOF, has been successfully implemented to predict the tensile moduli of PP/clay nanocomposites and evaluate the particle distribution effect on the deformation mechanism under the linear elastic condition. OOF modeling results and experimental data show the predominant intercalated structures with the best agreement with the Hui-Shia model. Both SEM and TEM morphological image approaches are found to be effective in the tensile modulus prediction in spite of the relative overestimate of TEM images due to the artifacts and phase distinction problem. The study of particle distribution in the microstructures suggests that uniform dispersion of clay particles is more favorable and clay clusters could lead to localized higher stress concentration, thus making the nanocomposites more prone to crack initiation.

The Taguchi DoE method is found to be a very effective approach for selecting suitable manufacturing conditions in order to enhance the mechanical properties of such nanocomposites. Recompounding at a relatively low screw speed can regain or improve their overall mechanical properties. Furthermore, the preferred combinations of factors to enhance each of tensile, flexural and impact properties are determined using Pareto ANOVA analysis. Evidently, the PP type used plays the most significant role in the enhancement of overall mechanical properties, particularly low viscosity PP-H380F being the most favorable grade. In addition, clay content is detected to be the second significant factor for enhancing tensile and flexural properties. The increase of the clay content mainly allows for moduli improvement, and the impact strength is conversely reduced due to clay aggregation. Both the clay type and MAPP content are determined as two non-significant factors in relation to the tensile and flexural properties. Due to the greater interlayer spacing for better intercalation, DK4 organoclays could be selected as the preferable clay type in future work.

Using the experience gained during the development of MFC, another type of polymer-polymer composite, the nanofibrillar composites could be developed. For this purpose, two immiscible polymers, PP and PET in the present case, were intimately melt blended in a twin screw extruder and further spun as textile fibers with a diameter of 30 μm. After thermal treatment of knitted samples above the T_m of PP but below the T_m of PET, one obtains a composite structure where the isotropic PP is reinforced with PET nanofibrils with diameters of 50–150 nm and an aspect ratio of up to 7000. The NFC shows an improvement of Young's modulus of 50% and tensile strength of 22% as compared to the neat PP.

SEM observations on cryofractures show a perfect distribution of nanofibrils in the matrix and complete absence of fibrillar aggregates, in contrast to the case of any other nanocomposites.

In the present study the composite materials are obtained only by means of compression molding, but the drawn blend comprising nanofibrils can be processed similarly to MFC *via* injection molding in order to obtain articles with NFC structure.

As far as the advantages of this approach for NFC preparation are concerned, one must mention the following peculiarities. In contrast to the other techniques used in composite preparation by means of textiles as reinforcement, in the present case the problems arising from the melt quality do not exist, since a perfect wetting has been achieved during melt blending. Even textiles from commingled yarn do not offer such superior melting and homogeneous distribution of the reinforcing nanofibrils. Also, this technique presumes the availability of yarn from nanofibrils in order to prepare a commingled material, which one cannot always easily obtain. Even the application of electrospun fibers having nanosizes does not solve the problem, since they have to be dispersed into the matrix just like any other nanomaterial that is used as a filler [61].

In summary, the advantages of polymer-polymer composites with MFC or NFC structure that have been developed include the following: reinforcement of polymer by polymer, no mineral additives, no need of extra compatibilizers (for condensation polymers), easy processing, reduction in weight, synergetic effect in mechanical properties, complete regeneration and renewing of process, applicability for recycling purposes, perfect distribution of nanofibrils in NFC (no aggregates), possibilities for isolation of nano- or microfibrils and articles thereof.

Acknowledgments

The authors would like to acknowledge the financial support of the Foundation for Research, Science and Technology of New Zealand. One of us (S. F.) appreciates also their support in making possible his sabbatical stay at the Department of Mechanical Engineering and the Centre for Advanced Composite Materials of The University of Auckland, Auckland, New Zealand, where this chapter was prepared.

References

1. Schulte K, Gojny F, Fiedler B, Sandler J and Bauhofer W (2005) Carbon nanotube-reinforced polymers: a state of the art review, in *Polymer composites: from nano- to macro-scale* (Eds. Friedrich K, Fakirov S and Zhang Z) Springer, New York, pp. 3–23.
2. Zilg C, Mulhaupt R and Finter J (1999) Morphology and toughness/stiffness balance of nanocomposites based upon anhydride-cured epoxy resins and layered silicates, *Macromol Chem Phys* **200**:661–670.
3. Zilg C, Thomann R, Finter J and Mulhaupt R (2000) The influence of silicate modification and compatibilizers on mechanical properties and morphology of anhydride-cured epoxy nanocomposites, *Macromol Mater Eng* **280**:41–46.
4. Lan T and Pinnavaia T J (1994) Clay-reinforced epoxy nanocomposites, *Chem Mater* **6**:2216–2219.
5. Sreekala M and Eger C (2005) Property improvements of an epoxy resin by nanosilica particle reinforcement, in *Polymer composites: from nano- to macro-scale* (Eds. Friedrich K, Fakirov S and Zhang Z) Springer, New York, pp. 91–106.
6. Yoshida M, Lal M, Kumar N D and Prasad P N (1997) TiO_2 nano-particle-dispersed polyimide composite optical waveguide materials through reverse micelles, *J Mater Sci* **32**:4047–4051.
7. Sumita M, Okuma T, Miyasaka K and Ishikawa K (1982) Effect of ultrafine particles on the elastic properties of oriented low-density polyethylene composites, *J Appl Polym Sci* **27**:3059–3066.
8. Shang S W, Williams J W and Soderholm K J M (1992) Using the bond energy density to predict the reinforcing ability of a composite, *J Mater Sci* **27**:4949–4956.
9. Rong M Z, Zhang Y X, Zheng H M, Walter R and Friedrich K (2001), Structure-property relationships of irradiation grafted nano-inorganic particle filled polypropylene composites, *Polymer* **42**:167–183.
10. Godovski D Y (1995) Electron behaviour and magnetic properties of polymer nanocomposites, *Adv Polym Sci* **119**:79–122.
11. Nakao Y (1995) Noble metal solid sols in poly(methyl methacrylate), *J Colloid Interf Sci* **171**:386–391.
12. Griffiths C H, O'Horo M P and Smith T W (1979) The structure, magnetic characterization, and oxidation of colloidal iron dispersions, *J Appl Phys* **50**:7108–7115.
13. Gonsalves K E, Carlson G, Chen X, Gayen S K, Perez R and Jose-Yacaman M (1995) Surface functionalized nanostructured gold/polymer composites, *Polym Mater Sci Eng* **73**:298–302.
14. Liu H (1997) Functional and structural investigation of nanometer particles/epoxy resin composites, *MSc Thesis*, Zhongshan University (in Chinese).
15. Giannelis E P (1992) A new strategy for synthesizing polymer-ceramic nanocomposites, *JOM* **44**:28-30.
16. Novak B M (1993) Hybrid nanocomposite materials – between inorganic glasses and organic polymers, *Adv Mater* **5**:422–433.
17. Zhang M Q, Rong M Z and Friedrich K (2005) Application of non-layered nanoparticles in polymer modification, in *Polymer composites: from nano- to macro-scale* (Eds. Friedrich K, Fakirov S and Zhang Z) Springer, New York, pp. 25–44.

18. Xu W, Huang R, Cai B and Fan W (1998) Nano-CaCO$_3$ filled HDPE composites, *Chin Plast* **12**:30-37 (in Chinese).
19. Okada A, Kawasumi M, Usuki A, Kojima Y, Kurauchi T and Kamigaito O (1990) Synthesis and properties of nylon–6/clay hybrids, in *Polymer based molecular composites* (Eds. Schaefer D W and Mark J E) MRS Symposium Proceedings, Pittsburgh, USA, vol. 171, pp. 45–50.
20. Giannelis E P (1996) Polymer layered silicate nanocomposites, *Adv Mater* **8**:29–35.
21. Vaia R A, Price G, Ruth P N, Nguyen H T and Lichtenhan J (1999) Polymer/layered silicate nanocomposites as high performance ablative materials, *Appl Clay Sci* **15**:67–92.
22. Giannelis E P (1998) Polymer-layered silicate nanocomposites: synthesis, properties and applications, *Appl Organomet Chem* **12**:675–680.
23. Messersmith P B and Giannelis E P (1995) Synthesis and barrier properties of poly(ε-caprolactone)-layered silicate nanocomposites, *J Polym Sci Part A* **33**:1047–1057.
24. Kojima Y, Usuki A, Okada A, Kurauchi T and Kamigaito O (1993) Mechanical properties of nylon 6-clay hybrid, *J Mater Res* **8**:1179–1184.
25. Gilman J W (1999) Flammability and thermal stability studies of polymer-layered silicate (clay) nanocomposites, *Appl Clay Sci* **15**:31–49.
26. Privalko V P, Shantalii T A and Privalko E G (2005) Polyimides reinforced by a sol-gel derived organosilicon nanophase: synthesis and structure-property relationships, in *Polymer composites: from nano- to macro-scale* (Eds. Friedrich K, Fakirov S and Zhang Z) Springer, New York, pp. 63–76.
27. Lertwimolnun W and Vergnes B (2005) Influence of compatibilizer and processing conditions on the dispersion of nanoclay in a polypropylene matrix, *Polymer* **46**:3462–3471.
28. Lei S G (2003) Formulation and mechanical properties of polypropylene nanocomposites, *MSc Thesis*, Concordia University, Department of Mechanical and Industrial Engineering.
29. Hasegawa N, Kawasumi M, Kato M, Usuki A and Okada A (1998) Preparation and mechanical properties of polypropylene-clay hybrids using a maleic anhydride-modified polypropylene oligomer, *J Appl Polym Sci* **67**:87–92.
30. Liu X H and Wu Q J (2001) PP/clay nanocomposites prepared by grafting-melt intercalation, *Polymer* **42**:10013–10019.
31. Wang Y, Cheng F B and Wu K C (2004) Twin-screw extrusion compounding of polypropylene/organoclay nanocomposites modified by maleated polypropylenes, *J Appl Polym Sci* **93**:100–112.
32. Perrin-Sarazin F, Ton-That M-T, Bureau M N and Denault J (2005) Micro- and nano-structure in polypropylene/clay nanocomposites, *Polymer* **46**:11624–11634.
33. Buryachenko V A, Roy A, Lafdi K, Anderson, K L and Chellapilla S (2005) Multi-Scale Mechanics of Nanocomposites Including Interface: Experimental and Numerical Investigation, *Compos Sci Technol* **65**:2435–2465.
34. Katti K S, Sikdar D, Katti D R, Ghosh P and Verma D (2006) Molecular Interactions in Intercalated Organically Modified Clay and Clay-Polycaprolactam Nanocomposites: Experiments and Modeling, *Polymer* **47**:403–414.
35. Zhu L J and Narh K A (2004) Numerical Simulation of the Tensile Modulus of Nanoclay-Filled Polymer Composites, *J Polym Sci* **42**:2391- 2406.
36. Fertig III R S and Garnich M R (2004) Influence of Constituent Properties and Microstructural Parameters on the Tensile Modulus of Polymer/Clay Nanocomposites, *Compos Sci Technol* **64**:2577–2588.

37. Sheng N (2002) Micro/Nanoscale Modeling of Anisotropic Mechanical Properties of Polymer/Layered-Silicate Nanocomposites, *MSc Thesis*, Massachusetts Institute of Technology, Cambridge, USA.
38. Sheng N, Boyce M C, Parks D M, Rutledge G C, Abes J I and Cohen R E (2004) Multiscale micromechanical modeling of polymer nanocomposites and the effective clay particle, *Polymer* **45**:487–506.
39. Liu Y J and Chen X L (2003) Evaluations of the Effective Material Properties of Carbon Nanotube-Based Composites Using a Nanoscale Representative Volume Element, *Mech Mater* **35**:69–81.
40. Dong Y and Bhattacharyya D (2008) Morphological-image analysis based numerical modelling of organoclay filled nanocomposites, *Mech Adv Mater Struc* (in press).
41. Langer S A, Reid A C E, Haan S I and Garcia R E (2007) The OOF2 Manual: Revision 3.6 for OOF2 Version 2.0.3, National Institute of Standards and Technology (NIST), Gaithersburg, March, [on-line]: http://www.ctcms.nist.gov/~langer/oof2man/index.html
42. Langer S A, Fuller Jr E R and Carter W C (2001) OOF: An Image-Based Finite-Element Analysis of Material Microstructures, *Comput Sci Eng* **3**:15–23.
43. Kornmann X, Lindberg H, Berglund L A (2001) Synthesis of epoxy-clay nanocomposites: influence of the nature of the clay on structure, *Polymer* **42**:1303–1310.
44. Dong Y and Bhattacharyya D (2008) Effects of clay type, clay/compatibiliser content and matrix viscosity on the mechanical properties of polypropylene/organoclay nanocomposites, *Composites A* **39**:1177–1191.
45. Park S H (1996) *Robust design and analysis for quality engineering*, Chapman & Hall, London.
46. Lochner R H, Matar J E (1990) *Design for quality: an introduction to the best of Taguchi and western methods of statistical experimental design*, Quality Resources and ASQC Quality Press.
47. Ghani J A, Choudhury I A and Hassan H H (2004) Application of Taguchi method in the optimization of end milling parameters, *J Mater Process Tech* **145**:84–92.
48. Gicía-López D, Picazo O, Merino J C and Pastor J M (2003) Polypropylene-clay nanocomposites: effect of compatibilizing agents on clay dispersion, *Eur Polym J* **39**:945–950.
49. Halpin J C and Kardos J L (1976) The Halpin-Tsai Equations: A Review, *Polym Eng Sci* **16**:344–352.
50. Dong Y, Bhattacharyya D and Hunter P (2008) Experimental characterization and object-oriented finite element modeling of polypropylene/organoclay nanocomosites, *Compos Sci Technol* (in press) (available online 30 October, 2007).
51. Wu Y P, Jia Q X, Yu D S and Zhang L Q (2004) Modeling Young's Modulus of Rubber-Clay Nanocomposites Using Composite Theories, *Polym Testing* **23**:903–909.
52. Hui C Y and Shia D (1998) Simple Formulae for the Effective Moduli of Unidirectional Aligned Composites, *Polym Eng Sci* **38**:774–782.
53. Shia D, Hui C Y, Burnside S D and Giannelis E P (1998) An Interface Model for the Prediction of Young's Modulus of Layered Silicate-Elastomer Nanocomposites, *Polym Compos* **19**:608–617.
54. Voigt W (1889) Über Die Beziehung Zwischen den Beiden Elasticitätsconstanten Isotroper Körper, *Ann Phys* **38**:573–587 (in German).
55. Reuss A (1929) Berechnung der fliebgrenze von Mischkristalen auf Grund der Plastizitätsbedingung für Einkristalle, *ZAMM* **9**:49–58 (in German).

56. Fornes T D and Paul D R (2003) Modeling Properties of Nylon 6/Clay Nanocomposites Using Composite Theories, *Polymer* **44**:4993–5013.
57. Dong Y, Bhattacharyya D and Hunter P J (2007) Characterisation and Object-Oriented Finite Element Modelling of Polypropylene/Organoclay Nanocomposites, *Key Eng Mater* **334–335**:841–844.
58. Luo J J and Daniel I M (2003) Characterization and Modelling of Mechanical Behaviour of Polymer/Clay Nanocomposites, *Compos Sci Tech* **63**:1607–1616.
59. Chawla N, Patel B V, Koopman M, Chawla K K, Saha R, Patterson B R, Fuller E R and Langer S A (2003) Microstructure-Based Simulation of Thermomechanical Behaviour of Composite Materials by Object-Oriented Finite Element Analysis, *Mater Charact* **49**:395–407.
60. Ganesh V V and Chawla N (2005) Effect of Particle Orientation Anisotropy on the Tensile Behavior of Metal Matrix Composites: Experiments and Microstructure-Based Simulation, *Mater Sci Eng A* **391**:342–353.
61. Greiner A, Wendorff J H (2007) Electrospinning: A fascinating method for the preparation of ultrathin fibres, *Angew Chem – Int Ed* **46**:5670–5703.
62. Chung D D L (2001) Comparison of submicron-diameter carbon filaments and conventional carbon, *Carbon* **39**:1119–1125.
63. Evstatiev M and Fakirov S (1992) Microfibrillar reinforcement of polymer blends, *Polymer* **33**:877–880.
64. Fakirov S, Evstatiev M and Schultz J M (1993) Microfibrillar reinforced composite from binary and ternary blends of polyesters and nylon 6, *Macromolecules* **34**:4669–4679.
65. Fakirov S, Evstatiev M and Petrovich S (1993) Microfibrillar reinforced composites, *Macromolecules* **26**:5219–5226.
66. Fakirov S, Evstatiev M and Friedrich K (2000) From polymer blends to microfibrillar reinforced composites, in *Polymer Blends* Vol 2: *Performance* (Eds. Paul D R and Bucknall C B) John Wiley & Sons, New York, pp. 455–475.
67. Evstatiev M, Nicolov N and Fakirov S (1996) Morphology of microfibrillar reinforced composites from polymer blends, *Polymer* **37**:4455–4463.
68. Evstatiev M, Fakirov S and Friedrich K (2000) Microfibrillar reinforced composite: another approach to polymer blends processing, in *Structure development during polymer processing* (Eds. Cunha A M and Fakirov S) Kluwer Academic Publisher, Dordrecht, pp. 311–325.
69. Fakirov S and Evstatiev M (1994) Microfibrillar reinforced composites – new materials from polymer blends, *Adv Mater* **6**:395–398.
70. Fakirov S, Evstatiev M and Friedrich K (2000) Nanostructured polymer composites from polyester blends: Structure-properties relationship, in *Handbook of thermoplastic polyesters* (Ed. Fakirov S) Wiley-VCH, Weinheim, pp. 1093–1129.
71. Evstatiev M, Fakirov S, Krasteva B, Friedrich K, Covas J and Cunha A (2002) Recycling of PET as polymer-polymer composites, *Polym Eng Sci* **42**:826–835.
72. Evstatiev M, Fakirov S and Friedrich K (2005) Manufacturing and characterization of microfibrillar reinforced composites from polymer blends, in *Polymer composites: from nano- to macro-scale* (Eds. Friedrich K, Fakirov S and Zhang Z) Springer, New York, pp. 149–167.
73. Friedrich K, Evstatiev M, Fakirov S, Evstatiev O, Ishii M and Harrass M (2005) Microfibrillar reinforced composites from PET/PP blends: processing, morphology and mechanical properties, *Comp Sci Technol* **65**:107–116.
74. Fakirov S and Bhattacharyya D (2005) Nanofibrillar-, microfibrillar- and microplates reinforced composites – new advanced materials from polymer blends for technical,

commodity and medical applications, *Proc 21st Ann Meet Polym Proc Soc (PPS 21)*, June 19–23, Leipzig, Germany.
75. Li J X, Wu J and Chan C M (2000) Thermoplastic nanocomposites, *Polymer* **41**:6935–6937.
76. Fakirov S, Bhattacharyya D and Shields R J (2008) Nanofibril reinforced composites from polymer blends, *Colloids Surf A: Physicochem Eng Aspects* **313–314**:2–8.
77. Fakirov S (2006) Modified Soxhlet apparatus for high temperature extraction, *J Appl Polym Sci* **102**:2013–2014.
78. Brandrup J and Immergut E H (1989) *Polymer handbook*, 3rd ed., Wiley, New York.
79. Fuchs C, Bhattacharyya D and Fakirov S (2006) Microfibril reinforced polymer-polymer composites: Application of Tsai-Hill equation to PP/PET composites, *Comp Sci Technol* **66**:3161–3171.
80. Friedrich K, Ueda E, Kamo H, Evstatiev M, Fakirov S and Krasteva B (2002) Direct microscopic observation of transcrystalline layers in drawn polymer blends, *J Mater Sci* **37**:4299–4305.
81. Friedrich K, Ueda E, Kamo H, Evstatiev M, Krasteva B and Fakirov S (2004) Microfibrillar reinforced composites from PET/LDPE blend: Morphology and mechanical properties, *J Macromol Sci – Phys* **B43**:775–789.
82. Krumova M, Michler G H, Evstatiev M, Friedrich K, Stribeck N and Fakirov S (2005) Transcrystallisation with reorientation of polypropylene in drawn PET/PP and PA66/PP blends. Part 2. Electron microscopic observations on the PET/PP blend, *Progr Colloid Polym Sci* **130**:166–171.
83. Friedrich K, Fakirov S and Zhang Z, Eds. (2005) *Polymer composites: from nano- to macro-scale*, Springer, New York.
84. Bousmina M (2005) Fundamental insight into polymer nanocomposites, *Abstr Book 21st Ann Meet Polym Proc Soc (PPS 21)*, June 19–23, Leipzig, Germany, p. 219.
85. Kee D E, Zhong Y, Zheng Y and Janes D (2005) Study on the mechanical properties and oxygen permeability of layered silicate-polyethylene nanocomposite films, *Abstr Book 21st Ann Meet Polym Proc Soc (PPS 21)*, June 19–23, Leipzig, Germany, p. 222.
86. Schaefer D W and Justice R S (2007) How Nano Are Nanocomposites, *Macromolecules*, **40**:8501–8517.
87. Brown J M, Anderson D P, Justice R S, Lafdi K, Belfor M, Strong K L and Schaefer D W (2005) Hierarchical morphology of carbon single-walled nanotubes during sonication in an aliphatic diamine, *Polymer* **46**:10854–10865.
88. Kalendova A, Peprnicek T, Kovarova L, Hrncirik J, Simonik J and Duchet J (2005) Poly(vinyl chloride)/montmorillonite nanocomposites: The effect of morphology on the barrier properties, *Abstr Book 21st Ann Meet Polym Proc Soc (PPS 21)*, June 19–23, Leipzig, Germany, p. 223.
89. Bismarck A, Mishra S and Lampe T (2005) Plant fibres as reinforcement for green composites, in *Natural fibres, biopolymers, and biocomposites* (Eds. Mohanty A, Misra M and Drzal L T) CRC/Taylor & Francis, pp. 37–108.
90. Fakirov S, Bhattacharyya D, Hutmacher D and Werner C (2008) Micro/Nanofibrillar and nanoporous scaffolds for tissue engineering, *Tissue Eng* (submitted).
91. Bhattacharyya D, Halliwell R, Fakirov S and Hutmacher D (2007) Nanoporous biocompatible and biodegradable free of contact with toxic substances polymeric materials for biomedical application and method for their preparation, *Patent Application*, The University of Auckland, New Zealand, Jan 15.

PART III
NANO- AND MICROCOMPOSITES: INTERPHASE

Chapter 6

Viscoelasticity of Amorphous Polymer Nanocomposites with Individual Nanoparticles

J. Kalfus

6.1. Introduction

The famous lecture on nanotechnology presented by Nobel Prize laureate Richard Feynman at Caltech in 1959 is generally considered as the beginning of the nanotechnology era [1]. He talked about miniaturization of engines, computers, small-scale recording of information, biological systems, and about nano-structured materials and forces determining the behavior of objects on the nanoscale. Currently, nano-materials are in the focus of interest of many researchers and nano-filled polymers have been used in a variety of applications for a long time (*e.g.*, tires).

The subject of this contribution is to address selected fundamental laws and principles governing the viscoelastic behavior of amorphous polymers filled with non-layered nanoparticles possessing high surface/volume ratio. Amorphous polymers represent systems composed of many long chain molecules with no long-range order. Although at first sight, one can consider the amorphous polymers a simple case of materials being well understood in all aspects, a closer look reveals that this statement is far from the truth in analogy to some other parts of physics. For instance, we observe various physical manifestations of the glass transition phenomenon but there is a lack of a widely accepted theory after tens of years of investigation [2]. Thus, "simple" amorphous polymers with disordered structure can be, in many regards, a more difficult issue for physicists than the crystalline solids [3].

Nevertheless, one may recognize that certain basic rules determine the behavior of amorphous polymer solids. Their properties are, in general, governed by the properties and structure of their components (chains, segments and monomer units), and by the nature of mutual interactions among them. Amorphous polymer structure is built up of several structural levels, from *ca.* 0.1 nm to several micrometers, which will be discussed later on. The structure, as well as the dynamics given by the inter- and intra-molecular interactions and fluctuations on the particular structural level, determine the microscopic processes and, consequently, the response of the entire amorphous material. One may wonder how a group of simple basic rules results in such a great complexity of physical properties. Furthermore, the amazing fact in this field is that even a "gentle" chemical or physical process on the molecular level can induce a far-reaching change of the macroscopic viscoelastic response.

Synthetic composite materials have passed through intensive progress in the last fifty years. Originally, synthetic polymers were blended with inorganic fillers for the sake of price reduction; on the other hand, fillers cause considerable change in many physical properties of polymers. Later, functional fillers bringing other effects to polymer composites, such as magnetic or electrical properties, bioactivity, reduced flammability, *etc.*, were introduced in the 80's and 90's. Nanocomposites can offer both unique properties and multi-functionality, which are well applicable in various fields. Undoubtedly, if one wants to use polymer nanocomposites in specific applications, the structure-property relationships have to be fully understood in these unique materials first.

6.2. Brief physics of amorphous polymer matrices

6.2.1. *Equilibrium structure of amorphous chains*

The structure of an amorphous polymer chain comprises various structural levels as it has been clearly presented by Jancar [4]: (i) constitution (atomic/molecular structure of monomer units), (ii) configuration (spatial arrangement of primary bonds independent of bond rotation), and (iii) conformation (spatial arrangement, which can be altered by bond rotation). The majority of amorphous polymer chains used as matrices contain the "carbon backbone" (ethylene/propylene rubber, EPR; poly(vinyl chloride), PVC; polystyrene, PS; poly(methyl methacrylate), PMMA; poly(vinyl acetate), PVAc; *etc.*). The approximate characteristics of the vinyl-type backbone are as follows: C-C bond length $b \cong 0.15$ nm, C–C bond angle $\theta \cong 109°$, C–C bond energy $U_{bond} \cong 350$ kJ/mol.

The addition of nanoparticles into the polymer matrix is a way to affect the chain conformation structure as well as conformation dynamics and, therefore, attention to the conformation statistics of macromolecules will be paid in this chapter at the expense of constitution and configuration structure.

Physically, it is most accurate to consider the polymer chain as a one-dimensional cooperative system [5]. This is treated *via* the one-dimensional Ising model, which takes into account the cooperative rotation of single bonds of a chain backbone [5–8]. The rotation isomerization in the polymer molecules means that the rotation around each single bond is more or less restricted. As a consequence, certain rotation positions (isomers) are preferred due to the local non-bonded potential field. The latter is caused, in general, by interactions among the electron clouds of atoms forming the polymer chain. The polymer chain flexibility is thus given by the number of available internal degrees of freedom determined by the rotations of backbone single bonds. Further contribution to the chain flexibility is the torsion oscillation. However, bond length and bond angle deformations are negligible in the majority of cases [5].

In the case of no restrictions imposed on the chain flexibility, the polymer chain represents a system without any directional memory and evokes a path of Brownian diffusive motion [9]. A suitable description of this self-similar process can be the random walk model (*rwm*) depending on Euclidean dimensionality, d. The *rwm*-based distribution function determines the probability of the root mean square chain end-to-end separation, $\langle r_0^2 \rangle^{1/2}$:

$$P(3d, r) = (3/2\pi Nb^2)^{3/2} \exp(-3r^2/2Nb^2), \tag{6.1}$$

where r is the end-to-end separation and b is the individual bond length. If one seeks the probability distribution function of the second chain-end lying in the spherical shell region $\langle r; r+dr \rangle$ one has to use the radial distribution function, $P(rdf)$, in the form of:

$$P(rdf) = 4\pi r^2 P(3d, r). \tag{6.2}$$

In this case, the maximum value appears rather far from the origin. Treloar [10] has considered this property of a Gaussian chain with a high number of bonds, N, to be an analogy of a target with high number of shots. The most probable position of one end relative to the other one lying in the origin is the origin, if it is related to the unit area. However, the end-to-end distance is $\sim N^{1/2}b$, when the end position is related to the shell with thickness dr. Note that the motional memory can be imposed to the random walk process using the Markov statistics [9], whereby each next step of the walk is characterized by an appropriate probability function.

Separation of chain ends in both the Gaussian and real chains is much lower than the value Nb ($N \gg 1$) and, therefore, the most probable spatial arrangement of the individual disordered macromolecule may exist: it is the random coil. Real unperturbed chains possess less flexibility and, hence, longer end-to-end distance, $\langle r_0^2 \rangle^{1/2}$, compared to the *rmw*-chain. A real chain is characteristic of a short-range order resulting from energetically non-equivalent rotation isomers and fixed bond angles. The one-dimensional Ising chain calculation considers the short-range interactions for near neighbors along the chain. It results in the necessity of calculation of the pair free energy of monomer units, $U(\Omega_{k-1}, \Omega_k)$. Consequently, the determination of the chain partition function and other thermodynamic quantities is possible [5]. Calculations according to the Ising model include procedures based on matrix or combinatorial methods.

The description of these mathematical procedures is beyond the scope of this contribution. Nevertheless, the one-dimensional Ising model is a method yielding reasonable values of the characteristic ratio, C_N, for a particular polymer chain. The parameter C_N can be simply used in the calculation of the real chain end-to-end distance:

$$\langle r_0^2 \rangle = C_N N b^2. \tag{6.3}$$

Obviously, the temperature dependent parameter C_N indicates the difference between the real chain and *rwm* structure. A review of the C_N values for a great number of polymers can be found in reference [11]. The unperturbed real chain exhibits "swelled" conformations (in contrast to the *rwm* approximation) due to the intra-molecular short-range interactions and almost fixed bond angles. This consequently leads to the temperature dependent end-to-end distance and other peculiarities such as the non-zero energy term in the rubber elastic response of real polymers [12].

6.2.2. *Microscopic relaxation modes and segmental mobility*

The entangled amorphous polymers are systems with macroscopic viscoelastic response given by processes on the microscopic level [13]. These microscopic modes represent particular molecular motions constrained to some characteristic volumes, V_c, varying from localized vibrations of 10^{-3} nm^3 to the non-local normal mode designating relaxation process connected with fluctuation of long chain parts and chain ends [13]. In the last case, the characteristic volume becomes dependent on the chain size, $V_c \propto N^{3/2}$. The time characteristic of each particular microscopic process, τ, varies from 10^{-13} s of local vibrations to infinitely long times at low temperatures. The macroscopic viscoelastic response is, thus, a manifestation of segmental or molecular fluctuations exhibiting extreme sensitivity to the local density of monomer units. These fluctuations are localized in some characteristic volume and the rate of the relaxation process is indirectly proportional to its locality (Figure 6.1).

The particular relaxation modes in polymers are a consequence of thermal agitation of monomer units. One may consider the polymer chain as a set of linearly linked Brownian particles, each executing thermally activated diffusive motion primarily *via* the torsion around the single bonds. Due to the intra- and inter-chain potential fields, kinetic energy of the monomer unit must be sufficiently high to overcome a certain energy barrier in the time coordinate of torsion motion to allow changes in the chain conformations [14]. One may consider the inter-molecular contribution an analogy to the real gas – "gas" of chain monomer units. Thus, certain activation energy, E_A, can be attributed to each particular relaxation mode. For instance, the reptation motion activation energy of polyethylene (PE) melt is of value $E_A = 20$–26 kJ/mol [15,16]. In the overwhelming majority of cases, PE local conformation dynamics occur through the *trans-gauche* and *gauche-trans* transitions (*gauche-gauche* jumps are negligible) with energy barrier of approximately $E_A^{intra} = 12$ kJ/mol [14,15].

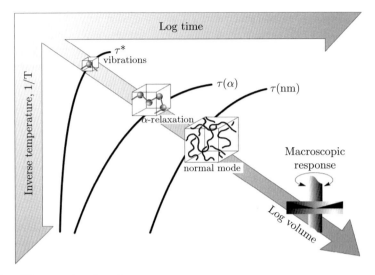

Figure 6.1. Microscopic relaxation processes

Hence, the rest of the activation energy can be attributed primarily to the intermolecular interactions. Lin [17] has demonstrated that around 18 chain strands are present in one cubed tube diameter in the melt of a linear PE ($V_{td} \cong 27$ nm^3). The term *tube* represents an approximation of the inter-molecular constraint of the chain in a melt imposed by the surrounding chains. Taking the range and shape of the inter-molecular interaction potential [15] into account, the above number of contributions to the inter-molecular interaction potential seems to be reasonable. Note, however, that these considerations represent approximations due to many simplifications involved.

In general, neglecting any secondary relaxation of Arrhenius type, the typical relaxation times of the cooperative single bond conformation dynamics may be taken as persistence times displaying temperature dependence. This can be described by the modified Vogel-Fulcher-Tamann (VFT) equation [18]:

$$\tau = \tau^* \exp\left(\frac{zE_A^{\text{intra}}}{k_B T}\right) \exp\left(-\frac{E_A^{\text{intra}}}{k_B T^*}\right), \tag{6.4}$$

where

$$z = \frac{T^* - T_0}{T^*} \cdot \frac{T}{T - T_0}. \tag{6.5}$$

The symbol τ^* represents the time characteristic of torsion vibration, k_B is the Boltzmann constant and T is temperature. T_0 is the Vogel temperature, typically expected approximately 50°C below the experimentally measured glass transition temperature, T_g. T^* is the empirical activation temperature denoting a hypothetical temperature, where the motional cooperativity vanishes (for PE at 160°C the $zE_A^{\text{intra}} \cong 25$ kJ/mol). Note that torsion vibration represents the mode being very localized, taking place without any energy barrier and, hence, it does not experience so strong temperature dependence compared to the other microscopic motions of Arrhenius and VFT types, $\tau^* = b(m_0/k_B T)^{1/2}$; m_0 is the monomer unit weight. According to Bueche [19], the monomer friction coefficient, ζ_0, can be estimated according to the following simple equation, which gives reasonable values for the temperature region $T > T_g$:

$$\zeta_0 = k_B T \tau / b^2, \tag{6.6}$$

where b is the length of a link between two Brownian particles (in the case of simple chains, b is of the order of the bond length). Consequently, one can express the monomer unit mobility, μ_0, as follows [20]:

$$\mu_0 = \zeta_0^{-1} \tag{6.7}$$

In a melt of long chains, each chain is entangled with its neighbors and the characteristic relaxation time of movement of the whole chain is given by the time of the chain to diffuse trajectory equal to its length. According to deGennes [20], the relaxation time of the chain reptation motion, τ_{rep}, can be expressed as:

$$\tau_{rep} \simeq \frac{NL^2}{D_0}, \tag{6.8}$$

where L is the chain length and D_0 is the monomer unit diffusion coefficient ($D_0 = k_B T/\zeta_0$). The reptation motion of an entangled chain can be written in the approximate form of the Doi-Edwards relation [21]:

$$\tau_{rep} = \frac{b^4 \zeta_0 N^3}{\pi^2 k_B T a_T^2}, \tag{6.9}$$

where a_T is the tube diameter. Supposing the a_T equal to the distance between the entanglement points yields the relation: $\tau_{rep} = b^2 \zeta_0 N^3 / \pi^2 k_B T N_e$ (N_e is the number of monomer units per entanglement strand). Unperturbed chain exhibits random coil structure in melt and its diffusion is retarded by the energy barrier of both intra- and inter-molecular origin included in the ζ_0. The rate of the chain reptation motion can be expressed as a path passed by the chain in the time τ_{rep}:

$$r_{rep} = L_{prim}/\tau_{rep}, \tag{6.10}$$

where L_{prim} is the primitive length equal to a sum of lengths of the chain Rouse segments.

6.2.3. Entropy vs. energy driven mechanical response

It is well known that a non-crystallizable polymer can pass through a certain transition during cooling from high temperature. Below and above this transition, the polymer material embodies pertinent properties given by the dynamics on the molecular level. On this basis, two general states of linear amorphous polymer can be detected: *equilibrium rubbery or liquid-like state* (depending on molecular weight) and *non-equilibrium solid-like glassy state*. A very general view on the difference between the equilibrium rubbery state and the "frozen" glassy state can be expressed using the Deborah number, representing a universal measure of equilibrium, $D = \tau_i/\tau_o$. The D denotes the relationship between intrinsic time of a relaxation process, τ_i, and external time of observation, τ_o. The equilibrium is attainable until $D \leq 1$.

In the case that the chain is able to retain its equilibrium structure, the single chain elasticity can be derived from the first and second law of thermodynamics. Assuming $pdV/fdl \approx 10^{-4}$, the retractive force can be expressed as [10]:

$$f = \left(\frac{dU}{dr}\right)_T - T\left(\frac{dS}{dr}\right)_T = f_E + f_S, \qquad (6.11)$$

where dU and dS is the change of internal energy and entropy, respectively. Hence, the symbols f_E and f_S depict the energy and entropy contribution to the chain force response. The change of the end-to-end distance, r, causes the conformation entropy decrease resulting in the chain retractive force increase. For the finite chain with $(dU/dr)_T = 0$, the retractive force can be expressed using the Langevin inverse function containing the Gaussian first-term parameter [10]:

$$f_S = \frac{k_B T}{b} L^{-1} \frac{r}{Nb} = \frac{k_B T}{b} \left\{ 3\left(\frac{r}{Nb}\right) + \frac{9}{5}\left(\frac{r}{Nb}\right)^3 + \frac{297}{175}\left(\frac{r}{Nb}\right)^5 + \frac{1539}{875}\left(\frac{r}{Nb}\right)^7 + \ldots \right\} \qquad (6.12)$$

In the case of real chains, the force response during deformation may exhibit a dependence more complex than presented in Eq. 6.12. During stretching, the chain energy change can follow the transition of coiled conformation to the elongated one [5,22]. Generally, the energy term depends on the intramolecular interactions and, hence, it can be extracted from experimental data using the equation [5]:

$$f_E = fT \frac{d \ln r^2}{dT}. \qquad (6.13)$$

The volume characteristic of torsion vibration corresponds approximately to $V_c^{vib} \cong b^3 \cong 10^{-3}$ nm^3. The characteristic volume corresponding to the entanglement strand is $V_c^{ent} \cong a_T^3 \cong 10^1$ nm^3. In high molecular weight polymers, the rubbery plateau can be detected because of the chain constraint caused by the entanglements acting as transient (physical) cross-links. Considering a purely entropic behavior in the rubbery region, the elastic modulus, G_{rubber}, of the entangled polymer can be expressed as follows [17]:

$$G_{rubber} = 4\rho RT/5M_e, \qquad (6.14)$$

where ρ is the polymer density, R is the universal gas constant, and M_e is the molecular weight of one entanglement strand. Because the characteristic volume of the entangled Gaussian network is $V_e^{ent} = (M_e C_N b^2/M_0)^{3/2} \pi/6$, the modulus of elasticity may be written in the form: $G_{rubber} = 4rRTC_N b^2 / 5 M_0 (6 V_c^{ent}/\pi)^{2/3}$; M_0 is the molecular weight of the monomer unit.

At low temperatures, the microscopic relaxation motions represent modes too slow compared to the experiment time scale, and are thus considered unable to maintain equilibrium values of the extensive thermodynamic quantities. The glassy chains can relax only due to the extremely rare conformation transitions of low-mobility chain segments trapped in the energy landscape. Hence, the glassy polymer response is considered as energy driven. The torsion vibrations represent the primary microscopic motion governing the response of the majority of segments in the glassy amorphous polymer solid. Thus, supposing the polymer glass composed of individually vibrating chain segments, the rotation isomeric state jumps represent infrequent cooperative events that are very sensitive to the local density of monomer units. Thereafter, one may very roughly estimate the glassy polymer modulus of elasticity, G_{glass}, in the approximate form:

$$G_{glass} = \gamma^* / V_c^{vib}, \qquad (6.15)$$

where γ^* is the single bond vibration force constant of the order $\gamma^* = 10^{-19} - 10^{-20}$ J [14]. This formula gives reasonable values of the elastic modulus $G_{glass} = 10^8 - 10^9$ Pa (note that Pa = J/m^3, because Pa = kg/ms^2 and J = kg×m^2/s^2). No doubt, polymer glasses represent more complex systems than is recognized in this consideration. Nevertheless, the idealized expressions for the rubbery and glassy modulus of elasticity demonstrate the fundamental difference between the entropy and energy driven elastic response of the polymer solid. While the first one is given by the huge number of accessible conformations, the second one is determined by the local energy landscape [23].

6.3. Basic aspects of amorphous polymer nanocomposites

First, it is necessary to emphasize that nanocomposite reinforcement comprises several contributions. Above the neat matrix T_g, immobilization of polymer chains becomes the primary effect given by the large nano-filler surface area open for polymer-filler contacts. Thus, the nanocomposite represents a system with high filler-polymer interface area. This is analogous in respect of surface interaction to another system produced by depositing thin polymer layers on solid substrates.

To justify this statement, let us consider a nanocomposite as a result of a certain reaction, where the degree of conversion depends on the fraction of chain segments adsorbed on filler surface:

$$xP + yF \xleftarrow{k(desorption) - k(adsorption)} P_n - F_m + (x-n)P + (y-m)F, \qquad (6.16)$$

where P and F depict polymer segments and filler adsorption sites, respectively. In general, *the filler surface-to-volume ratio* and *the filler-polymer interaction energy*, among other effects, primarily affect the amount of the P_n–F_m complex. Consequently, the P_n–F_m complex content has crucial effect on the nanocomposite reinforcement; its portion can be changed, for instance, during the deformation processes.

6.3.1. *Structure of surface adsorbed chains*

Pioneering attempts to solve the problem of the single one-dimensional *rwm* chain in the presence of the impermeable wall have been published by Chandrasekhar [9] and later by DiMarzio and McCrackin [24]. DiMarzio et al. [24] have verified the statistical-mechanical calculations by computer simulations utilizing the Monte-Carlo method (MC). They showed that the obtained results are universally valid even in the case of chains in the space. Furthermore, it was documented that the transition of the chain from the state of free random coil to the form of adsorbed chain is a second-order transition in the sense of the Ehrenfest definition. When the segment-wall interaction energy exceeds the value $-\varepsilon/k_B T \cong 0.7$ (one-dimensional model) or 0.4 (three-dimensional tetrahedral lattice model), the entropy repulsion is prevailed by the attractive forces and the single chain becomes multiply attached to the wall with fraction of the adsorbed units, N_A/N, approaching the limit $1/2$ ($N \to \infty$). The adsorbed chain exhibits the so-called train-loop-tail structure (Figure 6.2) and as the segment-wall interaction energy, $-\varepsilon$, increases, the size of loops and tails decreases in favor of trains.

Recently, various researchers [25–29] have performed precise molecular-dynamics (MD) and MC simulations of dense polymer melts in the presence of nanoparticles. Spatial arrangement of macromolecules near the surface has been investigated by Glotzer et al. [25] (MD) and Vacatello [26,27] (MC). Glotzer et al. [25] have simulated the dense polymer melt around one icosahedral nanoparticle in a cubic cell where attractive and non-attractive interactions between the filler particle and polymer chains were considered. They have clearly shown that the matrix chains, close to the attractive surface, displayed layered

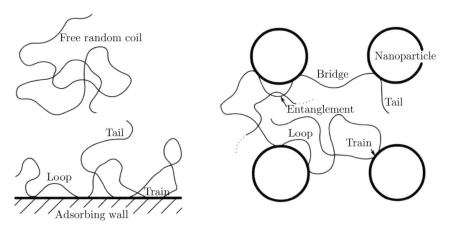

Figure 6.2. Sketch of polymer chain structure unperturbed (free random coil) and perturbed by the interaction with attractive wall and nanoparticle surfaces

density profile consisting of chain segments running parallel to the surface. The density of monomer units fluctuated in these layers by a factor of 0.5–1.5 compared to the bulk value. Moreover, the chains near the attractive nanoparticle surface exhibited asymmetric distribution of segments especially in the z-direction (perpendicular to the filler surface). This is in line with the early results achieved by DiMarzio and McCrackin [24].

Vacatello [26,27] simulated a dense polymer melt in a cubic cell with the adsorbing walls as well as randomly located nanoparticles using the MC method. His work offers many worthwhile results exceeding the scope of this contribution. Briefly, his simulations confirmed perturbation in the distribution function of the chain attached to the adsorbing surface. The distribution function depended on the chain stiffness and showed densely packed chain units in close vicinity to the attractive surface. In the case that the adsorbing surfaces are distributed through randomly placed filler nanoparticles with a diameter comparable or lower than two chain radii of gyration, $2R_g$, the R_g of chains in polymer nanocomposite decreased. This was in agreement with experimental data from the small angle neutron scattering (SANS) experiments [30]. However, more detailed results from SANS experiments and MC simulation presented by Mark *et al.* [28] revealed that the effect of nanoparticles on chain dimensions is difficult to be simply generalized and depends on a number of variables such as the particle diameter, particle volume fraction, *etc*. Furthermore, in contrast to a simple adsorbing wall, considered by DiMarzio and McCrackin [24], the polymer chain can form interparticle bridges in addition to trains, loops and tails (Figure 6.2). Thus, one can divide the average chain into a certain number of trains, loops, bridges, and tails.

Similarly to Vacatello [26,27], Kumar *et al.* [29] have carried out the MC simulation bringing more light into the problem of chain conformation structure in the presence of spherical nanoparticles. Their results, generally in agreement with those of Vacatello [26,27], show that the chain parts like bridges exhibit conformations different from the ideal Gaussian structure. The sub-chains connecting the adsorbed trains are in a stretched state compared to the Gaussian coil. For instance, nuclear magnetic resonance (NMR) measurements have revealed that PMMA chains adsorbed on the aluminum oxide surface exhibit lower population of the gauche conformations than the bulk PMMA [31]. Moreover, the bridges and loops have turned to be essential for the nano-reinforcing effect leading to the substantial solidification of amorphous polymers (Gaylord [32] and Keblinski *et al.* [33]).

Furthermore, Nowicki [34] carried out calculations on the chain entropy in the presence of nanoparticles using the MC method. He has shown that the loss in the conformational entropy is associated with spatial rearrangement of molecular segments caused by the adsorption process. The stronger the polymer-filler interaction is, the larger entropy loss can occur. As the particle diameter drops below the radius of gyration of the chain, the entropy perturbation starts to be more localized within the coil. Thus, it can be concluded that the nanoparticles, in contrast to microparticles, do not represent a filler in the common sense; they rather represent an agent able to modify the chain conformation structure and entropy as well as segmental dynamics.

Finally, note that some phenomena typical of real systems were neglected in the previous text. Such effects are, for instance, the distribution of activity of adsorption sites on filler surface as has been obtained from the inverse gas chromatography measurements [35] or higher equilibrium concentration of longer chains at the interface compared to the bulk [36]. However, they seem to be of second order effects according to experimental data reported.

6.3.2. *Segmental immobilization of chains in the presence of solid surfaces*

Generally, fluids constrained by the solid surfaces exhibit properties markedly differing from those of the bulk fluids. It has been clearly demonstrated that the layer of 3-methylpentane above its T_g deposited on Pt substrate exhibited spatial dependence of viscosity as inferred from measurements utilizing the ion movement [37]. The fluid viscosity increased by more than 6 orders of magnitude in a distance lower than 4 nm from the surface. One can expect that the effect of the surface confinement on the dynamics of polymer chains will be more pronounced. Chakraborty *et al.* [38] have summarized the kinetic constraints of chains adsorbed on a surface as follows: the nearest-neighbor interactions, the specific segment-surface interactions, the entropic constraint due to impermeable surface, the topological barriers due to other adsorbed chains, and the competition for surface sites. Whilst the first two effects are typical of strongly adsorbed chains, the other effects are generic to all polymer systems.

Many researchers [39–58] have observed considerably decreased segmental dynamics of chains in polymers filled with the high-specific surface area fillers utilizing various measurement methods. Experiments well documenting the immobilization phenomenon have been carried out by Asai *et al.* [52] using the pulse ^{1}H-NMR method for an oxidized carbon black filled natural rubber above the polymer matrix T_g. Asai *et al.* [52] identified three contributions to the free induction decay (FID) signal characterized by three T_2 relaxation constants (Figure 6.3). The first one originated from the chain segments closest to the surface exhibiting dynamics analogous to a polymer in the glassy state, i_1. The intermediate phase, i_2, was assigned to the loosely bounded chains, probably in the form of longer loops, bridges, tails and entanglements trapped by the adsorbed chains. The third phase was attributed to the bulk regions unaffected by the surface at all. Kaufman and his co-workers [43] came to similar conclusions. They studied the effect of carbon black on polybutadiene and ethylene/propylene/diene (EPDM) matrices by NMR measurements through the whole temperature range from $T < T_g$ to $T > T_g$. Unlike Asai *et al.* [52], Kaufman *et al.* [43] measured even samples pretreated by extraction in boiling solvent. They identified two phases in the bounded matrix, i_1 and i_2, from the resolve (FID) signal, see Figure 6.3. Below the matrix T_g, both phases, i_1

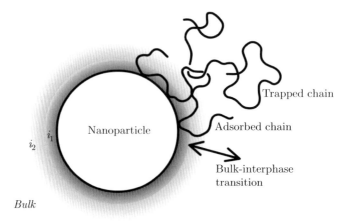

Figure 6.3. Idealized scheme of chain dynamics in the vicinity of nanoparticle above neat matrix T_g

and i_2, possessed the same relaxation times. In the glass transition region, two relaxation times appeared. Whilst the relaxation time of the i_1 phase remained approximately constant over the whole temperature range, the i_2 displayed an increase of T_2 by one order of magnitude and, therefore, it was assigned to the less bounded matrix.

Montes et al. [59] have done ^{1}H-NMR experiments on nano-silica filled acrylate based elastomer to estimate the extent of segmental immobilization. In this system, either covalent bonding or physical forces were the primary interaction between the nano-filler and elastomer matrix. In contrast to Asai et al. [52] and Kaufman et al. [43], Montes et al. [59] identified only two contributions to the transverse magnetization function – the immobilized interphase and bulk. If one approximates the region of reduced segmental mobility as a layer of finite boundary, its thickness is of the order of 1–2 nm. This layer thickness decreased with increasing temperature. This finding clearly demonstrates that the interphase is not characterized by a layer of constant thickness. The interphase thickness may increase or decrease under given external and internal conditions. In other words, the chains in the vicinity of the filler surface may show the same behavior as those of the neat polymer under certain conditions. When the conditions (e.g., temperature) are changed, the interphase may exhibit local relaxation dynamics in orders of magnitude slower than the neat polymer. One may consider this phenomenon as a *polymer bulk-interphase transition*. It is, however, necessary to note that this is only a schematic view. In real systems, no sharp interphase exists but rather a multi-layered one with diffuse boundaries from the point of view of chain dynamics is expectable.

Lin et al. [60] have tried to reveal the mobility course of PMMA chains in dependence on the distance from the adsorbing surface. They have used the neutron reflectivity measurements on the chain diffusion coefficient, D_c. PMMA was deposited on clean oxidized silicon substrate to document perturbation of D_c compared to the bulk after sufficient thermal annealing. Regardless of the PMMA molecular weight, the D_c started to decrease continuously at a distance from the surface of approximately $5R_g$. In addition, the desorption kinetics of PMMA chains was investigated utilizing the high-resolution neutron reflectivity method [61]. There, mechanical shear oscillations were applied to the sample resulting in a considerably increased rate of chain desorption from filler surface. This phenomenon has a serious consequence for the interpretation of the often peculiar, non-linear viscoelastic response of polymer nanocomposites above their T_g.

It is obvious from works devoted to the investigation of adsorption-desorption dynamics of chains on attractive surfaces that chains attached to the surface, even by strong specific bonding, are able to detach completely due to thermally activated processes [62–65]. Thus, the chain once attached to the surface can leave the surface after a period correlated to the time characteristic of cooperative multiple adsorption-desorption steps. For instance, Granick and co-workers [62,64] have shown that the time needed for complete desorption of PS chains from the oxygen-plasma treated silicon surface in cyclohexane solution may reach the value 10^4 s and higher (in dependence on molecular weight).

Furthermore, Rubinstein and co-workers [65] have demonstrated that the reptation model [21] can be used for description of the retarded normal mode dynamics of linear chains adsorbed on solid surface. In their experiments, a very thin layer of deuterated PS (d-PS) was deposited on attractive silica and less attractive poly(2-vinylpyridine) surface. The tracer diffusion coefficient of PS chains was measured by secondary ion mass spectroscopy after thermal annealing. The bulk diffusion coefficient was in the whole range of molecular weights considerably higher compared to the diffusion coefficient of chains on silica surface. According to Rubinstein et al. [65], the total friction coefficient of monomers belonging to a chain close to the surface, ζ_c, can be estimated by:

$$\zeta_c = \zeta_0 N + \zeta_a N^{1/2}, \tag{6.17}$$

where ζ_a is the adsorbed (train) monomer friction coefficient. Consequently, the reptation time of the chain adsorbed on the surface is in the approximate form:

$$\tau_{rep}^a = \frac{b^2 N^{5/2}}{2k_B T N_e}\left(\zeta_a + \zeta_0 N^{1/2}\right). \tag{6.18}$$

Thus, primarily the friction of monomer units attached to the adsorbing surface retards the chain reptation motion. The presented model agrees well with the experimental data and, as it will be shown later in the text, it can be used even for the interpretation of relaxation behavior of polymer nanocomposites above the matrix T_g.

6.4. Reinforcement of amorphous nanocomposite below and above matrix T_g

Researchers have investigated the effect of various micro-fillers on mechanical behavior of polymers in the last 40 years and comprehensive data are available for the general understanding of the structure-property relationships in these materials [66]. Adding stiff filler particles into the polymer matrix leads to the enhancement of modulus of elasticity. This phenomenon is caused, in principle, by one of the three main mechanisms: (i) substitution of soft matrix with stiffer filler, (ii) stress transfer from matrix to the surface of anisotropic particles, and (iii) segmental immobilization caused by the interaction of chain units with filler surface, respectively. The stress transfer mechanism depends on the inclusion aspect ratio and its orientation to the applied load and, thus, represents together with the substitution mechanism the size independent contribution. On the other hand, the segmental immobilization mechanism contributes to the overall composite reinforcement with the extent primarily affected by the size of inclusions. Surprisingly, the concept of mechanical and immobilization contribution to the polymer nanocomposite reinforcement has already been used by DiBenedetto in 1969 [40]; however, the term *nanocomposite* was not considered by researchers at that time. In the case of the shear modulus of elasticity, one can express the above contributions schematically as follows:

$$G_c/G_{matrix} \cong [f(v_f, A_r) + F(A_f)], \tag{6.19}$$

where G_c is the composite shear modulus of elasticity, G_{matrix} is the neat matrix shear modulus of elasticity, $f(v_f, A_r)$ regards the (i) + (ii) contribution given by the filler volume fraction, v_f, and the filler aspect ratio, A_r. $F(A_f)$ depicts the immobilization contribution (iii) being dependent on the filler-matrix contact area, A_f, among others.

In the case of micro-composites, specific filler surface area of common fillers is usually less than the value 10 m^2/g. In these systems, a very small portion of polymer molecules is in direct interaction with the filler surface. Moreover, dimensions of polymer chains and volumes characteristic of microscopic relaxation modes are orders of magnitude smaller compared to the dimensions of filler particles (see Table 6.1). Thus, continuum mechanics approaches, such as micromechanical models can be successfully used for description of their mechanical behavior. The continuum mechanics model, well usable for prediction of the mechanical contribution (i) to the modulus of elasticity of polymer composites, G_c, is the simple Kerner-Nielsen (K-N) model [66]:

$$\frac{G_c}{G_m} = \frac{1 + ABv_f}{1 - B\psi v_f}, \tag{6.20}$$

where
$$A = (7-5\mu_1)(8-10\mu_1)^{-1} \tag{6.21}$$

and
$$B = \left(\frac{G_f}{G_m} - 1\right)\left(\frac{G_f}{G_m} + A\right)^{-1}. \tag{6.22}$$

A is the factor depending on the matrix Poisson's ratio, μ_1, and B is the factor primarily related to the filler/matrix stiffness ratio, G_m and G_f are moduli of matrix and filler particles, respectively. The factor ψ represents a boundary condition and can be expressed using an empirical function taking into account the maximum filler volume fraction, v_{max}:

$$\psi = 1 + v_f(1-v_{max})v_{max}^{-2}. \quad (6.23)$$

The K-N model represents the substitution mechanism (i) of the composite reinforcement.

To include a further mechanical reinforcing effect – the stress transfer (ii) from matrix to the surface of anisometric particles, the Halpin-Tsai model [66] can be used providing reasonable predictions:

$$\left.\frac{G_c}{G_m}\right|_{L,T} = \frac{1+\zeta\eta v_f}{1-\eta v_f}, \quad (6.24)$$

and

$$\eta_{L,T} = \left(\frac{G_f}{G_m}-1\right)\left(\frac{G_f}{G_m}+\zeta_{L,T}\right)^{-1}. \quad (6.25)$$

The $\zeta_{L,T}$ is the semi-empirical shape related factor and $\eta_{L,T}$ is the factor related to the relative stiffness of the filler and the matrix. For ribbon shaped reinforcement with rectangular cross-section, $\zeta = 2A_r = 2\times length/thickness$. Elastic modulus for randomly oriented anisometric inclusions can be expressed using a simple Tsai randomization procedure [66]:

$$G_c^{random} = \frac{3}{8}G_{cL} + \frac{5}{8}G_{cT}. \quad (6.26)$$

In Eq. (6.26), G_{cL} and G_{cT} are the longitudinal and transverse elastic moduli of composite with unidirectionally aligned anisotropic inclusions, respectively.

Below the T_g, adding of nano-sized particles results in the increase of modulus of elasticity in accordance with the K-N micromechanics model (see Figure 6.4). Chains below T_g inhere in their non-equilibrium conformations exhibiting low conformational entropy regardless of the distance from the particle surface. Moreover, the dynamics of rotation isomerization is extremely low and most of the monomer units execute a vibration motion only in a given time scale. Hence, providing that torsion vibration is the primary microscopic motion in glassy polymer, the characteristic volume of local relaxation process is very low in comparison to the filler particle volume (Table 6.1). Thus, the glassy nanocomposite behaves as a true two-component system, where the effect of the nano-filler is merely the replacement of a portion of softer polymer matrix with more rigid filler nanoparticles. Hence, the dependence of the relative modulus on filler volume fraction is in accordance with the K-N model.

Nanocomposites represent systems, where specific filler surface area commonly exceeds 100 m^2/g and filler particle dimensions are easily comparable with the dimensions of polymer chains. Due to the large internal surface area exposed to the filler-polymer interaction, it is reasonable to assume that the

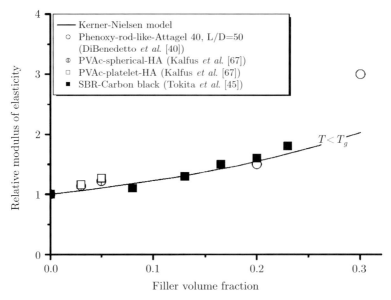

Figure 6.4. Relative storage modulus dependence on filler volume fraction for nano-filled amorphous polymers below T_g. HA depicts the hydroxyapatite nano-filler

Table 6.1. Ratios of the particle volume/polymer characteristic volume, r_v, and particle diameter/polymer characteristic length, r_l, for microcomposite and nanocomposite below and above the matrix glass transition [68]

	Filled glassy polymer ($T<T_g$)	Filled rubbery polymer ($T>T_g$)
Microcomposite ($d_{particle}$=10 μm)	$r_v=10^{15}$, $n=10^5$	$r_v=10^6$, $n=10^2$
Nanocomposite ($d_{particle}$=10 nm)	$r_v=10^6$, $n=10^2$	$r_v=10^{-3}$, $n=10^{-1}$

immobilization phenomenon can prevail over the substitution and stress transfer contribution to the composite reinforcement above the matrix T_g. Consequently, the filler surface interaction with polymer chains should be the first-order process governing the elastic behavior of polymer nanocomposite under given conditions (time, temperature and strain). Therefore, it is more convenient to use the filler-matrix contact area, A_f, rather than the filler volume fraction in order to express the dependence of modulus of elasticity on the filler content. Note, that the A_f was calculated simply from the filler weight fraction and known value of the specific filler surface area [67]. Selected data of the elastic modulus related to the neat matrix reported in literature are summarized in Figure 6.5. The relative modulus of elasticity of the nanocomposite is considerably higher than given by the Guth model [69]. Note that the latter is often used for prediction of the reinforcement of rubbery composites. The Guth model [69] is expressed by:

$$G_r = 1 + 2.5v_f + 14.2v_f^2, \tag{6.27}$$

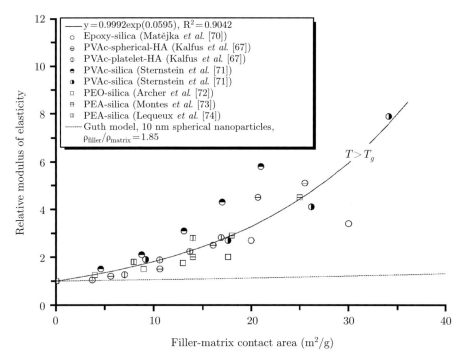

Figure 6.5. Relative storage modulus *vs.* filler volume fraction dependence for nano-filled amorphous polymers above T_g. The PEA and PEO depict cross-linked poly(ethyl acrylate) and poly(ethylene oxide) matrix, respectively

where G_r is the relative shear modulus of elasticity of polymer composite. The data in Figure 6.5 reveal that the reduced segmental mobility represents the primary contribution to the nanocomposite reinforcement. The experimental data in Figure 6.5 underlay some scatter most probably because these nanocomposites were obtained by different preparation procedures, exhibited various strength of filler-matrix interaction, differed in filler dispersion and matrix molecular characteristics. Nevertheless, the dependence clearly indicates the true nature of the nanocomposite reinforcement above the matrix T_g being driven by the extent of the filler-polymer interface area. Moreover, as it is seen from Table 6.1, the nanoparticle dimensions are comparable with those of polymer chains and volume characteristic of microscopic relaxation modes. Thus for low v_f, the nano-filler may be considered to be a kind of modifier or agent greatly affecting the properties of amorphous polymer matrix.

To address the physical merit of the immobilization phenomenon as the reinforcing mechanism in more detail it is necessary to recall the brief theory of amorphous polymers presented above. First, let us consider a general situation – the entangled polymer chain attached to the surface exhibiting the train-loop-bridge-tail structure. The chain is trapped by both entanglements with surrounding chains and by the trains situated on particle surfaces each with very different persistence time characteristic of the physical bond life-time. Moreover, the impermeable and adsorbing filler surface affects considerably the dynamics of rotation isomerization and normal mode relaxation. The first effect is rather localized to the regions close to the trains. Loops, bridges and tails should exhibit primarily perturbation of the Rouse modes and retardation of the chain reptation dynamics due to the effectively increased friction coefficient and probably decreased effective tube diameter.

The trains can act as additional semi-permanent cross-links above the matrix T_g at times and temperatures, where the train life time is substantially longer compared to the time of the experiment. In such a case, loops and bridges represent segments exhibiting the so-called Langevin effect, which has been precisely described by Sternstein et al. [71]. Such bridges and loops are much shorter than the entire chain clearly resulting in considerably increased incremental stiffness, $\mathrm{d}(f_S b/3k_B T)/\mathrm{d}(r/Nb)$. Furthermore, one may take into account the effect of the reference state bias (reference state = r_0/Nb) involving those parts of chains situated at the filler-polymer interface. They can be in a state $r/Nb > r_0/Nb$ leading to the further enhancement of the interface chain stiffness [71].

Lequeux et al. [74] and other authors [43,52] have shown that the nanocomposite contains a phase at the filler-polymer interface exhibiting dynamics substantially slower in comparison to the neat polymer above T_g. Davis and co-workers [43] have shown that the time constant T_2 (determined by the NMR method) belonging to the immobilized phase is comparable to the T_2 value of the neat matrix below its T_g. Thus, it seems necessary to extend the Langevin effect [71] by the dynamics effects, which are constrained to the diffuse shell of nanometer "thickness" surrounding each particle.

According to the simulations carried out by Ediger [75], the local bond conformation transition in neat PE-like chain exhibits cooperativity making distortion in the distance of 6–10 carbon atoms, which can gain in a distance of 0.5–1 nm along the chain. In that case, one can accept the extent of glassy dynamics measured by the NMR experiment [74] as reasonable due to the intrachain cooperative transmission of the motional constraint. The inter-molecular effect may be evoked, among others, due to the topological barriers caused by other adsorbed chains primarily [38].

At any rate, it is very difficult to analyze the retarded segmental dynamics of chains at the interface in a quantitative manner because it is connected with the glass transition phenomenon and, currently there is no generally accepted theory to interpret the T_g or α-relaxation process in amorphous polymers. Nevertheless, one can see that the elastic response of chains attached to the surface is far from being of purely entropy origin. Local energy effects may become crucial and the adsorbed chains can fall out from the equilibrium under given conditions [38]. Thus, the kinetics of conformation transitions of segments attached or close to the filler surface exhibit peculiar behavior. In systems of

Viscoelasticity of Amorphous Polymer Nanocomposites

higher filler-polymer interaction $(-\varepsilon/k_B T)$, the chain segments located at the interphase possess non-equilibrium structure due to the fact that their relaxation times reach a value many orders of magnitude higher compared to the unperturbed bulk under given conditions (the ergodicity is broken within the interphase).

The effect of perturbed segmental dynamics within the interphase is, most probably, responsible for the phenomenon presented in Figure 6.6, where the temperature dependence of the relative storage modulus, G'_r, of entangled and cross-linked matrix nanocomposite above T_g is presented. It is evident that the reinforcement displays temperature dependence proportional to the nature of the polymer network. As temperature increases, the segmental mobility of the interphase chains is continuously released resulting in a slow decrease of the interphase stiffness. Whilst the neat entangled network of poly(vinyl acetate) (PVAc) loses its elasticity with the increasing temperature due to the normal mode relaxation processes, the neat cross-linked poly(ethyl acrylate) elastomer slightly stiffens in accordance with the rubber elasticity theory. Hence, relating nanocomposite elastic modulus to that of neat matrix leads to either positive or negative slope of the temperature dependence, G_r–T dependence in Figure 6.6. Moreover, substantial effect of matrix-filler bonding can be recognized. In the case of covalent filler-matrix bonding, the mobility release is substantially slower

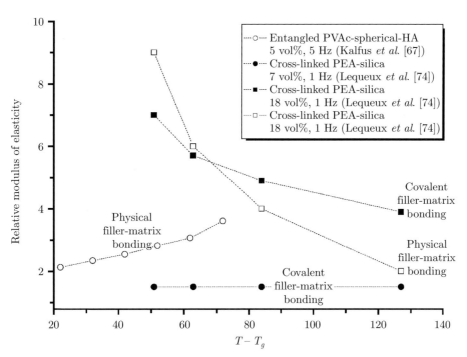

Figure 6.6. Temperature dependence of relative modulus of elasticity of an entangled and cross-linked nanocomposite

with increasing temperature compared to the system, where physical interaction between filler and matrix is dominating.

6.5. Strain induced softening of amorphous polymer nanocomposites

The strain induced softening of both entangled and cross-linked nanocomposites above their T_g, commonly termed the Payne effect [76], is a very important part of nanocomposite physics. Currently, it is in the center of interest of many researchers due to its principal relevance for the determination of a physically correct theory of nano-reinforcing mechanisms. First, it is necessary to emphasize that there is no consensus in literature on the origin of this phenomenon and, principally, two major concepts can be recognized [71,77,80]. The first concept is based on the assumption that the polymer matrix properties are modified by the filler surface corresponding to the immobilization mechanism as it has already been outlined above. During the large strain amplitudes, the chain desorption, slippage and subsequent segmental mobilization occur. Undoubtedly, this is modulated by other phenomena resulting from the structural inhomogeneity of polymer nanocomposite. The second concept utilizes the idea of filler-network structure exhibiting enough stiffness for rubbery matrix reinforcement. At high strain amplitudes, the breakage of the filler network causes partial loss of the composite stiffness. Increasing number of recently published experimental and simulation data considerably enhance validity of the first concept, which seems to be the only one able to bring a physically correct solution of the Payne effect and nanocomposite reinforcement above the matrix T_g. Nevertheless, it does not diminish the role of the inhomogeneous dispersion of filler nanoparticles. Poor dispersion of nanoparticles seems to be rather indirect by affecting local stress-strain field, accessibility of filler surface for adsorption processes, *etc.*

Schematic representation of the Payne effect for the entangled PVAc above its T_g filled with spherical and platelet hydroxyapatite nanoparticles (aspect ratio $A_r = 10$) is shown in Figure 6.7. Whilst neat PVAc exhibited constant modulus up to the maximum relative strain amplitude of 15%, the nano-filled samples softened even below 1–2% strain amplitude. Thus, a remarkable difference between linear low amplitude modulus (LAM) and non-linear high amplitude modulus (HAM) can be seen. The tangent delta increased at strain amplitudes of approximately 1–15%. At first sight, the Payne effect was more pronounced for the PVAc-platelet-HA nanocomposite. On closer look, this can be attributed to the higher extent of the filler-matrix interface area and not to the filler network and cluster percolation expectable at lower v_f compared to the spherical filler. This is clearly seen in Figure 6.8, where the low and high amplitude relative storage moduli are presented in dependence on the filler-matrix contact area. Obviously, as the strain amplitude increases the interphase is more and more perturbed. To reveal the difference of the Payne effect in an entangled and cross-linked nanocomposite above T_g, the dependence of the ratio LAM/HAM on temperature for different systems is shown in Figure 6.9. The related curve has a similar dependence as that shown in Figure 6.6.

Viscoelasticity of Amorphous Polymer Nanocomposites 229

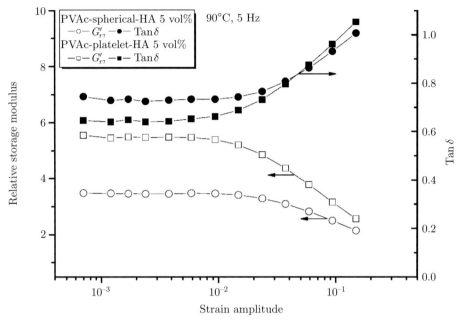

Figure 6.7. Example of dynamic strain softening of amorphous nanocomposite above neat matrix T_g. Hydroxyapatite spherical and platelet nanoparticles were used as filler. Experimental points were adopted from reference [79]

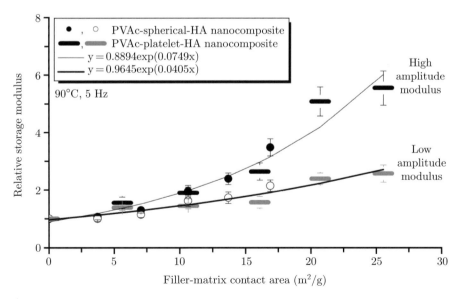

Figure 6.8. Effect of nanoparticle shape on the Payne effect; HAM – high amplitude modulus (non-linear) and LAM – low amplitude modulus (linear). Experimental points were adopted from reference [79]

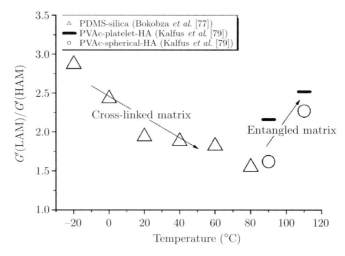

Figure 6.9. Temperature dependence of the LAM/HAM (low amplitude modulus/high amplitude modulus) for cross-linked polydimethyl siloxane (PMDS)-silica nanocomposite ($v_f = 0.15$) and entangled PVAc-HA nanocomposite ($v_f = 0.05$)

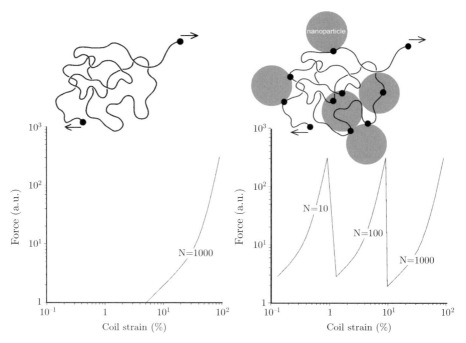

Figure 6.10. Sketch of force response of free random coil and adsorbed chain in dependence on the coil strain related to the maximum elongation, $100\% \approx 0.81 \, Nb$. Whilst free coil exhibits Gaussian response under given low strain approximately up to 20%, adsorbed loops and bridges, represented by $N = 10$ and 100, are highly Langevinean with incremental stiffness considerably increased. Strain induced perturbation of interphase accompanied by desorption of trains causes significant local stress and strain relief and, consequently, a lower stiffness

One can expect the chain desorption and slippage of chain segments at the filler-matrix interface at high strain amplitudes, which is schematically shown in Figure 6.10. The Langevin loops and bridges exhibit considerably higher stiffness compared to neat random coil. As the local high-strain limit is achieved, chain desorption from the filler surface followed by the stress release occurs [61]. The adsorption-desorption concept was utilized in the theory model by Maier and Goritz [81], where the chain desorption was considered as the primary source of the Payne effect. At high strain amplitudes, the portion of the adsorbed chains is too low to enhance nanocomposite modulus of elasticity. The above approach is further supported by the experimental works by Granick et al. [82] and Raviv et al. [83] for thin polymer layers being an analogy to the polymer interphase in polymer nanocomposites. Furthermore, the results by Keblinski et al. [33] and Raos et al. [84] indicate that the viscoelastic response of nanocomposites is primarily driven by the nature of the polymer matrix and filler surface induced immobilization rather than by the filler network. Note that Wang et al. [78] and Cassagnau [85] have published precise experimental results from viscoelastic measurements of polymer nanocomposites. Although they interpreted the measured results using the concept of the filler-network, their data fit fairly well into the concept of chain adsorption, segmental immobilization and strain induced desorption which all represent changes in the polymer matrix.

The last part of this chapter is devoted to discussion of the effect of non-uniform filler dispersion on the viscoelastic properties of nanocomposites. It seems that this phenomenon is rather inherent than just due to poor mixing of polymer and nano-filler. First, let us start with Figure 6.11, where a characteristic transmission electron microscopy (TEM) micrograph of PVAc-platelet-HA nanocomposite is presented in the upper part. Note that this is very similar to another nanocomposite system filled with spherical nano-fillers. On the micrometer level, it can be recognized that the nanocomposite is composed of heavily filled, low filled and unfilled domains and regions. In the lower part of Figure 6.11, a schematic drawing of the TEM picture is given resembling the morphology of blends composed of two immiscible polymers. In general, the observed poor nano-filler dispersion corresponds to the theoretical analysis proposed by Schweitzer et al. [86,87]. They used the microscopic polymer reference interaction site model theory to solve the miscibility phenomenon in polymer nanocomposites. For high molecular weight polymers strongly interacting with filler nanoparticles (holds for the PVAc-silica and PVAc-hydroxyapatite nanocomposites), phase separation occurs resulting in the formation of heavily filled transient polymer-filler clusters. Due to the interparticle distance within the heavily filled domains ranging from 0 to 30 nm, chain bridging may be at work. The chain end-to-end distance is commonly ranging from 10 to 30 nm in the case of flexible polymers. For PVAc presented in Figure 6.11, the end-to-end distance was about 15 nm. This is a very important finding suggesting that the filler arrangement is essential for the structure of polymer nanocomposites, but in a different manner than proposed for the filler network models. The author feels strongly that the transient character of the filler-polymer network should be considered properly.

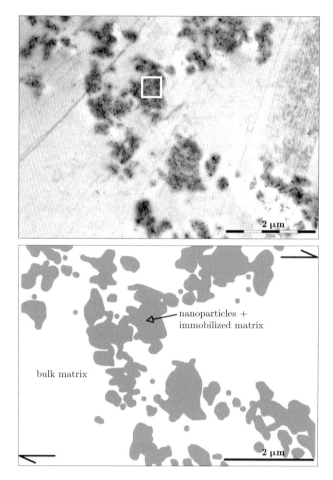

Figure 6.11. Transmission electron micrograph of local structure of PVAc-platelet-HA nanocomposite prepared by mixing matrix and filler in solution under intensive ultrasonic vibrations and stirring (upper) and an idealized drawing of the nanocomposite morphology (lower); scale bar = 2 μm. As seen, the applied shear strain may locally gain in very heterogeneous stress and strain field. The filler nanoparticles in the heavily filled domains cover 40–50% of the image area (within the white frame)

According to the image analysis of Figure 6.11, the filler particles occupy 40–50% of area in heavily filled domains. This is, however, difficult to generalize because a number of important factors affect the filler content in the heavily filled domains. They include the preparation procedure and thermal history of the nanocomposites, molecular weight of the matrix and its distribution, filler-polymer interactions, specific surface area of the filler. Nevertheless, the transient particle-polymer network in heavily filled domains affects most probably the local stress and strain field in polymer nanocomposite accompanied by extensive stress and strain concentration during mechanical loading. Moreover, occlusion of soft bulk matrix among stiff heavily filled domains can occur resulting in the strain screening effect [88]. Thus, the poor dispersion of nanoparticles is considered as the important factor affecting nanocomposite viscoelastic response. However, the principal mechanisms originate from the matrix itself; the experimental neutron scattering data recently published by Sternstein *et al.* [89] further support this.

6.6. Relaxation of chains in the presence of nanoparticles

It has been well documented that polymer dynamics is substantially affected by the presence of a nano-filler of high specific surface area. Archer *et al.* [72] have investigated the stress relaxation and frequency dependent viscoelastic behavior of PEO-silica nanocomposite. According to their results, the time characteristic of terminal normal mode relaxation is increased by many orders of magnitude with the addition of a few vol% of nano-filler. Another result has been presented by Sternstein *et al.* [76] for the case of frequency dependence of the Payne effect, which diminished at frequencies approximately below 10^{-5} Hz. Raviv *et al.* [83] have shown that the recovery time of the PEO thin layer after high strain perturbation is about 10^4 s. Moreover, Levresse *et al.* [90] have shown that the time needed for the bound layer formation in PDMS-silica nanocomposite reaches 10^6 s. No doubt, these results have been obtained for very different systems of varying matrix molecular parameters, temperatures, preparation procedure and history. On the other hand, these results highlight to what extent the relaxation behavior of the nanocomposites is affected by the large filler-matrix interface area.

Sternstein *et al.* [91] have shown that the Payne effect is a reversible process. The high strain amplitude perturbation is followed by long-time recovery of storage and loss moduli. An example of the recovery relaxation is presented in Figure 6.12. There, the experimental points were extrapolated to the equilibrium values, where the storage modulus related to the low amplitude value equals unity [79]. Relaxation behavior of this kind is attributed to multiple adsorption-desorption events occurring at a filler-polymer interface [92] leading to the recovery of the interphase structure after the Payne effect. In other words, one may approximate this process as driven by chain diffusion. In the vicinity of the surface, one can expect a higher monomer friction coefficient [93].

Kalfus *et al.* [93] have proposed a model describing modulus recovery after the Payne effect as retarded reptation motion of chains in the presence of nanoparticle surfaces. They adopted an approach, originally derived by Rubinstein *et al.* [65], for the reptation motion of PS chains deposited on SiO_2 surface. The extrapolated values of the terminal recovery relaxation time, $t_{recovery}$, after the Payne effect in the dependence on the filler-matrix contact area are shown in Figure 6.13. As it can be recognized, the transition from the lower bound (corresponding to neat polymer matrix) to the upper bound (where each single matrix chain is attached to the surface) exhibits a shape suggesting the onset of a percolation type transition. This is in agreement with the concept of transient filler-polymer network. In the region above the percolation threshold and

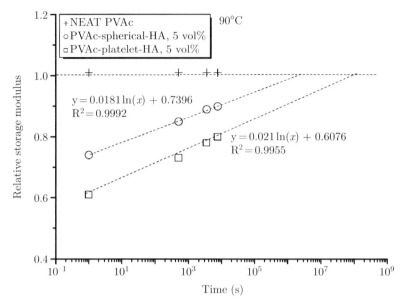

Figure 6.12. Example of recovery of storage modulus after high strain amplitude perturbation of PVAc filled with 5 vol% of nano-sized hydroxyapatite spheres and platelets [79]

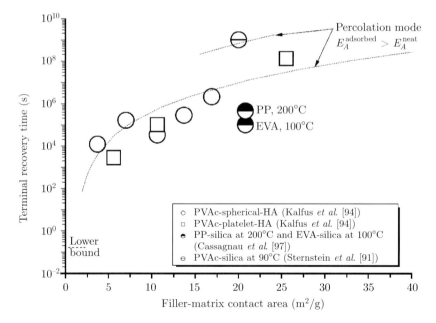

Figure 6.13. Terminal recovery times obtained from extrapolation of modulus recovery curves as it is shown in Figure 6.12. EVA and PP depict ethylene-vinylacetate copolymer and polypropylene, respectively

below the upper limit, the dependence of the recovery relaxation time on the filler content seems to follow a simple exponential law in the form often used for conducting composites [94,95]:

$$\tau_{recovery} \propto \left(\frac{v_{eff} - v_{eff}^*}{1 - v_{eff}^*}\right)^t, \tag{6.28}$$

where v_{eff} and v_{eff}^* are the effective filler volume fraction and critical effective filler volume fraction, respectively. The effective filler is considered here as a spherical particle composed of the interphase shell surrounding the true nano-filler particle core. The filler-matrix contact area is then calculated from the known filler weight fraction and specific surface area of a nano-filler. The exponent t in Eq. 6.28 was found to be around 3. For comparison, the dependence for PVAc-HA nanocomposite is supplemented with the value obtained from results by Sternstein et al. [91] for PVAc-silica nanocomposite and values obtained from the extrapolation of recovery data by Cassagnau et al. [96]. The activation energy of thermally activated movement of PVAc adsorbed on silica surface exhibited a slightly higher value compared to the PVAc-HA system most probably due to higher population of hydroxyl groups on silica surface.

6.7. `Conclusions and outlook

In this contribution, a brief review of selected results was presented on the elastic and viscoelastic behavior of amorphous polymer nanocomposites. The overview was done in a simplified manner, in order to support its understandability. Below the matrix T_g, a nanocomposite behaves like a two component system due to the low-entropy/low-mobility state of polymer matrix. Above the matrix T_g, the polymer chains near the nano-filler surface become perturbed in respect to their dynamics. These changes occurring on the molecular level cause severe effects observable on micro- and macroscopic levels. Due to the extensive nano-filler surface area, the filler nanoparticles are able to cause these effects even at a very low filler concentration. Interestingly, the immobilization phenomenon in polymers filled with high specific surface area fillers has already been addressed in the 60s by DiBenedetto [40], Lipatov [44,56] and others [39,43,45]. The cited authors interpreted the results properly although very poor computer simulation possibilities were available at that time.

Currently, no universal theory is available to predict quantitatively the effects of chain immobilization on the reinforcement of nanocomposites above their T_g. Most probably, the multi-length- and multi-time-scale computer models will be necessary for these quantitative predictions being able to account for the external effects (time, temperature, strain, etc.) and structural levels. However, this is connected with the more general, and still open, issue of the boundary between applicability of the continuum mechanics and the molecular approach. In addition, retarded and spatially heterogeneous segmental dynamics in polymer nanocomposites represents an issue principally connected with the glass transition phenomenon, which is still unresolved.

Hence, polymer nanocomposites represent a very attractive field of materials sciences, bringing new challenges for experimental, simulation, as well as theory scientists. Revealing those molecular mechanisms, which are responsible for the viscoelastic response of polymer nanocomposites, should be the first task to get a deeper understanding of the structure-property relationships in these materials. Afterward, one would be able to transfer the related knowledge in order to interpret other phenomena, such as creep, fatigue and impact behaviors accordingly.

Acknowledgements

The author is very thankful to Prof. J. Jancar, Prof. S. S. Sternstein and Dr. J. Kucera for many helpful discussions on the molecular level phenomena in polymer nanocomposites. Critical comments on the text by Dr. R. Balkova and K. Hynstova are also appreciated. The author is financially supported by the Czech Ministry of Education, Youth and Sports (grant MSM 0021630501).

References

1. Feynman R (2004) *The Pleasure of Finding Things Out*, Aurora, Prague (in Czech).
2. Ediger M D, Angell C A and Nagel S R (1996) Supercooled liquids and glasses, *J Phys Chem* **100**:13200–13212.
3. Zallen R (2004) *Physics of Amorphous Solids*, Wiley-Interscience, New York.
4. Jančář J (1999) Structure-property relationships in thermoplastic matrices, *Adv Polym Sci* **139**:2–65.
5. Birshtein T M and Ptytsin O B (1966) *Conformations of Macromolecules*, John Wiley & Sons, New York.
6. Flory P J (1969) *Statistical Mechanics of Chain Molecules*, John Wiley & Sons, New York.
7. Živný A (1968) *Statistics of Linear Cooperative Systems*, Czechoslovak Academy of Sciences, Prague (in Czech).
8. Boyd R H and Phillips P J (1996) *The Science of Polymer Molecules*, Cambridge University Press, Cambridge.
9. Chandrasekhar S (1943) Stochastic problems in physics and astronomy, *Rev Mod Phys* **15**:1–89.
10. Treloar L R G (2005) *The Physics of Rubber Elasticity*, Oxford University Press, Oxford.
11. Wu S (1992) Predicting chain conformation and entanglement of polymers from chemical structure, *Polym Eng Sci* **32**:823–830.
12. Flory P J, Hoeve C A J and Ciferri A (1959) Influence of bond angle restrictions on polymer elasticity, *J Polym Sci: Polym Symp* **34**:337–347.
13. Strobl G (2007) *Physics of Polymers*, Springer-Verlag, Berlin.
14. Skolnick J and Helfand E (1980) Kinetics of conformational transitions in chain molecules, *J Chem Phys* **72**:5489–5500.
15. Harmandaris V A, Mavrantzas V G and Theodorou D N (1998) Atomistic molecular dynamics simulation of polydisperse linear polyethylene melts, *Macromolecules* **31**:7934–7943.

16. Qiu X H and Ediger M D (2002) Entanglement effects in polyethylene melts: C–13 NMR relaxation experiments, *Macromolecules* **35**:1691–1698.
17. Lin Y H (2003) *Polymer Viscoelasticity, Basics, Molecular Theories, Experiments*, World Scientific, London.
18. Matsuoka S (1997) Entropy, free volume, and cooperative relaxation, *J Res Nat Inst Stand Tech* **102**:213–228.
19. Bueche F (1962) *Physical Properties of Polymers*, Interscience Publishers, New York.
20. deGennes P G (1979) *Scaling Concepts in Polymer Physics*, Cornell University Press, Cornell.
21. Doi M and Edwards S F (2003) *The Theory of Polymer Dynamics*, Oxford University Press, Oxford.
22. Kausch H H (1978) *Polymer Fracture*, Springer-Verlag, Berlin.
23. Lin Y H (2005) Whole range of chain dynamics in entangled polystyrene melts revealed from creep compliance: Thermorheological complexity between glassy-relaxation region and rubber-to-fluid region. 1, *J Phys Chem B* **109**:17654–17669.
24. DiMarzio E A and McCrackin F L (1965) One-dimensional model of polymer adsorption, *J Chem Phys* **43**:539–547.
25. Starr F W, Schroder T B and Glotzer S C (2002) Molecular dynamics simulation of a polymer melt with a nanoscopic particle, *Macromolecules* **35**:4481–4492.
26. Vacatello M (2002) Molecular arrangements in polymer-based nanocomposites, *Macromol Theory Simul* **11**:757–765.
27. Vacatello M (2003) Predicting the molecular arrangements in polymer-based nanocomposites, *Macromol Theory Simul* **12**:86–91.
28. Sharaf M A and Mark J E (2004) Monte Carlo simulations on the effects of nanoparticles on chain deformations and reinforcement in amorphous polyethylene networks, *Polymer* **45**:3943–3952.
29. Ozmusul M S, Picu C R, Sternstein S S and Kumar S K (2005) Lattice Monte Carlo simulations of chain conformations in polymer nanocomposites, *Macromolecules* **38**:4495–4500.
30. Nakatami A I, Chen W, Schmidt R G, Gordon G V and Han C C (2002) Chain dimensions in polysilicate-filled poly(dimethyl siloxane), *Int J Thermophys* **23**:199–209.
31. Konstadinidis K, Thakkar B, Chakraborty A, Potts L W, Tannenbaum R and Tirrell M (1992) Segment level chemistry and chain conformation in the reactive adsorption of poly(methyl methacrylate) on aluminum oxide surfaces, *Langmuir* **8**:1307–1317.
32. Gaylord R J (1979) The confined chain approach to the deformation behavior of bulk polymers, *Polym Eng Sci* **19**:955–965.
33. Sen S, Thomin J D, Kumar S K and Keblinski P (2007) Molecular underpinnings of the mechanical reinforcement in polymer nanocomposites, *Macromolecules* **40**:4059–4067.
34. Nowicki W (2002) Structure and entropy of a long polymer chain in the presence of nanoparticles, *Macromolecules* **35**:1424–1436.
35. Wang M J and Wolff S (1992) Filler-elastomer interactions, *Kaut Gum Kunst* **45**:11–17.
36. Santore M and Fu Z (1997) Direct measurement of molecular-weight driven competition during polymer adsorption, *Macromolecules* **30**:8516–8517.
37. Bell R C, Wang H, Iedema M J and Cowin J P (2003) Nanometer-resolved interfacial fluidity, *J Am Chem Soc* **125**:5176–5185.
38. Shaffer J S and Chakraborty A K (1993) Dynamics of poly(methyl methacrylate) chains adsorbed on aluminum surfaces, *Macromolecules* **26**:1120–1136.

39. Smit P P A (1966) The glass transition in carbon black reinforced rubber, *Rheol Acta* **5**:277–283.
40. Droste D H and DiBenedetto A T (1969) The glass transition temperature of filled polymers and its effect on their physical properties, *J Appl Polym Sci* **13**:2149–2168.
41. Waldrop M A and Kraus G (1969) Nuclear magnetic resonance study of the interaction of SBR with carbon black, *Rubber Chem Tech* **42**:1155–1166.
42. Kraus G and Gruver J T (1970) Thermal expansion, free volume, and molecular mobility in a carbon black-filled elastomer, *J Polym Sci: Part A-2* **8**:571–581.
43. Kaufman S, Slichter W P and Davis D D (1971) Nuclear magnetic resonance study of rubber-carbon black interactions, *J Polym Sci: Part A-2* **9**:829–839.
44. Lipatov Y S and Fabulyak F Y (1972) Relaxation processes in the surface layers of polymers at the interface, *J Appl Polym Sci* **16**:2131–2139.
45. Pliskin I and Tokita N (1972) Bound rubber in elastomers: analysis of elastomer-filler interaction and its effect on viscosity and modulus of composite systems, *J Appl Polym Sci* **16**:473–492.
46. Ziegel K D and Romanov A (1973) Modulus reinforcement in elastomer composites. I. Inorganic fillers, *J Appl Polym Sci* **17**:1119–1131.
47. O'Brien J, Cashell E, Wardel G E and McBrierty V J (1976) NMR investigation of interaction between carbon black and cis-polybutadiene, *Macromolecules* **9**:653–660.
48. Douglass D C and McBrierty V J (1979) Interfacial effects on the NMR of composite polymers, *Macromolecules* **19**:1054–1063.
49. Fabulyak F G, Lipatov Y S and Vovchuk D S (1981) Strain effect on dielectrical relaxation in filled rubber, *Vysokomol Soyed* A23 **11**:2449–2453.
50. Serizawa H, Nakamura T, Ito M, Tanaka K and Nomura A (1983) Effects of oxidation of carbon-black surface on the properties of carbon black-natural rubber systems, *Polym J* **15**:201–206.
51. Struik L C E (1987) The mechanical behavior and physical ageing of semicrystalline polymers. 2., *Polymer* **28**:1534–1542.
52. Asai S, Kaneki H, Sumita M and Miyasaka K (1991) Effect of oxidized carbon black on the mechanical properties and molecular motions of natural rubber studied by pulse NMR, *J Appl Polym Sci* **43**:1253–1257.
53. Jancar J, Kucera J and Vesely P (1991) Peculiarities of mechanical response of heavily filled polypropylene composites, *J Mat Sci* **26**:4878–4882.
54. Legrand A P, Lecomte N, Vidal A, Haidar B and Papirer E (1992) Application of NMR spectroscopy to the characterization of elastomer/filler interactions, *J Appl Polym Sci* **46**:2223–2232.
55. Shang S W, Williams J W and Soderholm K J M (1994) How the work of adhesion affects the mechanical properties of silica-filled polymer composites, *J Mat Sci* **29**:2406–2416.
56. Lipatov Y S (1995) *Polymer Reinforcement*, ChemTec Publishing, Toronto.
57. Xie X Q, Ranade S V and DiBenedetto A T (1999) A solid state NMR study of polycarbonate oligomer grafted onto the surface of amorphous silica, *Polymer* **40**:6297–6306.
58. Zhu A J and Sternstein S S (2001) Nanofiller-polymer interactions at and above the glass transition temperature, *Mat Res Soc Symp Proc* **661**:KK4.3.1.
59. Berriot J, Lequeux F, Monnerie L, Montes H, Long D and Sotta P (2002) Filler-elastomer interaction in model filled rubbers, a H–1 NMR study, *J Non-Cryst Solids* **307–310**:719–724.
60. Lin E K, Kolb R, Sajita S K and Wu W (1999) Reduced polymer mobility near the polymer/solid interface as measured by neutron reflectivity, *Macromolecules* **32**:3753–3757.

61. Lin E K, Kolb R, Sajita S K and Wu W (1999) Enhanced polymer segment exchange kinetics due to an applied shear field, *Macromolecules* **32**:4741–4744.
62. Wang Y, Rajagopalan R and Mattice W L (1995) Kinetics of detachment of homopolymers from solid surface, *Phys Rev Lett* **74**:2503–2506.
63. Frantz P and Granick S (1991) Kinetics of polymer adsorption and desorption, *Phys Rev Lett* **66**:899–902.
64. Johnson H S, Douglas and Granick S (1993) Topological influences on polymer adsorption and desorption dynamics, *Phys Rev Lett* **70**:3267–3270.
65. Zheng X, Sauer B B, Van Alsten J G, Schwarz S A, Rafailovich M H, Sokolov J and Rubinstein M (1995) Reptation dynamics of a polymer melt near an attractive solid surface, *Phys Rev Lett* **74**:407–410.
66. Nielsen L E and Landel R F (1994) *Mechanical Properties of Polymers and Composites*, Dekker, New York.
67. Kalfus J and Jancar J (2007) Elastic response of nanocomposite poly(vinylacetate)-hydroxyapatite with varying particle shape, *Polym Comp* **28**:365–371.
68. Kalfus J and Jancar J (2008) Reinforcing mechanisms in amorphous polymer nanocomposites, *Compos Sci Technol*, in press.
69. Guth E (1945) Theory of filler reinforcement, *J Appl Phys* **16**:20–25.
70. Matějka L, Dukh O and Kolařík J (2000) Reinforcement of crosslinked rubbery epoxies by *in-situ* formed silica, *Polymer* **41**:1449–1459.
71. Sternstein S S and Zhu A J (2002) Reinforcement mechanism of nanofilled polymer melts as elucidated by nonlinear viscoelastic behavior, *Macromolecules* **35**:7262–7273.
72. Zhang Q and Archer L A (2002) Poly(ethylene oxide)/silica nanocomposites: structure and rheology, *Langmuir* **18**:10435–10442.
73. Montes H, Lequeux F and Berriot J (2003) Influence of the glass transition temperature gradient on the nonlinear viscoelastic behavior in reinforced elastomers, *Macromolecules* **36**:8107–8118.
74. Berriot J, Montes H, Lequeux F, Long D and Sotta P (2002) Evidence for the shift of the glass transition near the particles in silica-filled elastomers, *Macromolecules* **35**:9756–9762.
75. Adolf A B and Ediger M D (1992) Cooperativity of local conformational dynamics in simulations of polyisoprene and polyethylene, *Macromolecules* **25**:1074–1078.
76. Chazeau L, Brown J D, Yanyo L C and Sternstein S S (2000) Modulus recovery kinetics and other insights into the Payne effect for filled elastomers, *Polym Comp* **21**:202–222.
77. Clement F, Bokobza L and Monnerie L (2005) Investigation of the Payne effect and its temperature dependence on silica-filled polydimethylsiloxane networks. Part II: Test of quantitative models, *Rubber Chem Tech* **78**:232–244.
78. Thu Z, Thompson T, Wang S Q, von Meerwall E D and Halasa A (2005) Investigating linear and nonlinear viscoelastic behavior using model silica-particle-filled polybutadiene, *Macromolecules* **38**:8816–8824.
79. Kalfus J and Jancar J (2007) Viscoelastic response of nanocomposite poly(vinyl acetate)-hydroxyapatite with varying particle shape-dynamic strain softening and modulus recovery, *Polym Comp* **28**:743–747.
80. Clement F, Bokobza L and Monnerie L (2005) Investigation of the Payne effect and its temperature dependence on silica-filled polydimethylsiloxane networks. Part I: Experimental results, *Rubber Chem Tech* **78**:211–231.
81. Maier P G and Goritz D (1996) Molecular interpretation of the Payne effect, *Kaut Gum Kunst* **49**:18–21.
82. Granick S, Hu H W and Carson G A (1994) Nanorheology of confined polymer melts. 2. Nonlinear shear response at strongly adsorbing surfaces, *Langmuir* **10**:3867–3873.

83. Raviv U, Klein J and Witten T A (2002) The polymer mat: arrested rebound of a compressed polymer layer, *Europ Phys J E* **9**:405–412.
84. Raos G, Moreno M and Elli S (2006) Computational experiments on filled rubber viscoelasticity: What is the role of particle-particle interactions?, *Macromolecules* **39**:6744–6751.
85. Cassagnau P (2003) Payne effect and shear elasticity of silica-filled polymers in concentrated solutions and in molten state, *Polymer* **44**:2455–2462.
86. Hooper J B and Schweizer K S (2007) Real space structure and scattering patterns of model polymer nanocomposites, *Macromolecules* **40**:6998–7008.
87. Hooper J B and Schweizer K S (2006) Theory of phase separation in polymer nanocomposites, *Macromolecules* **39**:5133–5142.
88. Pryamitsyn V and Ganesan V (2006) Origins of linear viscoelastic behavior of polymer-nanoparticle composites, *Macromolecules* **39**:844–856.
89. Narayanan R A, Thiyagarajan P, Zhu A J, Ash B J, Shofner M L, Schadler L S, Kumar S K and Sternstein S S (2007) Nanostructural features in silica-polyvinyl acetate nanocomposites characterized by small-angle scattering, *Polymer* **48**:5734–5741.
90. Levresse P, Feke D L and Manas-Zloczower I (1998) Analysis of the formation of bound poly(dimethylsiloxane) on silica, *Polymer* **39**:3919–3924.
91. Zhu A J and Sternstein S S (2003) Nonlinear viscoelasticity of nanofilled polymers: interfaces, chain statistics and properties recovery kinetics, *Comp Sci Tech* **63**:1113–1126.
92. Zhang Q and Archer L A (2003) Effect of surface confinement on chain relaxation of entangled *cis*-polyisoprene, *Langmuir* **19**:8094–8101.
93. Kalfus J and Jancar J (2007) Relaxation processes in PVAc-HA nanocomposites, *J Polym Sci: Part B: Polym Phys* **45**:1380–1388.
94. Dalmas F, Dendievel R, Chazeau L, Cavaille J Y and Gauthier C (2006) Carbon nanotube-filled polymer of electrical conductivity in composites. Numerical simulation of three-dimensional entangled fibrous networks, *Acta Mater* **54**:2923–2931.
95. Favier V, Dendievel R, Canova G, Cavaille J Y and Gilormint P (1997) Simulation and modeling of three-dimensional percolating structures: Case of a latex matrix reinforced by a network of cellulose fibers, *Acta Mater* **45**:1557–1565.
96. Cassagnau P and Melis F (2003) Non-linear viscoelastic behavior and modulus recovery in silica filled polymers, *Polymer* **44**:6607–6615.

Chapter 7

Interphase Phenomena in Polymer Micro- and Nanocomposites

J. Jancar

7.1. Introduction

In polymer matrix composites exhibiting heterogeneous structure at multiple length scales, the interphase phenomena at various length scales were shown to be of pivotal importance for the control of the performance and reliability of such structures [1]. Various models based on continuum mechanics were used to satisfactorily describe effects of the macro- and meso-scale interphase on the mechanical response of laminates and large fiber reinforced composite (FRC) parts. There is also a substantial body of literature dealing with the micro-scale interface/interphase phenomena in particulate filled polymers [2]. At the micro-scale, the interphase is considered a 3D continuum characterized by a set of average properties (elastic moduli, yield strength, toughness, *etc.*) and uniform thickness, t. A number of continuum mechanics models were derived over the last 50 years to describe the stress transfer between matrix and individual fiber with relatively good success. In these models, the interphase was characterized by some average shear strength, τ_a, and elastic modulus, E_a. On the other hand, models for transforming the properties of the micro-scale interphase around individual fibers into the mechanical response of macroscopic multifiber composites have not been generally successful. The anisotropy of these composite structures is the main reason causing the failure of these models. The strong thickness dependence of the elastic modulus of the micro-scale interphase suggested presence of its underlying sub-structure. At the same time, the micro-scale effects of interphase/interface phenomena have been modeled in particulate composites and translated into macromechanical models with significantly greater success compared to FRCs.

Over the last 20 years, polymer matrix nanocomposites came into the center of attention of the composite research community. According to information retrieved from the Web of Science database, more than 6000 research papers have been published on various aspects of polymer nanocomposites since the year 2000. More than 60% of these papers deal with polymers reinforced with montmorillonite or other layered silicates and another 20% were devoted to nanocomposites containing various types of carbon nanotubes (CNT). Only about 800 papers dealt with polymers containing other nanofillers such as silica, carbon black or hydroxyapatite. Among the matrices used in the published research, various polyamides and epoxies were prevailing, followed by polyolefins and rubbers. In most of the papers published, the interface/interphase phenomena are considered the most important for the physical and mechanical performance of the polymer nanocomposites. Unfortunately, more than 60% of the papers on layered silicate reinforced polymers published since the year 2000 dealt with the interface/interphase phenomena only as a problem related to various aspects of "exfoliation/intercalation". Moreover, in most cases, layered silicates are not true nano-sized fillers, since only one of their dimensions can be in the nanometer range provided cleavage of the stacks has been achieved, and thus their great effect on the mechanical response of polymers can easily be elucidated using existing micromechanics models such as the Halpin-Tsai equation [3–5]. In composites containing various types of CNT, the diameter of the tube falls within the nanometer range while the length is commonly in the micrometer range. Hence, only nanocomposites containing silica, carbon black, hydroxyapatite or other synthetic nanoparticles with all dimensions in the range similar to the polymer chain dimensions can be considered as *true nanocomposites**. Research of true nano-composites has resulted in renewed interest in basic principles of polymer physics capable of describing the changes in chain statistics and dynamics caused by true nanoparticles. Despite the fact that most of the effects exist also in micro-composites, they became prominent in nanocomposites due to the vast area of matrix-filler interface distinguishing the nano-composites from the micro-composites.

It has been recognized that, on the nano-scale, the discrete molecular structure of the polymer has to be considered. The term *interphase*, originally introduced in multiphase continuum models of composites, has to be re-defined to include the discrete nature of the matter at this length scale. The segmental immobilization resulting in retarded reptation of chains caused by interactions with solid surface seems to be the primary phenomenon which can be used to re-define the term interphase on the nano-scale. The very nature of chain and segment relaxations has to be re-examined since the presence of solid particles with size equal to the portion of the radius of gyration, R_g, resulted in perturbation of chain statistics and dynamics on a large scale. In most cases, there are no "bulk" chains in the nanocomposites containing more than 3–5 vol% of

---

* Editorial note: This term seems to be rather unusual since it is commonly accepted to consider as nanomaterials such ones having at least one dimension in the nano range.

the nano-sized filler, hence, all the chains are more or less in the "interphase" [6–10].

All the experiments measuring mechanical properties of nano-composites are performed using macro-scale specimens, however, attempts are made to interpret these data in terms of nano-structural features. One has to be very cautious in doing so, since the traditional continuum mechanics approach has to be modified substantially in order to account for the discontinuous nature of matter at the nano-scale. It has also been shown that for length scales below 5 nm, Bernoulli-Euler continuum elasticity becomes not valid and higher order elasticity has to be applied [11–13]. In order to find suitable means for bridging the gap between the models based on the mechanics of continuum matter at the micro-scale and mechanics of discrete matter at the nano-scale, molecular dynamics approach combined with higher order continuum elasticity can be used.

Despite the large volume of experimental data on the peculiarities of the deformation response of polymer nano-composites published over the last 20 years [14–16], models addressing the principal physical reinforcing mechanisms and taking into account appropriate length and time scales are scarce and not generally accepted [17]. The characteristic length and time scales in polymer nanocomposites [18–20] are strongly temperature dependent, which further complicates the theoretical treatment due to principal differences in the behavior of polymers below and above their glass transition temperature, T_g. One has to bear in mind, however, that the T_g is not a thermodynamic property as it is often assumed [21] and has rather a kinetic character representing various relaxation processes existing in glass forming materials [22]. In addition, the changes in chain mobility due to the presence of solid inclusions in the melt [10,23–25] also affect resulting structure and properties of a solid amorphous nanocomposite below T_g or crystalline nanocomposite below T_c due to effects of segmental immobilization on the formation of polymer structure during solidification [26,27] or crystallization [28–30]. It seems evident that the principal reinforcing mechanism above the T_g is related to the chain immobilization induced by the presence of rigid high specific surface area inclusions. In other words, the interphase/interface phenomena are of pivotal importance in systems with nano-sized solid inclusions. In small volume fraction (v_f) nanocomposites commonly considered ($v_f \leq 0.05$), the effect of filler volume becomes of second order importance due to the large surface to volume ratios and the surface effects are of primary importance. This should be reflected in modified predictive models where the v_f used as the main structural variable should be replaced with specific surface area per unit weight of composite, A_f [27].

In addition, when considering the molecular nature of the reinforcement at the nano-scale, the limits of validity of size independent continuum mechanics should be determined, since the discrete nature of the matter at the nano-scale prohibits simple scaling down of the micro-mechanics models [31–38]. In this contribution, the phenomena related to the behavior of chains in the immediate vicinity of nano-sized rigid inclusions residing in the polymer matrix are reviewed in view of the characteristic length and volume scales in polymers with nano-scale heterogeneities. The results are used to provide a conceptual framework for selecting a suitable molecular dynamics model accounting for molecular details relevant to the deformation response of true nanocomposites and identification of a possible bridging law between the micro- and nano-length scales.

Continuum mechanics can be used to describe effects of micro-scale interphase on the stress transfer in single fiber composites considering the system a three-phase material in which the individual phases, *i.e.*, solid inclusion, matrix and interphase, can be characterized by some average properties [39–47]. Unlike at the micro-scale, extreme caution has to be exercised when selecting a suitable modeling scheme at the nano-scale when the discrete molecular structure of the polymer becomes obvious [48–50]. One of the main difficulties when considering nano-scale in composites is to determine the size of the representative volume in which the discrete nature of the composite structure has to be taken into account [51]. Very little has been written so far on the laws governing the transition between the nano- and micro-length scales, especially on the reliability of classical continuum mechanics when scaled outside their validity range, *i.e.*, down to the nano-scale. Thus it seems desirable to perform a critical review of the current knowledge on the structure and properties of the micro-

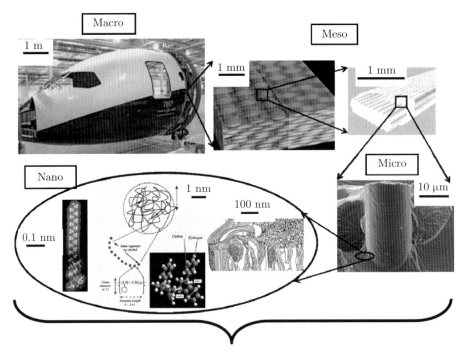

Figure 7.1. Part of the fuselage of Boeing 787 Dreamliner can serve as an example of a large manmade *multiscale* composite structure. This polymer composite structure has been designed using the engineering top-bottom methodology within the framework of continuum mechanics. No functional hierarchy exists between the various length scales [169]

and nano-scale interphases in polymer composites and the methodologies for their modeling in order to provide means for bridging the gap between continuum and discrete models useful for reliable design of future multiscale hierarchical composite structures.

The design of multi-length-scale composite structures, such as the fuselage of the Boeing 787 (Figure 7.1), represents the state-of-the-art engineering application of fiber reinforced composites (FRCs). These large structures are designed from top to bottom using continuum mechanics methodologies and the transitions between the individual length scales are treated simply by scaling down the structural features of the greater length scale. Such multiscale continuum mechanics modeling approach was demonstrated to provide reasonable means for transforming the mechanical response of polymer composites across several length and time scales from macro- down to micro-scale (Figure 7.2) [31].

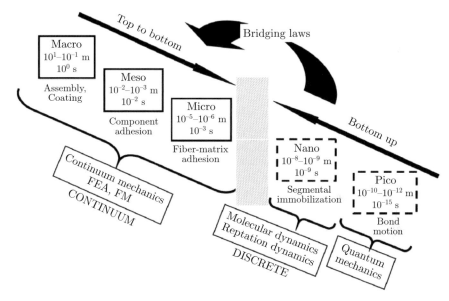

Figure 7.2. Schematic drawing of the two principal methodologies in designing multiscale composite structures, *i.e.*, the *top-bottom* engineering approach and the *bottom-up* approach observed in natural composites with the emphasis on the role of interphase phenomena at various length and time scales [169]

Since the FRCs exhibit heterogeneous structures at multiple length scales, the interphase phenomena are of pivotal importance for the control of the reliability and performance of multiscale FRC structures. There seems to be general agreement on using continuum mechanics models to account for interphase phenomena from macro- to micro-scale and the design schemes based on continuum mechanics, variational principles or finite element analysis (FEA) have been validated. The understanding of the translation of the properties of the microscale interphases into the response of macroscopic FRC parts is far less unam-

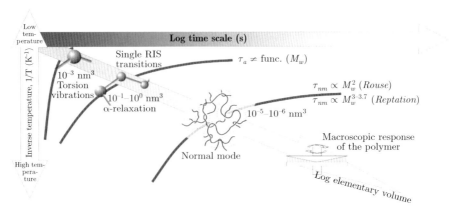

Figure 7.3. Molecular relaxation processes on the background of the time, temperature and elementary volume scale [101]

biguous. The greatest success has been achieved in understanding and modeling of the role of the micro-scale interphase in the stress transfer from the matrix to a single fiber in model composites. However, attempts to transfer properties of the micro-scale interphase in the performance of a multi-fiber FRC structures have generally failed. On the nano-scale, the discrete nature of the polymer has to be considered introducing additional length scale related phenomena, such as the chain relaxation heterogeneity. This results in more complex conditions which have to be taken into account when determining the size of the representative volume element in nanostructured composites (Figure 7.3).

In this chapter, the interphase phenomena in the polymer matrix composites at the micro- and nano-scales are briefly reviewed and their main differences are discussed. An approach for modeling mechanical properties of continuous macroscopic bodies considering peculiarities brought about by the discrete nature of the matter at the nano-scale is proposed based on the combination of gradient strain elasticity and chain reptation dynamics.

7.2. Micro-scale interphase in polymer composites

The research of the micro-scale interphase phenomena in composite materials has attracted considerable attention on the part of both the scientific and engineering communities over the last 50 years [1]. Good success has been achieved in describing the role of the interphase in stress transfer from the matrix to the fiber using model single fiber composites (Figure 7.4). From the simple Kelly-Tyson model, to the various lap shear models, to the numerical FEA models, the approach based on continuum mechanics has been employed [40,41,52–94]. Even though the molecular structure of the interphase has been anticipated in many papers, with some exceptions [48,49], the main effort has been devoted to the relationship between the type, thickness and deposition conditions of the fiber coating and the average shear strength of the interphase, τ_a, measured in a simple test employing model single fiber composite [52].

Interphase Phenomena in Polymer Micro- and Nanocomposites

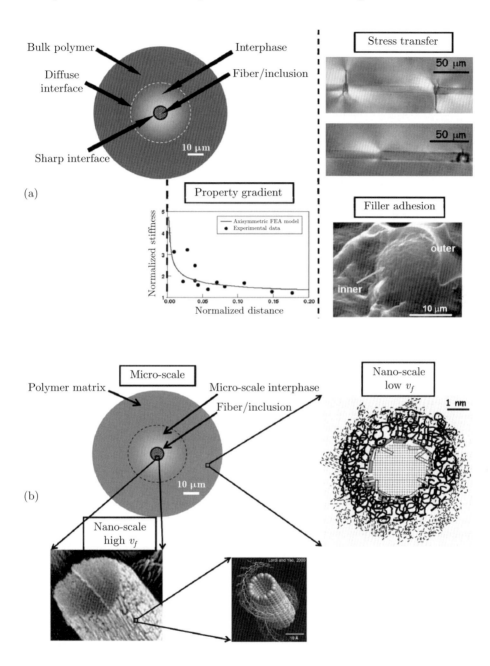

Figure 7.4. (a) Visualizing the interphase considering only the micro-scale. Interphase is a continuum layer with a gradient of properties reflecting variations in its structure. The main role of the micro-scale interphase is to provide stable and effective means for stress transfer between inclusions and polymer matrix even under adverse conditions. (b) Visualizing the structure of a micro-composite considering also the nano-scale structural features when the discrete structure of the matrix and inclusions becomes evident [169]

The mechanical properties and environmental stability of both fiber reinforced and particulate filled thermoplastic composites are strongly dependent upon the stability of the interfacial region between the matrix and fibers, especially when exposed to moist environments. This is of particular importance in glass fiber reinforced thermoplastic composites since the glass fibers are highly hygroscopic and the bond between the fibers and the thermoplastic matrix is usually weak. Hence, the tailoring of well-bonded, durable interphases between the thermoplastic matrix and glass reinforcement has become a critical concern. The use of coupling agents, chemically reactive with both matrix and reinforcement, and/or chemical modification of the surfaces of one or both constituents has been the most successful means of providing reasonably well controlled bonds between matrix and the encapsulated reinforcement [1,87,88].

From the published data, it seems clear that a monomolecular interphase layer with engineered molecular structure specific to the desired combination of resin and reinforcement should result in the most favorable mix of properties in thermoplastic matrix composites. Reactive end-capped polymers capable of chemically reacting with the fiber surface or various methods of grafting matrix molecules onto the reinforcement surface are the most promising candidates for further investigations [50,54,97]. Thickness of the interphase can be controlled *via* modification of the molecular weight and chain stiffness of the constituent molecules, its mechanical properties can be varied by selecting the backbone chain constitution and configuration, and its surface free energy can also be controlled by the chain constitution and by the polarity of the end-groups. Elastic properties of these layers are controlled by the attraction forces at the interface as well as the conformation entropy of the chains forming the layer.

Organofunctional silanes are so far the most widely used coupling agents for improvement of the interfacial adhesion in glass reinforced materials [87]. Upon application of a silane from either dilute solution or the vapor phase, a highly cross-linked multilayer siloxane "interphase" is presumably formed with thicknesses ranging from 1.5 to 500 nm. Unlike in thermosetting matrices with extensive interpenetration between organosilane layer and the matrix monomer, long chain molecules do not interpenetrate the organosilane layers significantly. On the other hand, immobilization phenomena are of greater importance in thermoplastic matrix composites. Moreover, processing temperatures of engineering thermoplastics ranging from 230 to 350°C can exceed the thermal stability of the commonly utilized organosilanes.

Elastic moduli of silane layers deposited in flat glass substrates decreased monotonically with increasing layer thickness, reaching a bulk value for layers thicker than 10^5 nm [95,96]. Reactive chlorine containing silane always formed stiffer layers compared to its alkoxy-analogues, most probably due to stronger interaction between the chlorine and glass surface and, most probably, due to less defective network structure (Figure 7.5). This hypothesis was further supported by the observed strong effect of deposition technique, controlling the layer supermolecular structure, on the layer elastic modulus. Solution deposition technique always yielded layers with lower elastic modulus compared to the

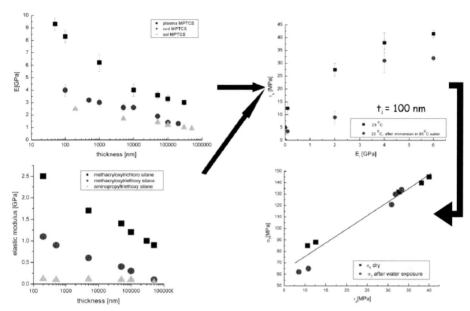

Figure 7.5. A typical example of the micro-scale organosilane interphases of various composition deposited on glass fibers using various deposition techniques and translation of the phenomenological properties of the interphases into mechanical response of multifiber composites with unidirectionally aligned fibers [96]

layers formed by low temperature radio-frequency-plasma (rf-plasma) or rf-plasma enhanced CVD deposition of the same substance. A qualitative explanation of the observed behavior was provided assuming formation of a strongly immobilized layer of constant thickness, t_i, and elastic modulus, E_i, near the bonded interface. Strength of the interfacial bond and network density of the polysiloxane interphase were proposed to be the factors determining t_i and E_i for the given external conditions. Experimental data showed that the contribution of this strongly immobilized layer started to play an important role for interphase thickness below 10^3 nm. This "inner" layer has been covered with a weaker "outer" layer with more defective network structure. The thickness of the "outer" layer was dependent on the concentration of the silane solution it was deposited from. The difference in E_i between the outer and inner interphase layers was increasing with strengthening the layer-surface interaction.

Similarly to the silane interphases, thickness dependence of the elastic modulus of thin polycarbonate (PC) layers deposited on a flat E-glass substrate was measured over the thickness interval ranging from 10^6 nm to 30 nm [39,44,46,95]. In all cases investigated, elastic moduli of the deposited layers, E_i, decreased monotonically with increasing layer thickness reaching a constant bulk value for layers thicker than 5×10^5 mm (Figure 7.4). Thermally annealed PC and $SiCl_4$ grafted oligo-PC interphases exhibited higher elastic moduli than the as received solution deposited PC interphase. No effect of thermal annealing on elastic modulus of strongly bonded oligo-PC interphase was observed. It has been shown that the shear strength of the interface, τ_a, measured in a single-fiber fragmentation test exhibited strong dependence on the interphase E_i.

In order to enhance the performance and reliability of the FRC structures at the macro-scale, the results obtained for the micro-scale interphase can be used to control the stress transfer between the matrix and the reinforcement. Stiff, well-bonded interphases provide very efficient stress transfer, less water diffusion; however, they promote brittle failure and thus limit the damage tolerance of the FRC structures. Low modulus tough interphases slightly reduce the effectiveness of stress transfer and may be less resistant to water attack; however, they provide significant enhancement of damage tolerance of the FRC part [98]. Moreover, the tough interphases are less sensitive to the direction of the external loading compared to the stiff ones. As it has already been shown, one can design hybrid FRCs with reinforcing fibers coated with both stiff and tough interphases in order to tailor the performance and reliability of the final FRC part.

Polymers filled with micrometer-sized particles form a commercially very important group of heterogeneous polymer matrix materials. In these materials, the role of interface/interphase phenomena is even more pronounced than in fiber reinforced composites. There has been a considerable effort put into understanding the role of these phenomena in controlling the mechanical response and thermal stability of filled polymers [2,91]. Analytical and finite element solutions of the stress fields around a particle under multiaxial thermomechanical load were derived as functions of the properties of the particle, the matrix and an interphase [92–94]. The calculations provide the stress field and show that an interphase with a Young's modulus lower than that of the matrix reduces stress concentration in the neighborhood of the particle. The superposition of loads in different directions results in the creation of a more or less uniform distribution of tensile stresses around the particle.

7.3. Nano-scale interphase

The role of the nano-scale interface/interphase in controlling the response of polymers containing nanometer-sized inclusions is in the center of attention of both academic and industrial research, since it is the single most important structural parameter which can be used to "tailor" properties of the nanocomposites. Research of these phenomena in true nanocomposites has resulted in renewed interest in basic principles of polymer physics capable of describing the changes in chain statistics and dynamics caused by the presence of nanoparticles with sizes similar to that of the polymer chain [99–136]. One has to keep in mind, however, that when the length scale considered reaches a few nanometers, which is equal to the size of individual polymer chains, the very term "interphase" becomes ill-defined due to the fact that the discrete nature of the matter has to be taken into account. Originally, the term *interphase* has been defined in the framework of continuum mechanics as a layer or series of layers, respectively, with some average property or gradient of properties, respectively, occurring between the matrix bulk and fiber or filler particle. However, the continuum elasticity in the Bernoulli-Euler form may no longer be valid on the nano-scale due to very large non-locality in elastic response of heterogeneous systems with coordinated movement of a large number of atoms, such as observed in polymer nanocomposites [137].

Nowicki [138] has shown that the spatial rearrangement of segments due to adsorption onto particle surface results in considerable drop of conformation entropy. The extent of entropy decrease is proportional to the particle-chain interaction energy [139–143]. In addition, as the particle size becomes smaller than the chain radius of gyration, the conformation change becomes more localized within the individual coil. Hence, the main effect of filler particles in the nanocomposite is to provide means for modification of segmental dynamics whose extent is proportional to the area of contact between particles and chain segments and thus, unlike in microcomposites where filler volume fraction is the main structural variable, the main structural variable in nanoparticle filled polymers is the area of the filler-chain contact per unit weight of the nanocomposite [144,145]. The deviation of conformation statistics of an adsorbed chain compared to the individual chain or entangled melt becomes even more complex when multi-particle and multi-chain problems are considered. Contrary to the traditional concept of continuum micro-scale interphase, the nano-scale "interphase" exhibits strong dependence of its "thickness" on temperature [146], supporting its claimed discrete molecular structure.

The role of the nano-scale "interphase" differs when considering the two limiting cases of (i) low v_f nanocomposites ($v_f < 0.05$) and (ii) high v_f nanocomposites ($v_f > 0.85$). The first case represents the majority of the published data and can be assumed as a direction to preparing new nano-structured advanced matrices, while the second case is more common in bio-nanocomposites [103,147] and its exploitation can result in designing new nano-structured advanced reinforcements. With a few exceptions, most of the published literature on synthetic nanocomposites deals with low v_f nanocomposites, while, on the other hand, most of the literature published on high v_f nanocomposites is related to the mechanics of bio-composites, such as bones, teeth and shells [148,149].

Most of the experimental evidence related to the interphase in the low v_f nanocomposites was obtained at temperatures below the polymer T_g using meso-scale test specimens. Assuming the chain immobilization to be the primary reinforcing mechanism on the nano-scale, spatial distribution of the conformation entropy within the polymer phase is of primary importance. Hence, experimental data for nano-composites above the matrix T_g has to be considered. Sternstein et al. [24,26,27] published an interpretation of the viscoelastic response of rubbery nanocomposite above the matrix T_g, i.e., the Payne effect. Kalfus and Jancar [144] analyzed the viscoelastic response of polyvinylacetate filled with nano-sized hydroxyapatite over the temperature range from −40 to +120°C and observed strain softening similar to the Payne effect [133,150].

The modulus recovery experiments allowed measuring the terminal relaxation time of reptation motion of bulk and surface immobilized chains, supporting the hypothesis that there is no "interphase" *per se* when nano-scale is considered. In order to bridge the gap between the continuum interphase on the micro-scale and the discrete molecular structure of the matrix consisting of freely reptating chains in the bulk and retarded reptating chains in contact with the inclusions, higher order elasticity combined with a suitable molecular dynamics model could be utilized [151–155].

It was demonstrated that the large specific surface area of the nano-sized filler is capable of immobilizing a large amount of entanglements, causing the steep increase of E' with a small addition of nanoparticles. This observation seemed to confirm the mainly entropic character of the reinforcement mechanism on the nano-scale. All the data published [7,23,27,48,50,104,105,116,119,126,128] support the dominant role of the chain immobilization as the main reinforcing mechanism above the matrix T_g. In addition, experimental results on the recovery of storage and loss moduli [26,27,99,144] in nanocomposites above T_g can be interpreted in terms of recovery of the "interphase" structure requiring chain reptation on the filler surface, hence, further proving the discrete molecular nature of the nano-scale "interphase". Due to the fact that the specific surface area of the true nano-fillers is two orders of magnitude greater, almost all the polymer chains are in contact with the surface at very low filler loadings above 2 vol%. In addition, continuum mechanics has only limited validity at this length scale and the discrete molecular structure prevails, resulting in a strong effect of the non-local character of viscoelastic response of the nano-composite matrix.

The interphases in high v_f nanocomposites were studied using abalone shells [103,142,143]. These shells represent a laminated sheet reinforced composite with more than 95 vol% of aligned 500 nm thin hexagonal aragonite sheets embedded in a protein matrix in apparently mesh-like fibrilar form. In the work of Fantner *et al.* [148] and Hansma *et al.* [149], the model of *sacrificial bonds* has been proposed to explain the observed high fracture resistance of nacreous composites. It has been shown [146] that the hypothesis of the sacrificial bonds can also be used to simulate the deformation response of lightly cross-linked long flexible chain network polymer fibrils. In order to apply the model to the behavior of an ensemble of chains in the vicinity of a rigid weakly attractive nanometer-sized inclusion, the immobilization phenomenon has to be investigated as the source of the drastic change in the viscoelastic behavior of polymers with the addition of a small amount of nano-scale inclusions.

7.4. Chain immobilization on the nano-scale

In a sense, the effects observed in nanocomposites resemble the behavior of colloidal systems and, in some cases, the behavior of thin polymer films. In comparison to colloids, the long chain character of the "liquid" phase, however, greatly complicates any theoretical treatment compared to the low molecular weight "liquid" phase in colloids. Reducing the size of rigid inclusions from micro- to nano-scale is accompanied by a 2–3 orders of magnitude increase in the internal contact area between the chains and the inclusions. At the same time, the particle diameter becomes comparable to the radius of gyration of a regular polymer matrix chain. Thus, the moments of inertia of the chain and the particle become similar, altering the chain dynamics substantially [6–9]. For a nanofiller with specific surface area of approximately 120 m^2/g, adding approximately 2 vol% of nanoparticles, the average interparticle distance is reduced below 2 radii of gyration, $2R_g$, of the chain [161]. It has been shown [144] that in the case of polyvinylacetate of $M_w = 9 \times 10^4$, almost all the chains are in contact with the solid surface and possess reduced segmental mobility at temperatures $T \geq T_g$ when the filler-matrix internal area reaches about 42 m^2/g [27,118]. Below T_g, main chain segmental mobility is frozen and only secondary low temperature side chain mobility can be affected by the nano-filler. In the molten state above T_g, the conformation statistics of chains in near solid surface can be altered from Gaussian random coil to Langevin coil above T_g and this phenomenon can be transformed into the behavior of nanocomposites also upon solidification below T_g.

In order to characterize the reduction in chain mobility in an entangled melt quantitatively, one can use the characteristic reptation relaxation time, τ_{rep}, introduced by deGennes [155] and Doi and Edwards [156]. The τ_{rep} is given for an entangled chain as:

$$\tau_{rep} \cong \frac{L^2}{D_c} \cong \frac{NL^2}{D_0}, \tag{7.1}$$

where L is the length of the reptation path, N is the number of monomer units in a chain, D_c and D_0 are diffusion constants of a chain and a monomer, respectively. The terminal relaxation time of a chain in a neat polymer melt can be expressed in a number of ways. Lin [157], Doi and Edwards have expressed τ_{rep} in the form:

$$\tau_{rep} = \frac{b^4 \zeta_0 N^3}{\pi^2 k_B T a_T^2} \tag{7.2}$$

where ζ_0 is the monomer friction coefficient, b is the length of the statistical segment, k_B is the Boltzmann constant, T is absolute temperature, a_T is the effective tube diameter.

In the case of a chain interacting with a filler surface and entangled with neighboring chains, the question of primary importance is how to establish a connection between the static conformation structure and the chain dynamics. Despite a certain intra-molecular order, the chain in a melt can be considered a random Gaussian coil. If such a chain approaches a solid surface, its conformation transfers to the train-loop-tail structure and the chain conformational entropy as well as the chain internal energy can alter very substantially depending on the surface-polymer interaction energy, e_{fp}, under given conditions. Assuming the chain friction coefficient, ζ_c, in the form [104,105,112,116,122]:

$$\zeta_c = \zeta_0 N + \zeta_a N^{1/2}, \tag{7.3}$$

where ζ_a is the friction coefficient of an adsorbed monomer unit and the number of monomer units in trains is $N_a = N^{1/2}$ for the weakly interacting surface, the terminal relaxation time is in the form:

$$\tau_{rep}^{ads} = \frac{L^2}{2D_c} = \frac{b^2 N^{5/2}}{2k_b TN_e}\left[\zeta_0 N^{1/2} + \zeta_a\right]. \tag{7.4}$$

Kalfus and Jancar [27] extended the use of the above Rubinstein model [104] to describe the reptation time of a linear chain weakly interacting with nano-filler surface.

The friction coefficient, ζ_a, is very difficult to measure and it is known just for the system silica-polystyrene [42] at 153°C. Estimation of ζ_a was based on the theoretical analysis given by Subbotin et al. [105]. Thus, one can establish a relation for the reptation time of a surface adsorbed chain as follows:

$$\tau_{rep}^{ads} = \frac{b^2 N^2}{\pi^2 k_B TN_e}\left[\zeta_0 (N - N_a) + \zeta_a N_a\right]. \tag{7.6}$$

In the case when $N_a = N^{1/2}$, the reptation time takes the form:

$$\tau_{rep}^{ads} = \frac{b^2 N^{5/2}}{2\pi^2 k_b TN_e}\left[\zeta_0 N^{1/2} + \zeta_a\right]. \tag{7.7}$$

To describe the change in reptation dynamics of the chains as a function of nanoparticle volume fraction, a percolation model was used. At the percolation threshold, a physical network formed by interconnection of immobilized chains on individual nanoparticles penetrates the entire sample volume. In this case, only physical "cross-links" are considered and the terminal relaxation time reaches the value characteristic for the life time of the physical filler-polymer bond. Thus, the relaxation time near the percolation threshold is expressed in the form [44]:

$$\tau_{composite}^{rec} = \tau_{rep}^{ads} v_{eff} \left(\frac{v_{eff} - v_{eff}^*}{1 - v_{eff}^*}\right)^b, \tag{7.8}$$

where v_{eff}^* is the critical effective filler volume fraction ($v_{eff}^* = 0.04$ for PVAc-HAP at 90°C) and b is the percolation exponent ($b = 4$ for the same system). The v_{eff}^* is a sum of the filler volume fraction and the volume fraction of immobilized chains and was shown to equal 0.04 for PVAc-HAP nanocomposites at 90°C. To simplify the percolation, random clustering of effective hard spheres was considered only in the way similar to that originally outlined by Jancar et al. [45] for micro-scale composites. Percolation threshold at approximately 2 m² of the filler-polymer contact area per 1 g of the composite was found in the PVAc-HA nanocomposite system (Figure 7.6). All the chains were immobilized at the internal contact area of 42 m² per 1 g of the nanocomposite.

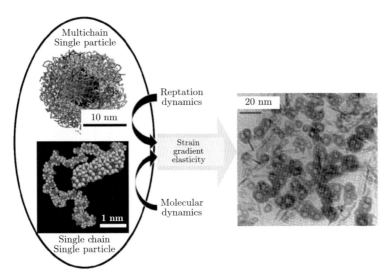

Figure 7.6. Schematic drawing of the nano-scale "interphase" and the models used to translate behavior of discrete matter consisting of molecules with segmental mobility reduced due to the presence of nano-scale solid inclusions. Crystalline inclusions surrounded by a number of long chain molecules can be modeled as a continuum, while amorphous inclusions with the extent of non-locality similar to polymers has to be considered discrete as well [170]

7.5. Characteristic length-scale in polymer matrix nanocomposites

The classical continuum mechanics is designed to be size-independent. For nanocomposites, however, size-dependent elastic properties have been observed which cannot be readily explained using continuum mechanics and, thus, prevent simple scaling down of the existing continuum elasticity models [23]. Polymers are unique systems with macroscopic viscoelastic response driven by the relaxation processes on the molecular level [46]. These relaxation processes represent particular molecular motions occurring in some characteristic volume, V_c. The V_c depends on the type of the relaxation process and temperature. The characteristic volumes vary from 10^{-3} nm^3 for localized bond vibrations to 10^6 nm for the non-local normal mode of relaxation (Figure 7.7) [46]. In the case of the non-local normal mode of relaxation, its characteristic volume is the upper limit for V_c displaying strong dependence on the chain size. Below this upper limiting characteristic volume, $V_c \sim R^3 \propto N^{3/2}$, where R is the chain end-to-end distance and N is the number of monomer units in the chain. The characteristic time, τ_c, for each particular relaxation process varies from 10^{-14} s for bond vibrations above T_g to the infinitely long times below T_g. Thus, the macroscopic viscoelastic response of a polymer is a manifestation of a range of molecular relaxations localized in some characteristic volume and the rate of the relaxation mode is indirectly proportional to its locality.

The physical reasons for the expected breakdown of continuum elasticity on the nano-scale include the increasing importance of surface energy due to

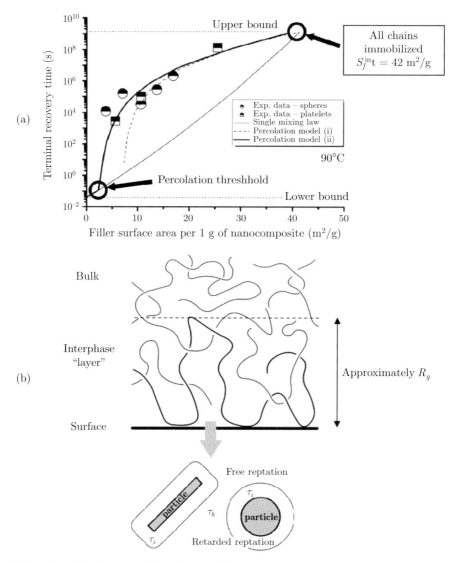

Figure 7.7. Simple approach combining the reptation dynamics and percolation model to describe the retarded reptation of chains in the vicinity of solid nano-sized inclusions representing the nano-scale "interphase" [144]. (a) The dependence of the terminal relaxation time from modulus recovery experiment plotted as a function of the specific filler-matrix interface area. Squares and circles represent experimental data for platelet and spherical particles, respectively. (b) Visualization of the molecular structure of the interphase (upper) and the 2-phase simplified structure of the nanoparticles with interphase considered in the percolation model.

appreciable surface to volume ratio [47], the discrete molecular nature of the polymer matrix resulting in non-local behavior contrary to the local character of classical elasticity [48], the presence of nano-scale particles with a length scale similar to the radius of gyration of the polymer chains [49,50] and internal strain due to molecular motion within a non-primitive lattice [51,52]. Quantum confinement effects can also play a role inducing a strain field on the nanoscale without presence of external loading, however, its importance is limited to the size range below 2 nm [53].

In order to estimate the length scale at which the classical elasticity becomes non-valid, the MD on a polyethylene chain in a cubic simulation cell was performed under periodic boundary conditions at 50 K [23]. The corresponding length scales for the longitudinal and transverse directions were 1.85 and 3.81 nm, respectively. A recent work by Nikolov *et al.* [54] estimated that rubbers above their T_g should have non-local length scale approximately 5 nm based on the calculation of the volume at which the ergodic principle is no longer valid and the network deformation is not affine.

The high non-locality in polymers most probably stems from a cooperative behavior of a large number of chain segments characteristic for entangled polymers. As a result, parts of the material system may undergo considerable non-affine deformation associated with the occurrence of high moment stresses. Consequently, for such systems, taking strain-gradient effects into account while investigating nano-scale elastic phenomena may impart significant size-dependent corrections to the results obtained from classical continuum elasticity [23]. The magnitude of corrections that strain-gradient effects may impart to results obtained by classical continuum elasticity has recently been proposed by Park and Gao [48]. The bending rigidity of a rectangular beam was calculated for various length scales [23]. For polyethylene, the bending rigidity increased with decreasing beam dimensions and was double that of the classical Bernoulli-Euler bending rigidity at beam dimensions of 5 nm. Some experimental data suggest that there is a different critical value of the characteristic length, L_c, for cross-linked and entangled polymers. Even though there are many aspects in calculating these critical characteristic lengths, considering the arguments put forward above, one can estimate L_c about 5–10 nm for cross-linked polymers and about 10–20 nm for entangled polymers.

Based on the arguments put forward above, one may consider to carefully revise published data on the elastic properties of polymer nanocomposites since the raw experimental data were usually analyzed to yield elastic constants using classical continuum elasticity.

7.6. Conclusions and outlook

In polymer matrix composites exhibiting heterogeneous structure at multiple length scales, the interphase phenomena are of pivotal importance for the control of their performance and reliability. In this chapter, a review of the current knowledge on the interphase phenomena at various length scales has been attempted, comparing multiscale manmade composite structures with natural multiscale functionally hierarchical composite structures. On the micro-scale, the interphase is considered a 3D continuum possessing some average properties such as elastic modulus, shear strength and fracture toughness. Existing continuum mechanics models provide satisfactory means to relate these properties to the stress transfer from matrix to the individual fiber. These models are far less successful in translating the properties of the interphase on an individual fiber into the performance of a multifiber composite part.

On the nano-scale, the discrete molecular structure of the polymer has to be considered. Segmental immobilization seems to be the primary reinforcing mechanism in true polymer nanocomposites at temperatures near and above the T_g. Reptation model and simple percolation model were used to describe immobilization of chains near solid nanoparticles and to explain the peculiarities in the viscoelastic response of polymers near solid surfaces of large polymer-inclusion contact areas. The "interphase" in the continuum sense does not exist at the nano-scale when relaxation processes in individual discrete chains are taken into account and the chains with retarded reptation can be considered forming the "interphase" analogue in the discrete matter. For a common polymer, all the chains in the composite are immobilized when the internal filler-matrix interface area reached about 42 m^2 per 1 g of the nanocomposite.

In polymers at very low temperatures, the classical Bernoulli-Euler continuum elasticity becomes not valid below approximately 5 nm. The size of this characteristic volume increases with increasing temperature and higher order strain elasticity along with molecular dynamics approach has to be used as the bridging law to connect behavior of the discrete matter at nano-scale with the mechanical response of continuous matter at larger length scales.

Acknowledgement

Financial support from the Czech Ministry of Education, Youth and Sports under grant MSM 0021630501 is greatly appreciated.

References

1. DiBenedetto A T (2001) Tailoring of interfaces in glass fiber reinforced polymer composites: A review, *Mater Sci Eng* **A302**:74–82.
2. Pukánszky B and Fekete E (1999) Adhesion and surface modification, in *Mineral fillers in thermoplastics* (Ed. Jancar J) Springer, Advances in Polymer Science, vol. 139, Heidelberg, Ch. 3, pp. 109–154.
3. Tsai J and Sun C T (2004) Effect of Platelet Dispersion on the Load Transfer Efficiency in Nanoclay Composites, *J Compos Mater* **38**:567–579.
4. Fornes T D and Paul D R (2003) Modeling properties of nylon 6/clay nanocomposites using composite theories, *Polymer* **44**:4993–5013.
5. Yung K C, Wang J and Yue T M (2006) Modeling Young's Modulus of Polymer-layered Silicate Nanocomposites Using Modified Halpin-Tsai Micromechanical Model, *J Reinf Plast Compos* **25**:847–861.
6. Vacatello M (2002) Chain Dimensions in Filled Polymers: An Intriguing Problem, *Macromolecules* **35**:8191–8193.
7. Vacatello M (2003) Phantom Chain Simulations of Polymer-Nanofiller Systems, *Macromolecules* **36**:3411–3416.

8. Vacatello M (2002) Molecular Arrangements in Polymer-Based Nanocomposites, *Macromol Theory Simul* **11**:757–765.
9. Vacatello M (2003) Predicting the Molecular Arrangements in Polymer-Based Nanocomposites, *Macromol Theory Simul* **12**:86–91.
10. Starr F W, Schrøder T B and Glotzer S C (2002) Molecular Dynamics Simulation of a Polymer Melt with a Nanoscopic Particle, *Macromolecules* **35**:4481–4492.
11. Maranganti R and Sharma P (2007) A novel atomistic approach to determine strain-gradient elasticity constants: Tabulation and comparison for various metals, semiconductors, silica, polymers and the (Ir) relevance for nanotechnologies, *J Mech Phys Solids* **55**:1823–1852.
12. Zhou D and Jin B (2003) Boussinesq-Flamant problem in gradient elasticity with surface energy, *Mech Res Commun* **30**:463–468.
13. Polizzotto C (2003) Gradient elasticity and nonstandard boundary conditions, *Int J Solids Struct* **40**:7399–7423.
14. Hussain F, Hojjati M, Okamoto M and Gorga R E (2006) Polymer-matrix nanocomposites, processing, manufactuting and applications: An overview, *J Compos Mater* **40**:1511–1575.
15. Ajayan P M, Schadler L S and Braun P V (2003) *Nanocomposite Science and Technology*, Wiley, New York.
16. Mai Y-W and Yu Z-Z (2006) *Polymer Nanocomposites*, Woodhead Publ. Ltd, Cambridge.
17. Zeng Q H, Yu A B and Lu Q B (2008) Multiscale modeling and simulation of polymer nanocomposites, *Prog Polym Sci* **33**:191–269.
18. Solunov H A (2006) The dynamic crossover temperature and the characteristic length of glass transition in accordance with the extended Adam-Gibbs theory, *J Non-Cryst Solids* **352**:4871–4876.
19. Plazek D J (2007) Anomalous viscoelastic properties of polymers: Experiments and explanations, *J Non-Cryst Solids* **353**:3783–3787.
20. Prevosto D, Capaccioli S, Lucchesi M and Rolla P (2005) Effect of temperature and volume on structural relaxation time: Interpretation in terms of decrease of configurational entropy, *J Non-Cryst Solids* **351**:2611–2615.
21. Mundra M K, Ellison C J, Behling R E and Torkelson J M (2006) Confinement, composition, and spin-coating effects on the glass transition and stress relaxation of thin films of polystyrene and styrene-containing random copolymers: Sensing by intrinsic fluorescence, *Polymer* **47**:7747–7759.
22. Quin Q and McKenna G B (2006) Correlation between dynamic fragility and glass transition temperature for different classes of glass forming liquids, *J Non-Cryst Solids* **352**:2977–2985.
23. Dionne P J, Picu C R and Ozisik R (2006) Adsorption and Desorption Dynamics of Linear Polymer Chains to Spherical Nanoparticles: A Monte Carlo Investigation, *Macromolecules* **39**:3089–3092.
24. Sternstein S S and Zhu A-J (2002) Reinforcement Mechanism of Nanofilled Polymer Melts as Elucidated by Nonlinear Viscoelastic Behavior, *Macromolecules* **35**:7262–7273.
25. Drozdov A D, Jensen E A and Christiansen J de C (2008) Thermo-viscoelastic response of nanocomposite melts, *Int J Eng Sci* **46**:87–104.
26. Kalfus J and Jancar J (2007) Immobilization of polyvinylacetate macromolecules on hydroxyapatite nanoparticles, *Polymer* **48**:3935–3938.
27. Kalfus J and Jancar J (2007) Viscoelastic response of nanocomposite poly(vynil acetate) hydroxyapatite with varying particle shape – Dynamic strain softening and modulus recovery, *Polym Compos* **28**:743–747.

28. Kalfus J, Jancar J and Kucera J (2008) Effect of weakly interacting nanofiller on the morphology and viscoelastic response of polyolefins, *Polym Eng Sci* **48**:889–894.
29. Nitta K, Asuka K, Liu B and Terano M (2006) The effect of the addition of silica particles on linear spherulite growth rate of isotactic polypropylene and its explanation by lamellar cluster model, *Polymer* **47**:6457–6463.
30. Liu X, Wu Q and Berglund L (2002) Investigation of Unusual Crystallization Behavior in Polyamide6/Montmorillonite Nanocomposites, *Macromol Mater Eng* **287**:515–522.
31. Cosoli P, Scocchi G, Pricl S and Fermeglia M (2008) Many-scale molecular simulation for ABS-MMT nanocomposites: Upgrading of industrial scraps, *Microporous Mesoporous Mater* **107**:169–179.
32. Mi C and Kouris D (2006) Atomistic calculations of interface elastic properties in noncoherent metallic bilayers, *J Mech Mater Struct* **1**:763–770.
33. Park S K and Gao X-L (2006) Bernoulli-Euler beam model based on a modified couple stress theory, *J Micromech Microeng* **16**:2355–2359.
34. Sharma P and Ganti S (2004) Size-Dependent Eshelby's Tensor for Embedded Nano-Inclusions Incorporating Sufrace/Interface Energies, *J Appl Mech* **71**:663–671.
35. Sharma P, Ganti S and Bhate N (2003) Effect of surfaces on the size-dependent elastic state of nano-inhomogeneties, *Appl Phys Lett* **82**:535–537.
36. Chen Y, Lee J D and Eskandarian A (2003) Examining the physical foundation of continuum theories from the viewpoint of phonon dispersion relation, *Int J Eng Sci* **41**:61–83.
37. Chen Y, Lee J D and Eskandarian A (2004) Atomistic viewpoint of the applicability of microcontinuum theories, *Int J Solids Struct* **41**:2085–2097.
38. Zang X, Sharma P and Johnsson H T (2007) Quantum confinement induced strain in quantum dots, *Phys Rev B* **75**:155319.
39. Pukánszky B (2005) Interfaces and interphases in multicomponent material: past, present, future, *Eur Polym J* **41**:645–662.
40. Hashin Z (2002) Thin interphase/imperfect interface in elasticity with application to coated fiber composites, *J Mech Phys Solids* **50**:2509–2537.
41. Nairn J A (2007) Numerical implementation of imperfect interfaces, *Comput Mat Sci* **40**:525–536.
42. Lauke B and Schuller T (2002) Calculation of stress concentration caused by a coated particle in polymer matrix to determine adhesion strength at the interface, *Comp Sci Technol* **62**:1965–1978.
43. Galiotis C (2005) Interfacial Damage Modelling of Composites In: *Multi-scale modelling of composite material systems* (Eds. Beaumont P W R and Soutis C) Woodhead Publ. Ltd, Cambridge, pp. 33–62.
44. Jancar J (2006) Effect of interfacial shear strength on the mechanical response of polycarbonate and PP reinforced with basalt fibers, *Comp Interfaces* **13**:853–864.
45. Kim J-K and Mai Y-W (1998) *Engineered interfaces in fiber reinforced composites*, Elsevier, Amsterdam, Ch. 3, pp. 43–85.
46. Jancar J (2006) Hydrolytic stability of PC/GF composites with engineered interphase of varying elastic modulus, *Comp Sci Technol* **66**:3144–3152.
47. Kim J-K and Mai Y-W (1998) in *Engineered interfaces in fiber reinforced composites*, Elsevier, Amsterdam, Ch. 4, pp. 93–164.
48. Droste D H and DiBenedetto A T (1969) The glass transition temperature of filled polymers and its effect on their physical properties, *J Appl Polym Sci* **13**:2149–2168.
49. Cave N G and Kinloch A J (1992) Self-assembling monolayer silane films as adhesion promoters, *Polymer* **33**:1162–1170.
50. Xie X-Q, Ranade S V and DiBenedetto A T (1999) A solid state NMR study of polycarbonate oligomer grafted onto the surface of amorphous silica, *Polymer* **40**:6297–6306.

51. Robertson C H and Palade L I (2006) Unified application of the coupling model to segmental, Rouse, and terminal dynamics of entangled polymers, *J Non-Cryst Solids* **352**:342–348.
52. Zinc P, Wagner H D, Salmon L and Gerard J-F (2001) Are microcomposites realistic models of the fiber/matrix interface? II. Physico-chemical approach, *Polymer* **42**:6641–6650.
53. Nairn J A (1997) On the use of shear-lag methods for analysis of stress transfer in unidirectional composites, *Mech Mater* **26**:63–80.
54. DiBenedetto A T, Huang S J, Birch D, Gomes J and Lee W C (1994) Reactive coupling of fibers to engineering thermoplastics, *Compos Struct* **27**:73–82.
55. Hodzic A, Kim J K, Lowe A E and Stachurski Z H (2004) The effects of water aging on the interphase region and interlaminar fracture toughness in polymer-glass composites, *Compos Sci Technol* **64**:2185–2195.
56. Ochiai S, Hojo M, Schulte K and Fiedler B (2001) Nondimensional simulation of influence of toughness of interface on tensile stress-strain behavior of unidirectional microcomposite, *Composites:Part A* **32**:749–761.
57. Johnson A C, Hayes S A and Jones F R (2005) An improved model including plasticity for the prediction of the stress in fibres with an interface/interphase region, *Composites: Part A* **36**:263–271.
58. Duan H L, Yi X, Huang Z P and Wang J (2007) A unified scheme for prediction of effective moduli of multiphase composites with interface effects. Part I: Theoretical framework, *Mech Mater* **39**:81–93.
59. Duan H L, Yi X, Huang Z P and Wang J (2007) A unified scheme for prediction of effective moduli of multiphase composites with interface effects. Part II: Application and scaling laws, *Mech Mater* **39**:94–103.
60. Wetherhold R C, Corjon M and Das P K (2007) Multiscale considerations for interface engineering to improve fracture toughness of ductile fiber/thermoset matrix composites, *Compos Sci Technol* **67**:2428–2437.
61. Hodzica A, Kimb J K and Stachurski Z H (2001) Nano-indentation and nano-scratch of polymer/glass interfaces. II. Model of interphases in water aged composite materials, *Polymer* **42**:5701–5710.
62. Wu Z J, Ye J Q and Cabrera J G (2000) 3D analysis of stress transfer in the micromechanics of fiber reinforced composites by using an Eigen-function expansion method, *J Mech Phys Solids* **48**:1037–1063.
63. Pisanova E, Zhandarov S and Mäder E (2001) How can adhesion be determined from micromechanical tests?, *Composites: Part A* **32**:425–434.
64. Goh K L, Aspden R L and Hukins D W L (2004) Review: Finite element analysis of stress transfer in short-fibre composite materials, *Compos Sci Technol* **64**:1091–1100.
65. Fu S-Y, Yue C-Y, Hua X and Mai Y-W (2000) On the elastic stress transfer and longitudinal modulus of unidirectional multi-short-fiber composites, *Compos Sci Technol* **60**:3001–3012.
66. Goutianos S, Galiotis C and Peijs T (2003) Mechanisms of stress transfer and interface integrity in carbon/epoxy composites under compression loading. Part II: Numerical approach, *Int J Solids Struct* **40**:5521–5538.
67. Xia Z, Okabe T and Curtin W A (2002) Shear-lag versus finite element models for stress transfer in fiber-reinforced composites, *Compos Sci Technol* **62**:1141–1149.
68. Nayfeh A H and Abdelrahman W G (1999) Dynamic stress transfer in fibrous composites with damage, *Composites: Part B* **30**:233–243.
69. Paipetis A and Galiotis C (1997) A study of the stress transfer characteristics in model composites as a function of material processing, fiber sizing and temperature of the environment, *Compos Sci Technol* **57**:827–838.

70. Anagnostopoulos G, Parthenios J, Andreopoulos A G and Galiotis C (2005) An experimental and theoretical study of the stress transfer problem in fibrous composites, *Acta Mater* **53**:4173–4183.
71. Guild F J, Vlattas C and Galiotis C (1994) Modelling of stress transfer in fiber composites, *Compos Sci Technol* **50**:319–332.
72. Nishiyabu K, Yokoyama A and Hamada H (1997) Assessment of the influence of interfacial properties on stress transfer in composites by a numerical approach, *Compos Sci Technol* **57**:1103-1111.
73. Hsueh C H, Young R J, Yang X and Becher P F (1997) Stress transfer in model composite containing a single embedded fiber, *Acta Mater* **45**:1469–1476.
74. Detassis M, Pegoretti A and Migliaresi C (1995) Effect of temperature and strain rate on interfacial shear stress transfer in carbon/epoxy model composites, *Compos Sci Technol* **53**:39–46.
75. Wu W, Jacobs E, Verpoest I and Varna J (1999) Variational approach to the stress-transfer problem through partially debonded interfaces in a three-phase composite, *Compos Sci Technol* **59**:519–535.
76. Thuruthimattam B J, Waas A M and Wineman A S (2001) Stress transfer modeling in viscoelastic polymer matrix composites, *Int J Non-Linear Mech* **36**:69–87.
77. Sretenovic A, Muller U and Gindl W (2006) Mechanism of stress transfer in a single wood fibre-LDPE composite by means of electronic laser speckle interferometry, *Composites:Part A* **37**:1406–1412.
78. Nishino T, Hirao K and Kotera M (2006) X-ray diffraction studies on stress transfer of kenaf reinforced poly(L-lactic acid) composite, *Composites: Part A* **37**:2269–2273.
79. You L H, You X Y and Zheng Z Y (2006) Thermomechanical analysis of elastic-plastic fibrous composites comprising an inhomogeneous interphase, *Comput Mater Sci* **36**:440–450.
80. de Lange P J, Mader E, Mai K, Young R J and Ahmad I (2001) Characterization and micromechanical testing of the interphase of aramid-reinforced epoxy composites, *Composites: Part A* **32**:331–342.
81. Gao S-L and Mader E (2002) Characterization of interphase nanoscale property variation in glass fiber reinforced polypropylene and epoxy resin composites, *Composites: Part A* **33**:559–576.
82. Mallarino S, Chailan J F and Vernet J L (2005) Interphase investigation in glass fibre composites by micro-thermal analysis, *Composites: Part A* **36**:1300–1306.
83. Pompe G and Mäder E (2000) Experimental detection of a transcrystalline interphase in glass-fibre/polypropylene composites, *Compos Sci Technol* **60**:2159–2167.
84. Li K and Saigal S (2007) Micromechanical modeling of stress transfer in carbon nanotube reinforced polymer composites, *Mater Sci Eng* **A457**:44–57.
85. Pegoretti A and DiBenedetto A T (1998) Measurement and analysis of stress transfer and toughness at fiber-matrix interface, *Composites: Part A* **29A**:1063–1070.
86. Hayes S A, Lane R and Jones F R (2001) Fibre/matrix stress transfer through a discrete interphase. Part 1: Single-fibre model composites, *Composites: Part A* **32**:379–389.
87. Pluedemann E P (1982) *Silane coupling agents*, Plenum Press, New York.
88. Park S-J and Jin J-S (2001) Effect of Silane Coupling Agent on Interphase and Performance of Glass Fibers/Unsaturated Polyester Composites, *J Colloid Interface Sci* **242**:174–179.
89. Wang J, Duan H L, Zhang Z and Huang Z P (2005) An anti-interpenetration model and connections between interphase and interface models in particle-reinforced composites, *Int J Mech Sci* **47**:701–718.

90. Olmos D and Gonzalez-Benito J (2007) Visualization of the morphology at the interphase of glass fibre reinforced epoxy-thermoplastic polymer composites, *Europ Polym J* **43**:1487–1500.
91. Keszei S, Matko S, Bertalan G, Anna P, Marosi G and Toth A (2005) Progress in interface modifications: From compatibilization to adaptive and smart interphases, *Europ Polym J* **41**:697–705.
92. Lauke B, Schuller T and Beckert W (2000) Calculation of adhesion strength at the interface of a coated particle embedded within matrix under multiaxial load, *Comput Mater Sci* **18**:362–380.
93. Erguney F M, Lin H, Mattice W L (2006) Dimensions of matrix chains in polymers filled with energetically neutral nanoparticles, *Polymer* **47**:3689-3695.
94. Wei P J and Huang Z P (2004) Dynamic effective properties of the particle-reinforced composites with the viscoelastic interphase, *Int J Solids Struct* **41**:6993–7007.
95. Jancar J (2008), Thickness dependence of elastic modulus of polycarbonate interphase, *Compos Interfaces* **15**:465-475.
96. Jancar J (2008), The thickness dependence of elastic modulus of organosilane interphases, *Polym Compos*, in print (available on line 26-04-08).
97. Polacek P (2007) *Reactive non-silane surface treatment of ceramic fibers,* PhD Thesis, Brno University of Technology.
98. DiAnselmo A, Jancar J, DiBenedetto A T and Kenny J M (1992) Finite element analysis of the effect of an interphase on the mechanical properties of polymeric composite materials, in *Composite Materials* (Eds. DiBenedetto A T, Nicolais L, Watanabe R) Elsevier Science Publ., pp. 49–59.
99. Narayanan R A, Thiyagarajan P, Zhu A-J, Ash B J, Shofner M L, Schadler L S, Kumar S K and Sternstein S S (2007) Nanostructural features in silica-polyvinyl acetate nanocomposites characterized by small-angle scattering, *Polymer* **48**:5734–5741.
100. Eitan A, Fisher F T, Andrews R, Brinson L C and Schadler L S (2006) Reinforcement mechanisms in MWCNT-filled polycarbonate, *Comp Sci Technol* **66**:1162–1173.
101. Kalfus J, Jancar J (2008), Reinforcing mechanisms in amorphous polymer nanocomposites, *Comp Sci Technol* **68**:3444–3447.
102. Kelarakis A, Giannelis E P and Yoon K (2007) Structure-properties relationships in clay nanocomposites based on PVDF/(ethylene-vinyl acetate) copolymer blends, *Polymer* **48**:7567–7572.
103. Bettye L. et al. (1999), Molecular mechanistic origin of the toughness of natural adhesives, fibres and composites, *Nature* **399**:761–763.
104. Zheng X, Sauer B B, van Alsten J G, Schwarz S A, Rafailovich M H, Sokolov J and Rubinstein M (1995) Repatation dynamics of a polymer melt near an attractive solid interface, *Phys Rev Lett* **74**:407–412.
105. Subbotin A, Semenov A and Doi M (1997) Friction in strongly confined polymer melts: Effect of polymer bridges, *Phys Rev E* **56**:623–630.
106. Li J, Peng C, Liu H and Hu Y (2005) Monte Carlo simulation for the adsorption of symmetric triblock copolymers. I. Configuration distribution and density profiles of adsorbed chains, *Europ Polym J* **41**:627–636.
107. Zheng Q, Xue Q, Yan K, Gao X, Li Q and Hao L (2008) Effect of chemisorption on the interfacial bonding characteristics of carbon nanotube polymer composites, *Polymer* **49**:800–808.
108. Tzavalas S, Drakonakis V, Mouzakis D E, Fischer D and Gregoriou V G (2006) Effect of Carboxy-Functionalized Multiwall Nanotubes (MWNT-COOH) on the Crystallization and Chain Conformations of Poly(ethylene terephthalate) PET in PET-MWNT Nanocomposites, *Macromolecules* **39**:9150–9156.

109. Jia Q-X, Wu Y-P, Lu M, He S-J, Wang Y-Q and Zhang L-Q (2008) Interface tailoring of layered silicate/styrene butadiene rubber nanocomposites, *Compos Interfaces* **15**:193–205.
110. Poole P H, Donati C and Glotzer S C (1998) Spatial correlations of particle displacements in a glass-forming liquid, *Physica* **A261**:51–59.
111. Dionne P J, Ozisik R and Picu C R (2005) Structure and Dynamics of Polyethylene Nanocomposites, *Macromolecules* **38**:9351–9358.
112. Hu X, Zhang W, Si M, Gelfer M, Hsiao B, Rafailovich M, Sokolov J, Zaitsev V and Schwarz S (2003) Dynamics of Polymers in Organosilicate Nanocomposites, *Macromolecules* **36**:823–829.
113. Ozmusul M S and Picu R C (2002) Structure of polymers in the vicinity of convex impenetrable surfaces: The athermal case, *Polymer* **43**:4657–4665.
114. Balijepalli S and Rutledge G C (2000) Conformational statistics of polymer chains in the interphase of semi-crystalline polymers, *Comput Theor Polym Sci* **10**:103–113.
115. Erguney F M, Lin H and Mattice W L (2006) Dimensions of matrix chains in polymers filled with energetically neutral nanoparticle, *Polymer* **47**:3689–3695.
116. Fragiadakis F, Pissis P and Bokobza L (2006) Modified chain dynamics in poly(dimethylsiloxane)/silica nanocomposites, *J Non-Cryst Solids* **352**:4969–4972.
117. Xia H and Song M (2005) Characteristic length of dynamic glass transition based on polymer/clay intercalated nanocomposites, *Thermochim Acta* **429**:1–5.
118. Sargsyan A, Tonoyan A, Davtyan S and Schick C (2007) The amount of immobilized polymer in PMMA/SiO$_2$ nanocomposites determined from calorimetric data, *Europ Polym J* **43**:3113–3127.
119. Lin E K, Kolb R, Satija S K and Wu W-L (1999) Reduced Polymer Mobility near the Polymer/Solid Interface as Measured by Neutron Reflectivity, *Macromolecules* **32**:3753–3757.
120. Sharaf M A and Mark J E (2004) Monte Carlo simulations on the effects of nanoparticles on chain deformations and reinforcement in amorphous polyethylene networks, *Polymer* **45**:3943–3952.
121. Smith J S, Bedrov D and Smith G D (2003) A molecular dynamics simulation study of nanoparticle interactions in a model polymer-nanoparticle composite, *Compos Sci Technol* **63**:1599–1605.
122. Harmandaris V A, Daoulas K C and Mavrantzas V G (2005) Molecular Dynamics Simulation of a Polymer Melt/Solid Interface: Local Dynamics and Chain Mobility in a Thin Film of Polyethylene Melt Adsorbed on Graphite, *Macromolecules* **38**:5796–5809.
123. Bhowmik R, Katti K S and Katti D (2007) Molecular dynamics simulation of hydroxyapatite-polyacrylic acid interfaces, *Polymer* **48**:664–674.
124. Cho J and Sun C T (2007) A molecular dynamics simulation study of inclusion size effect on polymeric nanocomposites, *Comput Mater Sci* **41**:54–62.
125. Adnan A, Sun C T and Mahfuz H (2007) A molecular dynamics simulation study to investigate the effect of filler size on elastic properties of polymer nanocomposites, *Compos Sci Technol* **67**:348–356.
126. Ozmusul M S, Picu C R, Sternstein S S and Kumar S K (2005) Lattice Monte Carlo Simulations of Chain Conformations in Polymer Nanocomposites, *Macromolecules* **38**:4495–4500.
127. Amalvy J I, Percy M J, Armes S P, Leite C A P and Galembeck F (2005) Characterization of the Nanomorphology of Polymer-Silica Colloidal Nanocomposites Using Electron Spectroscopy Imaging, *Langmuir* **21**:1175–1179.
128. Brown D, Marcadon V, Mélé V and Albérola N D (2008) Effect of Filler Particle Size on the Properties of Model Nanocomposites, *Macromolecules* **41**:1499–1511.

129. Reister E and Fredrickson G H (2004) Nanoparticles in a Diblock Copolymer Background: The Potential of Mean Force, *Macromolecules* **37**:4718–4730.
130. Fragiadakis D, Pissis P and Bokobza L (2005) Glass transition and molecular dynamics in poly(dimethylsiloxane)/silica nanocomposites, *Polymer* **46**:6001–6008.
131. Smith J S, Bedrov D and Smith G D (2003) A molecular dynamics simulation study of nanoparticle interactions in a model polymer-nanoparticle composite, *Compos Sci Technol* **63**:1599–1605.
132. Ciprari D, Jacob K and Tannenbaum R (2006) Characterization of Polymer Nanocomposite Interphase and Its Impact on Mechanical Properties, *Macromolecules* **39**:6565–6573.
133. Zhu A J and Sternstein S S (2003) Nonlinear viscoelasticity of nanofilled polymers: Interfaces, chain statistics and properties recovery kinetics, *Compos Sci Technol* **63**:1113–1126.
134. Yu T, Lin J, Xu J, Chen T, Lin S and Tian X (2007) Novel polyacrylonitrile/Na-MMT/silica nanocomposite: Co-incorporation of two different form nano materials into polymer matrix, *Compos Sci Technol* **67**:3219–3225.
135. Ha S R, Rhee K Y, Kim H C and Kim J T (2008) Fracture performance of clay/epoxy nanocomposites with clay surface-modified using 3-aminopropyltriethoxysilane, *Coll Surf A:Physicochem Eng Aspects* **313–314**:112–115.
136. Jiang L, Zhang J and Wolcott M P (2007) Comparison of polylactide/nano-sized calcium carbonate and polylactide/montmorillonite composites: Reinforcing effects and toughening mechanisms, *Polymer* **48**:7632–7644.
137. Cammarata R C (1997) Surface and interface stress effects on interfacial and nanostructured materials, *Mater Sci Eng* **A237**:180–184.
138. Nowicki W (2002) Structure and Entropy of a Long Polymer Chain in the Presence of Nanoparticles, *Macromolecules* **35**:1424–1431.
139. Tiller A R (1994) Towards first-principles prediction of polymer configurational statistics, *Polymer* **35**:4511–4520.
140. Rieger J (1998) Conformational statistics of polymers: A unifying approach comprising broken rods, blobs, and simple random walks, *Polymer* **39**:4477–4420.
141. Kholodenko A, Ballau M and Aguero Granados M (1998) Conformational statistics of semiflexible polymers: Comparison between different models, *PhysicaA* **260**:267–293.
142. Uhlherr A and Theodorou D N (1998) Hierarchical simulation approach to structure and dynamics of polymers, *Curr Opin Solid State Mater Sci* **3**:544–551.
143. Lang X Y, Zhu Y F and Jiang Q (2006) Size and interface effects on several kinetic and thermodynamic properties of polymer thin films, *Thin Solid Films* **515**:2765–2770.
144. Kalfus J and Jancar J (2007) Relaxation processes in PVAc-HA nanocomposites, *J Polym Sci: Part B: Polym Phys* **45**:1380–1388.
145. Asai S, Kaneko H, Sumita M and Miyasaka K (1991), Molecular examination of fracture toughness of amorphous polyesters as a function of copolymerization component, *Polymer* **32**: 2400–2405.
146. Ji B and Gao H (2004) Mechanical properties of nanostructure of biological materials, *J Mech Phys Solids* **52**:1963–1990.
147. Ji B and Gao H (2006), Elastic properties of nanocomposite structure of bone, *Comp. Sci. Technol.* **66**:1212–1218.
148. Fantner G E, Rabinovych O, Schitter G, Golde L S, Thurner P, Finch M M, Turner P, Gutsmann T, Morse D E, Hansma H and Hansma P K (2006) Hierarchical interconnections in the nano-composite material bone: Fibrillar cross-links resist fracture on several length scales, *Compos Sci Technol* **66**:1205–1211.
149. Hansma P K, Fantner G E, Kindt J H, Thurner P J, Schitter G, Turner P J, Udwin S F and Finch M M (2005) *J Muculoskelet Neuronal Interact* **5**:313–315.

150. Zidek J and Jancar J (2006) Simulation of inelastic stress-strain response of nanocomposites by a network model, *Key Eng Mater* **334–335**:857–860.
151. Cassagnau P (2003) Payne effect and shear elasticity of silica-filled polymers in concentrated solutions and in molten state, *Polymer* **44**:2455–2462.
152. Liu W K, Park H S, Qian D, Karpov E G, Kadowaki H and Wagner G J (2006) Bridging scale methods for nanomechanics and materials, *Comput Methods Appl Mech Eng* **195**:1407–1421.
153. Shenoy V B, Miller R, Tadmor E B, Rodney D, Phillips R and Ortiz M (1999) An adaptive finite element approach to atomic scale mechanics – the quasicontinuum method, *J Mech Phys Solids* **47**:611–642.
154. Miller R, Ortiz M, Phillips R, Shenoy V B and Tadmor E B (1998) Quasicontinuum models of fracture and plasticity, *Eng Fracture Mech* **61**:427–444.
155. deGennes P-G (1979) *Scaling Concepts in Polymer Physics*, Cornell University Press, London.
156. Doi M and Edwards S F (2003) *Theory of Polymer Dynamics*, Oxford University Press, London.
157. Lin Y-H (1985) Comparison of the pure reptational times calculated from linear viscoelasticity and diffusion motion data of nearly monodisperse polymers, *Macromolecules* **18**:2779–2781.
158. Riegleman R A, Douglas J F and dePablo J J (2007), Tuning polymer melt fragility with antiplasticizer additive, *J Chem Phys* **126**:234903.
159. Jancar J, Kucera J and Vesely P (1991) Peculiarities of mechanical response of heavily filled polypropylene composites, *J Mater Sci* **26**:4878–4882.
160. Bijsterbosch H D, Cohen Stuart M A and Fleer G J (1998) Adsorption of Graft Copolymers onto Silica and Titania, *Macromolecules* **31**:8981–8987.
161. Liu Z H, Li Y and Kowk K W (2001) Mean interparticle distances between hard particles in one to three dimensions, *Polymer* **42**:2701–2706.
162. Liu W K, Park H S, Qian S, Karpov E G, Kadowaki H and Wagner G J (2006) Bridging scale methods for nanomechanics and materials, *Comput Methods Appl Mech Eng* **195**:1407–1421.
163. Solomon M J and Lu Q (2001) Rheology and dynamics of particles in viscoelastic media, *Curr Opin Colloid Interface Sci* **6**:430–437.
164. Pozsgay A, Frater T, Szazdi L, Muller P, Sajo I and Pukanszky B (2004) Gallery structure and exfoliation of organophilized montmorillonite: Effect on composite properties, *Europ Polym J* **40**:27–36.
165. Liu H and Brinson K (2008) Reinforcing efficiency of nanoparticles: Simple comparison for polymer nanocomposites, *Compos Sci Tech* **68**:1502–1512.
166. Mendels D A, Leterriere Y and Manson J-A E (1999) Stress transfer model for single fibre and platelet composites, *J Compos Mater* **33**:1525–1543.
167. Buehler M J (2008) Molecular architecture of collagen fibrils: A critical length scale for tough fibrils, *Current Appl Phys* **8**:440–442.
168. Hague A and Ramasetty A (2005) Theoretical study of stress transfer in carbon nanotube reinforced polymer matrix composites, *Compos Struct* **71**:68–77.
169. Jancar J (2008), Review of the role of the interphase in the control of composite performance on micro- and nano-length scales, *J Mater Sci* **43**:6747–6757.
170. Jancar J (2009), Use of Reptation Dynamics in Modelling Molecular Interphase in Polymer Nano-Composite, in *IUTAM Book series*, Vol. 13 (Eds. Pyrz R and Rauhe J C), Springer, Berlin, pp. 293–301.

PART IV
NANO- AND MICROCOMPOSITES: CHARACTERIZATION

Chapter 8

Deformation Behavior of Nanocomposites Studied by X-Ray Scattering: Instrumentation and Methodology

N. Stribeck

8.1. Introduction

Overview. X-ray scattering from polymer composites during deformation has not yet become a routine method. Thus, a scientific group that intends to be successful in this field, still has to provide developer's qualifications both in the fields of engineering and computer programming. Therefore, this chapter does not focus on the documentation of results, but instead on the presentation of the state of development, of practical guidelines as an aid to avoid errors, and developing strategies for a broader application of this method. Because of this methodical approach, the examples are taken from demonstration experiments frequently carried out on semicrystalline polymers which may only be considered nanocomposites in a formal manner, because of their two-phase nanostructure (amorphous and crystalline domains). Nevertheless, this fact assures that the methods described here are readily applicable to true nanocomposites, as well.

Studies of deformation by X-ray scattering are presently experiencing a change from the so-called stretch-hold* technique [1] to an *in situ* study of nanostructure evolution during dynamical tensile tests. Thus, here the new perspectives, their technical foundations, and first results are discussed in com-

---

* In the stretch-hold technique the sample is first stretched to a desired elongation and, second, an X-ray scattering pattern is taken in the hold state.

parison to those of earlier studies. Ultimately, such studies are carried out in order to answer questions of some practical relevance: What is the constitution of the nanostructure of the composite that imprints tailored properties to the material such as special toughness or low fatigue? Such new materials are urgently sought after, *e.g.*, in the automotive industry, in order to become able to meet the politically requested goals of climate protection in the near future by weight reduction.

As research politics is presently urging science to step up the pace by utilizing computer modeling methods, the theoretician who tries to develop the model needs an accurate description of the complex nanostructure and understanding of the important evolution mechanisms. Section 8.6 presents an example that demonstrates that modeling the mechanical properties of such a simple material like polypropylene requires consideration of strain-induced crystallization and relaxation-induced melting. Unfortunately, to the best of my knowledge, modeling of such complexity is still beyond the development status of the computer modeling method. Thus, the practical modeling of the mechanics of a polymer nanocomposite is a huge challenge.

From statical to dynamical experiments. The investigation of deformation by the small-angle X-ray scattering (SAXS) method has been in a dilemma for decades. On the one hand, the X-ray method is promising to be able to reveal the structure of the material *in situ*; on the other hand, a deformation study implies that, in general, two-dimensional (2D) scattering patterns must be recorded. Until recently, one had to choose to either record noisy patterns while the mechanical test was dynamically performed, or to carry out the test stepwise, which is far from reality. Only by means of the latter technique it was possible to expose the material for a sufficiently long time at various constant elongations ("stretch-hold technique").

Noisy patterns are only appropriate for a qualitative interpretation of the structural changes in the material, which is afflicted with considerable uncertainties. In order to perform a quantitative analysis of the nanostructure, clear patterns are required. Nevertheless, if these clear images originate from an experiment that was carried out in a manner which is far from a representative loading of the material, it remains highly questionable how far the obtained results can be transferred to the behavior of the material during daily use.

Enhancement of the strain rate. Only recently scientists have been able to record good SAXS patterns with a cycle time of 30 s even from thin samples at synchrotron beamlines that are easily accessible*. Exploiting such short cycle times at slow strain rates ($\dot{\varepsilon} \approx 10^{-1} s^{-1}$) makes it possible to monitor a dynamic mechanical test by SAXS measurements, which are sufficiently accurate to study the nanostructure evolution inside the material. Admittedly, such strain rates

---

* For this purpose even a 2nd generation synchrotron is sufficient, if optimized beamline optics, a modern detector, and advanced data processing are combined. To my experience, it is impossible to set up a USAXS beamline for operation in the SAXS regime and to expect high throughput because of optical mismatch.

are still too low, but the technological gap is narrowing. By the time when the number of 3rd generation synchrotron sources has grown and the newly advertised high-power detectors will be available for materials research, the cycle time should decrease by a factor of 30 to 1 s. Reciprocally, the maximum $\dot\varepsilon$ will increase. At HASYLAB, Hamburg in Germany corresponding experiments will be enabled beginning in Fall 2008.

If noisy scattering patterns are accepted, even today mechanical tests performed at strain rates of practical relevance can be performed. This has been shown at the SRS Daresbury, Great Britain, where since 1995 structural evolution has been studied at cycle times as short as 40 ms [2–5].

Adapted extensometers and fatigue testing. It is still impossible or difficult to mount commercial tensile testers* at a synchrotron beamline, and for stretch-hold experiments even very simple designs are often sufficient. Only the perspective of genuine *in situ* experiments initiated the construction of extensometers adapted for operation in synchrotron beamlines with all the options of commercial devices. These options comprise the automatic recording of stress and strain to be assigned to the respective scattering patterns, and the possibility to run complex test programs.

At this stage of development, even a realization of fatigue tests comes into reach. The corresponding load-reversal experiments can be carried out, after the corresponding control programs for the mechanical tester have been written. From the results of such experiments one should be able to deduce how the topology of the nanocomposite is changing during the fatigue test. Thus, one should gain access to a deeper understanding of the mechanisms of materials fatigue.

Literature at a glance. Static investigations of strained polymer materials are, in fact, the earliest demonstrations of the SAXS method [7–9]. To this day, fibers and other nanocomposite materials are subjected to mechanical tests and investigated by X-ray scattering. Results of such experiments have fostered important developments in the field of SAXS theory. As an example, the fundamental work of Porod [10,11], Bonart [12–14] and Vonk [15] shall be mentioned.

Throughout the years, quasi-dynamical studies employing the stretch-hold principle [1] represent the predominant part of all the investigations on the nanostructure evolution of composite materials [6,16–35]. Since this technique means that the sample is held at a constant elongation until a low-noise scattering pattern has been exposed, the data can, in principle, be analyzed by quantitative evaluation, although most of the published papers do not proceed far beyond a presentation and interpretation of the accumulated images.

---

* The MiniMat® Tensile Tester of Rheometrics Scientific is no longer produced. An Instron® 4442 is operated on a regular basis at the polymer beamline X27C of the NSLS in Brookhaven, USA [6]. The small Zwicki® (Z1.0/TH1S, Zwick GmbH, Ulm, Germany) is of similar size, and space problems may arise in cramped experimental hutches.

Genuine dynamical *in situ* investigations of deformation have been carried out since the 1990s [36] using an instrument [37] at the SRS in Daresbury. Exploiting the shortest available cycle time of 40 ms with strongly scattering materials, it is possible to record 2D patterns that are sufficient for qualitative interpretation. In this environment, structure evolution can be studied at strain rates of $\dot{\varepsilon} \approx 10^{-1}\,\mathrm{s}^{-1}$, which are common in industrial processes. A selection of the studies performed with this instrument [2–5] demonstrates both its power and its limitations.

To our knowledge, the first paper based on high quality data with an exposure of 15 s and a cycle time of 30 s has recently been published by Chen *et al.* [38].

8.2. Scattering theory and materials structure

Reference to textbooks. In order to establish the relation between the structure of matter and the corresponding X-ray scattering pattern, a two-step deduction is carried out (cf. textbooks, *e.g.*, [39–43]).

In the first step, the interaction of X-rays with a single electron (in some absorbing matter) is considered, elaborating all the deviations of the practical interaction from the paradigm of a photon that is interacting elastically with a pseudo-electron that, in return, is emitting a spherical wave. As a result, several corrections should be applied to the measured raw data prior to a quantitative analysis* in order to make them satisfy the paradigm. These corrections comprise absorption correction, polarization correction, correction for Compton scattering, suppression of fluorescence and multiple scattering [44]. In practice, absorption correction should be applied in general. Polarization correction is important for the WAXS, as well as a subtraction of the Compton scattering background [41]. A test for multiple scattering [44] is important, if porous materials are studied by SAXS. It is very important in the field of USAXS (ultra-small angle X-ray scattering).

In the second step, the arrangement of the electrons (*i.e.*, the structure of matter) is added, and the Fourier relation of the structure to the scattered X-ray intensity, $I(\mathbf{s})$, is established by the so-called kinematic scattering theory.

The Fourier transform. According to the Fraunhofer approximation of kinematic scattering theory, the real space and the reciprocal space are related to each other by an integral transform known by the name of *Fourier transform*. It shall be indicated by the operator $\mathscr{F}()$. The *n-dimensional* (nD) Fourier transform of $h(r)$ is defined by

$$\mathscr{F}_n(h)(\mathbf{s}) := \int h(\mathbf{r}) \exp(2\pi i \mathbf{r}\mathbf{s})\, d^n r, \tag{8.1}$$

where i is the imaginary unit – and back-transformation simply yields

---

* Quantitative analysis means that the shape of the intensity function is considered or entered in a numerical evaluation procedure. For a simple qualitative discussion (peak positions, strong intensity changes) corrections are negligible.

$$\mathscr{F}_{-n}(H)(\mathbf{r}) := \int H(\mathbf{s})\exp(-2\pi i \mathbf{rs})\,d^n s, \tag{8.2}$$

where $(H)(\mathbf{s}) := \mathscr{F}_n(h)(\mathbf{s})$. In the field of scattering 1D-, 2D- and 3D-transforms are required. The kernel of the Fourier transform is called the harmonic function

$$\exp(2\pi i r s) = \cos(2\pi r s) + i\sin(2\pi r s), \tag{8.3}$$

and the Fourier transform is said to perform a harmonic analysis.

In the **q**-system, the pair of transformations

$$\mathscr{F}_n(h)(\mathbf{q}) := \int h(\mathbf{r})\exp(i\mathbf{rq})\,d^n r, \tag{8.4}$$

and

$$\mathscr{F}_{-n}(H)(\mathbf{r}) := \left(\frac{1}{2\pi}\right)^n \int H(\mathbf{q})\exp(-i\mathbf{rq})\,d^n q, \tag{8.5}$$

is asymmetric with different pre-factors that must be tracked in calculus.

Structure and scattering in a nutshell. The fundamental relations between the electron density distribution inside the sample, $\rho(\mathbf{r})$, and the observed scattering intensity, $I(\mathbf{s})$, are conveniently combined in a sketch

$$\begin{array}{ccc}
\rho(\mathbf{r}) & \overset{\mathscr{F}_3}{\Leftrightarrow} & A(\mathbf{s}) \\
\star 2 \Downarrow & & \Downarrow |\;|^2 \\
z(\mathbf{r}) \underset{\Delta}{\Leftrightarrow} P(\mathbf{r}) & \underset{\mathscr{F}_3}{\Leftrightarrow} & I(\mathbf{s})
\end{array} \tag{8.6}$$

from which the theoretically explored options for a quantitative analysis of the scattering, $I(\mathbf{s})$, can be accessed. According to the scheme, the real-space electron density, $\rho(\mathbf{r})$, is converted by 3D complex Fourier transform into the scattering amplitude, $A(\mathbf{s})$, in reciprocal space.

Stepping downwards in the scheme from the amplitude, we arrive at the scattering intensity

$$I(\mathbf{s}) = |A(\mathbf{s})|^2$$

by taking the square of the absolute value. The unidirectional downward arrow shows that this operation cannot be reversed. We are loosing the phase information of the structure. This means that we cannot reconstruct absolute positions of individual domains (*i.e.*, crystallites) in the material. Only relative distances among domains, *i.e.*, their correlations are readily determined. As a consequence of the last-mentioned operation, the intensity

$$I(\mathbf{s}) = I(-\mathbf{s})$$

is always an even function (point symmetry) with real (not complex) values. In practice, this means that we can reconstruct $I(\mathbf{s})$ from $I(-\mathbf{s})$ and vice versa, if only one of these points has been measured on the detector. Another consequence is relaxing the calculus, as long as we stay with the s-system: When switching back and forth between reciprocal and real space by means of the Fourier backtransform and the Fourier transform, even the switching sign in the harmonic kernel becomes negligible.

Proceeding to the left along the bottom edge of Eq. (8.6) we arrive back in real space at the *Patterson function*,

$$P(\mathbf{r}) = \mathscr{F}_3^{-1}(I(\mathbf{s})).$$

The physical meaning of the Patterson function is readily established by introduction and interpretation of the autocorrelation operation $\star 2$

$$P(\mathbf{r}) = \rho^{\star 2}(\mathbf{r}),$$

which turns the structure, $\rho(\mathbf{r})$, directly into the Patterson function. Because the autocorrelation integral expands into

$$\rho^{\star 2}(\mathbf{r}) = \int \rho(\mathbf{y})\, \rho(\mathbf{r}+\mathbf{y})\, d^3y, \tag{8.7}$$

it is identified by the overlap integral between the structure, $\rho(\mathbf{y})$, and its displaced ghost. Here, the vector $\mathbf{r}$ describes the amount and the direction of the displacement. In particular, in the field of SAXS a common synonym for the Patterson function is the *correlation function* (Debye (1949) [11], Porod (1951) [10]),

$$\gamma(\mathbf{r}) = \rho^{\star 2}(\mathbf{r})/\rho^{\star 2}(0) = P(\mathbf{r})/\rho^{\star 2}(0). \tag{8.8}$$

By its normalization, $\gamma(0) = 1$, it indicates that the correlation between structure and ghost is perfect, if there is no displacement at all.

Finally, the structure of soft matter frequently may be considered to be made from domains which can be distinguished from each other by a sufficient difference of their electron densities (contrast). This is the case with materials comprising soft and hard domains, voids, crystallites, or amorphous regions. In this case it has proven advantageous not to study the correlation function, but to perform edge enhancement

$$z(\mathbf{r}) = \Delta P(\mathbf{r})$$

by applying the Laplacian operator. A method for the computation of a 3D *chord distribution function* (CDF), $z(\mathbf{r})$, has been described in 2001 [45,46], but before that a one-dimensional chord distribution called *interface distribution function* (IDF), $g_1(\mathbf{r})$, had already been proposed by Ruland [47–49] for the study of lamellar systems. The basic idea of this approach reaches even back to 1965, when Méring and Tchoubar [50–53] proposed the (radial) *chord length distribution* (CLD), $g(\mathbf{r})$.

Comparing the IDF to the CDF, the 1D derivative d/dr_3 of the electron density $\rho(r_3)$ is replaced by the gradient $\nabla \rho(\mathbf{r})$, as well as the second derivative d^2/dx^2 in the IDF by the Laplacian Δ in the CDF [45]. In analogy to the particle-

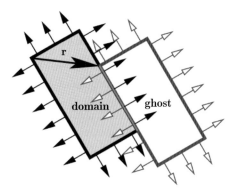

Figure 8.1. A particle–ghost autocorrelation of *gradient vectors* is generating the CDF. These vectors are emanating in normal direction from the surfaces of the particle and its displaced ghost. The ghost is displaced by the vector **r**. The scalar product of the gradient vectors is vanishing everywhere, except for the regions with *surface contact* between particle and ghost

ghost construction of the correlation function (cf. textbooks [43,44,54]) the construction of the CDF is readily demonstrated (Figure 8.1). In a multiphase material the gradient field, $\nabla \rho\left(\mathbf{r}\right)$, is vanishing almost everywhere. Exceptions are the domain surfaces. They are continuously populated with gradient vectors, the lengths of which are proportional to the heights of the density jumps. Thus, the autocorrelation among the edge-enhanced gradient field of the structure and its ghost as a function of ghost displacement, **r**, is computed from the scalar product of the gradient vectors. Approximately, it is proportional to the area of surface contact and the product of the density jumps at the contact surface. Its sign indicates if there is inner or outer contact. By definition of the CDF, outer contact is carrying the positive sign.

8.2.1. *Relation between a CDF and IDFs*

Every radial, 1D slice (cut, section) through the center of a CDF

$$[z]_1 \left(r_{\psi,\varphi}\right) = g_1 \left(r_{\psi,\varphi}\right)$$

is an IDF, by definition. In the above equation, the slicing direction is indicated by a polar and an azimuthal angle, ψ and φ, respectively. Of particular practical interest for the study of fibers is the cut of the CDF along the fiber axis,

$$[z]_1 \left(r_3\right) = z\left(0, r_3\right) = g_1 \left(r_3\right),$$

which describes the longitudinal structure of the material. In analogous manner, the transverse structure of the fiber is described by the 2D slice

$$[z]_2 \left(r_{12}\right) = z\left(0, r_{12}\right) = g_2 \left(r_{12}\right)$$

of the CDF.

8.3. Analysis options derived from scattering theory

8.3.1. *Completeness – a preliminary note*

If the structure of an investigated sample can be approximated by a lattice (crystal), the result of the scattering experiment is a diffraction pattern with many distinct peaks – the Fourier transform of an infinite lattice is, again, an infinite lattice of peaks and the success of crystallography is based on the fact that the lattice structure can be completely reconstructed from a limited number of diffraction peaks, if only their positions and strengths are measured.

In general, the structure of soft matter is imperfect and polydisperse. Only in rare cases it can be sufficiently described by a crystal. Thus, in most of the practical cases, an in-depth analysis of scattering intensity by means of the crystallographic approach is not permitted. Although this fact has been known for almost a century (Debye and Menke* (1931) [55]), far-reaching conclusions are still drawn, even if they are merely based on determinations of peak positions ("long periods") and peak shapes of scattering patterns.

Soft matter structure is characterized by a continuous density function, $\rho(\mathbf{r})$, which is subjected to Fourier transform by the scattering experiment yielding a continuous intensity function, $I(\mathbf{s})$, which does not drop to zero in long intervals between narrow peaks. Thus, for any in-depth analysis of distorted structure, either we have to model the complete shape of the pattern, or we have to switch to real space, again supplying the respective transform with a complete intensity function.

8.3.2. *Analysis options*

Options of data analysis can be deduced from Eq. (8.6) and our notions concerning the structure. As an example, let us consider the case of small-angle X-ray scattering. In this area, the structure is described by a continuous density function. Although there is no** way back from intensity to density, there are several options for data analysis:

1. Utilize theory and find out, how some structure parameters can be determined from the intensity directly;
2. Walk from the intensity along the lower edge half-way back to real space, where the transformed data are closer to human perception;
3. Model a structure and fit it to the intensity; or
4. In addition to item 2, carry out "edge enhancement" in order to visualize structure by means of the chord distribution function (CDF), $z(\mathbf{r})$, and interpret or fit it.

Effort of data analysis. The above-mentioned options are listed in the order of increasing complexity for the scientist. When scattering curves (isotropic

---

* For a translation of the note made by Debye and Menke cf. [44], p. 1.
** Except for the case of anomalous SAXS.

data) shall be analyzed, all the four listed options have proven to be manageable by many scientific groups.

In contrast, the analysis of scattering images from anisotropic materials is a real challenge, and in this subarea many scientists surrender and resort to the interpretation of peak positions and peak widths in raw data.

A shortcut solution for the analysis of anisotropic data is found by mapping scattering images to scattering curves as has been devised by Bonart in 1966 [13]. Founded on Fourier transformation theory, he has clarified that information on the structure "in a chosen direction" is not related to an intensity curve sliced from the pattern, but to a projection of the pattern on the direction of interest.

The barrier to the application of the shortcut is probably resulting from the need to pre-process the scattering data and to project the 3D scattering intensity. This task requires 3D geometrical imagination and knowledge of methods of digital image processing, a field that is quite new to the community of scatterers. Programmers, on the other hand, are rarely educated in the fields of scattering and multi-dimensional projections.

8.3.3. *Parameters, functions and operations*

The scheme of Eq. (8.6), which sketches the relation between structure and scattering intensity, contains many parameters, functions and operations which require explanation:

V	is the irradiated volume. It is defined by the sample thickness multiplied by the footprint of the incident primary beam on the sample.
$\rho(\mathbf{r})$	the electron density difference
$A(\mathbf{s})$	scattering amplitude
$k = \rho^{*2}(0)$	scattering power
$Q = k/V$	invariant
$P(\mathbf{r})$	Patterson function
$\gamma(\mathbf{r}) = P(\mathbf{r})/k$	SAXS correlation function
$z(\mathbf{r})$	chord distribution function (CDF)

The X-ray detector measures the intensity of electromagnetic waves, *i.e.*, the absolute square $|\ |^2$ of their amplitude. Thus, in combination, the upper path between density and intensity through the square is written as

$$I(\mathbf{s}) = \left|\mathscr{F}_3\left(\rho(\mathbf{r})\right)\right|^2.$$

In the lower path through the square we have an equivalent formulation

$$I(\mathbf{s}) = \mathscr{F}_3\left(\rho^{*2}(\mathbf{r})\right)$$

with the Patterson or correlation function, $\rho^{*2}(\mathbf{r})$, involved. $\rho^{*2}(\mathbf{r})$ is generated from the "inhomogeneities", $\rho(\mathbf{r})$, by means of the autocorrelation.

8.4. The experiment

8.4.1. *Principal design*

X-ray source: Laboratory *vs.* synchrotron. Concerning the output power, not only synchrotron beamlines, but also laboratory sources have experienced tremendous progress in the last decade. Details have been described in my textbook [44]. For a study of nanostructure evolution during continuous mechanical tests, the brilliance of a laboratory source will most probably never grow sufficiently high. As a consequence, only experiments following the stretch-hold technique [1] appear to be feasible in the laboratory, even if rotating anodes, Göbel mirrors and high-efficiency detectors are combined. In general, the results of stretch-hold experiments cannot be assigned to dynamic mechanical tests (cf. Figure 8.10).

Basic setup. Modern equipment is most frequently collecting small-angle scattering data from a large angular region at the same time by means of a two-dimensional (2D) X-ray detector*. In principle the basic setup resembles the traditional photographic X-ray cameras (pinhole camera, Kiessig camera). A sketch is presented in Figure 8.2. Such a setup comprises beam monitoring devices (ionization chamber, PIN-diode) and a detector. If the beam monitoring devices are not recording the primary beam intensity as a function of time, there is no chance to normalize the whole series of scattering patterns with

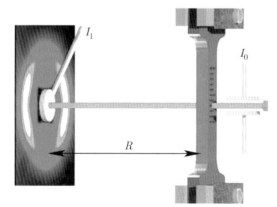

Figure 8.2. The basic X-ray setup for the study of deformation at a synchrotron. The intensity of the incident X-ray beam is measured in an ionization chamber (I_0). Thereafter it penetrates the sample which is subjected to mechanical load. At a distance, R, behind the sample, the detector is recording the scattering pattern. In its center the detector is protected by a beam stop. It is equipped with a PIN-diode (I_1) which records the intensity of the attenuated beam

---

* In the field of SAXS and USAXS (ultra-small angle X-rays cattering) the detector image is equivalent to a rectangular plane in reciprocal space, which is commonly addressed by either the symbol **s** or by the symbol $\mathbf{q} = 2\pi \mathbf{s}$.

respect to constant flux and a constant irradiated volume. In the horizontal direction there is ample space to place the tensile tester in the X-ray beam. Problems may arise from lack of space in the vertical direction, if the beamline construction is based on an optical bench with its bars running in the horizontal direction. A more flexible beamline design is "the dance floor" [44].

8.4.2. Engineering solutions

8.4.2.1. Setup for mechanical testing

The sample in the air gap. Figure 8.3 shows the mounting of tensile testers in the synchrotron beam. The close-up (Figure 8.3b) shows a polypropylene (PP) sample clamped between the cross-heads with fiducial marks applied to it. The tester is able to perform symmetrical drawing, so that the spot irradiated by the synchrotron beam is not drifting away too fast. On-line control of the spot position is not yet possible. A TV camera is monitoring the sample and the fiducial marks. The position of the X-ray beam is indicated on the TV screen by means of a cross-hair cursor (cf. Figure 8.6) which is positioned in advance by means of a scintillation screen mounted between the clamps.

Figure 8.3. Extensometers at synchrotron beamlines of HASYLAB, Hamburg: (a) Zwicki® Z1.0/TH1S mounted at beamline BW4; (b) Close-up view of a customized extensometer in the X-ray beam of beamline A2. Polypropylene test-bar (DIN 53504) between the cross-bars. TV camera monitoring the sample for the true elongation

Application of fiducial marks. Using a rubber stamp (Figure 8.4) appears to be an appropriate and easy way to apply marks to long samples which are uniaxially extended. If the samples are smaller or shall be subjected to biaxial deformation, varnish can be sprayed on the sample that is partly covered by

Figure 8.4. Rubber stamp with a line grid of 2 mm distance for marking polymer samples to be used in mechanical tests

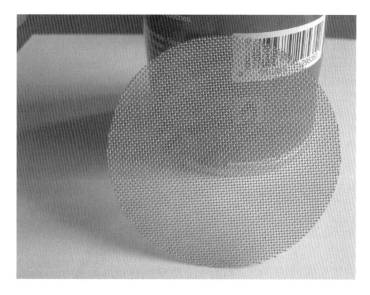

Figure 8.5. Commercial steel-wire sieve with a mesh width of 0.625 mm in front of a spray varnish

a woven steel-wire screen. Such screens are commercially available for mesh-widths down to 25 μm and can easily be cleaned (cf. Figure 8.5). Each of these methods has its drawback: The stamp ink will not dry on some materials, whereas dry spray varnish is prone to flaking off the sample upon straining.

During the experiment, its progress is observed on monitors (Figure 8.6) outside the experimental hutch. One of the monitors shows the TV-image of the sample with the penetration zone of the synchrotron beam indicated. For every SAXS pattern that is recorded, at least one image from the video stream is grabbed and stored together with the elapsed time. Further monitors display the actual SAXS pattern, the progress of the experiment reflected in the engineering stress-strain curve, and the testing program. Time, force and the distance of the cross-heads are continuously written into an ASCII file. From this file the force can directly be associated to each scattering pattern. The true elongation must be determined manually and after the end of the experiment from the distance between the fiducial marks on the images captured from the video.

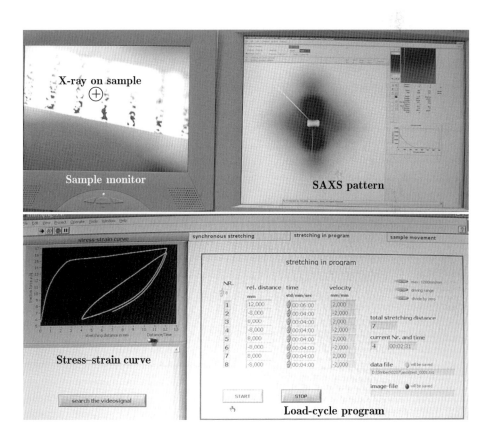

Figure 8.6. Mechanical testing of polypropylene in progress as watched on monitors at beamline A2, HASYLAB

8.4.2.2. *Setup for micro-tomography under bending load*

In order to study nanostructure variation as a function of position in the sample, there are synchrotron beamlines with a microbeam, *i.e.*, an X-ray beam with a diameter ranging from 100 μm down to hundreds of nanometers. An optical element by which some standard synchrotron beams can be turned into a microbeam is a stack of refracting beryllium lenses [56]. Figure 8.7 shows such an element mounted in the beam of beamline BW4 at HASYLAB, Hamburg. By means of this installation the ultra-small-angle X-ray scattering (USAXS) beamline with a beam diameter of 1 mm is turned into a microbeam small-angle X-ray scattering (SAXS) setup. Beam diameters down to 40 μm can presently be achieved with this setup.

Figure 8.7. Stacks of refracting Be-lenses are used to demagnify appropriate synchrotron beams. The resulting diameter is in the range of micrometers

If a polymer part is translated and rotated in such a microbeam (Figure 8.8), the three-dimensional (3D) spatial nanostructure variation in the part can be studied without slicing. For this purpose, tomographic reconstruction of the SAXS patterns that would have emerged from tiny volume elements (voxels) in the material is carried out [57,58]. Concerning a deformation study, bending load can be applied to a polymer part and the effect on the spatial variation of the nanostructure can be studied. The typical beamtime required for the collection of the complete set of projected scattering data is 24 h. A sample bender that has recently been used by us for test experiments is shown in Figure 8.9.

Deformation Behavior of Nanocomposites

(a) Side view

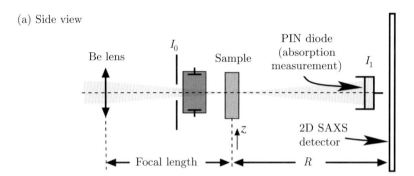

(b) Top view of sample region

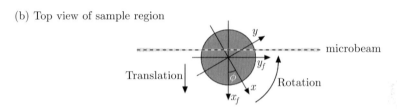

Figure 8.8. Sketch of a setup for X-ray SAXS micro computer tomography (SAXS μ-CT) and absorption μ-CT. The well-known absorption tomogram is constructed from the signal of the PIN diode, I_1/I_0. The SAXS μ-CT patterns are reconstructed from the projected scattering patterns recorded on the SAXS detector ($R \approx 2$ m) while the sample is translated and rotated in the beam

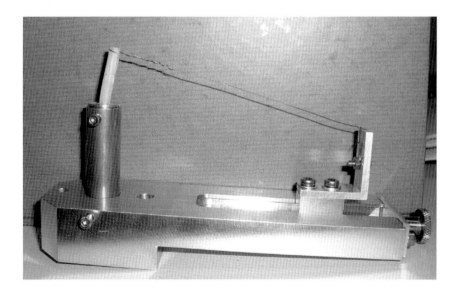

Figure 8.9. Sample bender for cylindrical polymer rods to be used in conjunction with microbeam SAXS experiments

8.4.3. *Scattering data and its evaluation*

Increasingly rarely scattering experiments are carried out by well-trained experts. Thus, there is a high risk of both experiment design flaws and of recording incomplete data. As a result, the analysis of the data is kept cursory or even becomes erroneous. In [44] the necessary steps concerning interfaces and electronical coupling, the design of the experiment, and the evaluation of the data are described in detail. Here, only the important issues concerning proper collection of data are presented.

How to collect complete data sets. After you have collected the first test patterns, check that the data files have arrived in the expected directories. Check the size of the files. Open the ASCII files or the ASCII headers in a text editor and check that the environmental data have arrived in the files. Vary environmental parameters in test measurements and check that the values in the ASCII files vary accordingly. If possible, calibrate environmental parameters (*e.g.*, sample temperature, straining force, cross-bar position). Ask the beamline staff to demonstrate what they tell you. Double check! Otherwise your effort may be wasted.

Before the first experiment and after each re-adjustment of the setup, the following static data should be collected:

– X-ray wavelength
– sample-to-detector distance, R
– technical description of the detector(s) used
– size of each pixel cell on the detector
– allocation of monitor channels to environmental parameters
– calibration of monitor channels for which absolute data values are required (*e.g.*, sample temperature, sample elongation)
– a measured primary beam profile or the integral breadths of the primary beam on the detector in the horizontal and the vertical directions
– in particular in studies of porous or fractal materials, assessment of multiple scattering should be carried out.

The parasitic scattering (*machine background*) should at least be recorded once within each synchrotron radiation run. You might consider measuring a background before every experiment, because there is some probability that the beam position jumps by a few pixels during refill of the storage ring.

The data evaluation bottleneck. The majority of users would benefit from a general and user-friendly data evaluation program. Nevertheless, as long as design rules of device interfacing are not respected and the data are not stored in a uniform and simple-to-read data format, this wish will remain a dream. Progress towards a uniform data format is not detectable. On the one hand, there is progress in subareas, but on the other hand, materials science is fed on experimental innovation, the realization of which is introducing new flaws and further complexity.

Flaws in engineering design and setup. If one intends to eliminate the machine background from a measured pattern, it is necessary to extract the exposure intervals, primary beam intensities, I_0, and attenuated intensities, I_1, from complex data sets which are different at different beamlines. Moreover, the sequence of the stored numbers changes as a function of the wiring during setup. Thus, if background correction and normalization are not done by the beamline staff, some users will not be able to carry out these steps.

On the other hand, if there is a normalization service, the result will be written into a set of files in another proprietary data format, and it may happen that during this process data are not copied which may be essential for further data evaluation (*e.g.*, the elapsed time since the start of the experiment, which is the most important parameter in any mechanical test – cf. Figure 8.11).

Even more severe flaws are encountered. For example, the student developer of an extensometer control does not hand over the signals from the load cell and from the transducer to channels of the voltage-to-frequency (VDC) converters of the standard beamline electronics, but carries out digitization on hardware of his own with a cycle time of his own. Finally, they are stored in an ASCII file on his own computer, using his computer clock as a second time normal. As long as the cycle time is longer than 10 s, a skilled programmer will be able to modify his data pre-evaluation module in such a way that it compensates this flaw.

Curing flaws by computer programming. In quintessence, the consistent evaluation of streams of scattering data remains a computer programming task, and thus, programming skills are required for a successful work with multi-dimensional scattering data. Scientists providing these skills will utilize and modify open sources (see for instance [59]). Without these skills, even the few popular data evaluation programs such as Fit2D [60] will probably not be employed in an efficient manner.

Automation – the next challenge. In a dynamical experiment, hundreds of 2D scattering patterns are collected which must be evaluated. Thus, efficient automation is essential. Instead of writing a thoroughly planned script and running it automatically on all patterns of the series, a user merely trained to operate programs with graphical user interfaces is easily misled. Having carried out the experiment, he will use the computer in accordance with his training: Install a user-friendly evaluation program and start to click buttons without resorting to manual or theory. Operated in such a way, a user-friendly evaluation program becomes data-hostile. Moreover, the processing of voluminous data sets turns into an unmanageable task.

For instance, Fit2D is frequently used to extract slices* from 2D scattering patterns, whereas either an azimuthal average or a projection would have been appropriate. Even such simple and inappropriate operation will turn into a nightmare, when Fit2D is operated by mouse clicks and many patterns are operated. Finally, noisy slices are extracted instead of low-noise integrated

---

* Slices are curves that show the intensity along a straight path in the pattern.

curves. As a consequence, further processing will produce low-significance results.

8.5. Techniques: Dynamic *vs.* stretch-hold

Only a few years ago, dynamic experiments with a good signal-to-noise (S/N) ratio of the scattering patterns were not possible. Therefore many of the studies listed in this chapter have been performed by means of the older stretch-hold technique. A recent comparative study [61] unexpectedly revealed pronounced differences after switching from stretch-hold to a slow dynamic experiment. Consequently, results on the nanostructure evolution retrieved in stretch-hold studies may be used only with a lot of cautiousness to explain the dynamic loading process that is relevant in practical applications.

Figure 8.10 shows in the top row (continuous straining experiment) scattering patterns with narrow and detailed peaks, although the technique implies integration over a considerable interval of the elongation ε ($\Delta\varepsilon \approx 0.05$), whereas the scattering patterns accumulated during the same beamtime in the stretch-hold technique at constant elongation appear much more blurred and show a stronger equatorial streak. Thus, this comparative experiment demonstrates that during the hold state in a stretch-hold experiment a considerable change of the nanostructure inside the composite may occur.

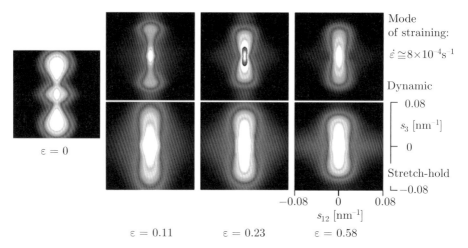

Figure 8.10. Comparison between SAXS patterns recorded during continuous straining (top row, $\dot{\varepsilon} \approx 10^{-1}\mathrm{s}^{-1}$) of hard-elastic PP films and results from the corresponding stretch-hold experiment (bottom row)

8.6. Advanced goal: Identification of mechanisms

One of the most important goals of the method is the elucidation of the mechanisms of nanostructure evolution during application of mechanical load. In the previous section it has been demonstrated that for this purpose not the stretch-hold technique, but the recording of scattering patterns during a dynamic materials test should be applied.

Moreover, concerning the mechanical test itself, not only the classical method of continuous straining should be considered, because of the fact that during service a component* made from a polymer nanocomposite will frequently be subjected to vibrations causing many cycles of alternating load – with materials fatigue becoming the most important reason for failure.

The scientific questions, which have recently come into reach, shall be demonstrated by results of a recent load-reversal study [62]. Figure 8.11 displays the macroscopic parameters and some of the nanostructure parameters as a function of the elapsed time. The parameters of the nanostructure have been evaluated from X-ray scattering data.

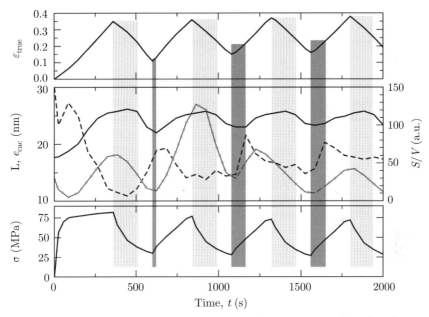

Figure 8.11. Dynamic load-reversal mechanical test of hard-elastic PP film. As a function of the elapsed time, t, nanostructural parameters (diagram in the middle) and the macroscopic parameters elongation, ε (top), and tensile stress, σ (bottom) are displayed. The middle diagram shows the long period, L (solid line), the lateral extension e_{cac} (dashed) of the layers, and a quantity S/V (solid line with dots) which is a measure of the number of lamellae in the material. Vertical bars indicate zones of strain-induced crystallization (dark gray) and relaxation-induced melting (light gray), respectively

Experimental. Commercial films of uniaxially oriented, hard-elastic PP were used to punch out tensile bars (DIN 53504) which were subjected to mechanical load parallel to the direction of the preferential orientation. The thickness of the films was 50 µm. Because of the fact that discrete X-ray scattering was observed

---

* Consider, for example, the spare-wheel hutch in a car that is supposed to be made from reinforced polypropylene in order to save weight.

in a wide angular range, every experiment was carried out both on an USAXS beamline and on a SAXS beamline. USAXS covers the region at extremely small scattering angles and was performed at beamline BW4 of HASYLAB. Measurements in the typical SAXS range were carried out at beamline A2 of HASYLAB. In order to record scattering patterns with good S/N-ratio, an exposure of 50 s was sufficient at both beamlines*. Before combination of the pairs of recorded scattering patterns, the SAXS pattern had been deconvolved from its primary beam profile.

Data analysis. From the scattering patterns the multidimensional chord distribution function (CDF) [44,45] was computed in order to extract the topological information from the SAXS data. Thereafter, the CDF was evaluated, and the curves in the central diagram of Figure 8.11 were obtained. The top diagram of Figure 8.11 shows the true elongation, ε, measured at the irradiation spot, as induced to the material by the extensometer. The bottom diagram shows the stress, ε, as measured by means of the load cell normalized to the actual cross-section of the material.

Results. In the central diagram the solid line presents the long period, L, i.e., the average distance between neighboring crystalline lamellae. Obviously, this distance remains constant for a while when relaxation sets in (gray bars). In the first straining stage, the long period is increasing monotonously with increasing ε. This finding is unspectacular, as macroscopic strain is almost proportional to nanoscopic strain. The behavior is changing during the subsequent straining stages, as at their beginning a plateau is formed, which is becoming broader with every cycle (dark vertical bars).

If the scattering patterns would have been interpreted directly and the CDF analysis would not have been performed, no further information on the nanostructure would have been obtained and we would have to leave it at the description of a strange phenomenon. Fortunately, from the CDF additional data on the nanostructure can be extracted with little effort. e_{cac} is the lateral extension of a sandwich made from two crystalline lamellae with an amorphous zone in between**. This quantity is a measure of the lateral extension of the average lamella and is reported by a dashed line in the diagram. It is observed that e_{cac} is decreasing considerably during the first straining phase – lamellae are fragmented into narrow blocks. Its extension remains almost constant during all the relaxation phases. However, in the beginning of each of the following

---

* After an adjustment of the USAXS beamline for the SAXS range the required exposure increased to 20 min, which is unacceptable. This shortcoming with respect to the claimed universality is probably a problem of the design of the beamline optics. After inserting Be-lenses, the SAXS performance of the USAXS beamline is much better.

** Granted – in principle it is possible to determine the lateral extension of the single lamella, as well. Concerning the PP studied here, the determination is difficult, because the corresponding peak is interfering both with the scattering effect of crystalline blocks and that of cross-hatched lamellae, but in this example we do not commit ourselves to the description of all the aspects of the complex nanostructure evolution of the studied material.

straining phases e_{cac} is increasing in a sudden burst, an effect that is readily explained by the fast lateral lamellae growth which is known from polymer crystallization. Upon further straining, the lateral extension of the crystallites is, again, decreasing. Even with this second nanostructure parameter there would be ample room for speculation on the fundamental mechanism by which the material reacts upon strain.

Before we start the interpretation, let us consider the total integral of the CDF which, to a first approximation, is proportional to the total surface, S, of the crystalline domains with respect to the volume, V, irradiated by the X-ray beam (curve marked with circular symbols). Now we see that in the beginning of the experiment, crystallites are "mechanically destructed" up to the point where the stress-strain curve reaches the yield plateau. From this point on strain-induced crystallization sets in and the number of crystals is strongly increasing. As the material is subjected to relaxation, the number of crystals is dropping again.

Interpretation. Conclusions can be drawn from the combination of these results. Only until the material enters the yield-plateau it can be considered as a nanocomposite solid body, the crystalline component of which is destroyed by the applied strain. Beyond this point, a description of the nanostructure in terms of mechanical categories appears to be inadequate, since the main mechanism by which the material reacts upon mechanical load is strain-induced crystallization and relaxation-induced melting. Moreover, the mechanism of fatigue is already indicated to some extent in the presented data: Concerning the extracted nanostructure parameters, the amplitude of their oscillations is decreasing from cycle to cycle. Thus, the material "is learning" a nanostructure with minimum resistance to load cycling, and favorable materials properties are continuously decreasing.

Outlook. At this point of understanding and with the dynamic X-ray method at hand, it appears promising to stabilize a favorable nanostructure of the polymer by fitting in a nanocomponent which is not melting and turns polypropylene into a low-fatigue material.

8.7. Observed promising effects from stretch-hold experiments

Even though the older stretch-hold experiments can hardly be used for a reliable description of nanostructure evolution under mechanical load, the observed general scattering effects indicate in which subareas of the structure analysis of nanocomposites there is a good chance for successful application of the X-ray method. Thus, some examples are presented.

8.7.1. *Orientation of nanofibrils in highly oriented polymer blends by means of USAXS*

The reinforcement of a fiber material with uniaxially microfibrils (a composite material) oriented parallel to the fiber axis suffers from a serious disadvantage, namely the extremely high anisotropy of the mechanical properties. This situation can be overcome by inclining the reinforcing microfibrils with respect to the principal axis of the material. This principle is observed in nature, where reinforcing fibrils are frequently inclined with respect to the principal axis of fibers found both in fauna [63] and flora [64,65].

Man-made model systems that show similar scattering patterns and in which the microfibrillar angle can easily be adjusted have been prepared and a feasibility study [66] has been performed utilizing USAXS at HASYLAB, beamline BW4 in different states of elongation. The studied materials are bundles of about 1000 highly oriented filaments, each having a diameter of 30 µm. The initial microfibrillar angle is adjusted by twisting the bundle more or less. Each filament is a blend of polypropylene/poly(ethylene terephthalate) (PP/PET) (80/20 by wt.) containing a huge amount of PET nanofibrils with diameters of only 60 nm to 100 nm and a length of many thousands of nanometers, as documented by scanning electron microscopy. These nanofibrils are homogeneously dispersed in the dominating PP matrix. It is worth noting that the studied bundles of filaments may easily be welded together to form a compact fiber without losing their unique orientation. This welding is achieved by annealing the bundles at a temperature between the melting points of PP and PET. Figure 8.12 presents some of the USAXS patterns. The left column displays the materials before the application of strain. The pattern of the twisted

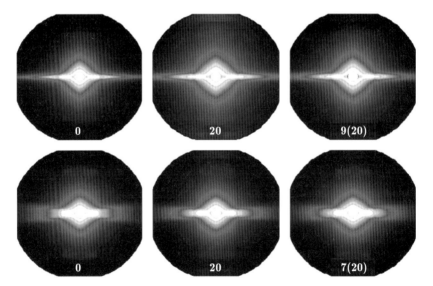

Figure 8.12. Nanofibrillar PET structure in USAXS patterns (pseudo-color, logarithmic scale) from bundles of fibers made from PP/PET blends by micro-die spinning at different states of elongation. Top row: Untwisted bundle. Bottom row: Twisted bundle. The elongation, ε, of the material in percent at which the pattern is taken is indicated at the bottom of each pattern. Right column: For the relaxed material the previous elongation is indicated in parentheses. Fiber axis s_3 is vertical. Each pattern shows a range of $-0.0625\,\text{nm}^{-1} < s_{12}, s_3 < 0.0625\,\text{nm}^{-1}$

sample is shown at the bottom. The double-tilted equatorial streak due to the twisting of the filaments is clearly visible. The preferential microfibrillar angle $\chi = 4°$ [64,65] is easily extracted from the inclination of the streak with respect to the equator of the bundle.

The middle column shows the patterns of both samples at an elongation of 20%. While the pattern of the straight-bundle sample appears virtually unaffected, the microfibrils in the twisted-bundle sample are straightened to $\chi = 1°$ under the load.

The right column displays the USAXS patterns of the two materials after relaxation ($\chi = 3°$). Obviously, in the twisted-bundle sample the microfibrillar angle has recovered to a considerable extent. It is interesting to note that the overall elastic recovery of the material with twisted bundles is better than that of the straight-bundle material.

8.7.2. USAXS studies on undrawn and highly drawn PP/PET blends

In an experiment performed at beamline BW4, HASYLAB [67], highly oriented fiber materials made from blends of the polymers poly(ethylene terephthalate) and poly(propylene) are studied. The materials are prepared in special processes aiming at dispersion of the two components on a nanometer scale. The initial aim has been to study the materials *in situ* during straining experiments. Because of poor conditions found at BW4 at that time, only static measurements of the materials at fixed elongations have been performed. Nevertheless, some conclusions on the materials properties as a function of preparation and elongational load can be drawn.

Dealing with polymer blends, the degree of dispersity is of prime importance because it determines to a great extent the final materials properties. Particularly, the present scientific focus on manufacturing of nanocomposites requires good knowledge of the real sizes of the dispersed particles. The study of domain sizes in the range of a couple of ten nanometers presumes the use of appropriate technique as, for example, the USAXS.

The samples are neat PP and its blend with PET in a composition PP/PET 80/20 by weight. Fibers are prepared by (1) conventional spinning followed by drawing at 130°C, and (2) by spinning from a 300 µm die without additional drawing resulting in filaments of 30 µm diameter. In the latter case, bundles of about 300 filaments are irradiated in the experiment.

For all the materials scattering patterns are taken at various deformations in the range between $\varepsilon = 0$ and $\varepsilon = 0.12$. After relaxation from the deformational steps, scattering patterns are taken again.

In Figure 8.13 (left column) some of the USAXS patterns from the neat PP filaments are shown, whereas the middle and the right columns show scattering patterns of the PP/PET blend, respectively.

The scattering patterns from the neat PP (Figure 8.13, left column) show diffuse scattering. In addition, the drawn and the elongated material show a

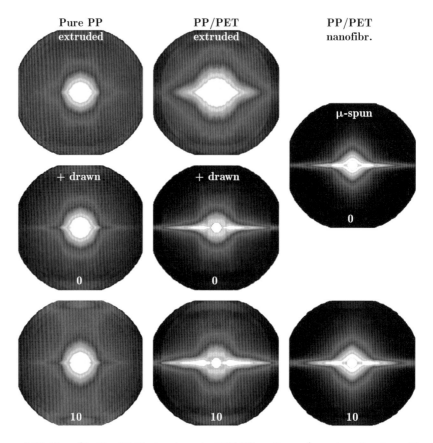

Figure 8.13. Nanofibrillar PET structure in USAXS patterns (pseudo-color, logarithmic scale) from fibers made from PP/PET blends after different steps of processing. The elongation, ε, of the material in percent at which the pattern is taken is indicated at the bottom. Fiber axis s_3 is vertical. Each pattern shows a range of -0.0625 nm$^{-1} < s_{12}, s_3 < 0.0625$ nm^{-1}

faint indication of void scattering (equatorial streak). This is typical for semi-crystalline highly oriented homopolymers. The long period reflection ($L = 17.5$ nm) is observed on the meridian just inside the accessible angular range of the USAXS. In the middle and the right columns the PP/PET blend is shown. Faint long periods are observed in the middle column as well, but not in the right column where the corresponding reflection is outside the USAXS range (i.e., $L < 16$ nm). After conventional extrusion (middle), the fine dispersion of the PET in the PP matrix on a nanometer scale is obvious from the size and shape of the central scattering. Drawing (central image) or micro-spinning (right column) leads to the formation of highly elongated needle-shaped PET domains in the PP fiber. Upon drawing of some of the PET, nanofibrillar domains appear to break (broadening of the equatorial streaks). The lengths of the nanofibrillar PET domains are much higher if the material is made by micro-die spinning (right column).

8.8. Choosing experiments

In case the experimental data shall be subjected to quantitative evaluation, the design of the experiment should be chosen carefully. In general, oriented materials with fiber symmetry should be preferred and the fiber axis should run parallel to one of the detector edges (cf. Figure 8.2) – otherwise the measured SAXS pattern is incomplete from the point of view of theory. Nanocomposites, which are studied at ambient temperature, frequently exhibit both discrete USAXS and discrete SAXS. In this case, the mechanical test should be run both at a USAXS beamline, and at a SAXS beamline. Before combining the pairs of patterns, the SAXS pattern must be deconvolved [44,68–70] from its primary beam profile (cf. Figure 8.14). At increased temperature or if the matrix material is polyethylene (PE), a single USAXS experiment may suffice.

8.8.1. *Experiments with a macrobeam*

Common synchrotron sources offer a beam with a breadth of approx. 2 mm (Figure 8.14). Since this is not a microbeam, it shall be called "macrobeam". As shown in the figure, the beam shape is frequently elongated in the horizontal direction. Bearing in mind quantitative data analysis, a normalization of the scattering patterns can be eased, if the shape and orientation of the primary beam is suitably combined with the shape and orientation of the sample to be studied.

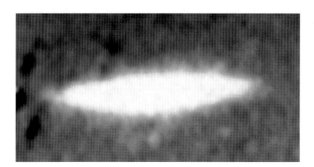

Figure 8.14. Primary beam profile of beamline A2, HASYLAB, Hamburg recorded on a scintillation screen after optical setup and before mounting of the beam stop and the detector. The horizontal extension of the bright spot is 3 mm

The usual normalization [44] involves several steps. First, the measured machine background is attenuated by the absorption factor. Second, this pattern is subtracted from the measured raw data of the sample. Third, the resulting pattern is normalized for constant V. V is the volume of the sample that is irradiated by the beam, and here the geometrical considerations step in. In mechanical tests, in particular, the irradiated volume is changing due to thinning and thickening of the sample – and the corresponding intensity changes must be compensated so that intensity changes resulting from a geometrical effect are not mistaken for crystallization or melting of domains.

Favorable conditions are present, if either the sample is flat and stays broader than the X-ray beam, or if the material is a fiber with a circular cross-section which is crossed in transverse direction by an oblong X-ray beam. The first mentioned case can easily be realized, if the sample is a tensile bar of sufficient breadth. In this case, only the sample thickness is changing as a function of load. In the second case, to a first approximation, the intersection between fiber and beam profile yields a truncated cylinder, V, and the cylinder diameter is varying upon mechanical loading of the fiber.

The variation of the sample geometry as a function of mechanical load is suitably measured in a separate experiment without X-ray beam and verified in the actual X-ray experiment utilizing the variation of the absorption factor, $I_1(t)/I_0(t)$, which can be computed from the data automatically recorded at every beamline which has been set up properly [44].

8.8.2. *Experiments with a microbeam*

Overview. Microbeam experiments are carried out if the structure in a small volume element of the sample is to be studied. Compared to macrobeam applications, the advantage of this method is the possibility to study spatial variations of nanostructure. An example of a matching scientific question is the analysis of the core-shell structure of polymer fibers, or the study of nanostructure variations in nanostructured gradient materials which are developed for medical applications. All the corresponding experiments are carried out in a "scanning microbeam setup".

Of course, there are also scientific questions which may be answered by time-resolved dynamic experiments with a microbeam. As an example, a dynamical study of the necking process could be performed and compared to the corresponding static experiment carried out in a scanning microbeam setup. In the dynamic experiment one would start probing a volume element outside the neck and monitor the passing of the necking front. To date, it should even be possible to vary the strain rate, $(\dot{\varepsilon})$, to some extent.

2D scanning of thin and flat material. Most of the scanning microbeam studies in which nanostructured material is scanned in 2 dimensions are not investigating material of homogeneous thickness but fibers. Thus, the superposition* of scattering patterns is neglected which, in fact, originate from regions that are different both with respect to orientation and to nanostructure. A comprehensive survey of the possibilities of the method has been published by Riekel *et al.* [71]. A paper by Wang and Cakmak [72] demonstrates the method in an investigation of microtomed slices of nanocomposites based on PVDF.

---

* In terms of mathematics this superposition is a *projection* of scattering patterns – and the fact that a complete set of projections can be turned into a complete set of *slices* (Fourier slice theorem) [44] is the basis of tomographic reconstruction.

SAXS microtomography. The setup for a SAXS μ-CT experiment has been sketched in Figure 8.8. Whereas most of the 2D scanning microbeam papers neglect the variation of nanostructure along the beam path, the μ-CT method makes proper use of the fact that the scattering pattern obtained when irradiating an inhomogeneous sample is a mathematical projection which should not be interpreted directly. Utilizing the well-known method of computer-aided back-projection [57], the scattering patterns originating from an almost homogeneous section (a voxel inside the material) are reconstructed. After that the nanostructure gradient inside the material can be studied [58]. The spatial resolution of the method is limited by the size of the microbeam.

For a practical application one should choose samples with macroscopic fiber symmetry. The axis of sample rotation should coincide with the fiber axis. If the sample is rotated about an axis different from the fiber axis, an effective fiber symmetry will be enforced by rotational averaging (smearing) – by analogy with the rotating crystal method [73,74] from the field of crystallography.

In the near future, a broad application of the method is unlikely, because the beamtime required for the recording of a complete set of projections is approximately one day for a single cross-section of a sample.

Fibers and reduced microtomography. In a discussion with K. Schneider (IPF, Dresden) we have developed the concept for a "Reduced SAXS μ-CT" method which saves a lot of time: If filaments of cylindrical shape are to be studied, the observed projected scattering pattern is not a function of a rotation angle of the fiber, and the step-wise rotation of the fiber can be saved. Only one translation of the fiber through the microbeam is performed within a short time. This is identical to the procedure carried out in the frequently published scanning microbeam studies of fibers. In contrast to the published studies, one would thoroughly choose the step-width for the translational step so that after the experiment the set of scattering patterns would be canalized through the backprojection algorithm in order to retrieve the true nanostructure gradient along the fiber diameter. In order to test the validity of the assumption of fiber symmetry, one could repeat the experiment with the fiber rotated by 90°.

8.9. Conclusion and outlook

Even if *in situ* investigations of nanocomposites by means of synchrotron radiation and the evaluation of the accumulated stream of scattering data are still a great challenge, the progress during the past decade has shown that now the questions of practical interest can be tackled with a perspective of resilient results. The development of detectors that had remained static for long has quickened considerably after its commercialization. Device developments are advanced by research groups worldwide. Obstacles for a successful exploitation of the methods by a broad community of users are the absence of standards concerning hard- and software interfaces as well as the shortage of scientists with good programming skills.

References

1. Wu J, Schultz J M, Yeh F, Hsiao B S and Chu B (2000) In-Situ Simultaneous Synchrotron Small- and Wide-Angle X-ray Scattering Measurement of Poly(vinylidene fluoride) Fibers under Deformation, *Macromolecules* **33**:1765–1777.
2. Butler M F, Donald A M, Bras W, Mant G R, Derbyshire G E and Ryan A J (1995) A Real-Time Simultaneous Small- and Wide-Angle X-ray Scattering Study of In-Situ Deformation of Isotropic Polyethylene, *Macromolecules* **28**:6383–6393.
3. Butler M F, Donald A M and Ryan A J (1997) Time resolved simultaneous small- and wide-angle X-ray scattering during polyethylene deformation: 1. Cold drawing of ethylene-α-olefin copolymers, *Polymer* **38**:5521–5538.
4. Hughes D J, Mahendrasingam A, Oatway W B, Heeley E L, Martin C and Fuller W (1997) A simultaneous SAXS/WAXS and stress-strain study of polyethylene deformation at high strain rates, *Polymer* **38**:6427–6430.
5. Blundell D J, Mahendrasingam A, McKerron D, Turner A, Rule R, Oldman R J and Fuller W (1994) Orientation changes during the cold drawing and subsequent annealing of PEEK, *Polymer* **35**:3875–3882.
6. Liu L Z, Hsiao B S, Fu B X, Ran S, Toki S, Chu B, Tsou A H and Agarwa P K (2003) Structure Changes during Uniaxial Deformation of Ethylene-Based Semicrystalline Ethylene-Propylene Copolymer. 1. SAXS Study, *Macromolecules* **36**:1929–1929.
7. Kratky O (1933) On the deformation mechanism of fiber materials. Part I, *Kolloid Z* **64**:213–222 (in German).
8. Kratky O (1933) On the deformation mechanism of fiber materials. Part II, *Kolloid Z* **68**:347–350 (in German).
9. Kratky O (1938) The computation of micellae dimensions of fiber materials from the interferences diffracted at extremely small angles, *Naturwiss* **26**:94 (in German).
10. Porod G (1951) The Small-Angle X-Ray Scattering from densely packed colloidal systems, *Colloid Polym Sci* **124**:83–114 (in German).
11. Debye P and Bueche A M (1949) Scattering by an Inhomogeneous Solid, *J Appl Phys* **20**:518–525.
12. Bonart R and Hosemann R (1962) Fibrillar structures in cold-drawn linear polyethylene, *Kolloid Z Z Polymere* **186**:16–29 (in German).
13. Bonart R (1966) Colloidal structures in strained polymers, *Kolloid Z Z Polymere* **211**:14–33.
14. Bonart R, Hosemann R and McCullough R L (1963) The Influence of Particle Size and Distortions upon the X-ray Diffraction Patterns of Polymers, *Polymer* **4**:199–211.
15. Vonk C G (1979) A SAXS study of PE fibers, using the two-dimensional correlation function, *Colloid Polym Sci* **257**:1021–1032.
16. Brandt M and Ruland W (1996) SAXS studies on the deformation of macro-lattices in block copolymers, *Acta Polym* **47**:498–506.
17. Wilke W and Bratrich M (1991) Investigation of the Superstructure of Polymers during Deformation by Synchrotron Radiation, *J Appl Cryst* **24**:645–650.
18. Grubb D and Bala V (1999) Simultaneous SAXS, WAXS and local strain measurement: deformation of row-structure in LLDPE, *Polym Mater Sci Eng* **81**:357–358.
19. Fu B X, Hsiao B S, Pagola S, Stephens P, White H, Rafailovich M, Sokolov J, Mather P T, Jeon H G, Phillips S, Lichtenhan J and Schwab J (2000) Structural development during deformation of polyurethane containing polyhedral oligomeric silsesquioxanes (POSS) molecules, *Polymer* **42**:599–611.
20. Yeh F, Hsiao B S, Sauer B B, Michel S and Siesler H W (2003) In-Situ Studies of Structure Development during Deformation of a Segmented Poly(urethane-urea) Elastomer, *Macromolecules* **36**:1940–1954.

21. Stribeck N, Polizzi S, Bösecke P and Zachmann H G (1989) Investigations of strained SBS block copolymers with dilute elastic networks, *Rev Roumaine Chem* **34**:635–648.
22. Stribeck N, Bösecke P and Polizzi S (1989) SAXS investigation on the influence of oil dilution on morphological changes in a SBS block copolymer during the first draw cycle, *Colloid Polym Sci* **267**:687–701.
23. Stribeck N (1999) The Equatorial Small-Angle Scattering During the Straining of Poly(ether ester) and its Analysis, *J Polym Sci Part B Polym Phys* **37**:975–981.
24. Stribeck N, Fakirov S and Sapoundjieva D (1999) Deformation Behavior of a Poly(ether ester) Copolymer. Quantitative Analysis of SAXS Fiber Patterns, *Macromolecules* **32**:3368–3378.
25. Fakirov S, Samokovlijsky O, Stribeck N, Apostolov A A, Denchev Z, Sapoundjieva D, Evstatiev M, Meyer A and Stamm M (2001) Nanostructure Deformation Behavior in Poly(ethylene terephthalate)/Polyethylene Drawn Blend as Revealed by Small-Angle Scattering of Synchrotron X-Radiation, *Macromolecules* **34**:3314–3317.
26. Stribeck N, Buzdugan E, Ghioca P, Serban S and Gehrke R (2002) Nanostructure evolution of SIS thermoplastic elastomers during straining as revealed by USAXS and multi-dimensional chord distribution analysis, *Macromol Chem Phys* **203**:636–644.
27. Stribeck N, Androsch R and Funari S S (2003) Nanostructure Evolution of Homogeneous Poly(ethylene-co-1-octene) as a Function of Strain, *Macromol Chem Phys* **204**:1202–1216.
28. Stribeck N, Fakirov S, Apostolov A A, Denchev Z and Gehrke R (2003) Deformation Behavior of PET, PBT and PBT-Based Thermoplastic Elastomers as Revealed by SAXS from Synchrotron, *Macromol Chem Phys* **204**:1000–1013.
29. Stribeck N and Funari S S (2003) Nanostructure Evolution in a Poly(ether ester) Elastomer during Drawing and the Displacement of Hard Domains from Lamellae, *J Polym Sci Part B Polym Phys* **41**:1947–1954.
30. Sauer B B, McLean R S, Brill D J and Londono D J (2002) Morphology and Orientation during the Deformation of Segmented Elastomers Studied with Small-Angle X-ray Scattering and Atomic Force Microscopy, *J Polym Sci Part B Polym Phys* **40**:1727–1740.
31. Murakami S, Yamakawa M, Tsuji M and Kohjiya S (1996) Structure development in the uniaxial–drawing process of poly(ethylene naphthalate), *Polymer* **37**:3945–3951.
32. Séguéla R and Prud'homme J (1988) Affinity of Grain Deformation in Mesomorphic Block Polymers Submitted to Simple Elongation, *Macromolecules* **21**:635–643.
33. Lee H S, Yoo S R and Seo S W (1999) Domain and segmental deformation behavior of thermoplastic elastomers using synchrotron SAXS and FTIR methods, *J Polym Sci Part B Polym Phys* **37**:3233–3245.
34. Welsh G E, Blundell D J and Windle A H (2000) A transient mesophase on drawing polymers based on polyethylene terephthalate (PET) and polyethylene naphthoate (PEN), *J Mater Sci* **35**:5225–5240.
35. Hernández J J, Garcìa Gutiérrez M C, Nogales A, Rueda D R, Sanz A, Sics I, Hsiao B S, Roslaniec Z, Broza G and Ezquerra T A (2007) Deformation behaviour during cold drawing of nanocomposites based on single wall carbon nanotubes and poly(ether ester) copolymers, *Polymer* **48**:3286–3293.
36. Bras W, Mant G R, Derbyshire G E, O'Kane W J, Helsby W I, Hall C J and Ryan A J (1995) Real-Time Simultaneous Wide- and Small-Angle Fibre Diffraction, *J Synchrotron Rad* **2**:87–92.
37. Hughes D J, Mahendrasingam A, Martin C, Oatway W B, Heeley E L, Bingham S J and Fuller W (1999) An instrument for the collection of simultaneous small and wide angle x-ray scattering and stress-strain data during deformation of polymers at high strain rates using synchrotron radiation sources, *Rev Sci Instrum* **70**:4051–4054.

38. Chen X, Yoon K, Burger C, Sics I, Fang D, Hsiao B S and Chu B (2005) In-situ x-ray scattering studies of a unique toughening mechanism in surface-modified carbon nanofiber/UHMWPE nanocomposite films, *Macromolecules* **38**:3883–3893.
39. Hosemann R and Bagchi S N (1962) *Direct Analysis of Diffraction by Matter*, North-Holland, Amsterdam.
40. Guinier A (1963) *X-Ray Diffraction*, Freeman, San Francisco.
41. Alexander L E (1979) *X-Ray Diffraction Methods in Polymer Science*, Wiley, New York.
42. Feigin L A and Svergun D I (1987) *Structure Analysis by Small-Angle X-Ray and Neutron Scattering*, Plenum Press, New York.
43. Baltá Calleja F J and Vonk C G (1989) *X-Ray Scattering of Synthetic Polymers*, Elsevier, Amsterdam.
44. Stribeck N (2007) *X-Ray Scattering of Soft Matter*, Springer, Heidelberg, New York.
45. Stribeck N (2001) Extraction of domain structure information from small-angle X-ray patterns of bulk materials, *J Appl Cryst* **34**:496–503.
46. Stribeck N and Fakirov S (2001) Three-Dimensional Chord Distribution Function SAXS Analysis of the Strained Domain Structure of a Poly(ether ester) Thermoplastic Elastomer, *Macromolecules* **34**:7758–7761.
47. Ruland W (1977) The evaluation of the small-angle scattering of lamellar two-phase systems by means of interface distribution functions, *Colloid Polym Sci* **255**:417–427.
48. Ruland W (1978) The evaluation of the small-angle scattering of anisotropic lamellar two-phase systems by means of interface distribution functions, *Colloid Polym Sci* **256**:932–936.
49. Stribeck N and Ruland W (1978) Determination of the Interface Distribution Function of Lamellar Two-Phase Systems, *J Appl Cryst* **11**:535–539.
50. Méring J and Tchoubar-Vallat D (1965) X-Rays. Small-Angle X-Ray Scattering in diluted suspensions. Computations of chord distributions, *C R Acad Sc Paris* **261**:3096–3099 (in French).
51. Méring J and Tchoubar-Vallat D (1966) Solid State Physics. Diffuse Small-Angle X-Ray Scattering from concentrated systems, *C R Acad Sc Paris* **262**:1703–1706 (in French).
52. Méring J and Tchoubar D (1968) Interpretation of the SAXS from porous systems. Part I, *J Appl Cryst* **1**:153–165 (in French).
53. Tchoubar D and Méring J (1969) Interpretation of the SAXS from porous systems. Part II, *J Appl Cryst* **2**:128–138 (in French).
54. *Small Angle X-ray Scattering* (1982) (Eds. Glatter O and Kratky O) Academic Press, London.
55. Debye P and Menke H (1931) Study of molecular order in liquids by means of X-rays, *Erg techn Rontgenkunde* **2**:1–22 (in German).
56. Lengeler B, Schroer C G, Kuhlmann M, Benner B, Günzler T F, Kurapova O, Zontone F, Snigirev A and Snigireva I (2005) Refractive x-ray lenses, *J Phys D Appl Phys* **38**:A218–A222.
57. Schroer C G, Kuhlmann M, Roth S V, Gehrke R, Stribeck N, Almendarez Camarillo A and Lengeler B (2006) Mapping the local nanostructure inside a specimen by tomographic small-angle x-ray scattering, *Appl Phys Lett* **88**:164102.
58. Stribeck N, Almendarez Camarillo A, Nöchel U, Schroer C, Kuhlmann M, Roth S V, Gehrke R and Bayer R K (2006) Volume-Resolved Nanostructure Survey of a Polymer Part by Means of SAXS Microtomography, *Macromol Chem Phys* **207**:1239–1249.
59. Stribeck N, Downloads, http://www.chemie.uni-hamburg.de/tmc/stribeck/dl.

60. Hammersley A P, FIT2D V12.012 Reference Manual, http://www.esrf.fr/computing/scientific/FIT2D/.
61. Stribeck N, Nöchel U, Funari S S and Schubert T (2008) Comparison of SAXS Patterns Obtained by the Stretch-Hold Technique and by X-Raying During Continuous Stretching of Polypropylene, *J Polym Sci Polym Phys* (in print).
62. Stribeck N, Nöchel U, Funari S S, Schubert T and Timmann A (2007) Nanostructure Evolution in Polypropylene during a Load-Reversal Mechanical Test, *Macromol Chem Phys* (submitted).
63. Putthanarat S, Stribeck N, Fossey S A, Eby R K and Adams W W (2000) Investigation of the nanofibril of silk fibers, *Polymer* **41**:7735–7747.
64. Müller M, Czihak C, Vogl G, Fratzl P, Schober H and Riekel C (1998) Direct Observation of Microfibril Arrangement in a Single Native Cellulose Fiber by Microbeam Small-Angle X-ray Scattering, *Macromolecules* **31**:3953–3957.
65. Reiterer A, Lichtenegger H, Fratzl P and Stanzl-Tschegg S E (2001) Deformation and energy absorption of wood cell walls with different nanostructure under tensile loading, *J Mater Sci* **36**:4681–4686.
66. Stribeck N, Almendarez Camarillo A, Roth S V, Fakirov S and Bhattacharyya D (2005) Highly Oriented Nanofibrillar Polymer Blends as Models for USAXS Studies, *Hasylab Ann Rep* **1**:897–898.
67. Stribeck N, Fakirov S, Heinemanns K, Friedrich K, Almendarez Camarillo A, Bhattacharyya D and Roth S V (2005) USAXS Studies on Undrawn and Highly Drawn PP/PET Blends, *Hasylab Ann Rep* **1**:997–998.
68. Glatter O (1981) Convolution Square Root of Band-Limited Symmetrical Functions and its Application to Small-Angle Scattering Data, *J Appl Cryst* **14**:101–108.
69. Pedersen J S (1997) Analysis of small-angle scattering data from colloids and polymer solutions: modeling and least-squares fitting, *Adv Coll Interf Sci* **70**:171–210.
70. Damaschun G, Müller J J and Pürschel H V (1971) Desmearing of noisy small-angle X-ray scattering curves, *Acta Cryst* **A27**:11–18 (in German).
71. Riekel C, Garcìa Gutiérrez M C, Gourrier A and Roth S V (2003) Recent synchrotron radiation microdiffraction experiments on polymer and biopolymer fibers, *Anal Bioanal Chem* **376**:594–601.
72. Wang Y D and Cakmak M (2001) Spatial variation of structural hierarchy in injection molded PVDF and blends of PVDF with PMMA. Part II. Application of microbeam WAXS pole figure and SAXS techniques, *Polymer* **42**:4233–4251.
73. Blake F C (1933) On the Factors Affecting the Reflection Intensities by the Several Methods of X-Ray Analysis of Crystal Structures, *Rev Mod Phys* **5**:169–202.
74. Hendershot O P (1937) Absorption Factor for the Rotating Crystal Method of Crystal Analysis, *Rev Sci Instr* **8**:324–326.

Chapter 9

Creep and Fatigue Behavior of Polymer Nanocomposites

A. Pegoretti

9.1. Introduction

Several types of nanofillers have been selected for the preparation of polymer nanocomposites (PNCs), with the aim to increase the mechanical properties of polymeric matrices commonly utilized in load-bearing applications [1]. Depending on the number of dimensions of nanometric size, the nanofillers can be classified into three main categories, *i.e.*, one-dimensional (1-D, layered materials), two-dimensional (2-D, fibrous materials), and three-dimensional (3-D, particulate materials) nanofillers [2]. Among the 1-D nanofillers various types of layered silicates (LS) that can be eventually exfoliated in dispersed nanoplates, have been considered, such as montmorillonite (MMT), bentonite (BT), fluorohectorite (FH), *etc.* The preparation and properties of LS based nanocomposites have been reviewed in great detail by Utracki [3], and more recently by Okada *et al.* [4] and Goettler *et al.* [5]. Regarding the 2-D nanofillers, most of the research efforts conducted so far have been focused on the use of carbon nanotubes (CNTs) [6–9] or carbon nanofibers (CNFs) [10,11]. 3-D nanoparticles comprise nearly spherical solid inorganic fillers [2] such as metal oxides (ZnO, Al_2O_3, $CaCO_3$, TiO_2, *etc.*), fumed silica (SiO_2), silicon carbide (SiC) and polyhedral oligomeric silsesquioxanes (POSS). Carbon black (CB) of very high surface area (up to 1350 m^2/g) has been often considered as a nanofiller since the dimensions of its primary particles are of the order of a few nanometers [12].

Focusing on the mechanical properties of the resulting PNCs, it can be noticed that most of the experimental information available in the scientific literature has been generated by constant deformation rate tests [13]. This type of tests are generally suitable to investigate the so-called *"short-term"* mechanical response, in terms of stiffness, strength and toughness. Depending on the testing configuration, elastic moduli and stress at break values are generally evaluated under tensile, compressive, flexure and shear conditions. The fracture toughness is measured in terms of (i) linear elastic approaches yielding a critical value of the stress intensity factor (K_{IC}) or a critical value of the strain energy release rate (G_{IC}), (ii) a non-linear elastic approach in terms of a critical value of the J integral (J_C), or (iii) an elastic-plastic approach such as the essential work of fracture (EWF) method. Conversely, information on the *"long-term"* mechanical response of PNCs under static (creep) and dynamic (fatigue) conditions are relatively scarce in the scientific literature. In fact, according to the ISI Web of Science® database [14], about 2900 papers published in peer-reviewed scientific journals in the period 1991–2007 contain the keywords "nanocomposite*", "polymer*" and "mechanical", where the asterisk is a wild character. Among this huge amount of articles dealing with various aspects of the mechanical response of PNCs, about 50 (*i.e.*, less than 2%) are entirely or partly focused on the long-term mechanical response of PNCs, such as the creep [15–44] or the fatigue [45–64] behavior. The oldest of them was published in 1999.

In this chapter, the behavior of PNCs under creep and fatigue loading conditions will be reviewed, including fundamental structure/property relationships emerging from the analysis of the scientific literature.

9.2. Generalities on the creep behavior of viscoelastic materials

Creep is a slow continuous deformational process occurring in materials under constant stress conditions [65]. Even if creep experiments can be performed under various testing configurations, such as tension, compression, flexure, shear, and indentation [13,66], the most frequently adopted configuration is the tensile one. It is important to underline that creep experiments are generally performed under a constant load, even when the specimen's cross-section reduces significantly with time. At a first approximation, the variation of specimen cross-section, ΔA, with respect to its initial value, A, can be estimated as:

$$\frac{\Delta A}{A} = 1 - 2\nu \varepsilon_a \qquad (9.1)$$

where ν is the Poisson's ratio and ε_a is the axial deformation. For isotropic materials, an upper limiting value for ν of 0.5 can be considered. Consequently, since the creep tests are most frequently performed at small strains (*i.e.*, ε_a lower than 5%), the expected area reduction is in any case lower than 5%, and consequently constant stress assumption is reasonable. A typical input-output data set for a creep experiment is represented in Figure 9.1a [67].

A sample is subjected to a constant tensile stress, σ_0, and its deformation, $\varepsilon(t, \sigma_0)$ is continuously monitored. The strain in isothermal tensile creep, $\varepsilon(t, \sigma_0)$, depending on time t and stress σ_0, is usually viewed as consisting of

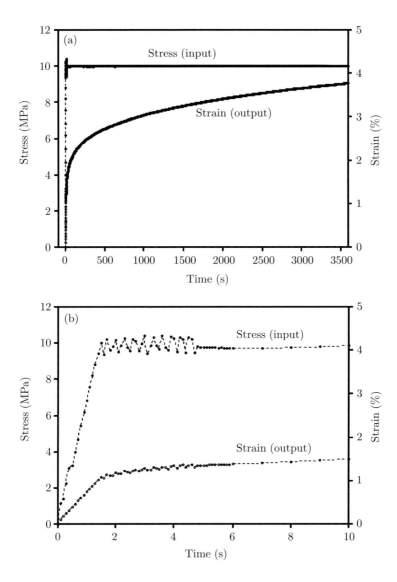

Figure 9.1. Example of a creep experiment performed under tensile conditions on high density polyethylene filled with 5 wt% of titania nanoparticles plotted over two different time intervals

three components [68–70]: (i) an elastic (instantaneous, reversible) $\varepsilon_E(\sigma_0)$ component; (ii) a viscoelastic (time-dependent, reversible) $\varepsilon_{VE}(t, \sigma_0)$ component; and (iii) a plastic (time-dependent, irreversible) $\varepsilon_P(t, \sigma_0)$ component. Linear viscoelastic behavior implies that the magnitude of the strain components is linearly proportional to the magnitude of the applied stress, so that a creep compliance $D(t) = \varepsilon(t)/\sigma_0$ can be defined as a function of time only. If no plastic deformation is produced in the course of the creep test, the tensile compliance for the isothermal creep can be written as:

$$D(t) = D_E + D_{VE}(t) \tag{9.2}$$

The loading step, evidenced in Figure 9.1b, is generally very short in comparison to the overall length of a creep experiment. Still, the input stress is not a true step function and therefore the creep compliance cannot be derived exactly at short times. A rule of thumb, supported by both experiment and analysis, is that creep data for elapsed times less than about ten times the loading period should be treated with reserve [13].

Several mechanical models, consisting of various combinations of purely elastic (spring) and purely viscous (dashpot) elements, have been developed to model the response of a viscoelastic solid under creep conditions. Among the most widely used ones, the Burgers (or four elements) model [65], and the generalized Kelvin model [65] can be mentioned. These two models are schematically represented in Figure 9.2.

Analytical expressions of the time-dependent creep compliance for Burgers and generalized Kelvin models are reported in Table 9.1.

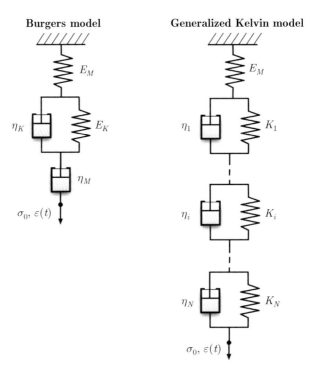

Figure 9.2. Combinations of elastic (spring) and viscous (dashpot) elements representing the Burgers and the generalized Kelvin models

Table 9.1. Most common equations adopted for modeling the time dependency of the creep compliance for linear viscoelastic solids

Burgers model	$D(t) = \dfrac{1}{E_M} + \dfrac{1}{E_K}\left[1 - \exp\left(-\dfrac{E_K}{\eta_K}t\right)\right] + \dfrac{t}{\eta_M}$
Generalized Kelvin model	$D(t) = \dfrac{1}{E_M} + \sum_{i=1}^{N}\dfrac{1}{K_i}\left[1 - \exp\left(-\dfrac{K_i}{\eta_i}t\right)\right]$
Kohlrausch-Watts-Williams equation	$D(t) = D_E + D_{VE}\left\{1 - \exp\left[-\left(\dfrac{t}{\tau}\right)^m\right]\right\}$
Power law equation	$D(t) = D_E + D_{VE}t^m$

The Burgers model consists of a Maxwell and a Kelvin unit connected in series [65]. The spring of the Maxwell unit (E_M) describes the instantaneous elastic deformation of the material occurring when the load is suddenly applied, the spring and the dashpot of the Kelvin units (E_K and η_K) describe the viscoelastic response in the early stage of creep, while the dashpot element of the Maxwell unit (η_M) represents the viscous flow of the material after a sufficiently long period of time. For the Kelvin unit, a retardation time defined as $\tau_K = \eta_K/E_K$ represents the time necessary to reach ($1 - e^{-1}$, corresponding to about 63.2%) of the total deformation (σ_0/E_K) allowed for this unit. A creep rate can be defined as $\dot{D} = dD/dt$. In the case of the Burgers model the creep rate assumes the following expression:

$$\dot{D} = \dfrac{1}{\eta_K}\exp\left(-\dfrac{E_K}{\eta_K}t\right) + \dfrac{1}{\eta_M} = \dfrac{1}{\eta_K}\exp\left(-\dfrac{t}{\tau_K}\right) + \dfrac{1}{\eta_M} \qquad (9.3)$$

On a sufficiently long time scale, i.e., when $t \gg \tau_K$, the creep rate asymptotically tends to a constant value given by $\dot{D}(\infty) = 1/\eta_M$.

Some other empirical functions, i.e., not directly derived from physical models, can be adopted to fit experimental creep data, such as the Kohlrausch-Williams-Watts [71,72] and the power law (or Findley) [73] functions. The expressions of the time-dependent creep compliance of these two functions are summarized in Table 9.1. In Figure 9.3, the ability of the above mentioned equations to fit the creep introduced in Figure 9.1 (with exclusion of the data within 10 times the loading period) is compared.

In all cases, a reasonably good fitting of the creep response of the investigated material (a high density polyethylene (HDPE)-titania nanocomposite) can be observed. The Chi Square (Chisq) values reported in the insert of Figure 9.3 represent the sum of the squared error between the original data and the calculated curve fit. In general, the lower the Chisq value, the better the fit. The fitting parameters of the curves have been optimized (by the software KaleidaGraph v.4) with the number of iterations required to minimize

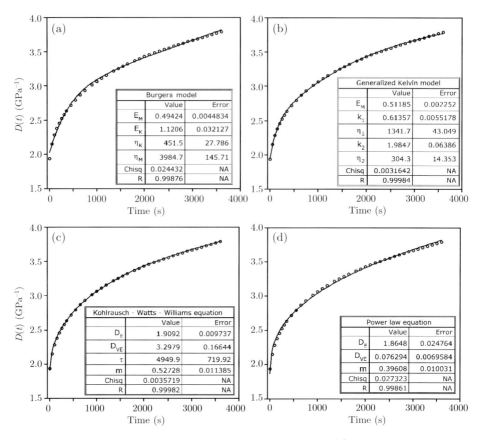

Figure 9.3. Fitting of the creep data of Figure 9.1 according to (a) the Burgers model, (b) the generalized Kelvin model, (c) the Kohlrausch-Watts-Williams equation, (d) the power law equation

Chisq parameter. For all the proposed equations, the Chisq value resulted to be very close to zero, with a somewhat better fitting ability of the generalized Kelvin model and of the Kohlrausch-Watts-Williams equation.

9.3. Generalities on the fatigue resistance of polymeric materials

Fatigue is a very general phenomenon in most engineering materials, and refers to microstructural damage and failure caused by applied cyclic loads or strains [74]. Many industrial components are subjected to load-controlled fatigue conditions, *i.e.*, cyclic loads periodically oscillating between a lower and an upper value. Such components include rotating shafts, pressure vessels under cyclic internal pressure, and tension rods or bars that must withstand fluctuating axial loads [75]. On the other hand, during their service life, other components such as coil and leaf springs, shoe soles and parts subjected to thermal cycling are cyclically deformed within certain limits of displacement or strain. Materials for these applications should be more appropriately tested under displacement- or strain-controlled fatigue conditions.

On the basis of considerations similar to those previously reported for creep experiments, the cross-section variations of unnotched specimens are generally ignored during fatigue experiments and, at a first approximation, tests performed under load-control are generally considered under stress-control. The typical key parameters associated with stress cyclic loading are represented in Figure 9.4.

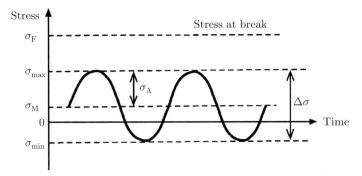

Figure 9.4. Key parameters for stress-controlled fatigue loading

The key stress variables shown in Figure 9.4 are defined as: σ_{min}, σ_{max} = minimum and maximum stress, respectively; $\Delta\sigma = \sigma_{max} - \sigma_{min}$ = stress range; $\sigma_A = (\sigma_{max} - \sigma_{min})/2$ = stress amplitude; $\sigma_M = (\sigma_{max} + \sigma_{min})/2$ = mean stress and R = $\sigma_{min}/\sigma_{max}$ = stress ratio.

The load history is completely defined when at least two of the above mentioned parameters are defined and when the waveform (sinusoidal, triangular, square, *etc.*) and the frequency (in Hz) of the cyclic load are specified. Analogous quantities can be defined for strain- (or displacement-) controlled fatigue experiments. Depending on the testing configuration, loads (or displacements) can be generated either by rotational bending, reciprocal bending, reciprocal torsion, or by pulsating strokes of an axial actuator.

Fatigue experiments can be broadly divided into two main classes [13]. The first and most common approach is aimed at exploring the relationships between the applied stress and the number of cycles to failure of unnotched specimens (S–N curves, or Wohler curves). The second is a more rigorous fracture mechanics approach, in which a crack growth rate is evaluated as a function of the applied stress intensity factor on specimens containing an initial sharp notch.

S-N curves are generally represented by plotting the stress amplitude $\Delta\sigma$ (or the maximum applied stress σ_{max}) as a function of the logarithm of the number of cycles to failure (N). Typical fatigue S–N curves for two injection molded short glass fiber (SGF) polymer composites are presented in Figure 9.5 [76].

The higher the amplitude of the stress, the smaller the number of cycles the material is capable of sustaining before failure. The fatigue strength

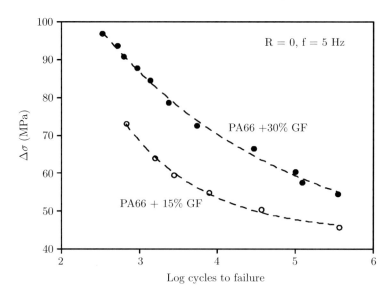

Figure 9.5. Typical fatigue S-N curves of injection molded polyamide–66 composites reinforced with (●) 15 and (○) 30 wt% of short glass fibers

represents the stress level at which failure is likely to occur for some given (for example 1 million) number of cycles. For some materials the S–N curves may become horizontal at high N values: the stress level beyond which the fatigue strength becomes independent of the number of cycles is termed the *endurance limit*. Several factors can affect the S–N curves of plastics and polymer composites [75], such as testing parameters (specimen geometry, loading method, test frequency, stress ratio, *etc.*) and materials characteristics (molecular weight, molecular orientation, crystallinity content, filler content, *etc.*).

Fatigue crack propagation (FCP) is an alternative approach used for studying the fatigue behavior of materials and it has been successfully applied also to plastics and composites [75]. A sharply notched test piece is used, such as compact tension (CT) or single-edge notched bending (SENB) specimens. Such specimen is tested under load control with a cyclic waveform of amplitude $\Delta P = P_{max} - P_{min}$. During the test, the crack length, a, is either measured using a video camera or calculated *via* the measured compliance. Paris found that the crack growth rate (da/dN) can be expressed in terms of the stress intensity factor amplitude ΔK with a relationship in the form (*Paris law equation*) [77]:

$$\frac{da}{dN} = A\Delta K^n \tag{9.4}$$

where A and n are constants, and the stress intensity factor amplitude ΔK is given by:

$$\Delta K = Y\Delta\sigma\sqrt{a} \tag{9.5}$$

where Y is a tabulated shape factor depending on the specimen geometry and loading configuration [78], and $\Delta\sigma$ is the fair-field stress remote from the notch.

A typical Paris plot is reported in Figure 9.6 for injection molded polypropylene (PP) and PP-SGF composites with three different fiber contents.

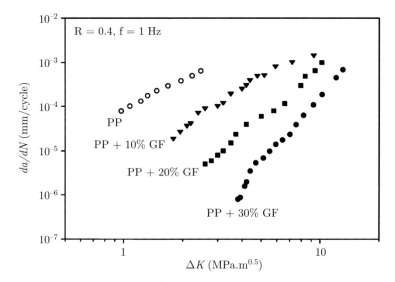

Figure 9.6. Typical Paris plot representing the fatigue crack propagation in (○) neat injection molded PP, and relative composites reinforced with (▼) 10 wt%, (■) 20 wt%, and (●) 30 wt% of short glass fibers

9.4. Creep behavior of polymer nanocomposites

Among the available scientific literature on the creep response of PNCs [15–43], the majority of papers are focused on the effect of 1-D layered nanofillers of the creep behavior of polymers such as poly(ethylene oxide) (PEO) [15], poly(ethylene terephthalate) (PET) [18], ethylene propylene rubber (EPR) [17], polypropylene [24], nylon-66 (PA66) [19], nylon-6 (PA6) [22,23], high density polyethylene (HDPE) [21,38,44], poly(ethylene-co-acrylic acid) copolymer [26], epoxy resin (EP) [28], polystyrene (PS) [29], polyurethane (PU) [30], and polystyrene-*block*-polybutadiene-*block*-polystyrene triblock copolymer (SBS) [37].

9.4.1. *Creep response of PNCs containing one-dimensional nanofillers*

In general, it has been proven that a substantial reduction of the creep compliance of the investigated polymeric matrices can be reached when layered silicates are introduced. Nevertheless, the level of intercalation and/or exfoliation of a layered silicate in a polymer matrix is a critical parameter determining the creep response of the resulting nanocomposite.

As reported by Pegoretti et al. [18], various amounts (1, 3 and 5 wt%) of a non-modified natural montmorillonite clay (Cloisite® Na+) or of an ion-exchanged clay modified with quaternary ammonium salt (Cloisite® 25A) were dispersed in a recycled poly(ethylene terephthalate) matrix (rPET) by a melt compounding process. Scanning transmission electron microscopy observations evidenced that Cloisite® 25A was much better dispersed in the rPET matrix than Cloisite® Na+. Concurrently, wide-angle X-ray scattering (WAXS) analysis showed that for both fillers a certain intercalation of rPET among the clay galleries occurred, whose extent appeared to be much more pronounced for Cloisite® 25A. The tensile creep compliance of the nanocomposites was only slightly lower than that of the neat rPET matrix, the reinforcing effect of Cloisite® 25A being somewhat higher. Moreover, both clays had a beneficial effect on the long-term dimensional stability of nanocomposites since – in contrast to the neat rPET – the creep rate did not rise at long creep periods.

Further indication of the importance of a proper dispersion of layered silicates in PNCs is provided by Ranade et al. [21] in a study reporting a comparison between the creep behavior of maleated and non-maleated polyethylene-montmorillonite layered silicate blown films. The authors claimed that maleated polyethylene (PE-g-MA) facilitated the dispersion of montmorillonite layered silicate in the polyethylene (PE) matrix. The creep experiments were performed at 25% and 50% of the yield stress and the resulting creep compliance was modeled with the Burgers model. The fitting parameters of the Burgers model for the creep behavior (evaluated at 50% of the yield stress) of neat PE matrix and relative PNCs are summarized in Table 9.2.

Table 9.2. Parameters of the Burgers model fitting the creep behavior of polyethylene – montmorillonite layered silicate (MLS) nanocomposites compatibilized with PE grafted maleic anhydride. Short-term creep tests were performed up to 3600 s at an applied stress of 50% of the yield stress [21]

Material composition			Best fit parameters of the Burgers model				
PE (wt%)	PE-g-MA (wt%)	MLS (wt%)	E_M (MPa)	E_K (MPa)	η_M (MPa s)	η_K (MPa s)	τ_K (s)
100	–	–	5.5	7.0	33 000	2 300	329
99	1	–	1.8	1.6	7 300	500	313
99	–	1	5.0	5.0	26 500	1 300	260
97.5	–	2.5	7.0	7.4	32 000	1 900	257
95	–	5	4.9	5.0	18 000	1 500	300
98	1	1	11.5	15.0	54 000	2 800	187
95	2.5	2.5	11.0	15.7	66 200	2 500	159
90	5	5	9.6	12.0	69 000	2 000	167
92.5	5	2.5	12.0	15.0	75 000	1 900	127

From the data reported in Table 9.2 it clearly emerges that the addition of PE-g-MA has a detrimental effect on the creep behavior of the PE matrix itself. On the other hand, the nanocomposites display a better creep stability when compared to the PE matrix provided that a proper dispersion of the layered silicate is promoted by the addition of PE-g-MA compatibilizer. The better creep response of the MLS+PE-g-MA nanocomposites compared to the compositions without PE-g-MA is documented by the increasing value of the fit parameters of the Burgers model. The decrease of the retardation time for MLS+PE-g-MA nanocomposites indicates that the occurrence of a stable creep deformation stage (secondary creep) is anticipated with respect to neat PE.

As reported by Joshi and Viswanathan [24], the creep deformation of PP filaments subjected to 20% of their breaking load can be reduced by a factor of 2 by the introduction of small amounts of clay (0.25, 0.5 and 1.0 wt%) and with up to 3 wt% of anhydride grafted polypropylene (PP-g-MA) as compatibilizer. The importance of using a grafted polymer as a compatibilizer for improving the dispersion of organoclays in non-polar matrices turned out to be crucial also for the creep response of HDPE-clay nanocomposites. In fact, as reported by Pegoretti et al. [38], an improvement of the creep resistance of an HDPE matrix filled with a commercial (Cloisite® 20A) organoclay was reached only through the addition of a significant (10 wt%) amount of a PE-g-MA compatibilizer.

The creep behavior of various types of PA6-clay nanocomposites was investigated by Vlasveld et al. [23]. For both commercial (Unitika M1030D) and experimental materials they showed that the creep compliance above the glass transition temperature (T_g) can be reduced by more than 80% by filling PA6 with 5 wt% of exfoliated layered silicates.

The advantages of nanofillers over traditional microfillers were recently evidenced by Siengchin and Karger-Kocsis [29] who compared the tensile creep behavior of PS/fluorohectorite micro- and nanocomposites. Direct melt compounding resulted in microcomposites, whereas a latex-mediated (masterbatch) technique had been proven to be suitable for the production of PS-fluorohectorite nanocomposites. Both X-ray diffraction (XRD) and transmission electron microscopy (TEM) demonstrated that direct melt blending of PS with fluorohectorite produced microcomposites since no delamination of the pristine clay occurred. On the other hand, the latex-mediated technique furnished intercalated and partly exfoliated nanocomposites. A deeper insight into the creep response of the investigated materials was obtained by the construction of creep compliance master curves over an extended time interval on the basis of a time-temperature superposition principle. The creep master curves of unfilled PS and relative micro- and nanocomposites are compared in Figure 9.7 at a reference temperature of 40°C.

The beneficial effect of fluorohectorite on the creep response of PS clearly emerges from Figure 9.7, together with the better performance of nanocomposites over microcomposites. Moreover, it is interesting to note that the creep

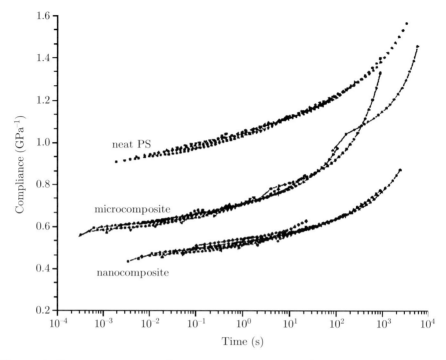

Figure 9.7. Creep master curves referred at a temperature of 40°C for neat PS and its micro- and nanocomposites filled with 4.5 wt% of fluorohectorite [29]

curves resulted almost parallel to one another, thus indicating that fluorohectorite did not significantly influence the creep rate of the PS matrix.

Also, the creep behavior of rubbers and elastomers can be improved by the addition of clays. EPR-clay nanocomposites were prepared by Hasegawa et al. [17] at Toyota Central Research and Development Laboratories by melt compounding maleic anhydride modified EPR (EPR-MA) with organophilic clay. Transmission electron microscopy observations evidenced that silicate layers of organophilic clay were exfoliated and homogeneously dispersed at a nanometric level in the EPR matrix. As reported in Figure 9.8, the creep resistance of EPR-MA-clay nanocomposites was much improved in comparison to the pure EPR-MA matrix.

In particular, the elongation of the nanocomposite containing 6.1 wt% of clay is restricted to less than 1% at 30 h, while the unfilled EPR-MA matrix elongates over 50% at 1 h and breaks within 2 h. It is worthwhile remarking that the creep elongation of EPR-MA nanocomposite filled with 6.1 wt% of clay is similar to the creep elongation of the same matrix filled with 30 wt% of carbon black. The possibility to reach the same mechanical properties of microcomposites with a much lower nanofiller content probably represents one of the most attractive features of polymer nanocomposites over traditional microcomposites.

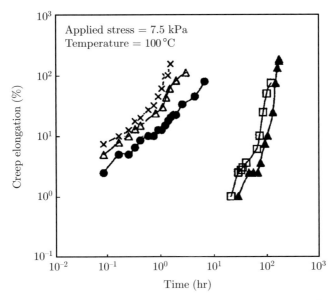

Figure 9.8. Creep elongation of maleic anhydride modified EPR matrix (X) and relative composites filled with (△) 5 wt% or (□) 30 wt% of carbon black, and nanocomposites filled with (●) 2.9 wt% or (▲) 6.1 wt% of clay [17]

As recently reported by Lietz et al. [37], the creep resistance of thermoplastic elastomers such as SBS block copolymers can also be markedly increased by the introduction of small amounts of a commercial organoclay (Cloisite® 20A) by twin-screw extrusion. Table 9.3 summarizes the parameters of the Burgers model obtained by fitting the creep compliance curves of neat SBS and relative clay-nanocomposites tested at room temperature for a period up to 200 h under a constant stress of 8.5 MPa (corresponding to 28% of the tensile strength of neat SBS). The parameters of the fitting model reported in Table 9.3 clearly show that the creep behavior of SBS thermoplastic elastomer is greatly improved by the addition of only 1.5 wt% of clay. Surprisingly enough, the addition of a higher quantity of clay (3.0 wt%) does not further improve the

Table 9.3. Parameters of the Burgers model fitting the creep behavior of SBS block copolymer and relative nanocomposites. Long-term creep tests were performed up to 225 h at a constant applied stress of 8.5 MPa [37]

Material composition	Best fit parameters of the Burgers model				
	E_M (MPa)	E_K (MPa)	η_M (GPa h)	η_K (GPa h)	τ_K (h)
SBS	2349	4749	682	19.1	4.02
SBS + 1.5 wt% Cloisite® 25A	8026	7582	1653	9.7	1.28
SBS + 3.0 wt% Cloisite® 25A	1930	5807	289	29.7	5.11

creep resistance of the material, which proved to be even lower than that of neat SBS. The short retardation time for the nanocomposite containing 1.5 wt% of organoclay indicates that the transition period from the primary to the secondary creep stage is anticipated for this particular nanocomposite, consistently with its improved dimensional stability. This unusual behavior has been explained by the authors on the basis of the microstructural information collected by WAXS and TEM analyses that concurrently indicated a worsening of the clay dispersion as its concentration in the nanocomposite increased. Therefore, the creep resistance of the nanocomposite loaded with 3.0 wt% of clay is reduced due to the generally worse dispersion and distribution of the intercalated clay stacks and to the presence of aggregates acting as stress concentrators.

Microhardness [26] and nanoindentation [19,22,28] techniques were also adopted to obtain experimental information on the creep behavior of polymer-clay nanocomposites. Even if a creep compliance can hardly be derived from indentation experiments [66], some useful indications can be inferred from the comparison of the creep displacement curves. In particular, Shen and co-workers investigated the creep behavior of PA66-clay [19], PA6-clay [22], and epoxy-clay [28] nanocomposites by measuring the displacement of a three-side pyramid (Berkovich) diamond nanoindenter pressed on the sample surface under a constant load (in the range from 40 to 100 mN) for a very short time period (60 s). Nanoindentation tests evidenced that the addition of 1, 2 and 5 wt% of organoclay (Nanomer® I.34TCN by Nanocor, USA) in a PA66 matrix induced a deterioration of its creep resistance. These results were explained by considering that, as calculated by deconvolution of XRD patterns, a monotonous decrease of PA66 crystallinity occurred as the organoclay content increased. In semicrystalline thermoplastics, the reduction of the polymer mobility caused by well dispersed, stiff, high aspect ratio nanoclay platelets is counterbalanced, and in this case even overcome, by changes of crystalline morphology owing to heterogeneous nucleation on clay particles that could lower or destroy the crystal perfection or crystallinity of the matrix. A completely different behavior was reported for PA6-clay nanocomposites whose displacement under nanoindentation creep resulted to be markedly reduced by the presence of 2.5, 5, 7.5 and 10 wt% of organoclay (Nanomer® I.30TC by Nanomer, USA) [22]. Even in this case, as evidenced by differential scanning calorimetry (DSC) analysis, the crystallinity of the PA6 matrix declined as the clay content increased, but the creep stabilizing effect of the well dispersed organoclay platelets prevailed. A slight improvement of the resistance to creep penetration of a nanoindenter was also reported for an epoxy matrix filled with various amounts of an organoclay (Cloisite® 30A by Southern Clay Products, USA) [28]. Figure 9.9 summarizes the effects of clay content on the nanoindentation creep displacement of epoxy-clay nanocomposites. On the same plot, the variation of the glass transition temperature of the epoxy matrix with the clay content is reported. To maximize the creep resistance of the epoxy matrix, an optimum

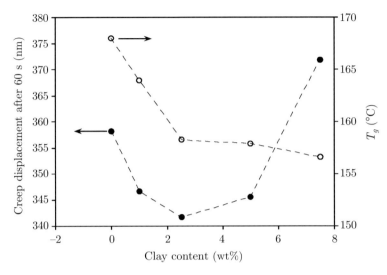

Figure 9.9. Effect of clay content on (●) the creep displacement after 60 s of nanoindentation under a load of 100 mN and on (○) the glass transition temperature of epoxy-clay nanocomposites [28]

clay content was found to be 2.5 wt%. The worsening of the creep resistance with higher clay loadings could be attributed to the reduced crosslinks density and the plasticizing effect caused by the pending groups of alkyl ammonium chains of the organoclay.

9.4.2. Creep response of PNCs containing two-dimensional nanofillers

The literature data on the effect of 2-D nanofillers on the creep behavior of polymers are relatively scarce [35,36,42]. Yang *et al.* [35] and Ganss *et al.* [36] almost simultaneously reported on the tensile creep behavior of PP nanocomposites obtained by melt-compounding multi-walled carbon nanotubes (MWCNTs) in PP by twin-screw extrusion followed by injection molding of the extruded granules.

Yang *et al.* [35] investigated the creep behavior of nanocomposites with 1 vol% of two types of MWCNTs having different aspect ratios. The creep strain and strain rate of the resulting PP-MWCNTs nanocomposites were markedly lower than that of pure PP matrix, this positive effect being more enhanced at higher (20 MPa) in comparison to lower (14 MPa) applied stress values. Moreover, the reinforcing effect resulted to be more evident as the aspect ratio of the MWCNTs increased. Additionally, the creep lifetime of the nanocomposites, *i.e.*, the time to failure under creep conditions, resulted to be considerably extended by 1000% compared to that of neat PP.

Ganss *et al.* [36] produced PP-based nanocomposites containing 0.5, 1.0, 2.5, and 5.0 wt% of MWCNTs. On the basis of a time-temperature superposi-

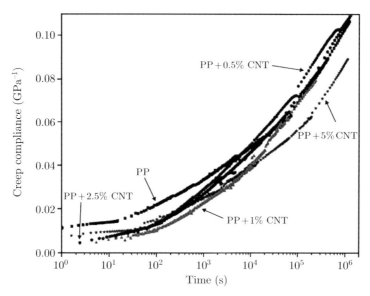

Figure 9.10. Creep compliance master curves at a reference temperature of 50°C for neat PP and relative nanocomposites with various contents of carbon nanotubes (CNT) [36]

tion principle, short term creep tests at low stress (2.5 MPa) performed at various temperatures allowed them to construct the creep compliance master curves reported in Figure 9.10.

As expected, the creep compliance master curves indicated a lower compliance for all the nanocomposites at shorter times (up to 10^4 s) as compared to neat PP. However, it is interesting to note that the creep compliance curves for nanocomposites with 0.5, 1, and 2.5 wt% of MWNT showed a cross over at about 10^3 s. The nanocomposite with 5 wt% of MWNTs showed the lowest compliance over the entire time range when compared to all the other compositions. This peculiar behavior was tentatively explained by the authors considering that two competing factors may concur in determining the overall creep response: (i) the size of the spherulites/crystallites (which is a consequence of incorporation of nanotubes and hence inter-phase effects), (ii) the thermal expansion coefficient (which depends on the amount of nanotubes).

Zhang et al. [42] demonstrated that very low amounts (0.1–0.25%) of single-walled carbon nanotubes (SWCNTs) may significantly improve the creep response of epoxy based nanocomposites. As reported in Figure 9.11, they observed that the improvement of the material's resistance to creep strain shows an optimum value for a given amount of SWCNTs. For very small SWCNTs contents (e.g., 0.001 wt%) no noticeable reduction of the creep strain is observed. When the SWCNTs weight fraction is increased a reduction of creep strain is observed with a plateau value of about 30% for weight fractions from 0.1 to 0.25 wt%. When the SWCNTs content was raised to 1 wt% the creep strain after 50 h of testing showed a 90% reduction. SEM investigation of the fracture

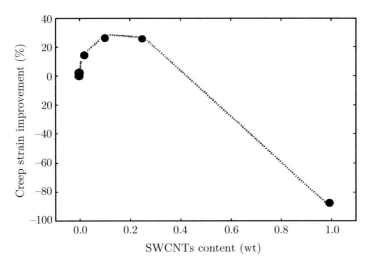

Figure 9.11. Percentage of creep strain reduction for nanocomposites compared to pure epoxy after 50 h of testing at 10 MPa for various SWCNTs contents [42]

surfaces of nanocomposites evidenced that the dispersion level of SWCNTs is quite uniform up to a nanotubes content of 0.25 wt%. In contrast, for the 1 wt% SWNT-epoxy sample, a significant agglomeration of SWNTs had been observed. These agglomerations are believed to decrease the number of nanotubes-epoxy interfacial contacts and create voids/defects in the material, which could cause a worsening of the creep response, as observed in the experiments. The authors also showed that for the same weight fraction (0.1 wt%) SWCNTs are much more effective in reducing the creep strain of the epoxy matrix when compared to MWCNTs and C_{60} fullerene.

9.4.3. *Creep response of PNCs containing three-dimensional nanoparticles*

The incorporation of 3-D nanoparticles such as titania [20,32,33,41,44], silica [34,43], and alumina [39] was proven to be a promising road to be traveled to enhance the creep resistance of thermoplastics.

The creep performances of PA66 nanocomposites reinforced with TiO_2 nanoparticles were extensively investigated by Friedrich and co-workers [20,32, 33]. In particular, they melt compounded 1 vol% (3.4 wt%) of 21 nm titania nanoparticles (Aeroxide P25 by Degussa, Germany) in a PA66 matrix by a twin-screw extruder and dog-bone specimens for tensile creep tests were formed using an injection-molding machine [20]. The quasi-static tensile properties of PA66 matrix at room temperature and at 50°C were only slightly modified by the presence of titania nanoparticles. At the same time, as documented in Figure 9.12, the titania nanoparticles had a dramatic impact on the room temperature creep strain at very high applied stress such as 80 and 90% of the quasi-static strength.

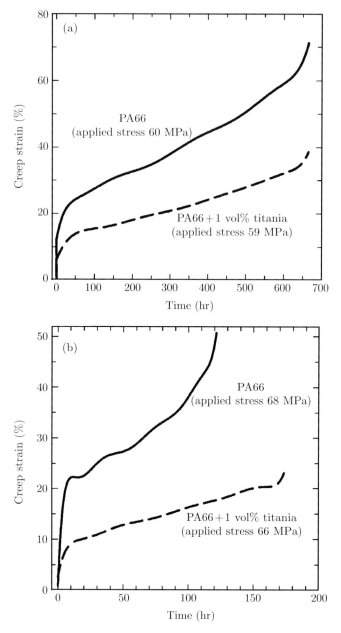

Figure 9.12. Creep strain of neat PA66 and of relative nanocomposites reinforced with 1 vol% of titania nanoparticles under an applied stress of (a) 80% and (b) 90% of the quasi-static stress at break [20]

Also at 50°C, the creep resistance of PA66 resulted to be markedly improved by the presence of titania nanoparticles. According to the authors' hypothesis, nanoparticles may restrict the slippage, reorientation and motion of polymer chains. The same authors also investigated the creep response of PA66 and relative nanocomposites under moderately low stress levels (20, 30 and 40 MPa) at various temperatures (23, 50 and 80°C) over a time interval of up to 200 h [32,33]. In particular, the authors compared the creep behavior of nanocomposites containing the same amount (1 vol%) of (i) titania nanoparticles of 21 nm diameter (Aeroxide P25 by Degussa, Germany), (ii) titania nanoparticles of 300 nm diameter (Kronos 2310 by Kronos International, Germany), (iii) surface modified titania nanoparticles of 21 nm diameter (Aeroxide T805 by Degussa, Germany), and (iv) an organo-modified clay (Nanofil 919 by Süd-Chemie, Germany). The resulting creep compliance curves were fitted by the Burgers model and with a power law equation. The parameters of the Burgers model fitting the creep curves at the highest investigated stress level (40 MPa) are summarized in Table 9.4.

As expected, for each sample the parameter E_M decreased as the test temperature rose. At the same time, the parameter E_M of nanocomposites was higher than that corresponding to neat PA66 matrix, thus reflecting the fact that the instantaneous elasticity was increased by the addition of nanofillers.

Table 9.4. Parameters of the Burgers model fitting the creep behavior of neat PA66 and relative nanocomposites filled with 1 vol% of various nanofillers. Long-term creep tests were performed up to 200 h at various temperatures at a constant applied stress of 40 MPa [32]

Material	Temperature (°C)	Best fit parameters of the Burgers model				
		E_M (MPa)	E_K (MPa)	η_M (GPa h)	η_K (GPa h)	τ_K (h)
PA66	23	2490	1616	404	23.7	14.7
	50	357	1348	429	10.4	7.7
	80	156	232	207	0.75	3.2
PA66 + 21 nm TiO$_2$	23	3708	4652	1033	91.7	19.7
	50	732	2418	988	24.9	10.3
	80	289	1428	793	14.2	9.9
PA66 + 21 nm TiO$_2$ surface treated	23	3442	3584	896	60	16.7
	50	692	2302	774	23.4	10.2
	80	290	1128	573	12.3	10.9
PA66 + 300 nm TiO$_2$	23	2647	2646	681	38.1	14.4
	50	525	1773	570	16.0	9.0
	80	181	586	322	5.65	9.6
PA66 + clay	23	3130	3900	696	78.1	20.0
	50	551	1712	544	15.2	8.9
	80	190	619	319	7.47	12.1

Among the tested fillers, 21 nm titania nanoparticles turned out to be the most efficient in improving the instantaneous elastic response. As concerns the short-term viscoelastic response, represented by the Kelvin unit, it can be pointed out that all nanofillers showed a reinforcing effect documented by the increase of both E_K and η_K parameters. According to the observed retardation time values, the small-sized particles (21 nm titania) turned out to be the most effective in retarding the deformation of the Kelvin unit. Moreover, it can be noticed that the creep behavior of PA66 filled with surface treated titania nanoparticles was slightly weakened with respect to isodimensional unmodified particles, probably because of the more compliant interphase caused by modified surfaces of particles. Nanocomposites reinforced with 300 nm titania and organoclay resulted to be much less efficient in improving the PA66 creep behavior when compared to nanocomposites obtained with the 21 nm titania nanoparticles. The long-term creep response of nanocomposites, represented by the parameter η_M, was also improved by the presence of nanofillers. Similar effects have been observed by the same authors also on polypropylene filled with 1 vol% of titania nanoparticles with diameters of 21 and 300 nm [41]. In particular, the dimensional stability of the nanocomposites resulted to be significantly improved, especially for those containing small-sized nanoparticles. The creep strain and creep rate of nanoparticle-filled polypropylene are reduced by 46% and 80%, respectively, compared to those of the neat matrix. Additionally, creep lifetime is extended by a factor of 3.3 due to the addition of the smaller nanoparticles.

Also, the creep resistance of high-density polyethylene was successfully improved, as reported by Bondioli *et al.* [44], through melt compounding of 1 vol% of amorphous or crystalline submicron titania particles with diameters ranging from 20 to 350 nm.

A novel route for improving creep resistance of thermoplastic polymers using nanoparticles has been proposed by Zhou *et al.* [43]. Silica nanoparticles with an average diameter of 7 nm (Aerosil 200 by Degussa, Germany) were pre-treated by a silane coupling agent to introduce C=C double bonds on their surfaces. Afterwards, during melt mixing with polypropylene, butyl acrylate monomers reacted with the silane coupling agent on the nanoparticles and crosslinked to interconnect the particles with each other, while the PP chains penetrated into the networks, forming a semi-interpenetrating polymer network (semi-IPN) structure. The time-dependent creep compliance of neat PP and relative nanocomposites containing 1.36 vol% of silica is reported in Figure 9.13.

Various amounts (0.5, 1, 2, 3 and 6 wt%) of both unmodified (Aerosil 200 by Degussa, Germany) and silane-modified silica nanoparticles (Aerosil R974 by Degussa, Germany) have also been melt compounded in a commercial segmented thermoplastic elastomer composed of a hard and crystallizable poly(tetramethylene terephthalate) and a flexible poly(tetramethylene-ether-glycol-terephthalate) (Hytrel, DuPont) [34]. The creep compliance, evaluated under a constant stress of 13 MPa (corresponding to 41% of the tensile strength of pure Hytrel), decreased clearly and steadily with both unmodified and modifi-

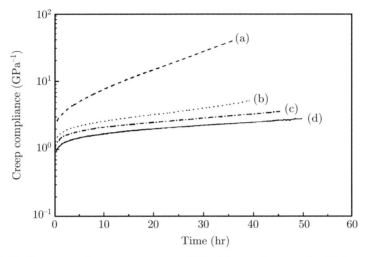

Figure 9.13. Creep compliance curves under 80% of the static strength of (a) neat PP, and nanocomposites containing (b) as-received silica nanoparticles, (c) pretreated silica without the crosslinking agent (butyl acrylate monomer), (d) pretreated silica with the crosslinking agent [43]

ed SiO_2 content. After 84 h, the creep compliance of Hytrel was reduced by a factor of 0.64 in nanocomposites filled with 6 wt% of unmodified silica and by a factor of 0.48 in nanocomposites filled with the same amount of silane-treated silica. This difference is likely to be the consequence of the existence of a thin, flexible interfacial silane layer on modified silica nanoparticles that may contribute to deformation.

Alumina fillers were incorporated in polystyrene in 4.5 wt% by melt blending with and without latex precompounding, as reported by Siengchin et al. [39]. SEM and TEM observations concurrently indicated that direct melt mixing of alumina nanoparticles with PS resulted in microcomposites, whereas the masterbatch technique gave origin to nanocomposites. The stiffness and resistance to creep (summarized in master curves) of the nanocomposites were improved compared to those of the microcomposites.

9.5. Fatigue resistance of polymer nanocomposites

Despite the relatively scarce amount of available literature data, some scientific papers contain information on the fatigue resistance of various polymeric matrices filled with 1-D nanofillers such as layered silicates [45,46,49,50,55–58,61], 2-D nanofillers such as carbon nanotubes [47,48,51,59,62] or carbon nanofibers [63], and 3-D nanoparticles such as silicon carbide [54], titania [60], and alumina [60]. In most cases the fatigue resistance was investigated through the classical S-N curves [45–49,51,52,54,56,57,59,61–63], while in a limited number of papers the subject was afforded by a fracture mechanics point of view [50,60].

9.5.1. Fatigue behavior of PNCs containing one-dimensional nanofillers

The first literature information on the fatigue resistance of PNCs is a study published in 1999 by Yamashita et al. [45] on a commercial nylon-6 nanocomposite (1015C2 by Ube Industries, Japan). Nanocomposites were prepared by in situ polymerization of ε-caprolactam in the presence of 2 wt% of montmorillonite (MMT) whose sodium ions were exchanged with ammonium ions of 12-aminolauric acid. TEM observation of the injection molded specimens indicated that MMT was fully exfoliated and dispersed in the PA6 matrix as single layers oriented along the flow direction. As reported in Figure 9.14, the fatigue lifetime of the nanocomposites was compared with that of traditional composites consisting of the same PA6 matrix reinforced with 30 wt% of short glass fibers (SGFs). SGFs were surface treated with an aminosilane coupling agent.

It is worthwhile noting that the fatigue lifetime of PA6-MMT nanocomposites was much longer than that of PA6-SGF composites tested under identical fatigue conditions. Moreover, it is important to underline that a small amount (2 wt%) of well dispersed 1-D nanofillers turned out to be much more efficient in improving the fatigue resistance of PA6 than a larger amount (30 wt%) of a traditional reinforcing agent such as SGFs. The authors also verified that

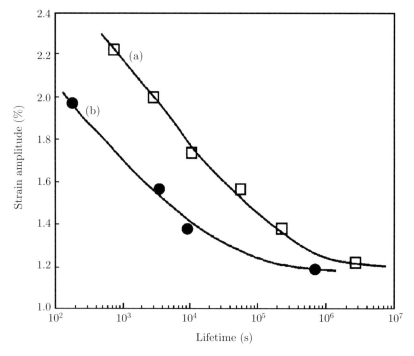

Figure 9.14. Fatigue lifetime for (a) PA6 – 2 wt% MMT nanocomposites, and (b) PA6 – 30 wt% SGFs composites, tested at 30°C under strain-control at a frequency of 6.91 Hz [45]

the fatigue resistance of both nano- and microcomposites was significantly higher than that of neat PA6 (1015B by Ube Industries, Japan).

The fatigue resistance of nano- and microcomposites was also compared by Mallick and co-workers [46,56]. In particular, they performed stress-controlled tension-tension fatigue tests on PP nanocomposites reinforced with 5 wt% of organoclay and on 40 wt% talc-filled PP. As reported in Figure 9.15, the S–N diagram of nanocomposites was slightly better than that of neat PP and much higher than that of talc-filled composites.

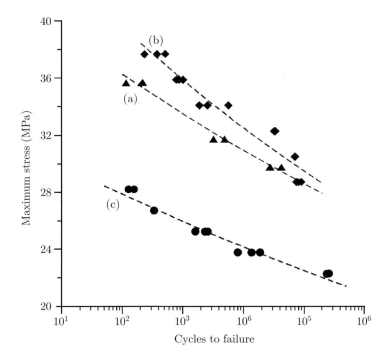

Figure 9.15. Fatigue S-N curves of (a) neat PP, (b) PP – 5 wt% organoclay, and (c) PP – 40 wt% talc. Tests have been performed at room temperature under stress-control with R = 0.1 and a frequency of 1 Hz [56]

The effect of layered silicates on the fatigue resistance of PA6 was also investigated from a fracture mechanics point of view. Bellemare *et al.* [50,53] studied the fatigue resistance of neat PA6 (1015B by Ube Industries, Japan) and relative commercial nanocomposites with 2 wt% of organoclay (1015C2 by Ube Industries, Japan) prepared by *in situ* polymerization. Unfilled PA6 had the same molecular weight (M_n = 22 200 g/mol) as the PA6 matrix of the nanocomposite. Preliminary stress-controlled fatigue tests performed on unnotched dogbone specimens confirmed the results previously obtained by Yamashita *et al.* [45] on the same materials tested under strain control. In fact, from the 5 Hz S–N curves the fatigue life of PA6-clay nanocomposites resulted to be improved in

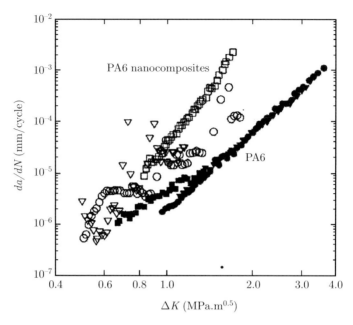

Figure 9.16. Fatigue crack propagation rate as a function of the applied stress intensity factor range for neat PA6 (full points) and PA6 nanocomposites (open points) [50]

comparison to that of the neat PA6 matrix. At the same time, fatigue crack propagation experiments were performed on compact tension specimens under load control at a frequency of 5 Hz with R = 0.1. The fatigue crack propagation rate (da/dN) was monitored as a function of the applied stress intensity range (ΔK) and the results are summarized in Figure 9.16 [50]. Surprisingly enough, at any given ΔK value, the crack growth rate was comparatively higher in PA6-clay nanocomposites than in neat PA6. The fracture surfaces generated during the fatigue propagation tests and transverse cross-section of cracked specimens were observed by SEM. Both near the fatigue initiation site and in an intermediate portion of the fatigue crack propagation zone, the fracture surfaces of neat PA6 specimens did not present fibrils similar to those observed in PA6-nanocomposites, thus indicating a lower tendency for the nanocomposite to form microvoids ahead of the fatigue crack tip under fatigue loading conditions. The effect of the nanoparticles is to increase the yield stress, thus facilitating the formation of microvoids ahead of the crack tip: this could be an important factor to consider for explaining the observed decrease in fatigue propagation resistance. The fatigue life is given by the summation of the time spent in initiating the crack and in propagating it to a critical length, which depends on the fracture toughness of the material. Since both the fatigue crack propagation resistance and the fracture toughness [79] of PA6-clay nanocomposites are lower than for neat PA6 matrix, the increased fatigue life induced by the layered silicate can be attributed to an increase in the fatigue crack initiation resistance.

A dispersion of layered silicates was also proven to have beneficial effects on the fatigue endurance of an epoxy resin, as documented by Juwono and Edward [58]. In fact, their results showed that the bending fatigue life of an epoxy resin improved significantly with the addition of a commercial MMT (Nanomer® I.30E by Nanocor, USA) at strain amplitudes below a threshold value. The failure mode of neat epoxy and relative epoxy-clay nanocomposites was similar (brittle failure). However, the morphology and the striation pattern on the fracture surfaces as well as the direction of propagation changed.

The effect of layered silicates on the fatigue resistance of several different elastomers such as polyurethanes [49,57], poly(styrene-co-butadiene) rubber [55,61] and polybutadiene rubber (BR) [55] was also investigated. In particular, Song et al. [49] measured the fatigue life of PU-nanocomposites based on poly(propylene glycol), 4,4'-methylene bis(cyclohexyl isocianate), 1,4-butandiol and various amounts (1, 3, 5 and 7 wt%) of a commercial organoclay (Cloisite® C20 by Southern Clay Products, USA). The fatigue life was determined as the number of cycles to failure in strain-controlled tests performed at 2 Hz and 18°C on dogbone specimens with a constant maximum strain of 200% and R = 0. The results are summarized in Figure 9.17.

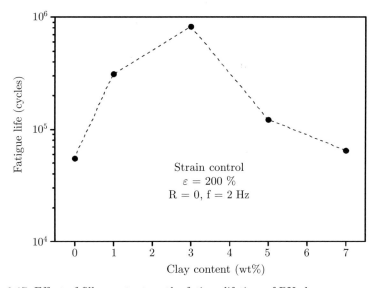

Figure 9.17. Effect of filler content on the fatigue lifetime of PU-clay nanocomposites [49]

The data reported in Figure 9.17 clearly show that MMT can effectively improve the fatigue life of the PU elastomeric matrix. The occurrence of an optimal clay content value of 3 wt% was not explained by the authors. It is worthwhile noting that a similar non-monotonic trend with the filler content was previously reported for other ultimate mechanical properties of PU-clay nanocomposites, such as the tensile stress and strain at break [80–84]. It could be postulated that after a certain filler loading (of about 3–5 wt%) the layered silicates tend to form aggregates in the PU-clay nanocomposites that would cause a deterioration of the ultimate properties, including their fatigue resistance. An analysis on the enhancement of fatigue durability of polyurethane by organoclay nanofillers was performed by Jin *et al.* [57] by WAXD measurements and computer simulations. They concluded that organoclay nanoplatelets show reversible deformation behavior during a fatigue test when properly dispersed (*i.e.*, at an optimal content of 3 wt%). This behavior could help in releasing concentrated stress and stopping crack propagation, thus improving the fatigue resistance. By a simple mechanical blending process Song *et al.* [55] prepared both SBR and BR nanocomposites filled with various amounts of a natural (Cloisite® Na+) and an organo-modified (Cloisite® 20C) montmorillonites by Southern Clay Products, USA. The results of the strain-controlled fatigue tensile tests performed at 20°C with a constant maximum strain of 60%, and a frequency of 1.5 Hz are summarized in Figure 9.18. Both clays had a positive effect on the fatigue lifetime of SBR, with somewhat better performances for the natural (unmodified) Cloisite® Na+ clay. The fatigue life of SBR-clay nanocomposites increased with the clay content. For the BR matrix the better performances of nanocomposites prepared with natural clay was even more marked. However, as previously observed for PU-clay nanocomposites, the effect of organoclay Cloisite® C20 started to diminish after a certain clay content (7 wt% in this case). A positive effect of natural montmorillonite on the flexural fatigue life of SBR/carbon-black/clay nanocomposites was also reported by Wu *et al.* [61]. Testing specimens with a semicircular groove on a De Mattia flexing machine (ASTM D430), they found that the fatigue life of carbon-black filled SBR dramatically improved with the incorporation of 4–5 phr of nano-dispersed clay. Moreover, they observed that clay addition did not decrease the degree of crosslinking of the material but improved the hysteresis and tearing energy.

9.5.2. *Fatigue behavior of PNCs containing two-dimensional nanofillers*

Some attempts were made to improve the fatigue resistance of polymers by adding 2-D nanofillers such as carbon nanotubes or nanofibers, with particular attention to thermosetting resins, such as epoxies, having a potential interest as matrices for structural composites [47,48,51,63].

In particular, Ren *et al.* [47,48,51] investigated the tension-tension fatigue behavior of an epoxy resin reinforced with single-walled carbon nanotube (SWCNT) ropes. The SWCNT ropes were synthesized by an hydrogen/argon electric arc discharge method with lengths up to 100 mm [85]. The SWCNT ropes were then aligned by a slight tension, subsequently laid onto an epoxy layer, and finally covered by more epoxy up to a final thickness of 0.4–0.6 mm. Due to the fabrication procedure the authors were not able to keep constant the SWCNT ropes content, as it varied in the range of 0.1–0.9 wt% depending

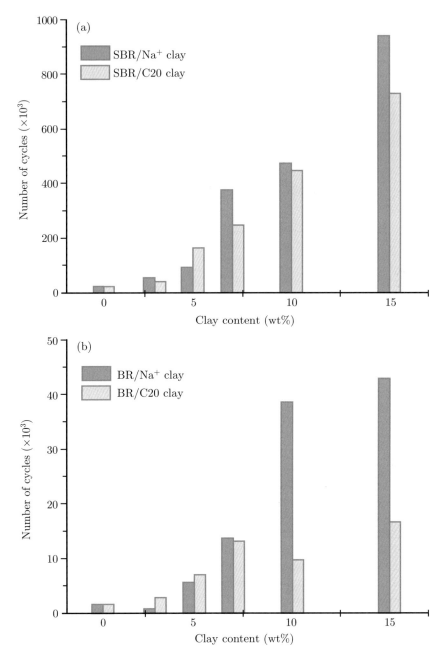

Figure 9.18. Effect of clay content on fatigue life of (a) SBR, and (b) BR [55]

on the specimens. Nine unidirectional SWCNT/epoxy specimens were fatigue tested in tension at 5 Hz under stress control with R = 0.1 and the resulting S–N data are reported in Figure 9.19.

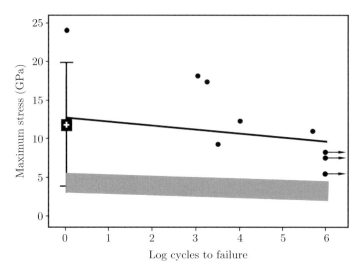

Figure 9.19. Tension-tension S–N fatigue data. Filled circles refer to SWCNT-epoxy composites, while the grey band covers most S–N data for unidirectional carbon fiber reinforced composites [48]

The data reported in Figure 9.19 indicate that SWCNT-epoxy composites possess a fatigue resistance at least twice that of unidirectional carbon fiber reinforced composites, whose typical S–N data are represented by the grey band. Macroscopically, SWCNT-epoxy composites exhibited a brittle failure with smooth fracture surfaces. However, SEM observations indicated ductile-like local failure modes around the SWCNT ropes with bridging of matrix cracks. Moreover, observed damage and failure modes included [51]: (i) splitting of SWCNT bundles, (ii) kink formation and subsequent failure in SWCNT bundles, (iii) fracture of SWCNT bundles.

The effect of addition of small amounts (1–3 wt%) of vapor grown carbon nanofibers (VGCNF) on fracture toughness and fatigue resistance of an epoxy matrix was investigated by Zhou *et al.* [63]. The S-N curves reported in Figure 9.20 were obtained from stress-controlled tension-tension fatigue tests, performed at 21.5°C with R = 0.1 and a frequency of 1 Hz. From Figure 9.20 it clearly emerges that at any given stress level, the fatigue life of nanophased epoxy was significantly higher than that of the neat epoxy matrix. As the authors pointed out, the experimental data were fitted reasonably well by a power law equation in the form:

$$\sigma_{max} = \sigma_f N^b \qquad (9.6)$$

where σ_{max} is the maximum applied fatigue stress, N is the number of cycles to failure, σ_f is a fatigue strength coefficient, and b is a fatigue strength exponent. Figure 9.21 summarizes both the fatigue strength coefficient and exponent for the neat epoxy matrix and relative VGCNF-filled nanocomposites, whose S–N curves are reported in Figure 9.20.

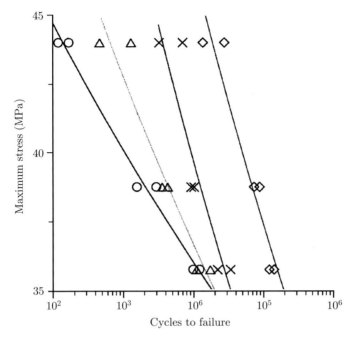

Figure 9.20. Tension-tension fatigue S-N curves of (○) neat epoxy matrix and relative nanocomposites reinforced with (△) 1 wt%, (◇) 2 wt%, and (×) 3 wt% of VGCNFs [63]

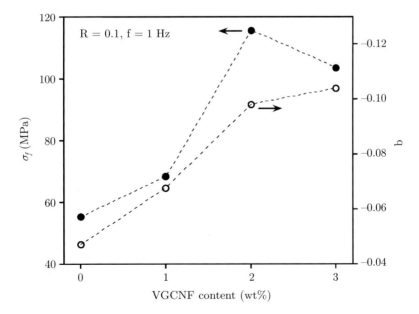

Figure 9.21. Effect of VGCNF content on the strength coefficient (σ_f) and the fatigue strength exponent (b) of epoxy-VGCNF nanocomposites [63]

The fatigue strength exponent decreases as the CNF content increases, while the fatigue strength coefficient increases as the CNF content is increased up to 2 wt%; however, at 3 wt%, the fatigue strength coefficient decreases. It is interesting to note that a similar trend with the VGCNF content is followed by the fracture toughness of the materials evaluated by the K_{IC} parameter [63].

Marrs et al. [59] investigated how MWCNTs may alter the fatigue resistance of an acrylic bone cement based on methyl metacrylate-styrene copolymer (MMA-co-Sty), that is a proven polymer having important applications in medicine and dentistry. MWCNTs were synthesized on a quartz substrate in an argon-hydrogen atmosphere and dispersed throughout the molten matrix of pre-polymerized commercial bone cement powder in the heated (220°C) chamber of an internal mixer. Fatigue tests in 4-point bending configuration were performed at room temperature between constant maximum (40 N) and minimum (4 N) load values at a frequency of 5 Hz. The effect of MWCNT content on the fatigue cycles to failure is summarized in Figure 9.22 that also contains the quasi-static flexural strength data.

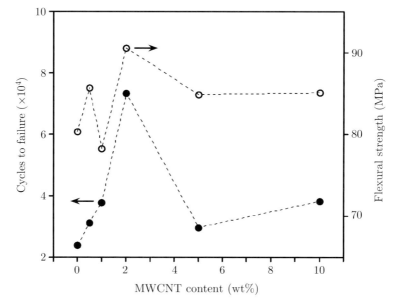

Figure 9.22. Effect of MWCNT content on the flexural (●) cycles to failure and (○) quasi-static strength of MMA-co-Sty–MWCNT nanocomposites [59]

Unlike the small but significant improvement of quasi-static flexural strength, a 2 wt% loading of MWCNTs was associated with a substantial enhancement of bone cement's fatigue performance. In fact, specimens with a 2 wt% content of MWNTs showed a 3.1-fold increase in the mean actual fatigue life. This result is even more interesting considering that previous work showed that the addition of 2 wt% untreated, milled carbon fibers (150 μm lengths; 8 μm diameters) did not substantially improve the fatigue performances of bone cement containing an untreated radiopacifying agent [86]. Moreover, it is interesting to observe how fatigue endurance passed through a maximum for an optimal MWCNT content of 2 wt%. Although the nanotube-matrix interactions are not yet fully understood, the authors speculate that the observed decreases in material property enhancement with increasing MWNT loading could be related to beneficial effects of MWNT material augmentation competing against the adverse effects of sporadic inadequately dispersed (still agglomerated) "clumps" of MWNTs. These MWNT clumps, whose number and extension is naturally increasing with the MWCNTs content, may act as fracture initiation sites. The same authors provided experimental data also on the fully reversed (R = −1) tension-compression fatigue resistance of these materials under three different peak stress amplitudes (20, 30 and 35 MPa) in a 37°C saline environment [62]. By using a three-parameters Weibull model, the cycles to failure (N) for each sample were converted into a probability of failure $P(N)$, as follows:

$$P(N) = 1 - \exp\left[-\left(\frac{N - N_0}{\alpha - N_0}\right)^\beta\right] \quad (9.7)$$

where N_0 is the minimum fatigue life (the baseline number of cycles for the sample at a given load amplitude), α is the location parameter (the number of cycles to failure below which 63.2% of the specimens fail), and β is the Weibull shape parameter (an indicator of the variance within the sample). In order to compare the fatigue resistance of the investigated materials, the authors proposed to adopt the Weibull mean (N_{WM}) parameter calculated as follows:

$$N_{WM} = N_0 + (\alpha - N_0)\Gamma\left(1 + \frac{1}{\beta}\right) \quad (9.8)$$

The higher the N_{WM}, the better the material fatigue resistance. For the investigated nanocomposites Figure 9.23 summarizes how N_{WM} values depend on the MWCNTs content. The data reported in Figure 9.23 clearly indicate that the fatigue resistance of MMA-co-Sty–MWCNT nanocomposites stressed under physiologically relevant conditions depends on the concentration of MWCNTs, the dispersion of the MWCNTs, and the peak stress of the dynamic loading cycle. In particular, testing at peak stress amplitudes of 20 and 30 MPa confirmed the existence of an optimal MWCNTs content (of about 2 wt%). Higher amounts of MWCNTs had a detrimental effect on the fatigue resistance. SEM observations evidenced that MWCNTs protruded from the crack faces into the crack wake normal to the direction of fracture growth, while observation of the secondary crack showed some of the MWCNTs bridging the growing crack. Testing at the highest peak stress amplitude of 35 MPa revealed that N_{WM} decreased slightly with increasing the concentration of MWCNTs up to and including 2 wt%. The authors postulated that at these higher stress levels the

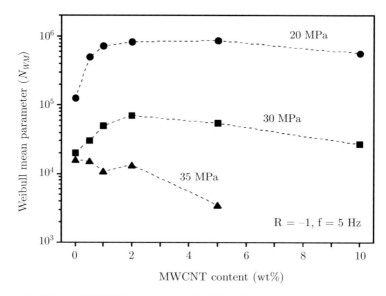

Figure 9.23. Effect of MWCNT content on the Weibull mean parameter (N_{WM}) representing the fatigue resistance of MMA-co-Sty–MWCNT nanocomposites tested at various amplitudes of the peak stress: (●) 20 MPa, (■) 30 MPa, and (▲) 35 MPa [62]

onset of damage may overcome the ability of MWCNTs to slow crack growth *via* the aforementioned mechanisms.

9.5.3. *Fatigue behavior of PNCs containing three-dimensional nanoparticles*

Very little literature information exists on the effect of inorganic 3-D nanoparticles on the fatigue resistance of polymeric matrices [60] or composites [54].

Wetzel *et al.* [60] performed a comprehensive study on the fracture and toughening mechanisms in epoxy nanocomposites containing various amounts (in the range from 1 to 10 vol%) of either titania (TiO_2) or alumina (Al_2O_3) nanoparticles. Various nanocomposites were prepared by dispersing alumina (Aeroxide Alu C by Degussa, Germany) nanoparticles with a primary particle size of about 13 nm, and titania (Kronos 2310 by Kronos International, Germany). They found that both kinds of nanoparticles can markedly improve the fracture toughness of the selected epoxy matrix with K_{IC} values steadily increasing with the nanoparticles content. At any given volume fraction, the finer alumina nanoparticles resulted to have a more pronounced toughening effect when compared to the coarser titania nanoparticles. Compact tension specimens of epoxy-alumina nanocomposites were also fatigue tested under stress control in order to assess their fatigue crack propagation (FCP) curves. Tension-tension fatigue tests at a frequency of 5 Hz and R = 0.2 allowed the authors to obtain the crack propagation rate curves as a function of the stress intensity factor range, reported in Figure 9.24.

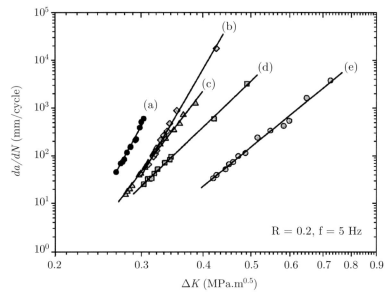

Figure 9.24. Fatigue crack propagation curves of (a) neat epoxy and relative nanocomposites containing (b) 1 vol%, (c) 3 vol%, (d) 5 vol%, and (e) 10 vol% of Al_2O_3 nanoparticles [60]

It clearly emerges that titania nanoparticles were able to significantly increase the resistance to fatigue crack propagation of epoxy under dynamic loading conditions. In fact, the FCP curves shifted towards higher ΔK values as the nanoparticle content rose. Through SEM analysis of the plastic zone on the fracture surface of epoxy/Al_2O_3 nanocomposites close to the initial starting crack, the authors were able to obtain some indications on the fracture mechanisms. The observed features cautiously indicated the occurrence of (i) plastic deformation, (ii) formation of cavities/debonding, (iii) shear banding, and (iv) crack bridging (pinning). All these mechanisms may account for the improved fracture toughness and fatigue crack propagation resistance of the epoxy/Al_2O_3 nanocomposites.

It is worthwhile noting that epoxy resins are widely used for the fabrication of structural composites reinforced with continuous glass or carbon fibers. The possibility to improve the mechanical properties of epoxies by nanofillers could provide a way to enhance the mechanical properties of epoxy-matrix structural composites, including their fatigue resistance. Crisholm et al. [54] conducted a study on the effect of 1.5 and 3 wt% of spherical SiC nanoparticles about 29 nm in diameter (MTI Corporation, USA) on the fatigue resistance of epoxy matrix-carbon fibers laminates. Nanosized SiC particles were ultrasonically mixed in the diglycidylether of Bisphenol A (DGEBA) epoxy resin and the reaction mixture was used in a vacuum assisted reaction transfer molding process with satin weave carbon fiber preforms to fabricate carbon/epoxy nanocomposite panels. Test coupons were extracted from each category of panels and subjected to mechanical tests. Under both flexural and tensile quasi-static tests, an average of 20–30% increase in mechanical properties was observed with an optimal SiC content of 1.5 wt%. Flexure fatigue tests were conducted at a stress ratio of 0.1 and a frequency of 3 Hz, and the resulting S–N diagrams are shown in Figure 9.25. The maximum applied stress is normalized to the flexural strength (relative stress level) and plotted as a function of the number of cycles to failure.

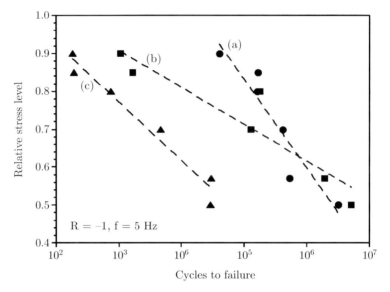

Figure 9.25. S–N curves of carbon/epoxy composites realized with (a) neat epoxy matrix, and epoxy matrix filled with (b) 1.5 wt% and (c) 3.0 wt% of SiC nanoparticles [54]

A critical stress level of about 60% of the ultimate flexural strength seems to exist. Below this threshold stress level, the system filled with 1.5% of SiC nanoparticles performed better than the neat system, whereas the situation reversed above this stress level. The authors believe that below this threshold stress level the fatigue failure mechanisms such as matrix cracks, filament splitting, and delamination are significantly slowed. On the other hand, throughout the entire loading range, the 3% system filled with 1.5% of SiC nanoparticles resulted to be inferior to the other two systems. The formation of nanoparticle aggregates acting as stress concentrators could be a reason for the observed worsening of the fatigue resistance of this latter system.

9.6. Conclusions and outlook

In this chapter the experimental investigations regarding the mechanical response of PNCs under creep or fatigue loading conditions have been reviewed.

In general, from the available literature information, it emerges that both creep stability and fatigue resistance of several polymeric matrices can be significantly improved by the addition of various types of nanofillers, such as layered silicates, carbon nanotubes or nanofibers, and inorganic nanoparticles. Some experimental evidences indicate that small well-dispersed amounts of such nanofillers may have reinforcing effects comparable to or better than those induced by much larger amounts of traditional microfillers, such as short glass or carbon fibers.

However, unlike in traditional microcomposites, in PNCs the observed reinforcing effects are often not directly proportional to the filler content. In fact, the existence of an optimal filler concentration was experimentally observed in several cases, depending on the type of matrix and the selected nanofiller. At a nanometric level, the filler-matrix interactions may be very complex since the nanofiller may induce modifications of the matrix structure, such as the crystals content and morphology in thermoplastic matrices, or the crosslinking density in thermosetting matrices. Consequently, the reinforcing effects of nanofillers and the matrix modifications (such as reduction of its crystallinity content or crosslinking degree) induced by their presence may play opposite roles on the creep and fatigue performances of the resulting nanocomposites. Moreover, after a certain amount, nanofillers tend to form aggregates in the polymeric matrices that would cause a deterioration of the ultimate properties, including their fatigue resistance.

In conclusion, even if some very promising results have already been reached, a deeper understanding of the nanofiller-matrix interactions is certainly required in order to optimize the creep stability and fatigue resistance of both thermoplastic and thermosetting polymeric nanocomposites.

References

1. Tjong S C (2006) Structural and mechanical properties of polymer nanocomposites, *Mater Sci Eng R-Reports* **53**:73–197.
2. Hussain F, Hojjati M, Okamoto M and Gorga R E (2006) Review article: Polymer-matrix nanocomposites, processing, manufacturing, and application: An overview, *J Compos Mater* **40**:1511–1575.
3. Utracki L A (2004) *Clay-containing polymeric nanocomposites*, Rapra Technology Ltd, Shawbury, UK, Vol. 1 and 2.
4. Okada A and Usuki A (2006) Twenty years of polymer-clay nanocomposites, *Macromol Mater Eng* **291**:1449–1476.
5. Goettler L A, Lee K Y and Thakkar H (2007) Layered silicate reinforced polymer nanocomposites: Development and applications, *Polym Reviews* **47**:291–317.
6. Coleman J N and Khan U (2006) Mechanical reinforcement of polymers using carbon nanotubes, *Adv Mater* **18**:689–706.
7. Miyagawa H, Misra M and Mohanty A K (2005) Mechanical properties of carbon nanotubes and their polymer nanocomposites, *J Nanosci Nanotechnol* **5**:1593–1615.
8. Thostenson E T, Ren Z F and Chou T W (2001) Advances in the science and technology of carbon nanotubes and their composites: A review, *Compos Sci Technol* **61**:1899–1912.
9. Xie X L, Mai Y W and Zhou X P (2005) Dispersion and alignment of carbon nanotubes in polymer matrix: A review, *Mater Sci Eng R-Reports* **49**:89–112.

10. Choi Y K, Sugimoto K, Song S M, Gotoh Y, Ohkoshi Y and Endo M (2005) Mechanical and physical properties of epoxy composites reinforced by vapor grown carbon nanofibers, *Carbon* **43**:2199–2208.
11. Manocha L M, Valand J and Manocha S (2007) Development of composites incorporating carbon nanofibers and nanotubes, *J Nanosci Nanotechnol* **7**:1845–1850.
12. Traina M, Pegoretti A and Penati A (2007) Time-temperature dependence of the electrical resistivity of high density polyethylene – carbon black composites, *J Appl Polym Sci* **106**:2065–2074.
13. Moore D R and Turner S (2001) *Mechanical evaluation strategies for plastics*, Woodhead Publishing Ltd, Cambridge, UK.
14. http://portal.isiknowledge.com/portal.cgi.
15. Beake B D, Chen S, Hull J B and Gao F (2002) Nanoindentation behavior of clay/poly(ethylene oxide) nanocomposites, *J Nanosci Nanotechnol* **2**:73–79.
16. Dutta A K, Penumadu D and Files B (2004) Nanoindentation testing for evaluating modulus and hardness of single-walled carbon nanotube-reinforced epoxy composites, *J Mater Res* **19**:158–164.
17. Hasegawa N, Okamoto H and Usuki A (2004) Preparation and properties of ethylene propylene rubber (EPR)–clay nanocomposites based on maleic anhydride-modified EPR and organophilic clay, *J Appl Polym Sci* **93**:758–764.
18. Pegoretti A, Kolarik J, Peroni C and Migliaresi C (2004) Recycled poly(ethylene terephthalate)/layered silicate nanocomposites: morphology and tensile mechanical properties, *Polymer* **45**:2751–2759.
19. Shen L, Phang I Y, Chen L, Liu T X and Zeng K Y (2004) Nanoindentation and morphological studies on nylon 66 nanocomposites. I. Effect of clay loading, *Polymer* **45**:3341–3349.
20. Zhang Z, Yang J L and Friedrich K (2004) Creep resistant polymeric nanocomposites, *Polymer* **45**:3481–3485.
21. Ranade A, Nayak K, Fairbrother D and D'Souza N A (2005) Maleated and non-maleated polyethylene-montmorillonite layered silicate blown films: creep, dispersion and crystallinity, *Polymer* **46**:7323–7333.
22. Shen L, Tjiu W C and Liu T (2005) Nanoindentation and morphological studies on injection-molded nylon-6 nanocomposites, *Polymer* **46**:11969–11977.
23. Vlasveld D P N, Bersee H E N and Picken S J (2005) Creep and physical aging behavior of PA6 nanocomposites, *Polymer* **46**:12539–12545.
24. Joshi M and Viswanathan V (2006) High-performance filaments from compatibilized polypropylene/clay nanocomposites, *J Appl Polym Sci* **102**:2164–2174.
25. Li K, Gao X L and Roy A K (2006) Micromechanical modeling of viscoelastic properties of carbon nanotube-reinforced polymer composites, *Mech Adv Mater Struct* **13**:317–328.
26. Peneva Y, Tashev E and Minkova L (2006) Flammability, microhardness and transparency of nanocomposites based on functionalized polyethylenes, *Europ Polym J* **42**:2228–2235.
27. Rivera I, Martin D and Truss R (2006) Viscoelastic studies on thermoplastic polyurethane nanocomposites using a nano-hardness tester, *J Polym Eng* **26**:903–918.
28. Shen L, Wang L, Liu T and He C (2006) Nanoindentation and morphological studies of epoxy nanocomposites, *Macromol Mater Eng* **291**:1358–1366.
29. Siengchin S and Karger-Kocsis J (2006) Creep behavior of polystyrene/fluorohectorite micro- and nanocomposites, *Macromol Rapid Commun* **27**:2090–2094.
30. Xia H, Song M, Zhang Z and Richardson M (2006) Microphase separation, stress relaxation, and creep behavior of polyurethane nanocomposites, *J Appl Polym Sci* **103**:2992–3002.

31. Xue Y, Wu W, Jacobs O and Schadel B (2006) Tribological behavior of UHMWPE/HDPE blends reinforced with multi-wall carbon nanotubes, *Polym Testing* **25**:221–229.
32. Yang J L, Zhang Z, Schlarb A K and Friedrich K (2006) On the characterization of tensile creep resistance of polyamide 66 nanocomposites. Part II: Modeling and prediction of long-term performance, *Polymer* **47**:6745–6758.
33. Yang J L, Zhang Z, Schlarb A K and Friedrich K (2006) On the characterization of tensile creep resistance of polyamide 66 nanocomposites. Part I. Experimental results and general discussions, *Polymer* **47**:2791–2801.
34. Aso O, Eguiazábal J I and Nazábal J (2007) The influence of surface modification on the structure and properties of a nanosilica filled thermoplastic elastomer, *Compos Sci Technol* **67**:2854–2863.
35. Yang J L, Zhang Z, Friedrich K and Schlarb A K (2007) Creep resistant polymer nanocomposites reinforced with multiwalled carbon nanotubes, *Macromol Rapid Commun* **28**:955–961.
36. Ganß M, Satapathy B K, Thunga M, Weidisch R, Potschke P and Janke A (2007) Temperature dependence of creep behavior of PP-MWNT nanocomposites, *Macromol Rapid Commun* **28**:1624–1633.
37. Lietz S, Yang J L, Bosch E, Sandler J K W, Zhang Z and Altstadt V (2007) Improvement of the mechanical properties and creep resistance of SBS block copolymers by nanoclay fillers, *Macromol Mater Eng* **292**:23–32.
38. Pegoretti A, Dorigato A and Penati A (2007) Tensile mechanical response of polyethylene-clay nanocomposites, *Express Polym Lett* **1**:123–131.
39. Siengchin S, Karger-Kocsis J and Thomann R (2007) Alumina-filled polystyrene micro- and nanocomposites prepared by melt mixing with and without latex precompounding: structure and properties, *J Appl Polym Sci* **105**:2963–2972.
40. Starkova O, Yang J and Zhang Z (2007) Application of time-stress superposition to nonlinear creep of polyamide 66 filled with nanoparticles of various sizes, *Compos Sci Technol* **67**:2691–2698.
41. Yang J L, Zhang Z, Friedrich K and Schlarb A K (2007) Resistance to time-dependent deformation of nanoparticle/polymer composites, *Appl Phys Lett* **91**:011901-1–011901-3.
42. Zhang W, Joshi A, Wang Z, Kane R S and Koratkar N (2007) Creep mitigation in composites using carbon nanotube additives, *Nanotechnology* **18**:1–5.
43. Zhou T H, Ruan W H, Yang J L, Rong M Z, Zhang M Q and Zhang Z (2007) A novel route for improving creep resistance of polymers using nanoparticles, *Compos Sci Technol* **67**:2297–2302.
44. Bondioli F, Dorigato A, Fabbri P, Messori M and Pegoretti A (2008) High-density polyethylene reinforced with submicron titania particles, *Polym Eng Sci* **48**:448–457.
45. Yamashita A, Takahara A and Kajiyama T (1999) Aggregation structure and fatigue characteristics of (nylon 6/clay) hybrid, *Compos Interf* **6**:247–258.
46. Mallick P K and Zhou Y (2003) Yield and fatigue behavior of polypropylene and polyamide-6 nanocomposites, *J Mater Sci* **38**:3183–3190.
47. Ren Y, Li F, Cheng H M and Liao K (2003) Tension-tension fatigue behavior of unidirectional single-walled carbon nanotube reinforced epoxy composite, *Carbon* **41**:2177–2179.
48. Ren Y, Lil F, Cheng H M and Liao K (2003) Fatigue behavior of unidirectional single-walled carbon nanotube reinforced epoxy composite under tensile load, *Adv Compos Lett* **12**:19–24.
49. Song M, Hourston D J, Yao K J, Tay J K H and Ansarifar M A (2003) High performance nanocomposites of polyurethane elastomer and organically modified layered silicate, *J Appl Polym Sci* **90**:3239–3243.

50. Bellemare S C, Bureau M N, Denault J and Dickson J I (2004) Fatigue crack initiation and propagation in polyamide-6 and in polyamide-6 nanocomposites, *Polym Compos* **25**:433–441.
51. Ren Y, Fu Y Q, Liao K, Li F and Cheng H M (2004) Fatigue failure mechanisms of single-walled carbon nanotube ropes embedded in epoxy, *Appl Phys Lett* **84**:2811–2813.
52. Song M and Yao K J (2004) X-ray diffraction detection of compliance in polyurethane-organoclay nanocomposites, *Mater Sci Technol* **20**:989–992.
53. Bellemare S C, Dickson J I, Bureau M N and Denault J (2005) Bulk fatigue damage evolution in polyamide-6 and in a polyamide-6 nanocomposite, *Polym Compos* **26**:636–646.
54. Chisholm N, Mahfuz H, Rangari V K, Ashfaq A and Jeelani S (2005) Fabrication and mechanical characterization of carbon/SiC-epoxy nanocomposites, *Compos Struct* **67**:115–124.
55. Song M, Wong C W, Jin J, Ansarifar A, Zhang Z Y and Richardson M (2005) Preparation and characterization of poly(styrene-co-butadiene) and polybutadiene rubber/clay nanocomposites, *Polymer Intern* **54**:560–568.
56. Zhou Y X, Rangari V, Mahfuz H, Jeelani S and Mallick P K (2005) Experimental study on thermal and mechanical behavior of polypropylene, talc/polypropylene and polypropylene/clay nanocomposites, *Mater Sci Eng A* **402**:109–117.
57. Jin J, Chen L, Song M and Yao K J (2006) An analysis on enhancement of fatigue durability of polyurethane by incorporating organoclay nanofillers, *Macromol Mater Eng* **291**:1414–1421.
58. Juwono A and Edward G (2006) Mechanism of fatigue failure of clay-epoxy nanocomposites, *J Nanosci Nanotechnol* **6**:3943–3946.
59. Marrs B, Andrews R, Rantell T and Pienkowski D (2006) Augmentation of acrylic bone cement with multiwall carbon nanotubes, *J Biomed Mater Res Part A* **77A**:269–276.
60. Wetzel B, Rosso P, Haupert F and Friedrich K (2006) Epoxy nanocomposites – fracture and toughening mechanisms, *Eng Fracture Mech* **73**:2375–2398.
61. Wu Y P, Zhao W and Zhang L Q (2006) Improvement of flex-fatigue life of carbon-black-filled styrene-butadiene rubber by addition of nanodispersed clay, *Macromol Mater Eng* **291**:944–949.
62. Marrs B, Andrews R and Pienkowski D (2007) Multiwall carbon nanotubes enhance the fatigue performance of physiologically maintained methyl methacrylate-styrene copolymer, *Carbon* **45**:2098–2104.
63. Zhou Y X, Pervin F and Jeelani S (2007) Effect of vapor grown carbon nanofiber on thermal and mechanical properties of epoxy, *J Mater Sci* **42**:7544–7553.
64. Blackman B R K, Kinloch A J, Sohn Lee J, Taylor A C, Agarwal R, Schueneman G and Sprenger S (2007) The fracture and fatigue behavior of nano-modified epoxy polymers, *J Mater Sci Lett* **42**:7049–7051.
65. Findley W N, Lai J S and Onaran K (1976) *Creep and relaxation of nonlinear viscoelastic materials*, Dover Publications Inc., Toronto.
66. Kolarik J and Pegoretti A (2004) Indentation creep of heterogeneous blends poly(ethylene-terephthalate)/impact modifier, *Polym Testing* **23**:113–121.
67. Pegoretti A, unpublished results.
68. Crawford R (1999) *Plastics Engineering*, Pergamon Press, Oxford, UK.
69. Lakes R S (1999) *Viscoelastic solids*, CRC Press, Boca Raton, FL.
70. Ward I M and Hadley D W (1993) *An introduction to the mechanical properties of solid polymers*, Wiley, Chichester, UK.

71. Kohlrausch F (1863) Ueber die elastische Nachwirkung bei der Torsion, *Poggendorff's Ann Phys* **119**:337–368 (in German).
72. Williams G and Watts D C (1970) Non-symmetrical dielectric relaxation behavior arising from a simple empirical decay function, *Trans Faraday Soc* **66**:80–85.
73. Findley W N (1971) Combined stress creep of non-linear viscoelastic material, in *Advances in Creep Design* (Eds. Smith A I and Nicolson A M) Applied Science Publication, London, Ch. 14.
74. Ritchie R O (1999) Mechanisms of fatigue-crack propagation in ductile and brittle solids, *Int J Fracture* **100**:55–83.
75. Hertzberg R W and Manson J A (1980) *Fatigue of engineering plastics*, Academic Press, New York.
76. Pegoretti A and Ricco T (2000) Fatigue fracture of neat and short glass fiber reinforced polypropylene: effect of frequency and material orientation, *J Compos Mater* **34**:1009–1027.
77. Paris P C (1962) *PhD Dissertation*, Lehig University.
78. Murakami Y (1987) *Stress intensity factors handbook*, Pergamon, Oxford.
79. Bureau M N, Denault J, Cole K C and Enright G D (2002) The role of crystallinity and reinforcement in the mechanical behavior of polyamide–6/clay nanocomposites, *Polym Eng Sci* **42**:1897–1906.
80. Ma J S, Zhang S F and Qi Z N (2001) Synthesis and characterization of elastomeric polyurethane/clay nanocomposites, *J Appl Polym Sci* **82**:1444–1448.
81. Chang J H and An Y U (2002) Nanocomposites of polyurethane with various organoclays: Thermomechanical properties, morphology, and gas permeability, *J Polym Sci, Part B: Polym Phys* **40**:670–677.
82. Choi W J, Kim S H, Kim Y J and Kim S C (2004) Synthesis of chain-extended organifier and properties of polyurethane/clay nanocomposites, *Polymer* **45**:6045–6057.
83. Ni P, Li J, Suo J S and Li S B (2004) Novel polyether polyurethane/clay nanocomposites synthesized with organic-modified montmorillonite as chain extenders, *J Appl Polym Sci* **94**:534–541.
84. Wang J C, Chen Y H and Wang Y Q (2006) Preparation and characterization of novel organic montmorillonite-reinforced blocked polyurethane nanocomposites, *Polym Polym Compos* **14**:591–601.
85. Liu C, Cheng H M, Cong H T, Li F, Su G, Zhou B L and Dresselhaus M S (2000) Synthesis of macroscopically long ropes of well-aligned single-walled carbon nanotubes, *Adv Mater* **12**:1190–1192.
86. Kim H Y and Yasuda H K (1999) Improvement of fatigue properties of poly(methyl methacrylate) bone cement by means of plasma surface treatment of fillers, *J Biomed Mater Res* **48**:35–142.

Chapter 10

Deformation Mechanisms of Functionalized Carbon Nanotube Reinforced Polymer Nanocomposites

S. C. Tjong

10.1. Introduction

The demand for lightweight, high performance and functional materials for advanced structural applications is ever increasing in industrial sectors. Polymer nanocomposites offer significant potential in the development of novel materials for advanced engineering applications. An ideal high-performance polymer nanocomposite is a lightweight material with a high Young's modulus, high yield/tensile strength, good ductility and toughness. Polymer nanocomposites having thermoplastic or thermoset matrices reinforced with stiff inorganic nanofillers have attracted considerable attention in recent years. Inorganic nanofiller materials include silicate clay, silica, alumina and calcium carbonate [1–7]. The nanoclay additions cause significant embrittlement despite the fact they can enhance the tensile stiffness and strength of polymer nanocomposites. In contrast, spherical silica, alumina and calcium carbonate nanoparticles are beneficial to enhance the tensile stiffness, strength and toughness of polymer nanocomposites simultaneously [3,8]. Further improvement in these properties can be achieved by adding one-dimensional nanofillers with large aspect ratio and flexibility such as carbon nanotubes (CNTs) and carbon nanofibers (CNFs).

Carbon nanotubes with high aspect ratio, large surface area, low density together with extraordinary mechanical, electrical and thermal properties make them particularly attractive reinforcing fillers for polymers. CNTs are reported to exhibit an extremely high Young's modulus of up to 1 TPa and tensile strength approaching 60 GPa [9]. Depending on their structural configurations, CNTs can be classified into single-walled and multi-walled materials. *Single-walled CNT* (SWNT) is formed from a single graphene layer rolled into a cylinder in which hemispherical caps seal both ends of the tube. *Multi-walled CNT* (MWNT) comprises many graphene layers wrapped onto concentric cylinders. The diameters of CNTs range from a few (for single-walled) to several nanometers (for multi-walled tubes). By comparison with expensive CNTs, particularly with SWNTs, cost-effective CNTs can be mass-produced by a catalytic chemical vapor deposition using gaseous hydrocarbons as the feed stock [10,11]. Vapor grown CNTs generally have larger diameter than MWNTs, typically in the range of 50 to 200 nm.

In general, CNT/polymer nanocomposites can be prepared by solution blending [12–14], *in situ* polymerization [15–19] and melt compounding processes [20–23]. The solution processing method is the most common method to prepare CNT/polymer nanocomposite thin films due to its simplicity. The technique involves mixing of the nanotubes and polymer in solution *via* energetic agitation or dispersion of nanotubes in either a solvent or polymer solution under vigorous stirring. High-power sonication is needed because it is difficult to disperse pristine nanotubes in a solvent by simple stirring. After sonication, thin nanocomposite films are produced by evaporating the solvent. The evaporation time can be shortened by drop casting the nanotube-polymer suspension on a hot substrate. *In situ* polymerization involves the dispersion of CNTs in a suitable monomer followed by polymerization [15]. This technique is especially suitable for the fabrication of insoluble and thermally unstable polymers. Melt compounding is a versatile commercial process capable of producing a wide variety of polymer composites on large volume scales. It uses high temperature for the melting of polymer pellets, and large shear force to disperse nanotubes in the polymer melt. However, melt compounding tends to reduce the length of CNTs in certain cases. Bulk samples can be prepared *via* typical techniques such as extrusion, injection molding or compression.

Poor dispersion and weak interfacial bonding remain the major challenges for effectively incorporating CNTs into polymer matrix. An intrinsic van der Waals force between nanotubes tends to promote cluster formation rather than homogeneous dispersion. In particular, MWNTs produced by the arc-discharge method are highly crystalline and defect free. Their surfaces are inert and incompatible with most polymers. The agglomeration of nanotubes is detrimental to the mechanical properties of nanocomposites due to inferior load-transfer mechanism. The interfacial bonding between the polymer and CNTs can be enhanced *via* chemical functionalization of nanotubes. The nanotube caps are more reactive because of their high degree of curvature. Oxidative or acid treatments are used to develop functional groups such as carboxylic ($-COOH$) and hydroxyl groups at the cap open ends [24,25]. These functional groups at the tube ends interact directly with the polymer *via* special reactions such as etherification and amidization, thereby improving interfacial bonding between the CNT and the polymer matrix. Alternatively, direct addition of chemical species onto the conjugated carbon framework at the nanotube sidewalls can also improve the solubility of CNTs in polymers [26,27]. Thus, chemical functionalization is considered to be an effective way to achieve homogeneous dispersion of CNTs in polymers. The enhancement of the mechanical properties of CNT/polymer composites depends greatly on the interfacial filler-matrix interactions, distribution of nanofillers and types of polymer matrix employed. One of the main features of polymer nanocomposites is the capability of tailoring their performances based upon the properties of their parent components. A wide variety of polymers ranging from brittle glassy thermoplastics and epoxies to tough elastomers can be selected as matrices of the nanocomposites. This chapter reviews the influences of nanotube dispersion on the deformation behavior of CNT/polymer nanocomposites having different polymer matrix materials, and the mechanisms responsible for mechanical deformation of such materials.

10.2. Deformation characteristics

The mechanical reinforcement of CNT/polymer nanocomposites relies mainly on an effective load-transfer between the polymer matrix and the nanotubes. Several factors such as aspect ratio, homogeneous dispersion and alignment of nanotubes play decisive roles for effective reinforcement. The lengths of CNTs synthesized by chemical vapor deposition (CVD) vary from several micrometers up to millimeters, forming a coiled feature that is undesirable for practical applications. Both physical and chemical approaches have been adopted to reduce the length of CNTs to certain values that are suitable for blending. The physical dispersion route includes ball milling, grinding and high speed shearing [28,29]. Such treated CNTs generally have a large aspect ratio, typically in the range of ∼1000–2000. Chemical treatment of CNTs in a concentrated sulfuric/nitric mixture solution also activates the nanotubes at the open-ends to form carboxyl functional groups. This in turn improves the dispersion of nanotubes in polymers. The CNTs are known to orient randomly within the polymer matrix during the composite fabrication. The physical and mechanical properties of the CNT/polymer nanocomposites can be further enhanced by aligning the nanotubes. Alignment of CNTs in the matrix can be achieved by melt drawing, electrospinning and a range of force field methods such as electric and magnetic [30,31].

There are three main *mechanisms of load transfer* from the polymer matrix to nanotubes [32]. These are weak van der Waals bonding between the fillers and the matrix, micromechanical interlocking as well as chemical bonding between the nanotubes and the matrix. Molecular dynamics simulations show that shear strength and critical length required for effective load transfer in the CNT/polymer nanocomposites can be enhanced significantly by forming cross-links *via* chemical functionalization. The shear strength of a functionalized CNT/polymer composite can be increased by over an order of magnitude in comparison to weak nonbonded interactions. Such enhancement arises from

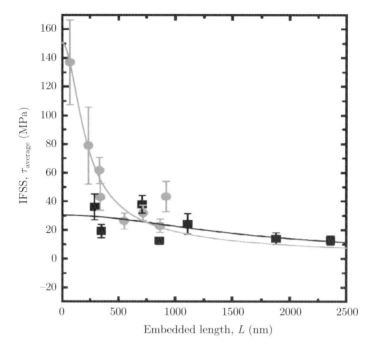

Figure 10.1. The average interfacial shear strength vs. embedded length for pristine and chemically modified nanotubes pulled from the epoxy matrix. The symbols indicate pristine nanotube pullout (■) and modified nanotube pullout (●) [34]

the introduction of chemical bonds between the nanotubes and the matrix [33]. More recently, Wagner and coworkers [34] determined experimentally the interfacial strength of epoxy composites reinforced with pristine and chemically modified MWNTs (Figure 10.1). The interfacial strength was evaluated by pulling individual nanotubes, embedded at different lengths, out of the polymer matrix using atomic force microscopy. Both pristine and modified show an average interfacial shear strength (IFSS) that decreases with increasing embedded length of nanotubes in epoxy resin. This implies that most of the interfacial shear stresses develop at the ends of the nanotubes.

The CNTs generally exhibit well-defined Raman peaks. It is possible to use Raman peak shift to characterize the load transfer mechanism of the CNT/polymer composites. When a strain is applied to a material, the interatomic distances change, leading to variations in the frequency of vibrational modes. Accordingly, Raman spectroscopy can provide useful information related to the load transfer between the polymer matrix and CNTs. The Raman spectrum of SWNTs generally shows characteristic peaks located at 1350 cm^{-1} (D band), ~1550–1605 cm^{-1} (G band) and ~2700 cm^{-1} (D* band) [35–37]. The D-band derives from the disorder-induced mode and its second-order harmonic is D* (G′) band. The G-band is associated with the graphite-like in-plane mode. Dresselhaus *et al.* have provided an in-depth review of the Raman spectra of isolated SWNTs [36]. Generally, the G-band for SWNTs is a doublet, having peak positions at 1553 and 1589 cm^{-1}, respectively. The peak at 1553 cm^{-1} is normally considered as a shoulder peak, indicating the presence of SWNTs. However, the G-band for MWNTs is a single peak at 1589 cm^{-1}.

10.2.1. *CNT/glassy thermoplastic nanocomposites*

Glassy polymers such as polystyrene (PS), poly(methyl methacrylate) (PMMA) and polycarbonate (PC) are widely used in many industrial applications owing to their superior transparency, high stiffness and relative ease of processing. However, PS and PMMA are brittle materials that fail at small strains. PC displays ductile mechanical behavior but is highly notch-sensitive. A large number of studies have attempted to improve the mechanical properties of glassy polymers by adding nanofillers of different shapes and aspect ratios [3]. The additions of CNT to glassy thermoplastics can lead to improvements in tensile stiffness and strength, or even ductility, depending on the dispersion of CNTs in the polymer matrix. This in turn relates to the processing methods used to fabricate the polymer nanocomposites. Qian *et al.* used ultrasonic assisted solution blending to prepare the MWNT/polystyrene nanocomposite film [38]. Tensile tests showed that the addition of 1 wt% pristine MWNT to PS results in 36–42% increase in the elastic modulus and about 25% increase in the tensile strength. Moreover, TEM observation revealed that nanotube fracture and pullout are responsible for the failure of the composite. The fracture of MWNTs in a PS matrix demonstrates that a certain degree of load transfer has taken place. However, the pullout of MWNTs from the PS matrix implies that the interfacial strength is not strong enough to resist debonding of the fillers from the matrix. Wong *et al.* indicated that the interfacial stress of solution blended MWNT/PS nanocomposite arises from mechanical interlocking of atoms at nanometer scale [39]. It is considered that local non-uniformity of MWNTs embedded in the polymer matrix such as kinks and bends contribute to the CNT-polymer adhesion by mechanical locking of atoms as shown by transmission electron microscopy (TEM) (Figure 10.2). To further enhance the interaction between pristine CNTs and polymer, Zhang *et al.* combined both solution mixing and melt blending to prepare the MWNT/PS nanocomposites. The precipitate from the MWNT/PS suspension was dried and then melt mixed with MWNTs [40]. They reported that the melt mixing led to enhanced interaction between PS and MWNTs *via* chemical bonding.

More recently, Lee *et al.* also used combined solution mixing and melt blending, *i.e.,* injection molding to fabricate the MWNT/PMMA nanocomposites [41]. This process allows elimination of shortcomings of the solution mixing such as difficulty of solvent removal and re-aggregation of CNTs after solvent evaporation. Figure 10.3 shows tensile strength *vs.* MWNT content plot for the MWNT/PMMA nanocomposites prepared by the combined process. The tensile strength of the nanocomposites prepared solely by injection molding is also shown for comparison purposes. Apparently, the tensile strength of the nanocom-

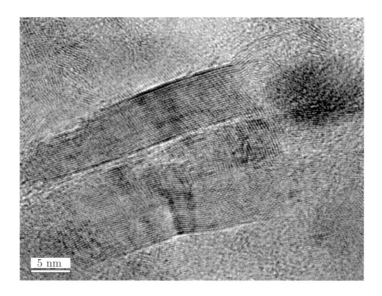

Figure 10.2. TEM micrograph showing details of the PS-MWNT interface. The kink and change in diameter of the MWNT is believed to promote mechanical interlocking [39]

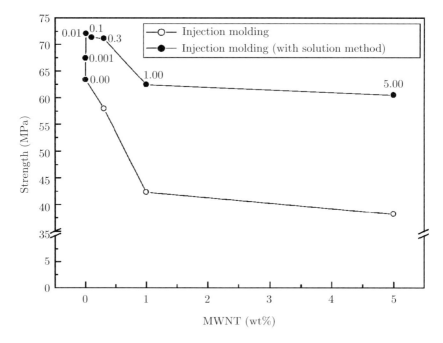

Figure 10.3. Comparison of strength of MWNT/PMMA nanocomposites prepared by different methods [41]

posites prepared by the combined process is much higher than that fabricated by injection molding. For the former composites, small MWNT additions from 0.001 to 0.01 wt% are beneficial in improving the tensile strength of PMMA. On the contrary, the tensile strength of injection molded nanocomposites decreases with increasing filler content as expected. Melt blending is generally less effective at dispersing CNTs in polymers and is confined to lower nanotube contents due to high viscosities of the composites at higher filler loadings. As aforementioned, functionalized CNTs or surfactant additions enable a better dispersion of nanotubes during the composite fabrication. The effect of surfactant sodium dodecyl sulfate (SDS) additions on the tensile strength of the 0.01 wt% MWNT/PMMA nanocomposite is shown in Figure 10.4. It is obvious that the incorporation of 0.01 wt% SDS can further enhance the tensile strength of the 0.01 wt% MWNT/PMMA from 72 to 73 MPa.

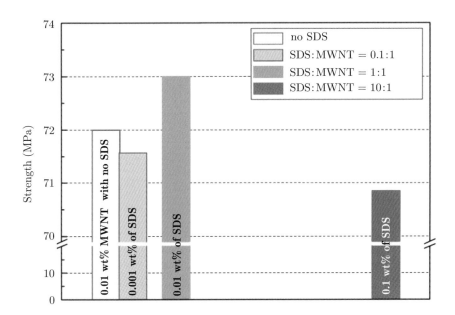

Figure 10.4. Changes of strength with MWNTs/SDS concentrations [41]

In general, oriented CNTs in polymer matrix can yield better mechanical performances compared to randomly oriented nanotubes. Gorga and Cohen studied the effect of nanotube orientation on the mechanical properties of extruded MWNT/PMMA nanocomposites [31]. The nanotubes were oriented by drawing the extrudates at different draw ratios. Figure 10.5 shows the effect of extrudate ratios on the tensile toughness and modulus for PMMA and 1% MWNT/PMMA nanocomposite. The tensile toughness is determined from the area of the stress-strain curve. It can be seen that the tensile toughness and modulus increase with increasing draw ratio for both PMMA and filled

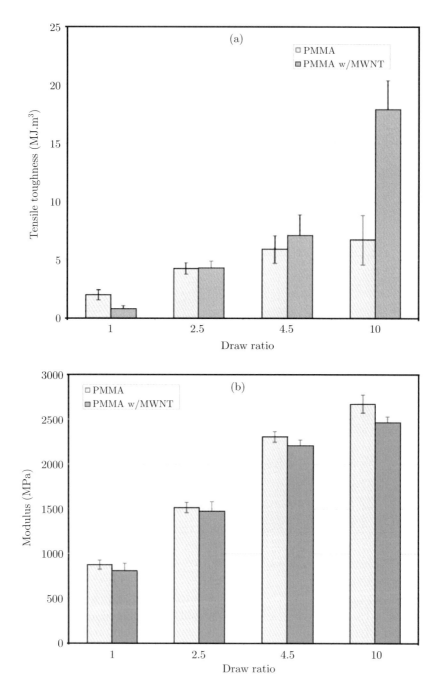

Figure 10.5. The effect of extrudate draw ratio on (a) tensile toughness and (b) modulus for PMMA and PMMA-1wt%MWNT samples [31]

composite. The addition of 1 wt% MWNT to PMMA (oriented nanocomposite) leads to the largest increase in tensile toughness with a 170% improvement over oriented PMMA. The tensile toughness of the nanocomposite is lower than that of PMMA under a draw ratio of 1; it is equivalent to PMMA at both draw ratios of 2.5:1 and 4.5: 1. The tensile toughness of the nanocomposite improves significantly under a draw ratio of 10:1. At such a high draw ratio, the nanotubes tend to bridge the microcracks in the nanocomposites during tensile loading as evidenced from the scanning electron microscopy (SEM) fractographs.

The *in situ* polymerization method is anticipated to induce strong covalent bonding between functionalized nanotubes and the polymer matrix, leading to large interfacial shear strength. Jia *et al.* synthesized MWNT/PMMA nanocomposites *via in situ* polymerization using a radical initiator, 2,2'-azobisisobutyronitrile (AIBN) [15]. Both pristine and acid-treated nanotubes were used as nanofillers. They demonstrated that the π-bonds of nanotubes could be open by the initiator. The nanotubes then participate in PMMA polymerization. A bonding between the MWNTs and the PMMA matrix is then formed. Velasco-Santos *et al.* also used AIBN initiator to incorporate both pristine and carboxyl functionalized MWNTs into PMMA matrix during *in situ* polymerization [42]. The tensile strength and stiffness of composites filled with 1 wt% pristine and functionalized MWNT (f-MWNT) are markedly improved compared to those of PMMA. The 1 wt% f-MWNT/PMMA nanocomposite exhibits higher tensile ductility than the composite filled with 1 wt% neat MWNT (Figure 10.6). For the composite with f-MWNTs, the storage modulus at 40°C is increased by

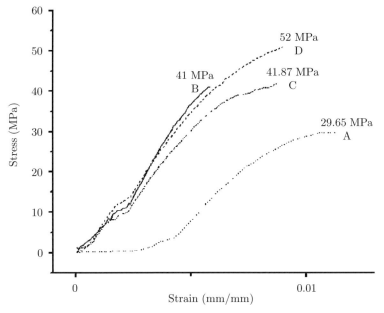

Figure 10.6. Stress strain curves: (A) – PMMA, (B) – 1 wt% pristine MWNT/PMMA, (C) – 1 wt% f-MWNT, and (D) – 1.5 wt% f-MWNT/PMMA specimens [42]

66% and the glass transition temperature (T_g) by about 40°C as compared to 50% and 6°C in the composite with pristine MWNTs (Figure 10.7).

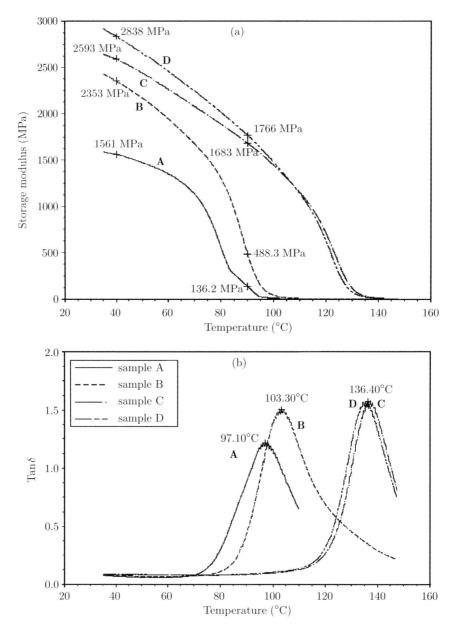

Figure 10.7. Dynamical mechanical analysis of PMMA and carbon nanotube composites: (a) storage modulus, and (b) loss tangent results: (A) – PMMA, (B) – 1 wt% pristine MWNT/PMMA, (C) – 1 wt% f-MWNT/PMMA, and (D) – 1.5 wt% f-MWNT /PMMA specimens [42]

Deformation of polymer/CNT nanocomposites 351

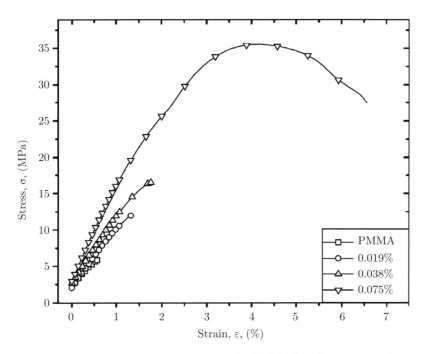

Figure 10.8. Representative stress-strain curves for PMMA-based composites for a range of nanotubes volume contents [43]

Very recently, Coleman and coworkers investigated the mechanical behavior of f-MWNT/PMMA nanocomposites prepared by *in situ* polymerization [43]. The stress-strain curves of neat PMMA and its nanocomposites are shown in Figure 10.8. It is interesting to see that small additions of f-MWNTs to PMMA are beneficial to enhance the tensile strength, stiffness and strain to break. Neat PMMA displays brittle behavior with a strain of break of 0.57%. Addition of 0.075 vol% MWNT to PMMA leads to a drastic increase of the strain of break. The variations of Young's modulus and mechanical strengths with volume fraction of MWNTs are shown in Figure 10.9 and 10.10, respectively. Both figures reveal that the Young's modulus, ultimate tensile strength and breaking strength increase linearly with volume fraction before falling off at higher volume fractions due to filler agglomeration. This trend is also observed in the plot of strain at break *vs.* filler volume fraction. Young's modulus increases from 0.71 GPa to 1.38 GPa by adding MWNT at a volume fraction of 1.5×10^{-3}. This corresponds to an improvement of 94%. Obvious improvements in ultimate tensile strength (360%), breaking strength (373%) and elongation at break (526%) are also associated with the MWNT additions. These enhancements in mechanical properties are attributed to a homogeneous dispersion of nanotubes in the polymer matrix, thereby promoting effective load transfer from the polymer to the nanotubes. SEM examination of the tensile fractured surface reveals the bridging of matrix cracks by MWNTs. In this respect, the nanotube

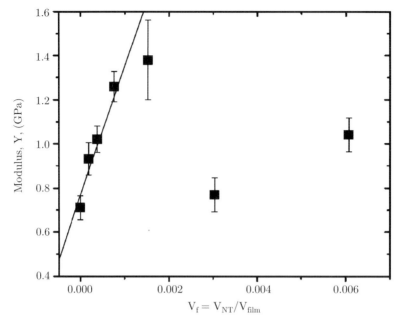

Figure 10.9. Young's modulus as a function of volume fraction of MWNTs for PMMA-based composites [43]

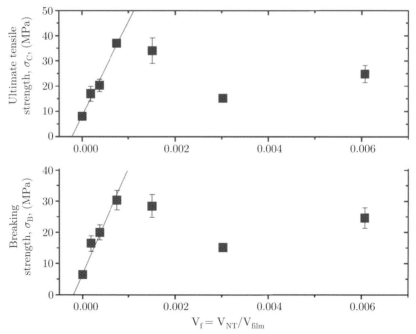

Figure 10.10. Ultimate tensile strength and breaking strength as a function of volume fraction of MWNTs for PMMA-based composites [43]

reinforcement also influences the crack-tip deformation mechanisms such as crack bridging and crack closure. This leads to the improvement of tensile ductility of brittle PMMA (Figure 10.8). In general, brittle materials such as ceramics and epoxies can be toughened by adding short fibers as a result of crack deflection and delamination mechanisms. In the absence of crack deflection, a crack can propagate very fast through the matrix without interruption when it encounters the fibers. Relatively little energy is dissipated during deformation, leading to little or no toughening. On the contrary, crack deflection *via* matrix-crack bridging by intact fibers dissipates energy considerably. This contributes to toughening of the composite materials [44]. It is worth noting that the so-called "sword-sheath" morphology or telescopic pullout of inner tube from MWNT has been observed in the f-MWNT/PMMA nanocomposite thin film subjected to applied stresses. This indicates the formation of strong interfacial bonding between the nanotubes and the polymer matrix as a result of chemical functionalization of the nanotubes, capable of transferring the stress load [45]. The telescopic pullout of nanotubes dissipates a considerable amount of energy and contributes to toughening of brittle PMMA.

In the case of tougher PC, crack bridging by intact nanotubes in the 0.06 wt% SWNT/PC nanocomposite prepared by solution blending has been reported. The SWNTs were purified in nitric acid [13]. Figure 10.11 is the SEM

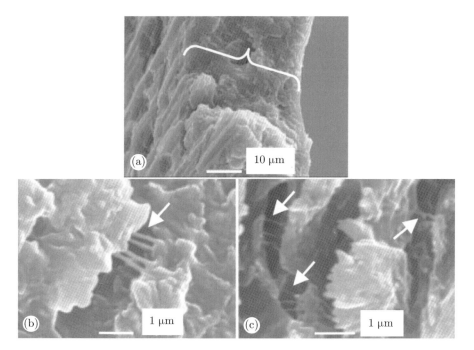

Figure 10.11. SEM images of SWNT/PC room temperature fracture: (a) low magnification showing the whole fracture surface; (b) and (c) close-up of embedded SWNTs indicated by arrows, aligned by the stretching process [13]

micrograph showing the SWNTs bridge the matrix cracks effectively. Moreover, the nanotube surfaces are covered with a polymer layer, indicating the formation of strong interactions between the nanotube and the polymer. Recent research by Schadler and coworkers [46–48] has also shown the formation of polymer sheaths in nanotubes of the MWNT/PC composites (Figures 10.12a–12c). To study this layer, an atomic force microscopy (AFM) tip of a nanomanipulator is brought into contact with the PC coated nanotubes (Figure 10.12a). After contact, the polymer coating balls up, indicating a strong adhesion of PC layer to the nanotube surface [46]. They subsequently modified the MWNT surfaces chemically by means of epoxide-based functional groups [47]. In the

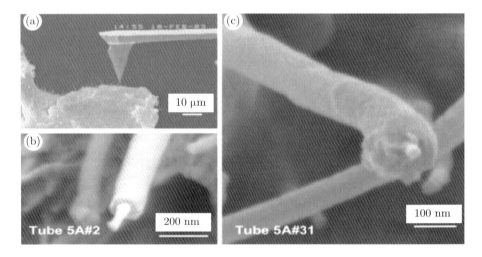

Figure 10.12. High-resolution SEM images of MWNT/PC fracture surface: (a) far-field image of the nanomanipulation experiment; (b, c) Nanotube structures coated with a polymer sheath protruding from the fracture surface [46]

process, the MWNTs were initially treated in nitric acid, forming carboxyl groups along their walls. The carboxylated nanotubes were further reacted with di-epoxide terminated resin as shown in Scheme 10.1. More recently, they fabricated the MWNT/PC nanocomposites using pristine MWNTs and f-MWNTs *via* combined solution blending and injection molding [48]. The tensile properties of these nanocomposites are listed in Table 10.1. The stiffness and yield stress of PC are improved by adding 2 and 5 wt% pristine MWNTs. A higher increase is observed for the surface modified MWNT. The stiffness is improved by ∼90% and the yield stress is enhanced by 32% at 5 wt% loading compared to pure PC. This demonstrates that there is a load transfer from the matrix to the MWNTs. They attributed such improvements in mechanical properties to the formation of immobilized polymer layer at the filler-matrix interface. The efficiency of the load transfer is improved by surface modification of the MWNT as evidenced by Raman spectroscopy (Figure 10.13). Raman

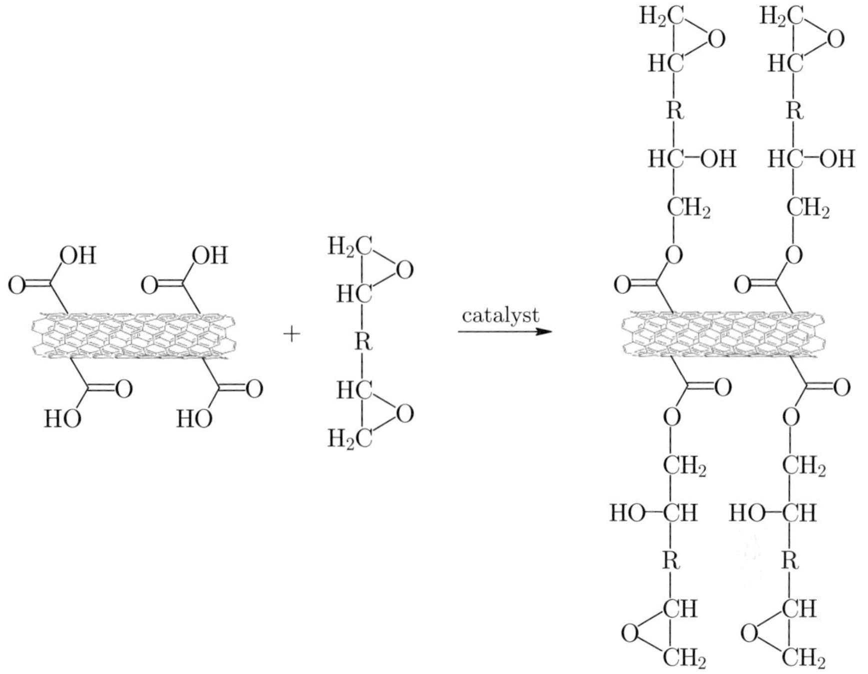

Scheme 10.1. Reaction scheme of carboxylated nanotube with a di-epoxide molecule [47]

Table 10.1. Comparison of mechanical properties for MWCNT/PC and f-MWCNT/PC nanocomposites [48]

Sample	Young's modulus (GPa)	Yield stress (MPa)	Yield strain (%)	Strain to failure (%)
Polycarbonate	2.0±0.1	59	6.5±0.2	>100
2 wt% -MWCNT	2.6±0.1	67	6.5±0.2	80±30
5 wt% -MWCNT	3.3±0.1	70	4.8±0.2	25±10
2 wt% f-MWCNT	2.8±0.1	69	5.0±0.2	50±20
5 wt% f-MWCNT	3.8±0.1	78	4.0±0.2	10±3

spectroscopy is an effective tool to analyze microscopic load transfer from the matrix to the CNTs. It has been demonstrated that the position of the second-order Raman band at 2610 cm^{-1} (D*-band) is sensitive to stress in the CNTs. This peak tends to shift down in frequency when the nanotubes are subjected to tension [37]. From Figure 10.13, the shift of the Raman peak indicates that load is transferred from the matrix to the MWNT. The steeper slope of the f-MWNT/PC composite (6.0 cm^{-1}/%) compared to 3.4 cm^{-1}/% of the MWNT/PC composite reveals a more efficient load transfer for the epoxide-modified samples.

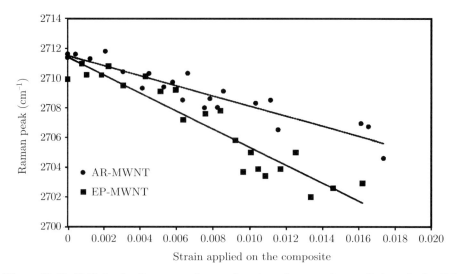

Figure 10.13. Shift in the Raman peak as a function of composite applied strain for PC nanocomposites filled with pristine (AR, %) and epoxide-modified MWNTs (EP, %) [48]

10.2.2. *CNT/semicrystalline thermoplastic nanocomposites*

Semicrystalline thermoplastics such as polyethylene (PE), polypropylene (PP), polyamide (PA), poly(ethylene terephthalate) (PET) and poly(butylene terephthalate) (PBT) are an important commodity and engineering thermoplastics widely used in the industrial sectors. Their mechanical properties can be further upgraded by adding flexible CNTs with large aspect ratios [49–57]. It is well recognized that the mechanical properties of a semicrystalline polymer are affected by its morphology, crystalline structure and the degree of crystallinity. The CNTs normally act as nucleating agents for semicrystalline polymers during the crystallization process, thereby forming finer spherulites [58,59].

The dispersion of CNTs in PE and PP is rather poor due to their polar nature, resulting in weak adhesion between the filler surface and the polymer matrix. Poor mechanical behavior is therefore expected for the PE nanocomposites filled with pristine MWNTs [60]. In this respect, solubilizing CNTs with suitable chemical groups attached to their surface can improve their dispersion in polar polyolefins. Chemical modification of the CNTs sidewalls by fluorination is an attractive method because it does not require a solvent for the functionalization reaction to take place [61]. Studies carried out by Barrera's group have explicitly shown that when the SWNT is fluorinated, an improved interfacial bonding between the SWNT and polyolefins is obtained [49,50]. The fluorinated SWNT/PP nanocomposites were prepared by shear mixing (Scheme 10.2). The partial removal of a fluorine moiety from the f-SWNTs during shear mixing with polyolefins provides the opportunity for direct covalent bonding between the nanotubes and the polymers. This results in remarkable increases in storage modulus, Young's modulus, and tensile strength as compared with the com-

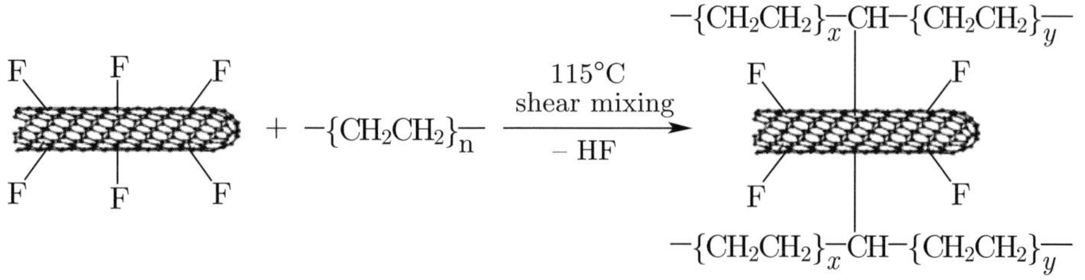

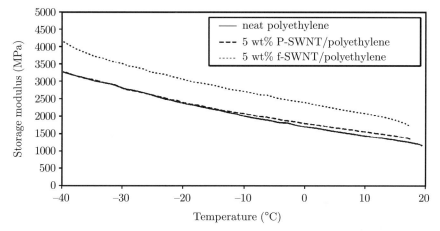

Scheme 10.2. Proposed covalent bonding of polyethylene to f-SWNTs during shear mixing processing [49]

posites prepared from pristine CNTs (Figures 10.14 and 10.15). However, the tensile ductility of the f-SWNT/PP nanocomposite fiber is inferior to that of pure PP (Figure 10.15c).

Figure 10.14. Storage modulus vs. temperature for neat polyethylene, 5 wt% SWNT/PE and 5 wt% f-SWNT/PE nanocomposites [49]

The fluorinated nanotubes can be further reacted with alkyl lithium, and the resulting products can be used as initiators of anionic polymerization for *in situ* synthesis of PS-grafted nanotube composites [62,63]. Using a similar approach to modify the nanotubes, Blake *et al.* prepared ultra strong and tough f-MWNT/PP nanocomposites [51]. In the process, *n*-butyllithium-functionalized MWNTs reacted with chlorinated polypropylene (CPP) to give nanotubes covalently bonded to CPP (CPP-MWNT) *via* a coupling reaction with elimination of LiCl (Scheme 10.3). The CPP-MWNT was then dispersed in tetrahydrofuran, blended with a CPP polymer solution followed by casting to form composite disks. Figure 10.16 shows the stress-strain curves of CPP and its nanocomposites. It is obvious that the elastic modulus, yield strength, ultimate tensile strength and strain to failure of CPP improve dramatically with increasing nanotube volume content. The largest improvement in these properties can be observed for the nanocomposite with 0.6% filler.

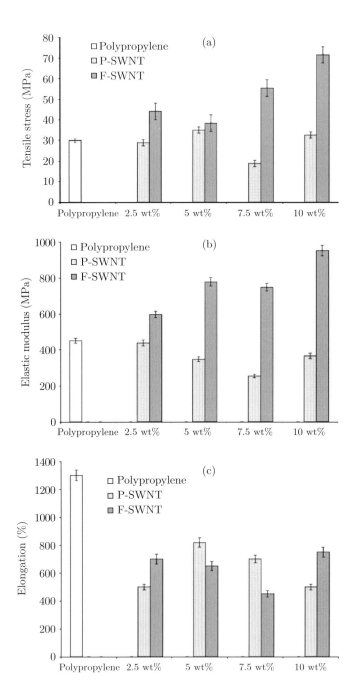

Figure 10.15. Comparison of tensile strengths (a), elastic moduli (b) and percent elongations (c) for 130 μm fibers of PP, SWNT/PP and f-SWNT/PP [50]

Deformation of polymer/CNT nanocomposites 359

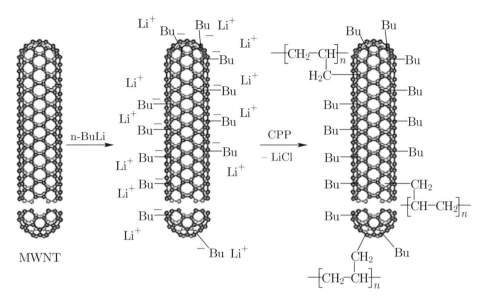

Scheme 10.3. Schematic presentation for functionalization of MWNTs with chlorinated polypropylene (CPP) [51]

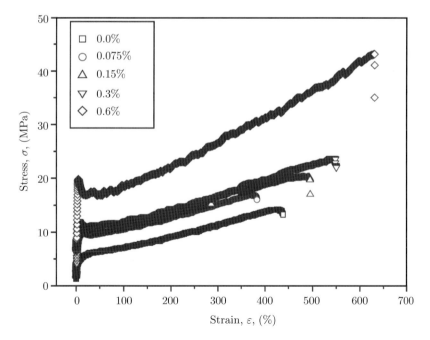

Figure 10.16. Stress-strain curves of CPP and the f-MWNT/CPP nanocomposites containing different nanotube volume contents [51]

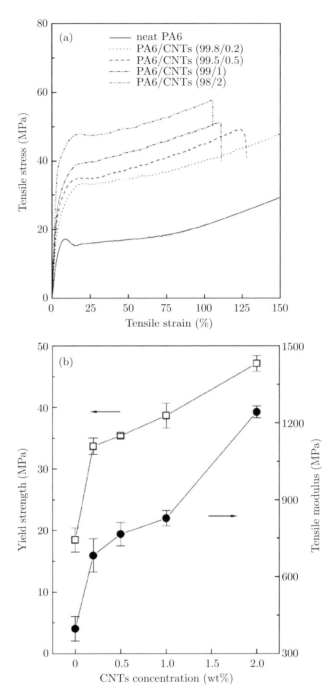

Figure 10.17. Typical stress-strain curves (a), yield stress and tensile modulus (b) for PA6 and its nanocomposites as a function of MWNTs concentration [55]

Liu *et al.* studied the morphology and mechanical properties of the melt-compounded f-MWNT/PA6 nanocomposites [55]. Functionalized carboxyl and hydroxyl groups were bonded to MWNT surfaces by means of acid treatments. Figure 10.17a shows the stress-strain curves for PA6 and its nanocomposites. The variations of tensile modulus and yield strength with the MWCNT content are depicted in Figure 10.17b. At 1 wt% MWNTs, the tensile modulus and the tensile strength are greatly improved by ~115% and 120%, respectively, compared to the neat PA6. The strain to failure reduces from 150% to 115%. However, the significant increases in ultimate and yield stresses combined with a small decrease in tensile elongation leads to an apparent increase in toughness as characterized by the area of the stress-strain curve. The improvements of these mechanical properties are attributed to a better dispersion of MWNTs within the PA6 matrix and to a strong interfacial adhesion between the nanofillers and the PA6 matrix. Furthermore, the MWNTs are sheathed by several small polymer beads. Such polymer sheathed nanotubes bridge the matrix cracks effectively during tensile loading (Figure 10.18).

The improvements in tensile elongation and toughness of glassy and semicrystalline thermoplastics reinforced with low loadings of functionalized carbon nanotubes are very impressive. An important issue for toughening in functionalized CNT-reinforced polymer composites is the nature of the interface between the nanotube and the matrix as discussed above. The high flexibility and large

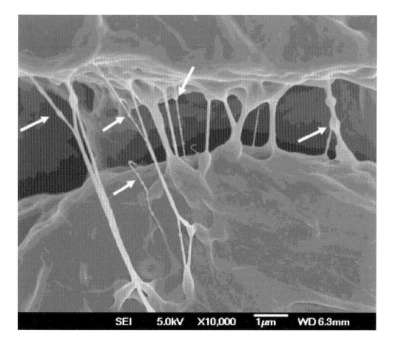

Figure 10.18. SEM image showing microcracks linked by stretched nanotubes. Some MWNTs are sheathed by several small polymer beads as indicated by the arrows [55]

aspect ratio of functionalized nanotubes can lead to crack deflection *via* matrix-crack bridging during mechanical loading. The bridging of the nanotubes restrains the crack from opening and reduces the driving force for crack propagation in the polymer matrix. This bridging behavior is believed to be the main source of energy dissipation, and contributes to toughening of functionalized CNT/polymer nanocomposites. Another possible toughening mechanism is the nanotube pullout effect [34,64–66]. The work required for pulling the embedded broken nanotubes out against the interfacial adhesion and frictional stresses at the nanotube/matrix region can also contribute to toughness of the polymer nanocomposites [64]. For weak interfacial CNT-polymer bonding, pullout of the nanotubes from the matrix occurs more readily. For a strong interfacial bonding, telescopic pullout associated with fracture of the outer layer and telescopic pullout of the inner tube of CNT would occur [67]. Molecular dynamics has been used extensively to study nanotube deformation and pullout from the polymer matrix [33,64–66]. Frankland and Harik analyzed the pullout force of CNT from a polymer matrix by the interface friction model based on the sliding of a nanotube and the surrounding polymer chains [64]. Qian *et al.* indicated that the surface tension forces like friction and adhesion and intertube corrugation are the main factors governing the interfacial strength [66]. Chowdhury and Okabe demonstrated that the presence of chemical cross-links in the nanotube-matrix interface increases the ISS of CNT/PE composites [65]. Frankland *et al.* also reported large increase in the ISS of the CNT/polymer composite due to addition of cross-links at the interface [33]. It seems that a strong nanotube-matrix bonding is essential to improve the toughness of the CNT/polymer nanocomposites.

The deformation behavior of CNT/thermoplastic nanocomposites is now compared and contrasted with that of conventional polymer microcomposites reinforced with short fibers. In general, fiber pullout dissipates energy substantially in conventional polymer microcomposites during mechanical loading. Thus, fiber pull-out is mainly responsible for toughening of conventional microcomposites. Moreover, a strong interfacial fiber-matrix bonding can have an adverse effect on fracture toughness of microcomposites as it restrains the pullout of fibers. This means that weak interfacial bonding favors fiber pullout during mechanical deformation [68,69]. It is worth noting that conventional microcomposites are reinforced with a large volume content of stiff short fibers, ca 25–30% in comparison with the CNT/polymer nanocomposites having very low loading levels of nanofillers capable of stretching during deformation. The flexible nanotubes can bridge the matrix cracks readily under a strong interfacial bonding condition. It appears explicitly that crack bridging, pullout or sword-sheath deformation could occur in CNT/thermoplastic nanocomposites, depending greatly on the interfacial bonding.

10.2.3. *CNT/epoxy nanocomposites*

Epoxy resin is widely used in industrial sectors because of its low shrinkage in cure, excellent dimensional stability, outstanding adhesion and good corrosion resistance. One of the disadvantages in the use of epoxy for structural applications is its brittle nature. The toughness of brittle epoxies can be improved by adding rigid particles, elastomers and nanoparticles. The nanoparticles are of particular interest because only low loading levels of nanofillers are needed to achieve desired mechanical properties. Recently, several researchers attempted to enhance the strength, stiffness and even the toughness of epoxy resin by adding suitable functionalized CNTs [70–75]. For example, Zhu et al. [70] improved the dispersion of SWNTs in an epoxy matrix through functionalization of the nanotubes in mixed sulfuric/nitric acid (3:1 volume ratio) followed by fluorination. The combination of acid treatment and fluorination caused both end-tip and sidewall functionalization (Scheme 10.4). Carboxyl and fluorine groups covalently attached to the nanotubes offer the opportunity

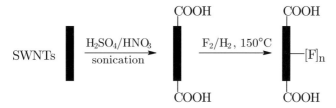

Scheme 10.4. Functionalization of single-walled carbon nanotubes [70]

for chemical interactions with the epoxy. The carboxyl moieties can react directly with the epoxy to form esters. The fluoronanotubes can also react with diamine agent during the high temperature curing process (Scheme 10.5). As a result, functionalized nanotubes can disperse homogeneously within the epoxy matrix.

$$\text{F-SWNT} + \text{H}_2\text{N-Y-NH}_2 \xrightarrow{\text{Py, 130°C}} \text{SWNT-NH-Y-NH}_2$$

Scheme 10.5. Reaction of fluoronanotube with a diamine curing agent [70]

This leads to enhanced mechanical strength and stiffness of functionalized nanocomposite over the neat epoxy (Figure 10.19). The epoxy composite with 1 wt% f-SWNT exhibits a tensile strength of 95 MPa and a modulus of 2632 MPa, showing 18 and 24% improvements over the epoxy composite filled with pristine SWNT. However, the elongation at break of the functionalized epoxy composite is slightly lower than that of neat epoxy.

To improve the tensile toughness, Tseng et al. used plasma treatment to graft maleic anhydride (MA) molecule onto the MWNTs, yielding functionalized nanotubes, *i.e.*, CNTs-MA [76]. The functionalized nanotubes can readily react with diamine curing agent followed by further chemical bonding with the epoxy *via* an amidation reaction. This leads to the formation of a highly cross-linked structure (Scheme 10.6). Figures 20a–20e show SEM images of fractured surfaces of the CNTs-MA/epoxy nanocomposites loaded with different MWNT contents. A homogenous dispersion of nanotubes can be readily observed in

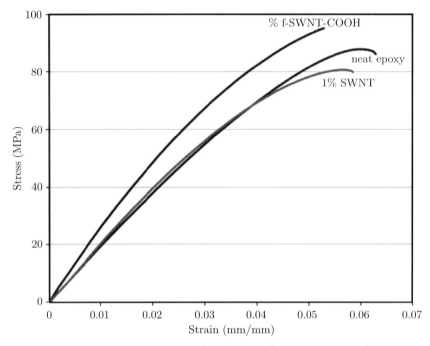

Figure 10.19. Tensile stress-strain curves for nanotube/epoxy composites [70]

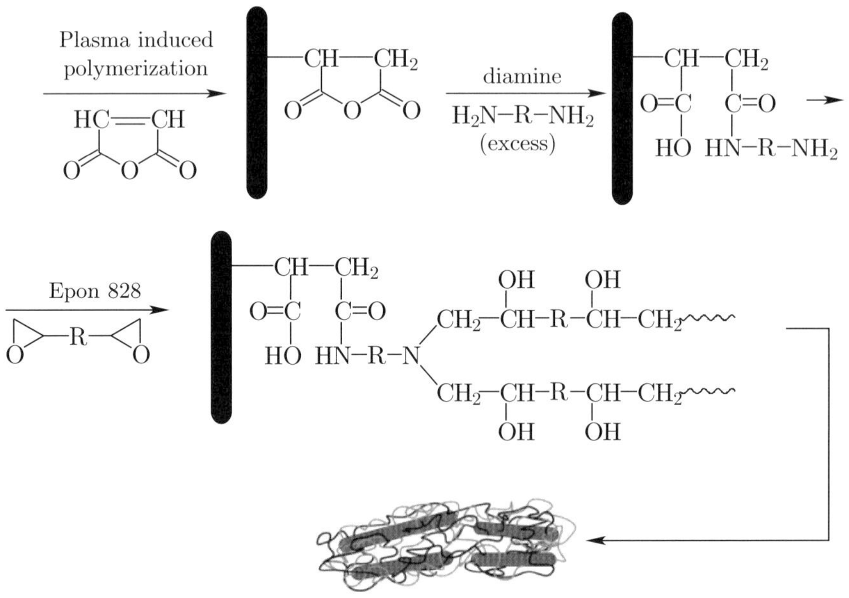

Scheme 10.6. Overall procedure for preparation of CNTs-MA/epoxy nanocomposites [76]

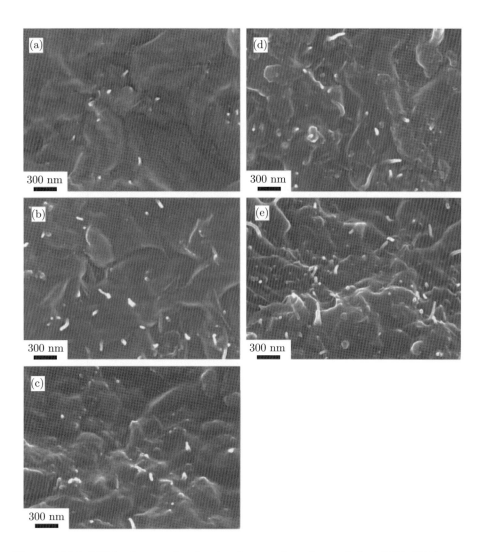

Figure 10.20. SEM micrographs showing fractured surfaces of the CNTs-MA/epoxy nanocomposites filled with: (a) – 0.1 wt%, (b) – 0.3 wt%, (c) – 0.5 wt%, (d) – 0.7 wt% and (e) – 1.0 wt% functionalized nanotubes [76]

these micrographs. The tensile properties of the CNTs-MA/epoxy nanocomposites as a function of filler content are shown in Figures 10.21a–21c. For the purposes of comparison, the tensile properties of nanocomposites filled with untreated MWNTs are also shown. It can be seen that the tensile strength, elongation at break and tensile modulus of functionalized nanocomposites increase with increasing filler content. In sharp contrast, the tensile strength and elongation at break of untreated epoxy nanocomposites decrease with increasing filler content as expected.

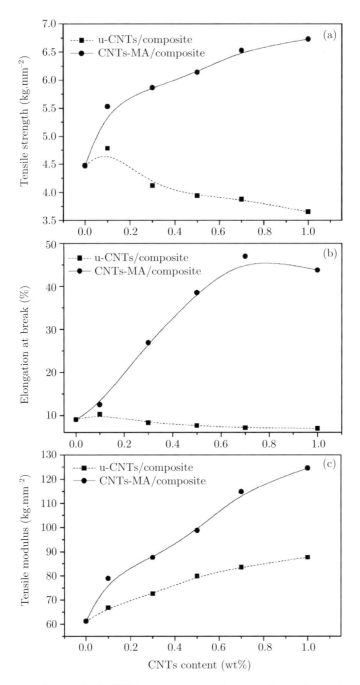

Figure 10.21. Effect of the MWNT content on tensile strength (a), elongation at break (b) and tensile modulus (c) of the epoxy nanocomposites reinforced with untreated (u-CNT) and funtionalized nanotubes (CNTs-MA) [76]

More recently, Kim et al. investigated the effects of surface modification on the mechanical properties of MWNT/epoxy nanocomposites [75]. The MWNTs were treated with an acid solution to remove impurities and modified subsequently by amine treatment or plasma oxidation to improve interfacial bonding and dispersion of nanotubes in the epoxy matrix. Tensile measurements revealed that the modified MWNT composites exhibit higher stiffness, tensile strength and strain to failure than the epoxy resin (Figure 10.22 and Table 10.2). The surface modified MWNTs are well dispersed in the epoxy matrix and have strong interfacial bonding with the matrix. This leads to the plasma treated

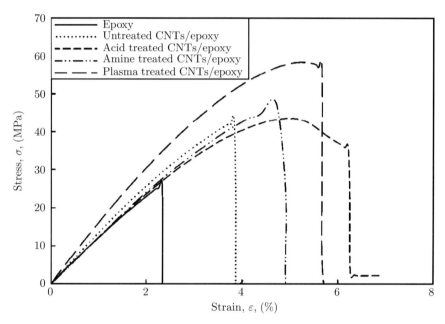

Figure 10.22. Stress-strain curves of cure epoxy and its nanocomposites containing 1 wt% untreated and surface modified MWNTs [75]

Table 10.2. Tensile properties of neat epoxy, untreated and functionalized CNT/epoxy composites [75]

Sample	Young's modulus (GPa)	Tensile strength (MPa)	Elongation at break (%)
Epoxy	1.21	26	2.33
Untreated CNT/epoxy	1.38	42	3.83
Acid treated CNT/epoxy	1.22	44	4.94
Amine treated CNT/epoxy	1.23	47	4.72
Plasma treated CNT/epoxy	1.61	58	5.22

MWNT/epoxy nanocomposite demonstrating the most significant improvements in tensile stiffness, strength and toughness.

Schulte and coworkers used TEM to observe the crack morphologies developed in carboxyl- and amino-functionalized MWNT/epoxy composites [71]. TEM observation revealed typical nanotube pullout morphology in the carboxyl-functionalized composite, indicating a weak interfacial bonding between the nanotube and the epoxy resin. However, crack bridging of the MWNT can be readily seen in the TEM micrograph of amino-functionalized MWNT/epoxy composite (Figure 10.23a). Furthermore, telescopic pullout of the outer tube layer associated with improved interactions has occurred. (Figure 10.23b). The amino functional group is highly reactive and reacts readily with epoxide resin, leading to the opening of the epoxide ring to form a cross-link. Accordingly, the MWNTs are incorporated into the epoxy resin and become an integral part of the composite. Telescopic pullout is caused by the fracture of the outer MWNT layer due to a strong interfacial layer and pullout of the inner tube. The outer shell of the tube often remains embedded in the matrix following pullout. These effects render the amino-functionalized CNT/epoxy nanocomposites exhibiting good tensile strength, stiffness and elongation as well as excellent fracture toughness (K_{IC}) [67]. On the basis of these results, a schematic diagram showing possible nanomechanical deformation mechanisms of the CNT embedded in an epoxy resin is proposed. Several mechanisms, *e.g.*, pullout, rupture or crack bridging of nanotubes could occur, depending on the nature of interfacial adhesion (Figure 10.24).

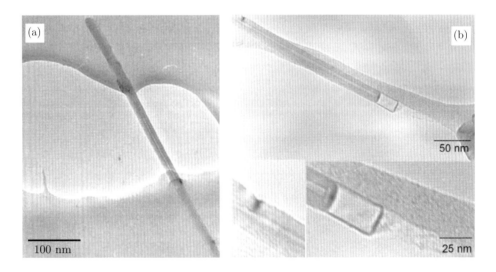

Figure 10.23. TEM images of amino-functionalized MWCNT/epoxy composites showing crack bridging (a) and telescopic pullout (b) of nanotubes [71]

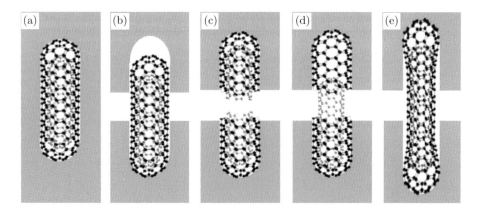

Figure 10.24. Schematic description of possible fracture mechanisms of CNTs: (a) – ideal state of the CNT; (b) – pullout caused by CNT/matrix debonding in case of weak interfacial adhesion; (c) – rupture of CNT- strong interfacial adhesion in combination with extensive and fast local deformation; (d) – telescopic pullout-fracture of the outer layer due to strong interfacial bonding and pullout of the inner tube and (e) – bridging and partial debonding of the interface – local bonding to the matrix enables crack bridging and interfacial failure in the non-bonded regions [67]

10.2.4. *CNT/elastomer nanocomposites*

Elastomers generally exhibit low mechanical strength and stiffness. In this respect, high loadings of microparticles such as carbon blacks and metal flakes are incorporated into elastomers to improve their mechanical and electrical properties. In this respect, organoclay and alumina nanoparticles have been considered as substitutes of carbon blacks in elastomeric materials recently [77,78]. The interest in using CNTs as reinforcing fillers in elastomers arises from their remarkable electrical and mechanical properties. Among various types of elastomers, polyurethane (PU) is particularly attractive because it is widely used in coatings, adhesives and biomaterials. PU copolymer consists of hard segments based on diisocyanate and low molecular weight diol or diamine, and soft segments having a high molecular weight polyester or polyether diol. These segments aggregate into microdomains resulting in a structure consisting of glassy and hard domains, as well as rubbery and soft domains, having their glass transition temperatures below and above room temperature. The extent of hard segments primarily determines the modulus of PU.

Xiong *et al.* prepared a 2 wt% f-MWNT/PU nanocomposite by solution mixing [79]. They reported that the tensile stress at break of the nanocomposite is considerably higher than that of a neat PU at the expense of the elongation at break. Xia and Song synthesized polyurethane grafted SWNT (SWNT-g-PU) through a two-step reaction [80]. The SWNT-g-PU/polycaprolactone diol dispersion was obtained by ball milling, and SWNT-g-PU/PU nanocomposites were prepared by further *in situ* polymerization. At 0.7 wt% SWNT-g-PU con-

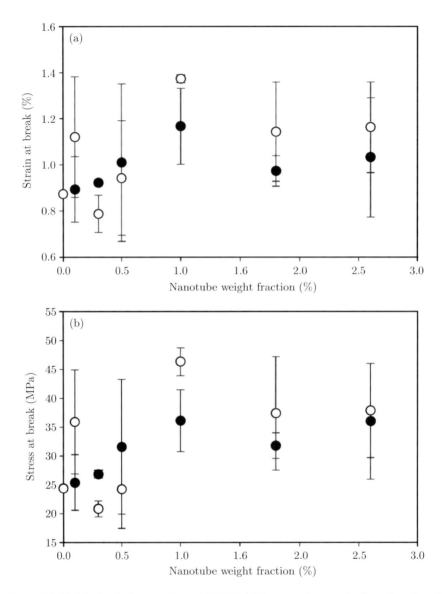

Figure 10.25. Mechanical properties of SWNT/PU composites made from functionalized (•) and unfunctionalized (○) nanotubes [81]

centration, Young's modulus increases from 7.94 MPa (neat PU) to 22.10 MPa, which is 278% improvement over neat PU. The elongation at break is nearly unchanged for the nanocomposite specimens (Figure 10.25). Very recently, Buffa et al. introduced pristine and carboxyl-functionalized SWNTs into polyurethane with a high hard-segment content by solution mixing [81]. The modulus, tensile strength and strain at break tend to increase with increasing nanofiller content up to 1 wt%. It can be concluded that the nanotube additions are also beneficial in improving the mechanical performances of elastomers.

10.3. Conclusions

The deformation behavior of CNT/polymer nanocomposites is summarized and analyzed in this chapter. At present, CNT is one of the most promising reinforcing nanofillers for high performance polymer nanocomposites. Of great importance is the ability of CNT to blend with a wide variety of polymers, including semi-crystalline and glassy thermoplastics, epoxies and elastomers. Nanocomposites with high elastic modulus, high mechanical strength and good ductility are important in aerospace, automotive and naval applications. However, efforts to improve the mechanical properties of CNT-based nanocomposites are essential for these applications. A homogeneous dispersion of nanotubes in the polymer matrix and effective load transfer from the matrix to the nanotubes are the primary key parameters in achieving superior mechanical properties of the nanocomposites. This can be realized *via* chemical functionalization of the CNTs as well as proper selection of suitable processing techniques. Appropriate functionalization of nanotube surfaces is found to enhance chemical interactions with the polymer molecules, to assist the dispersion and optimize the mechanical properties of the nanocomposites. *In situ* polymerization is considered to be an effective route to fabricate CNT/polymer nanocomposites with high tensile strength, stiffness and ductility due to a better dispersion of nanotubes in the polymer matrix. The main deformation mechanism responsible for such superior mechanical performances is the crack bridging effect of polymer sheathed nanotubes. In certain cases, telescopic pullout is also responsible for the strengthening and toughening of CNT/polymer nanocomposites having brittle polymer matrices (*e.g.*, PMMA and epoxy) but with a strong interfacial bonding.

References

1. Reichert P, Kressler J, Thomann R, Mulhaupt R and Stoppelmann G (1998) Nanocomposites based on synthetic layer silicate and polyamide-12, *Acta Polym* **49**:116–122.
2. Alexandre M and Dubois P (2000) Polymer-layered silicate nanocomposites: preparation, properties and uses of a new class of materials, *Mater Sci Eng R* **28**:1–63.
3. Tjong S C (2006) Structural and mechanical properties of polymer nanocomposites, *Mater Sci Eng R* **53**:73–197.
4. Reynaud E, Jouen T, Gauthier C, Vigier G and Varlet J (2001) Nanofillers in polymeric matrix: a study on silica reinforced PA6, *Polymer* **42**:8759–8768.
5. Musto P, Ragosta G, Scarinzi G and Mascia L (2004) Toughness enhancement of polyimides by in situ generation of silica particles, *Polymer* **45**:4265–4274.
6. Chen C H, Teng C C, Su S F, Wu W C and Yang C H (2006) Effects of microscale calcium carbonate and nanoscale calcium carbonate on the fusion, thermal and

mechanical characterizations of rigid poly(vinyl chloride)/calcium carbonate composites, *J Polym Sci Part B Polym Phys* **44**:451–460.
7. Ash B J, Rogers D F, Wiegand C J, Schadler L S, Siegel R W, Benicewicz B C and Apple T (2002) Mechanical properties of Al_2O_3/polymethylmethacrylate nanocomposites, *Polym Eng Sci* **23**:1014–1025.
8. Cotterell B, Chia J Y and Hbaieb K (2007) Fracture mechanisms and fracture toughness in semicrystalline polymer nanocomposites, *Eng Fract Mech* **74**:1054–1078.
9. Treacy M M, Ebbesen T W and Gibson J M (1996) Exceptional high Young's modulus observed for individual carbon nanotubes, *Nature* **381**:678–680.
10. Tibbetts G G (1989) Vapor–grown carbon fibers: status and prospects, *Carbon* **27**:745–747.
11. Hammel E, Tang X, Trampert M, Schmitt T, Mauthner K, Eder A and Potschke P (2004) Carbon nanofibers for composite applications, *Carbon* **42**:1153–1158.
12. Benoit J M, Corraze B, Lefrant S, Blau W J, Bernier P and Chauvet O (2001) Transport properties of PMMA-carbon nanotube composites, *Synth Met* **121**:1215–1216.
13. Singh S, Pei Y, Miller R and Sundararajan P (2003) Long-range, entangled carbon nanotubes networks in polycarbonate, *Adv Funct Mater* **13**:868–871.
14. Badaire S, Poulin P, Maugrey M and Zakri C (2004) In situ measurements of nanotubes dimensions in suspensions by depolarized dynamic light scattering, *Langmuir* **20**:10367–10370.
15. Jia Z, Wang Z, Xu C, Liang J, Wei B, Wu D and Zhu S (1999) Study on poly(methyl methacrylate)/carbon nanotube composites, *Mater Sci Eng A* **271**:395–400.
16. Haggenmueller R, Du F, Fischer J E and Winey K I (2006) Interfacial in situ polymerization of single wall carbon nanotube/nylon 6,6 nanocomposites, *Polymer* **47**:2381–2388.
17. Zhao C, Hu G, Justice R, Schaefer D W, Zhang S, Yang M and Han C C (2005) Synthesis and characterization of multi-walled carbon nanotubes reinforced polyamide 6 via in situ polymerization, *Polymer* **46**:5125–5132.
18. Park C, Qunaies Z, Watson K A, Crooks R E, Joseph S J, Lowther S E, Connell J W, Siochi E J, Harrison J S and Clair T L (2002) Dispersion of single wall carbon nanotubes by in situ polymerization under sonication, *Chem Phys Lett* **364**:303–308.
19. Vaisman L, Maron G and Wagner H D (2006) Dispersions of surface-modified carbon nanotubes in water-soluble and water-insoluble polymers, *Adv Funct Mater* **16**:357–363.
20. Tang W, Santare M H and Advani S G (2003) Melt processing and mechanical property characterization of multi-walled carbon nanotube/high density polyethylene composite films, *Carbon* **41**:2779–2785.
21. Chen L, Pang X J and Yu Z L (2007) Study on polycarbonate/multi-walled carbon nanotubes composite produced by melt processing, *Mater Sci Eng A* **457**:287–291.
22. Tibbetts G G and McHugh J J (1999) Mechanical properties of vapor–grown carbon fiber composites with thermoplastic matrices, *J Mater Res* **14**:2871–2880.
23. Finegan I C, Tibbetts G G, Glasgow D G, Ting J M and Lake M L (2003) Surface treatments for improving the mechanical properties of carbon nanofiber/thermoplastic composites, *J Mater Sci* **38**:3485–3490.
24. Hamon M A, Hu H, Bhowmik P, Niyogi S, Zhao S, Itkis M E and Haddon R C (2001) End-group and defect analysis of soluble single-walled carbon nanotubes, *Chem Phys Lett* **347**:8–12.
25. Sun Y P, Fu K and Huang W (2002) Functionalized carbon nanotubes: properties and applications, *Acc Chem Res* **35**:1096–1104.
26. Dyke C A and Tour M J (2004) Overcoming the insolubility of carbon nanotubes through high degrees of sidewall functionalization, *Chem Eur J* **10**:812–817.

27. Balasubramanian K and Burghard M (2005) Chemically functionalized carbon nanotubes, *Small* **1**:180–192.
28. Wang Y, Wu J and Wei F (2003) A treatment to give separated multi-walled carbon nanotubes with high purity, high crystallization and a large aspect ratio, *Carbon* **41**:2939–2948.
29. Konya Z, Zhu J, Niesz K, Mehn D and Kiricsi I (2004) End morphology of ball milled carbon nanotubes, *Carbon* **42**:2001–2008.
30. Xie X L, Mai Y W and Zhou X P (2005) Dispersion and alignment of carbon nanotubes in polymer matrix: a review, *Mater Sci Eng R* **49**:89–112.
31. Gorga R E and Cohen R E (2004) Toughness enhancements in poly(methyl methacrylate) by addition of oriented multiwall carbon nanotubes, *J Polym Sci Part B Polym Phys* **42**:2690–2702.
32. Schadler L S, Giannaris S C and Ajayan P M (1998) Load transfer in carbon nanotube epoxy composites, *Appl Phys Lett* **73**:3842–3844.
33. Frankland S J, Caglar A, Brenner D W and Grieble M (2002) Molecular simulation of the influence of chemical cross-links on the shear strength of carbon nanotube-polymer interfaces, *J Phys Chem B* **106**:3046–3048.
34. Barber B A, Cohen S R, Eitan A, Schadler L S and Wagner H D (2006) Fracture transitions at a carbon nanotube/polymer interface, *Adv Mater* **18**:83–87.
35. Holden J M, Ping Z, Bi X X, Eklund P C, Bandow S J, Jishi R A, Chowdhury K D, Dresselhaus G and Dresselhaus M S (1994) Raman scattering from nanoscale carbons generated in a cobalt-catalyzed carbon plasma, *Chem Phys Lett* **220**:186–191.
36. Dresselhaus M S, Dresselhaus G, Jorio A, Souza Filho A S and Saito R (2002) Raman spectroscopy on isolated single wall carbon nanotubes, *Carbon* **40**:2043–2061.
37. Cooper C A, Young R J and Halsall M (2001) Investigation into the deformation of carbon nanotubes and their composites through the use of Raman Spectroscopy, *Composites A* **32**:401–411.
38. Qian D, Dickey E C, Andrews R and Rantell T (2000) Load transfer and deformation mechanisms in carbon nanotube-polystyrene composites, *Appl Phys Lett* **76**:2868–2870.
39. Wong M, Paramsothy M, Xu X J, Ren Y, Li S and Liao K (2003) Physical interactions at carbon nanotube-polymer surface, *Polymer* **44**:7757–7764.
40. Zhang Z, Zhang J, Chen P, Zhang B, He J and Hu H H (2006) Enhanced interactions between multi-walled carbon nanotubes and polystyrene induced by melt mixing, *Carbon* **44**:692–698.
41. Lee W J, Lee S F and Kim C G (2006) The mechanical properties of MWNT/PMMA nanocomposites prepared by modified injection molding, *Compos Struct* **76**:406–410.
42. Velasco–Santos C, Martinez–Hernandez A L, Fisher F T, Ruoff R and Castano V M (2003) Improvement of thermal and mechanical properties of carbon nanotube composites through chemical functionalization, *Chem Mater* **15**:4470–4475.
43. Blond D, Barron V, Ruether M, Ryan K P, Nicolosi V, Blau W J and Coleman J N (2006) Enhancement of modulus, strength, and toughness in poly(methyl methacrylate)-based composites by the incorporation of poly(methyl methacrylate)-functionalized nanotubes, *Adv Funct Mater* **16**:1608–1614.
44. Hwang G L, Shieh Y T and Hwang K C (2004) Efficient load transfer to polymer-grafted multiwalled carbon nanotubes in polymer composites, *Adv Funct Mater* **14**:487–491.
45. Parmigiani J P and Thouless M D (2006) The roles of toughness and cohesive strength on crack deflection at interfaces, *J Mech Phys Solids* **54**:266–287.
46. Ding W, Eitan A, Fisher F T, Chen X, Dikin D A, Andrews R, Brinson L C, Schadler L S and Ruoff R S (2003) Direct observation of polymer sheathing in carbon nanotube-polycarbonate composites, *Nano Lett* **3**:1593–1597.

47. Eitan A, Jiang K, Dukes D, Andrews R and Schadler L S (2003) Surface modification of multiwalled carbon nanotubes: Toward tailoring of the interface in polymer composites, *Chem Mater* **15**:3198–3201.
48. Eitan A, Fischer F T, Andrews R, Brinson L C and Schadler L S (2006) Reinforcement mechanisms in MWNT-filled polycarbonate, *Compos Sci Technol* **66**:1162–1173.
49. Schoner M L, Khabashesku V N and Barrera E V (2006) Processing and mechanical properties of fluorinated single-walled carbon nanotube-polyethylene composites, *Chem Mater* **18**:906–913.
50. McIntosh D, Khabashesku V N and Barrera E V (2006) Nanocomposite fiber systems processed from fluorinated single-walled carbon nanotube and a polypropylene matrix, *Chem Mater* **18**:4561–4569.
51. Blake R, Gun'ko Y K, Coleman J, Cadek M, Fonseca A, Nagy J B and Blau W J (2004) A generic organometallic approach toward ultra-strong carbon nanotube polymer composites, *J Am Chem Soc* **126**:10226–10227.
52. Dondero W E and Gorga R E (2006) Morphological and mechanical properties of carbon nanotube/polymer composites via melt compounding, *J Polym Sci Part B Polym Phys* **44**:864–878.
53. Funck A and Kaminsky W (2007) Polypropylene carbon nanotube composites by *in situ* polymerization, *Compos Sci Technol* **67**:906–915.
54. Zhang W D, Shen L, Phang I Y and Liu T X (2004) Carbon nanotubes reinforced nylon 6 composite prepared by simple melt-compounding, *Macromolecules* **37**:256–259.
55. Liu T X, Phang I Y, Shen L, Chow S Y and Zhang W D (2004) Morphology and mechanical properties of multiwalled carbon nanotubes reinforced nylon-6 composites, *Macromolecules* **37**:7214–7222.
56. Lee H J, Oh S J, Choi J Y, Kim J W, Han J, Tan L S and Baek J B (2005) In situ synthesis of poly(ethylene terephthalate) (PET) in ethylene glycol containing terephthalic acid and functionalized multiwalled carbon nanotubes (MWNTs) as an approach to MWNT/PET nanocomposites, *Chem Mater* **17**:5057–5064.
57. Ania F, Broza G, Mina M F, Schulte K, Roslaniec Z and Balta–Calleja F J (2006) Micromechanical properties of poly(butylene terephthalate) nanocomposites with single- and multiwalled carbon nanotubes, *Compos Interf* **13**:33–45.
58. Assouline E, Lustiger A, Barber A H, Cooper C A, Klein E, Wachtel E and Wagner H D (2003) Nucleation ability of multiwall carbon nanotubes in polypropylene composites, *J Polym Sci Part B Polym Phys* **41**:520–527.
59. Li J, Fang Z, Tong L, Gu A and Liu F (2006) Polymorphism of nylon-6 in multiwalled carbon nanotubes/nylon-6 composites, *J Polym Sci Part B Polym Phys* **44**:1499–1512.
60. McNally T, Potschke P, Halley P, Murphy M, Martin D, Bell S E, Brennan G P, Bein D, Lemoine P and Quinn J P (2005) Polyethylene multiwalled carbon nanotube composites, *Polymer* **46**:8222–8232.
61. Khabashesku V N, Billups W E and Margrave J L (2002) Fluorination of single-wall carbon nanotubes and subsequent derivatization reactions, *Acc Chem Res* **35**:1087–1095.
62. Boul P J, Liu J, Mickelson E T, Huffman C B, Ericson L F, Chiang I W, Smith K A, Colbert D T, Hauge R H, Margrave J L and Smalley R E (1999) Reversible sidewall functionalization of buckytubes, *Chem Phys Lett* **310**:367–372.
63. Viswanathan G, Chakrapani N, Yang H, Wei B, Chung H, Cho K, Ryu C Y and Ajayan P M (2003) Single-step in situ synthesis of polymer-grafted single-wall nanotube composites, *J Am Chem Soc* **125**:9258–9259.
64. Frankland S J and Harik V M (2003) Analysis of carbon nanotube pull-out from a polymer matrix, *Surf Sci* **525**:L103–L108.

65. Chowdhury S C and Okabe T (2007) Computer simulation of carbon nanotube pullout from polymer by the molecular dynamics method, *Composites A* **38**:747–754.
66. Qian D, Liu W K and Ruoff R S (2003) Load transfer mechanisms in carbon nanotube ropes, *Compos Sci Technol* **63**:1561–1569.
67. Gojny F H, Wichmann M H, Fiedler B and Schulte K (2005) Influence of different carbon nanotubes on the mechanical properties of epoxy matrix composites – A comparative study, *Compos Sci Technol* **65**:2300–2313.
68. Tjong S C, Xu S A, Li R K Y and Mai Y W (2002) Mechanical behavior and fracture toughness evaluation of maleic anhydride compatibilized short glass fiber/SEBS/polypropylene hybrid composites, *Compos Sci Technol* **62**:831–840.
69. Tjong S C, Xu S A, and Mai Y W (2003) Impact fracture toughness of short glass fiber-reinforced polyamide 6,6 hybrid composites containing elastomer particles using essential work of fracture concept, *Mater Sci Eng A* **347**:338–345.
70. Zhu J, Kim J D, Peng H, Margrave J L, Khabashesku V N and Barrera E V (2003) Improving the dispersion and integration of single-walled carbon nanotubes in epoxy composites through functionalization, *Nano Lett* **3**:1107–1113.
71. Gojny F H, Nastalczyk J, Roslaniec Z and Schulte K (2003) Surface modified multi-walled carbon nanotubes in CNT/epoxy composites, *Chem Phys Lett* **370**:820–824.
72. Gojny F H and Schulte K (2004) Functionalisation effect on the thermo-mechanical behavior of multi-walled carbon nanotube/epoxy composites, *Compos Sci Technol* **64**:2303–2308.
73. Liu L and Wagner H D (2005) Rubbery and glassy epoxy resins reinforced with carbon nanotubes, *Compos Sci Technol* **65**:1861–1868.
74. Wang J, Fang Z, Gu A, Xu L and Liu F (2006) Effect of amino-functionalization of multi-walled carbon nanotubes on the dispersion with epoxy resin matrix, *J Appl Polym Sci* **100**:97–104.
75. Kim J A, Song D G, Kang T J and Young J R (2006) Effects of surface modification on rheological and mechanical properties of CNT/epoxy composites, *Carbon* **44**:1898–1905.
76. Tseng C H, Wang C C and Chen C Y (2007) Functionalizing carbon nanotubes by plasma modification for the preparation of covalent-integrated epoxy composites, *Chem Mater* **19**:308–315.
77. Arroyo M, Lopez-Manchado M A and Herrero B (2003) Organo-montmorillonite as substitute of carbon black in natural rubber compounds, *Polymer* **44**:2447–2453.
78. Donnet J B (2003) Nano and microcomposites of polymer elastomers and their reinforcement, *Compos Sci Technol* **63**:1085–1088.
79. Xiong J, Zheng Z, Qin X, Li M, Li H and Wang X (2006) The thermal and mechanical properties of a polyurethane/multi-walled carbon nanotube composite, *Carbon* **44**:2701–2707.
80. Xia H and Song M (2006) Preparation and characterization of polyurethane grafted single-walled carbon nanotubes and derived polyurethane nanocomposites, *J Mater Chem* **16**:1843–1851.
81. Buffa F, Abraham G A, Grady B P and Resasco D (2007) Effect of nanotube functionalization on the properties of single-walled carbon nanotube/polyurethane composites, *J Polym Sci Part B Polym Phys* **45**:490–501.

Chapter 11

Fracture Properties and Mechanisms of Polyamide/Clay Nanocomposites

A. Dasari, S.-H. Lim, Z.-Z. Yu, Y.-W. Mai

11.1. Introduction

It is well-known that materials can show significantly improved properties if they possess multi-component phase separated morphology at the nanoscale [1]. Polymer nanocomposites are a good example of this class of nanostructured materials that provide unique combinations of mechanical, physical, optical and thermal properties at relatively low filler loading compared to traditional micro-composites. This is due to (i) the exceptionally large surface area-to-volume ratio of the nano-additives available for interaction with the polymer matrix; (ii) nanoscale arrangement and presence of a large number of reinforcements; and (iii) the confinement of polymer matrix chains at the nano-level [2–11]. These fundamental characteristics of the nano-reinforcements, if fully exploited, will affect the microstructure, crystallinity, glass transition and degradation temperatures, and other inherent properties of the resulting nanocomposites, which in turn enable enhanced multi-functional properties to be achieved. But the extent of improvement is determined by the microstructure represented by the size, shape, and homogeneity of the reinforcement in the polymeric matrix [12]. Recent advances in the formulation and evaluation of the energetics and interatomic interactions in materials combined with the development and implementation of computational methods and simulation techniques have led to investigations of the microscopic origins of complex nano-domains in materials [13,14]. However, there are still many unanswered and critical issues to be addressed, particularly in polymer/clay nanocomposites, which have wide applications.

Polymer/clay nanocomposites originated from pioneering research conducted at the Toyota Central Research Laboratories in the late 80s and early 90s, where the replacement of inorganic exchange cations in the galleries of native clay was effected by using alkylammonium surfactants that compatibilized the surface chemistry of nanoclay and the hydrophobic polymer matrix [12,15–18]. This concept was later expanded to many different polymers and enhanced properties were obtained. However, as will be shown in this chapter, the results and conclusions from different studies appear to vary widely, particularly fracture properties, with only subtle changes of testing conditions, material/filler or even characterization techniques. These apparent contradictions in many studies are mainly due to the poor characterization of polymer nanocomposites and lack of quantitative descriptions of observed phenomena. These include detailed depictions of the orientation and dispersion of fillers, local dynamics of the polymer at the interface, quantifiable and controllable interfacial interactions, polymer crystallization and crystallite morphology, *etc.* Hence, fundamental knowledge of the nature of different processes from fabrication to final structure is crucial in determining the ultimate performance of the polymer nanocomposites. In this chapter, we will briefly discuss some of these issues and then focus on the efforts that have been put forth to understand the fracture properties and failure mechanisms in polyamide/clay nanocomposites. In addition, we will outline our recent and current research findings on the same topics for the same materials.

Before discussing the fracture properties and mechanisms of polyamide/clay nanocomposites, a brief review is provided on both direct and indirect effects of adding clay or organically modified clay in the matrix, on the nanocomposites structure and ultimate properties.

11.2. Dispersion of clay in polymers

Layered silicates, or more commonly known as *clay* (minerals), belong to the family of 2:1 phillosilicates. Their crystal structure is made up of layers of two tetrahedrally coordinated silicon atoms fused to an edge-shared octahedral sheet of either aluminum or magnesium hydroxide (Figure 11.1) [2–4,6–10,17]. The layer thickness is ~0.94 nm, has a stiffness of ~170 GPa, and the lateral dimensions of the layers vary from 30 nm to several microns, depending on the particular layered silicate (*e.g.*, saponite ~50–60 nm; montmorillonite ~100–150 nm; hectorite ~200–300 nm) [10]. This provides a large surface area ~750 m^2/g of silicate material. These layers organize themselves into stacks leading to a regular van der Waals gap between the layers known as the interlayer or intragallery. Isomorphic substitution within the layers (*e.g.*, Al^{3+} replaced by Mg^{2+} or by Fe^{2+}, or Mg^{2+} replaced by Li$^+$) generates negative charges that are, generally, counterbalanced by alkali or alkaline earth cations located in the interlayer. Therefore, this type of clay is characterized by a moderate negative surface charge, cation exchange capacity (CEC), that is an important factor during the fabrication of nanocomposites as it determines the amount of sur-

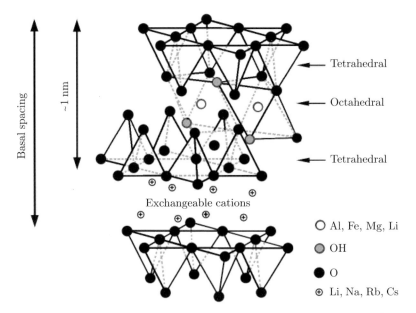

Figure 11.1. Schematic of crystal structure of 2:1 phyllosilicates [9]

factants that can be intercalated between the silicate layers (into the intragallery) to render the hydrophilic phyllosilicates more organophilic and compatible with organic polymers. This is needed since the complete exfoliation of clay into individual layers optimizes the number of available reinforcing elements and results in better improvement in elastic modulus/tensile strength and other properties than the intercalated structures.

From a thermodynamic viewpoint, both the entropic and enthalpic factors are important in controlling the dispersion of clay layers in a polymer matrix [19–21]. It was reported that the confinement of the polymer chains inside the silicate galleries results in a decrease in the overall entropy of the macromolecular chains [10,20]; this is, however, compensated by the increase in conformational freedom of the tethered alkyl surfactant chains as the inorganic layers separate due to the less confined environment (Figure 11.2). It was also shown that apolar interactions are generally unfavorable and so in the case of nonpolar polymers, there is no favorable excess enthalpy to promote the dispersion of clay platelets and it is hence necessary to improve the interactions between the polymer and clay so as to become more favorable than the alkylammonium-clay interactions. This can be achieved by functionalization of the polymer matrix or addition of compatibilizers [3,18,22–29]. For polar polymers, an alkylammonium surfactant is adequate to offer sufficient excess enthalpy and promote the formation of exfoliated nanocomposites.

Besides the thermodynamic considerations, it is important to realize that the source or type of clay mineral [30], charge density [31,32], inherent size of clay platelets, matrix molecular weight [33,34], surfactant type and length [11,

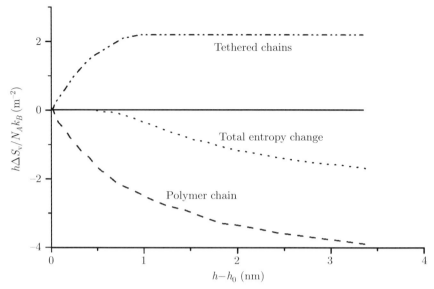

Figure 11.2. Entropy change per unit area as a function of the change in gallery height for an arbitrary polymer and a silicate functionalized with octadecylammonium groups [10]

33,35–39], processing conditions [40,41], melt rheology [34], *etc.*, are crucial in determining the dispersion extent of clay in polymers and thus the magnitude of property enhancements. Paul and his co-workers [30] examined the effect of sodium montmorillonite source (from Yamagata, Japan and Wyoming, USA) on the morphology and properties of polyamide 6 nanocomposites. The related clays were made organophilic by ion exchange using the same alkylammonium chloride compound. They found that the Yamagata clay comprised of platelets that were slightly larger than those of the Wyoming clay. Hence, the larger particle lengths, in addition to a slightly higher degree of platelet exfoliation, observed for nanocomposites made from Yamagata clay gave a higher particle aspect ratio (∼69) than that of nanocomposites made from Wyoming clay (∼57). The trends in tensile properties appear to reflect the nanocomposite structure in that higher stiffness and strengths are obtained for the Yamagata nanocomposite per unit mass of montmorillonite.

Reichert *et al.* [42] studied how the length of the alkyl group on the amine used to modify sodium fluoromica and addition of maleic anhydride grafted polypropylene (PP-g-MA) affected the morphology and mechanical properties of polypropylene nanocomposites formed by melt processing. A critical alkyl length of 12 carbons or more was found necessary for promoting exfoliation in conjunction with PP-g-MA. Similarly, Usuki *et al.* [18] also found that swelling of montmorillonite modified with ω-amino acids by ε-caprolactam increased significantly when the carbon number of the amino acid was greater than 8. Apart from the length of the alkyl group, the number of alkyl tails is another important factor influencing the extent of swelling of clay in a polymer.

Specifically, organic modifiers consisting of one long alkyl tail led to considerably higher levels of organoclay exfoliation than those having two alkyl groups [38]. The dependence of exfoliation on alkyl chain length and tails is believed to stem from the amount of silicate surface that the alkylammonium cation allows to be exposed to the polymer. That is, organic modifier consisting of two alkyl tails shields more silicate surface than one alkyl tail and, thus, precludes desirable interactions between polymer and clay surface, which ultimately limits the degree of organoclay exfoliation (Figure 11.3). However, nanocomposites derived from an organoclay having no alkyl tails in the quaternary cation have an immiscible morphology, consisting primarily of agglomerated clay particles as it contributes the least reduction of cohesive forces between adjacent platelets. As expected, the extent of mechanical reinforcement follows the degree of exfoliation.

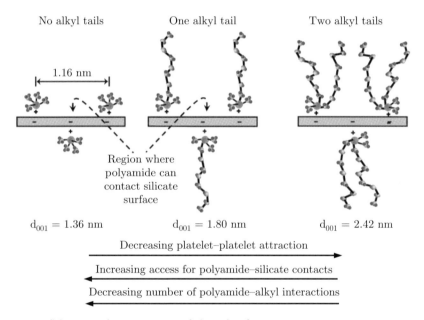

Figure 11.3. Schematic demonstration of the role of quaternary ammonium cations with different tails on the intercalation and exfoliation of organoclay by polyamide 6 [38]

High molecular weight polymers are also shown to result in better exfoliation of organoclay due to higher shear stresses generated during processing caused by their high melt viscosities [33,34,43]. The same is also thought to be true even with normal molecular weight polymers, if high shear stresses are generated during processing by changing the screw configuration or rotation speed. However, whilst the applied external shear stresses are important to achieve a finely dispersed structure, if the matrix polymer lacks sufficient compatibility with the nanoparticles, stress alone cannot achieve fine dispersion of the particles.

Weon and Sue [44] studied the effect of aspect ratio and orientation of clay layers on the toughness and ductility of polyamide 6 nanocomposites (see Table 11.1). By changing the shear using the equal channel angular extrusion (ECAE) process, they altered the aspect ratio and orientation of clay layers.

Table 11.1. Effect of aspect ratio and degree of orientation of clay layers on the mechanical properties and heat distortion temperature of polyamide 6 with 2 wt% of organoclay [44]

	Aspect ratio	Degree of orientation	Young's modulus, GPa	Elongation-at-break, %	K_{IC}, MPa.m$^{0.5}$	Heat distortion temperature, °C
Sample 1	131±37	8	4.67±0.20	4.7	2.7±0.07	110
Sample 2	89±24	7	4.09±0.18	5.3	3.1±0.13	90
Sample 3	80±25	25	3.80±0.14	5.8	3.0±0.11	84

They found that low aspect ratio clay layers along with minimum orientation show improvements in toughness and elongation-at-break despite reductions in elastic modulus, strength and heat distortion temperature (HDT) of the nanocomposites. The major drawback in this study is the assumption of the thickness of clay layers as 0.94 nm irrespective of the system. As indicated in our work [45] and those of others [46,47] on the quantification of clay layers in polyamide matrix, although qualitatively the dispersion state of clay appears to be perfectly exfoliated, quantitative analysis showed that, in reality, it is very difficult to produce a perfectly exfoliated structure of clay in a polymer matrix. In addition, different processing conditions lead to different length and thickness (and therefore, aspect ratio). This in turn affects the mechanical properties of the nanocomposites so that the conclusions derived therein may not always be accurate. It is also reported that clay layers are oriented at some precise angles to the flow direction in different samples; but actually, it is very difficult to identify the flow direction from transmission electron microscopy (TEM) micrographs, particularly at low clay loadings and when the images are taken from the core where the influence of shear is negligible and so the clay layers are randomly oriented.

In addition to the above-mentioned parameters, preferred orientation and spatial distribution of nanoclay layers affect the clay reinforcement efficiency. The absence of Bragg diffraction peaks in the X-ray diffraction (XRD) patterns should not be used as sole evidence for the formation of an exfoliated structure. Morgan and Gilman [48] and Eckel et al. [49] pointed out that XRD analysis alone can lead to false interpretations of the extent of exfoliation. Several factors such as clay dilution, peak broadening and preferred orientation make XRD characterization of polymer nanocomposites susceptible to errors. Clay dilution and peak broadening can yield false result that exfoliation has occurred. Conversely, preferred orientation effects can yield false conclusion that exfoliation has not occurred. Hence, for accurate interpretation of the nanostructures of polymer nanocomposites, both direct and reciprocal space methods [50] should be applied. Visualization of atomic-scale structures is possible by using direct space methods; while interference and diffraction effects of lattice planes are utilized in reciprocal space methods for structural characterization. The usage of either method alone cannot fully identify the structure and can lead to false conclusions.

It is also important to note that despite the many advantages of organic modification of clay, there is a major disadvantage associated with it, particularly when using long chain alkyl ammonium surfactants. Based on the cation-exchange capacity of clay, the alkyl ammonium surfactant content in organic clay is usually over 30 wt% [34,40,43]. It should be noted that not all clay platelets can be fully ion exchanged because of pre-existing natural defects/charge heterogeneities. Therefore, some of the organic surfactant may not be ionically bound and only physiosorbed onto the clay surface [51]. Also, alkyl ammonium organic treatments are thermally unstable and decompose usually from ~180°C, which is lower than the processing temperature of most engineering polymers including polyamides. Organoclay decomposes following Hofmann's elimination reaction that depends on the basicity of the anion, the steric environment around the ammonium, temperature and its product, other than clay itself, which can catalyze the degradation of the polymer matrix [52,53]. Therefore, the presence of such a large amount of low molecular weight bound/unbound surfactant may adversely affect their performance and application.

Many efforts have been directed towards developing alternative routes to produce nanocomposites without using the conventional alkyl ammonium surfactants [54–59]. These include modifying clay layers with thermally stable (like aromatic compounds) surfactants and water-assisted approach. Thermally stable surfactants (such as phosphonium, pyridinium, iminium, and imidazolium salts) decompose at higher temperatures than conventional alkyl ammonium surfactants and are stable, particularly during processing. In the water-assisted approach, water is used as a substitute to organic agents. The underlying principle is that water is a powerful swelling agent for pristine clay *and* a natural plasticizer of polyamide and hence will assist intercalation/exfoliation of the clay layers in the polyamide matrix, improving its thermal stability. Based on this concept, we prepared polyamide 6/clay nanocomposites where the clay layers were finely dispersed even without using any organic surfactants [54]. The degradation temperature of the final nanocomposite is higher than the conventionally prepared organoclay nanocomposite (Figure 11.4), confirming the advantage of nanocomposites based on clay which does not contain alkyl ammonium surfactants.

Despite evidential advantages, the water-assisted approach is too new to apply commercially. The difficulty of applying this method to other polymers is another major disadvantage. For example, Kato *et al.* [58] prepared polypropylene-clay nanocomposites using a modified clay-slurry method in which clay-slurry was obtained by pumping water into the screw extruder containing the polypropylene melt and clay mixture. Although exfoliated structure of clay in

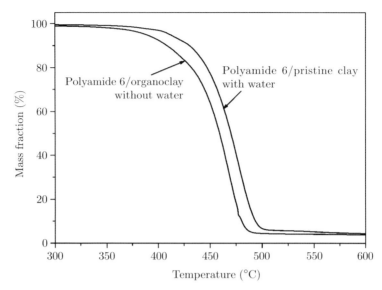

Figure 11.4. Thermogravimetric analysis (TGA) curves for polyamide 6/clay nanocomposites (95/5) prepared conventionally and with the aid of water [54]

polypropylene was successfully obtained, these authors had to use two types of organic compatibilizers to stabilize the nano-morphology. Hence, the original concern of thermal stability returns.

11.3. Crystallization behavior

Polyamides are known for their strong hydrogen bonding ability and seek to maximize the number of H-bonds within and between the polymer chains [60,61]. These strong H-bonding characteristics dominate some physical behavior of these materials. Maximization of H-bonds in its crystalline state requires the polyamide chains to adopt either a fully extended or a twisted configuration, which determines the resulting crystal form (either in monoclinic α- or pseudo-hexagonal γ-form). The essential difference between them is the molecular packing: in the α-form (fully extended configuration) hydrogen bonds are formed between anti-parallel chains, while the molecular chains have to twist away from the zigzag planes to form hydrogen bonds among the parallel chains in the γ-form (twisted configuration) giving rise to lesser inter-chain interactions compared to the α-form.

There are several factors that influence the crystallization kinetics of polyamide materials. It was shown that isothermal crystallization behavior of even as-received pellets and extruded polyamide materials are different and the latter materials exhibit faster crystallization, higher crystallization temperatures and narrower peak widths than the former [60]. Impurities, memory effects (thermal and stress histories), along with generation of low molecular weight chains during processing (through degradation) [62] were attributed to this. Therefore, in the additional presence of clay layers in polyamide materials, the crystallization behavior is obviously more complex.

It has been generally reported that the introduction of clay induces polymorphism, increases crystallization temperature and rate, alters crystal fraction and percentage crystallinity, and promotes formation of small, irregular crystallites of polymers owing to their heterogeneous nucleation effect [59,60,63–71]. For example, the crystallization temperature of polyamide 66/organoclay nanocomposites is shown by Zhang *et al.* [71] to increase for all organoclay loadings by more than 5°C compared to neat polyamide 66 due to heterogeneous nucleation (Figure 11.5). This is because the heterogeneous nuclei form simultaneously once the sample reaches the crystallization temperature; but homogeneous nucleation (by chain aggregation below the melting point) requires a low

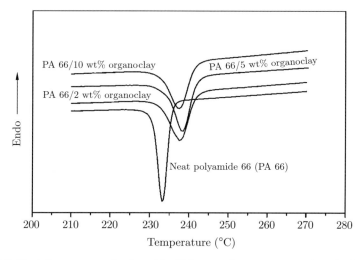

Figure 11.5. Heat-flow curves of polyamide 66/organoclay nanocomposites during crystallization at a cooling rate of 10°C/min [71]

temperature to form stable nucleation. Therefore, the temperature to reach the maximum crystallization rate in neat polyamide 66 was lower than that of its nanocomposites. Maiti and Okamoto [70] also showed the overall crystallization rate to increase in polyamide 6/clay nanocomposites (*in situ* polymerized) when compared to the neat polymer, and the rate increased with increasing clay content at a given crystallization temperature. It was even reported (by considering the Avrami indices) that the spherulites in a nanocomposite were crystallized in a 2D fashion [63,72]. Another investigation stated the formation of axialites rather than spherulites [73]. Clearly, the extent of these effects depends on the processing history.

Many reasons and issues were considered and debated for this behavior and include: bias of chain conformation near interface [63,64,67,74], decrease of activation barrier to nucleation [75,76], and change of chain mobility

[27,77,78]. Both enhancement of interfacial mobility (due to increased free volume at the polymer-layer interface) and retardation of interfacial mobility (due to hydrogen bonding, coordination or electrostatic interactions between the polymer and surface) were also discussed [63]. Nevertheless, lack of proper quantification of the extent of dispersion and an accurate measurement of polymer-clay interfacial area hinders proper description of the impact of layered silicates and comparison with different studies.

Moreover, incorporation of clay layers in some polymers is shown to induce the formation of new crystalline forms, which in turn may affect significantly the ultimate properties of the nanocomposites. Most common examples are poly(vinylidene difluoride) (PVDF) and polyamides. PVDF normally crystallizes in the α-form; but in the presence of clay, because of the crystallization of PVDF in a confined state, transition from large spherulites made of α-crystallites in neat PVDF (Figure 11.6a) to disordered, fiber-like β-crystallites (Figure 11.6b) was observed [79,80]. Shah *et al.* [80] also suggested that the large flat surface of nanoclay and similar crystal lattices between clay and β-crystallites may allow intimate interaction between polymer and inorganic components and stabilize the structures. With polyamides, normally, clay addition favors a γ-crystalline phase instead of α-form as a result of chain conformation change, limiting the formation of hydrogen-bonded sheets of polyamides [60,63].

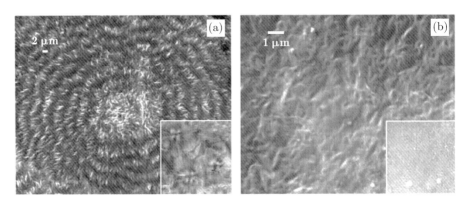

Figure 11.6. Scanning electron microscopy (SEM) micrographs with polarized optical microscope image insets of crystal morphologies of (a) neat PVDF and (b) PVDF/clay nanocomposite [80]

Ma *et al.* [69] used three types of swelling agents with different end groups (-CH_3, -COOH, and -NH_2) to modify montmorillonite and then blended with polyamide 6. They showed that irrespective of the swelling agent, all types of clay resulted in γ-crystalline form of polyamide 6 in the nanocomposites and also enhanced the crystallization rate. Additionally, as the clay content increased, the content of γ-form crystals increased (based on the endothermic peak). However, it is not entirely correct to attribute the complete formation of γ-crystalline form of polyamides to the presence of clay layers. It has been well established that rapid cooling and low temperature crystallization promotes γ-form of polyamides, while higher crystallization temperatures or slow cooling yields predominantly α-form [81–84]. Toyota researchers [12] demonstrated that the γ/α ratio in polyamide 6/organoclay nanocomposite decreased towards the monolith center of injection-molded bars. This implies that the formation of γ depends not only on layered silicate content, but also on the thermal and shear histories [85].

Many other physical and mechanical properties of polyamides, such as glass transition and heat distortion temperatures, melt viscocity, dimensional stability, flame retardancy, *etc.*, are affected directly or indirectly by the addition of clay nano-layers but are not discussed here due to the defined scope and focus of this chapter.

11.4. Fracture properties and mechanisms

It is generally accepted that the major contribution to toughness of polymers comes from the plastic deformation mechanisms. As the material adjacent to the crack-tip is usually under plane-strain condition, it is therefore subjected to high plastic constraint. Without a constraint relief mechanism, the material under this triaxial stress tends to fail in brittle manner with a low toughness [86–90]. In conventional polymer/rubber binary blends, cavitation of rubber particles occurs, relieves the high constraint and triggers large scale plastic deformation of the surrounding matrix material. But there is general agreement based on the majority of studies that polymer/clay nanocomposites are brittle compared to their neat polymeric matrices (see Section 11.4.2 below). It is thought that the brittleness of polyamide/clay nanocomposites may arise from degradation of molecular weight due to high compounding temperature and poor dispersion of clay [91]. However, contrary to these expectations, many investigations showed minimal (or no) degradation of the matrix during compounding with clay and dispersion was never a problem [3,54,92]; and even when the clay layers were uniformly dispersed, they showed brittle failure [47,93–96]. In fact, brittle fracture occured particularly when clay was well exfoliated in the matrix and it is believed that the presence of a stiff nano-filler confines the mobility of the surrounding chains, thus limiting their ability to undergo plastic deformation. But little evidence can be found in the published literature validating the fundamental causes of embrittlement in these polymer-based nanocomposites.

11.4.1. *Improved toughness in polymer/clay nanocomposites*

Nonetheless, in a few studies, toughness (or notched impact strength or ductility) data for some binary polymer (particularly, thermosets)/clay nanocomposites are inspiring. Zerda and Lesser [97] reported an increase in toughness of epoxy/clay nanocomposites, particularly at high concentrations of clay (up to 10 wt%), Figure 11.7. By investigating the fracture surface roughness and crack growth under subcritical loading, they hypothesized that the creation of

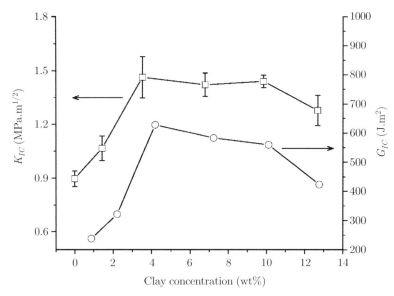

Figure 11.7. Variations of fracture toughness of epoxy nanocomposites with clay loading [97]

additional surface areas is the primary means for toughening in the intercalated systems. It is noted that the fracture surface roughness was characterized by using an atomic force microscope (AFM), which obviously does not have flexibility of more than ∼1.5–1.7 μm in z-direction and limited x- and y-scan resolutions; but the fracture surfaces seem to be very rough from their SEM images. In another epoxy/clay system, Wang et al. [57] have shown an increase in toughness through the initiation and development of many microcracks due to the delamination of clay layers and the increase of fracture surface area due to crack deflections. Fracture toughness (K_{IC}) value increased from 0.70 MPa.m$^{1/2}$ in neat epoxy to 1.26 MPa.m$^{1/2}$ in epoxy nanocomposite with 2.5 wt% of clay and then decreased with further increase in clay loading (Figure 11.8). This apparently contradicts the results of Zerda and Lesser [97] where they show improvements in fracture toughness at higher clay loadings. Fröhlich et al. [98] synthesized anhydride cured epoxy nanocomposites in which two types of phenolic imidazolineamides were used for clay modification. A steady increase in fracture toughness by ∼50% was observed with increasing silicate content but with no loss of glass transition temperature. Again, the creation of additional surface area due to crack growth was assumed to be the primary cause for the improved fracture toughness. Zilg et al. [99] and Becker et al. [100] also showed increases in toughness and stiffness by the addition of organoclay in epoxy matrix. It was theorized that the exfoliated structure mainly leads to improvement in modulus while the remaining tactoids of the intercalated organoclay serve as a toughening phase, possibly by an energy-absorbing mechanism of shearing of the intercalated clay layers.

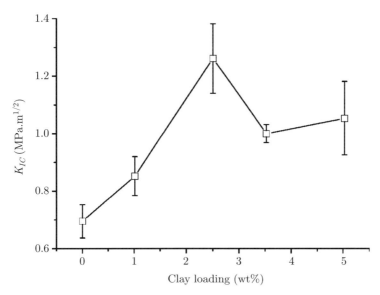

Figure 11.8. Fracture toughness of neat epoxy and epoxy/clay nanocomposites as a function of clay loading [57]

Liu and Wu [101] fabricated polyamide 66/organoclay nanocomposites by melt compounding where the organoclay was obtained through co-intercalation of epoxy resin and quaternary ammonium into Na-montmorillonite so as to improve the interaction between matrix and clay. Impact strength was improved at a clay content of 5 wt% by ∼50% (from ∼96 J/m to ∼146 J/m); though the reasons for this improvement were unclear. When organoclay was added to PVDF, an intercalated structure was observed (along with a transition from α-crystallites in neat PVDF, Figure 11.6a, to disordered, fiber-like β-crystallites, Figure 11.6b, as discussed earlier) and the elongation-at-break of these nanocomposites increased from 20% to 140% [80]. The authors suggested that the fiber-like β-phase on the faces of individual silicate layers led to a structure that promoted plastic flow under the applied stress thus giving rise to more energy dissipation in the nanocomposites. Thus, based upon these results, it is understood that clay is not directly responsible for the enhanced ductility; but it does so indirectly by changing the structure of the neat PVDF from α- to β-crystallites. However, the authors did not investigate the toughening mechanisms, making it difficult to realize the direct influence of clay on toughening.

Kim et al. [102,103] reported that the toughness of polyamide 12/clay and polyamide 6/clay nanocomposites was enhanced due to the occurrence of some energy dissipating events. They performed *in situ* tensile tests on thin films of these nanocomposites in a transmission electron microscope and observed that the major deformation process in these materials was micro-void formation inside the stacked silicate layers. Since the clay interlayer strength is weaker than the clay/matrix interfacial strength, delamination or interlayer debonding of clay tactoids occurs. Depending on the orientation of intercalated clay, delamination under tensile mode of deformation resulted in splitting, opening or sliding and any combination thereof of the high surface area clay layers. However, it is important to note that unlike spherical nano-fillers, clay platelets, due to their 2D nanostructure, show significant orientation differences on any one plane varying from a fully disc shape to a fine layer. Due to its high aspect ratio, during injection molding into a tensile or rectangular bar, only translation motion of clay platelets is possible and rotational motion is often negligible. So generally, along a plane parallel to the flow direction, large percentages of disc shaped particles are present. In the study of Kim et al. [102,103], the observations were made along this plane, which makes it difficult to identify those micro-voids formed by the real deformation processes and presence of discs or those removed during microtoming (which can be easily removed if microtomed along this plane). This is obvious from the TEM micrographs (Figure 11.9), where it is hard to identify whether those are real voids or thin planar discs (since they are present on different planes).

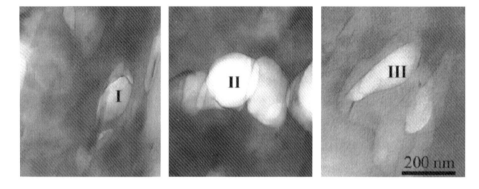

Figure 11.9. Deformation structures around silicate layers in polyamide 6 nanocomposite. Depending on the orientation of layered silicates, splitting, opening and sliding modes of deformation process may occur [103]

Giannelis and co-workers [104] have shown that nanoclay layers in PVDF reorient and align after being subjected to a tensile stress above the glass transition temperature T_g (Figure 11.10) and suggested that this process is responsible for the energy dissipation. These observations were also supported by the molecular dynamics simulations conducted by Gersappe [105] where it was suggested that "mobility" of nanoparticles in the polymer is crucial to enhance energy absorption mechanisms that results in improved toughness. The "mobility" concept was later applied to non-layered fillers like silica and it was found that low particle-particle interaction, strong filler-matrix adhesion, and adequate polymer matrix mobility (usually achieved above T_g) are required for this mechanism to occur [106]. For the purpose of weakening the interaction between nanoparticles while enhancing nanofiller/matrix adhesion, poly(dodecafluoroheptyl

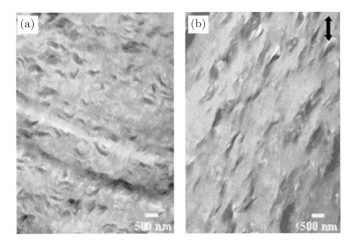

Figure 11.10. TEM micrographs before (a) and after (b) the sample has been subjected to tensile stress. The double-headed arrow in (b) points to the tensile stretching direction [104]

acrylate) (PDFHA) was grafted onto silica particles. It is expected that fluoride facilitates the relative sliding of the grafted nanoparticles under applied stress, and the grafting would result in improved filler/matrix adhesion. Figure 11.11 clearly reveals that when the polymer matrix is in the rubbery state (above T_g), the grafted nanoparticle agglomerates in the composites (strong filler-matrix adhesion) were aligned and reoriented along the tensile stretching direction. On the contrary, when untreated nano-silica particles were used, the agglomerated clusters in the polymer matrix do not align or reorient under tension (Figure 11.11d). These results are also well reflected in the tensile stress-strain curves of these materials at different temperatures above and below T_g of a polypropylene matrix (~18°C) (Figure 11.11e). Nevertheless, an important question that needs to be addressed is the role (acting as a plasticizer in the matrix?) of the grafting agent as the ratio of grafting monomer to nanoparticles was kept at 0.5 by weight.

It is also necessary to distinguish whether it is the particle mobility *per se* or it is facilitated by the mobility of polymer chains during tensile testing; that is, during tensile deformation (especially above T_g), the polymer chains will stretch and align along the tensile direction causing the particles to rotate with the plastically deformed matrix. We have revealed similar observations during scratching of polyamide 6/organoclay nanocomposites underneath the sliding indenter where the clay layers are reoriented from nearly parallel to the sliding direction (Figure 11.12a) to an angle with the contact zone (Figure 11.12b) [107]. This is possible under large loads since the scratch grooves are formed beneath the indenter and so the clay platelets must also rotate with the plastically deformed matrix. This is different from micro-composites, such as carbon fiber/polymer, where extensive cracking of fibers is generally observed with no fiber rotation [108]. So, these observations suggest that plastic deformation

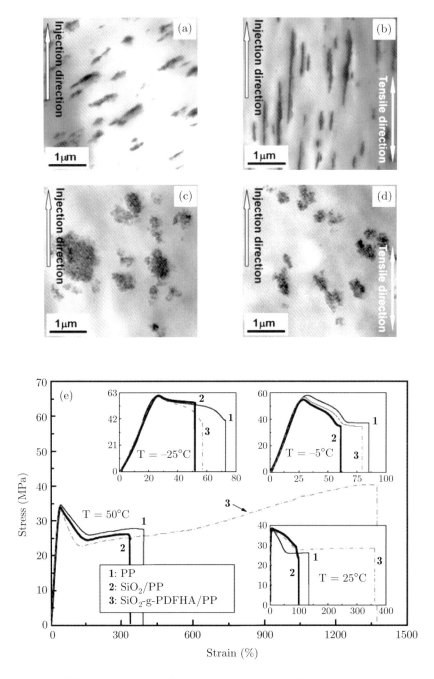

Figure 11.11. TEM micrographs before (a, c) and after (b, d) the samples, polypropylene (PP)/silica-g-PDFHA (a, b) and PP/untreated silica (c, d) have been subjected to tensile stress at 25°C; (e) tensile stress-strain curves for neat PP and PP/silica composites at different temperatures above and below T_g of polypropylene [106]

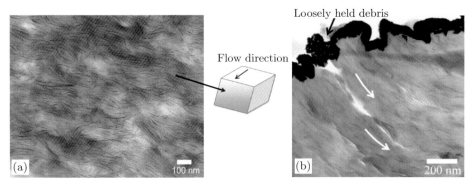

Figure 11.12. (a) TEM micrograph showing the original silicate layer orientation in the sample; and (b) micrograph taken after the scratch test showing subsurface damage beneath scratch track. White arrows point to the rotation of clay layers under the applied stress field. Scratch test conditions: normal load ∼60 mN, scratch velocity ∼5 μm/s, spherical indenter and parallel to flow direction [107]

or "mobility" of the polymer matrix leads to the "mobility" of nano-fillers and that the nano-fillers (unlike micron-sized fillers) are able to actively participate in the mechanical response of the matrix polymer under an applied stress field.

In summary, even though an increase in toughness of *some* polymer/clay nanocomposites was achieved, the results are not always consistent. Especially, the role of polymer matrix plastic deformation *versus* nano-filler mobility is a critical issue that needs further detailed studies.

11.4.2. *Brittleness of polymer/clay nanocomposites*

Most studies have shown a drop in toughness of polymers and a sharp increase in the ductile-brittle transition temperature with the addition of nanoclay layers. Using three-point bending and tensile tests, we have shown that the plane strain fracture toughness K_{IC} and elongation-at-break, respectively, of polyamide 66 decrease with organoclay loading (up to 10 wt%) despite improvements in both Young's modulus and tensile strength [95,109]. Likewise, Nair *et al.* [110] reported a significant reduction in the initiation toughness represented by the J-integral, J_{IC}, of polyamide 66/organoclay nanocomposites as a function of organoclay content. Also, for clay loadings less than 5 wt%, Wong and co-workers [96] observed large reductions in toughness K_C (by ∼1/3) and specific work of fracture G_C (by ∼1/7) in the nanocomposites compared to neat polyamide 66 (Figure 11.13). By using double-notched four-point-bending tests to examine the deformation near the crack tip during sub-critical crack growth, they found a few broken fibrils and ligaments connecting the two surfaces in the crack opening region of polyamide 66/organoclay nanocomposite with 5 wt% of organoclay (Figure 11.14a); and a devious crack path with some voids in the crack tip region (Figure 11.14b). Although processes such as micro-voiding and crazing would help release the triaxial stress state and dissipate energy, they mentioned that the micro-voids observed near the crack tip were

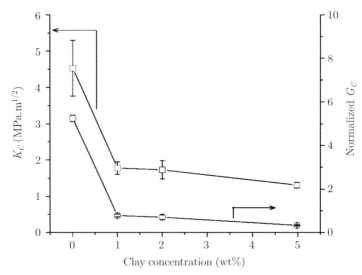

Figure 11.13. Variation of fracture toughness with organoclay loading for polyamide 66/organoclay nanocomposites [96]

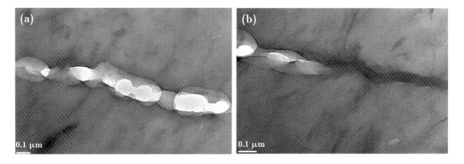

Figure 11.14. TEM micrographs showing the (a) crack opening region and (b) crack tip region in polyamide 66 reinforced with 5 wt% of organoclay [96]

minimal and small, which finally resulted in the plastic deformation being constrained and a low toughness value. We also reported similar observations on binary polymer/clay nanocomposite where only a limited number of microvoids were found along the crack path in the vicinity of the arrested crack-tip (Figure 11.15) [111]. It was also noted that these voids mostly originated from the clay layers due to the weak forces between the stacked silicate layers so that delamination occurred readily and expanded under the applied crack-tip stress field, eventually forming nano- to micron-sized voids.

More recently, He *et al.* [91] showed that K_{IC} decreases (while Young's modulus E increases) monotonically with clay loading in polyamide 6 nanocomposites (Figure 11.16a). Transmission optical microscope (TOM) images of the damage zones obtained from double notch–4-point bending test (Figure 11.16b)

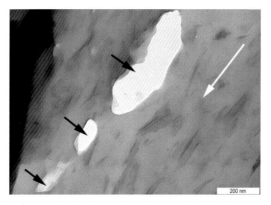

Figure 11.15. TEM micrograph of polyamide 6/organoclay (95/5) nanocomposite showing voids extension along the path of the arrested crack tip. White arrow indicates crack growth direction and black arrows show delamination of clay layers [111]

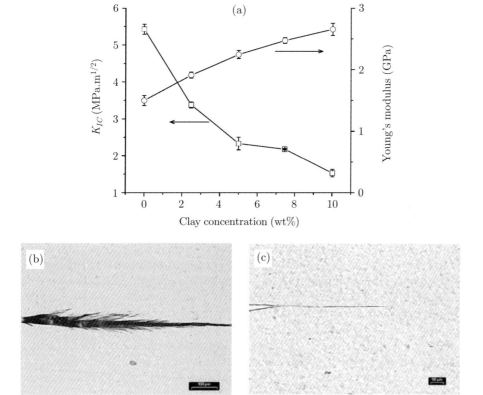

Figure 11.16. (a) Young's modulus and fracture toughness K_{IC} for polyamide 6/organoclay nanocomposites with various organoclay concentrations [91]; transmission optical microscope images of double notch–4-point bending tested nanocomposites with (b) 2.5 wt% and (c) 7.5 wt% organoclay [91]

and small angle X-ray scattering (SAXS) results of tensile deformed samples show that the craze density is highest in polyamide 6 with 2.5 wt% clay loading. These crazes initiate in the matrix adjacent to the interface between clay platelets and polymer, which develop into microcracks upon continued deformation to failure. Further, TEM studies of the crack-tip process zone reveal multiple crazing as the major toughening mechanism for the 2.5 wt% clay nanocomposite. Note that the toughness obtained is still lower than of the neat PA 6. At higher clay loadings, contrary to expectation, SAXS results show much lower density of crazes, supporting the brittle fracture evidenced in TOM (Figure 11.16c). It is suggested that the large number of microcracks (due to the higher clay loading) has limited the crazing mechanism to be fully operative before failure thus leading to even lower toughness.

From earlier discussions in Section 11.3, it is clear that the clay layers serve as heterogeneous nucleation agents and initiate crystallization along the interface; while at the same time, they also restrict the mobility of the surrounding chains and limit the plastic deformation of the polymer matrix. Thus, it is important to understand whether the crystallized nuclei and grown lamellae along the clay interface (*i.e.*, preferentially organized lamellae) are beneficial for shear stress transfer at the interface, hence improving the mechanical properties; or if they have a negative or negligible effect on the ultimate mechanical properties.

With rigid particles, especially clay, there were a few attempts to understand the preferential orientation behavior of polyamide lamellae induced by the presence of nanoclay and its contributions to the overall mechanical properties of the nanocomposite [70,102,103,112]. It has been shown that in intercalated systems polyamide lamellae are oriented perpendicular to the intercalated clay layers and the flow direction, extending over several microns [102]. Although the authors suggested this to be the transcrystallization process, it raises a critical question whether the preferred orientation is a direct result of the presence of the clay layers or a combination of the presence of clay layers (acting as nucleation agents) and flow-induced crystallization. Based on these observations and assuming the entire material to be transcrystalline, Sheng *et al.* [113] included the effect of the transcrystallized matrix layers to calculate the enhancement of the composite modulus using finite element simulation. They assumed that each particle has a transcrystalline matrix layer of thickness ~50 nm on either side. While the transcrystallization of the matrix does contribute to the enhancement of the overall composite modulus, however, this effect was found to be very minor in comparison with the "composite-level" effects of stiff particles in the matrix. With soft elastomeric additives, Muratoglu *et al.* [114] proposed that in the inter-particle regions of closely spaced particles, polyamide lamellae are arranged close to each other and perpendicular to the rubber-matrix interface. Therefore, the hydrogen-bonded planes of low slip resistance are aligned parallel to these interfaces. Consequently, under an external applied load, upon cavitation of the rubber particles, the local deformation will be easily initiated on these planes, and eventually allows deformation to large strains without initiating any critical fracture process, hence enhancing the toughness. While Corte et al. [115] have recently shown that in injection molded polyamide 12/ethylene-propylene rubber (EPR) systems, the crystalline lamellae are aligned in the same direction over tens of microns pointing to the absence of any transcrystallization zone in the vicinity of the amorphous rubber fillers. However, it is important to note that they examined the intermediate regions of the injection molded bar, i.e., in-between the surface and the core, where the influence of shear exists and so the lamellae may be aligned normal to the flow direction consistent with the theories of flow-induced crystallization; while Muratoglu et al. [114] investigated spin-coated thin films in order to remove any such shear-induced effects.

In an attempt to clarify and understand the role of clay layers and their constraint effect on the adjacent polymer matrix, we studied the influence of preferentially organized lamellae in the vicinity of clay layers on the toughening processes and compared with conventional binary rubber toughened polyamide 6 [116]. For binary polyamide 6/maleic anhydride grafted polyethylene-octene elastomer (POE-g-MA) blend, in the core region of the injection-molded sample, the prevalence of lamellae impinging on particle interfaces was apparent for all the particles and when the adjacent particles are intimately near to each other, the inter-particle regions of these particles showed a different morphology whereby the lamellae appeared to be "interpenetrating" caused by the fact that they are very closely spaced. Even in polymer/clay nanocomposites, it was shown that the crystalline lamellae are aligned normal to the lateral interface (on both sides) of each clay layer and matrix, and closely organized to each other depending on the orientation of clay layers (Figure 11.17a). These preferentially organized layers are around 30–40 nm (including both sides) for each clay layer and confirmed that nucleation occurs at the silicate surface during crystallization of the polyamide matrix. Hence, as the inter-platelet distance is smaller, the entire lamellae in the region are highly constrained.

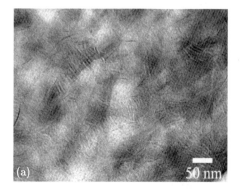 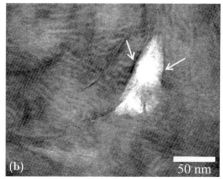

Figure 11.17. TEM micrographs in the core region of an injection-molded polyamide 6/organo-clay nanocomposite (90/10) along a plane normal to the flow direction (a) before and (b) after subjected to a tensile extension of 60%. The preferred orientation of lamellae is seen even after delamination, pointed with arrows [116]

Comparing the crystal forms of polyamide 6 in polyamide 6/rubber (α-form) and polyamide 6/clay (mixture of α- and γ-forms) systems, it can be seen that transcrystallinity may not be an effect of polymorphism. Generally, transcrystallization is believed to occur only on solid, crystalline surfaces and is governed by epitaxy [117–119]. Besides the surface topographical match, there are, however, other possibilities through which transcrystallization occurs. These include differences in thermal conductivity, surface energy, chemical composition, and thermal expansion coefficients between filler and matrix [120–124]. For example, if a temperature gradient is generated due to the thermal conductivity mismatch (clay $\sim$1.14 W/m-K; polyamide 6 $\sim$0.25 W/m-K), the clay surface can be cooler than the matrix. The lower surface temperature can result in a larger super-cooling at the filler-matrix interface and subsequently increasing the nucleation rate and result in transcrystallization [124]. This phenomenon was used to explain the formation of transcrystallized layers in polyacrylonitrile (PAN)-based carbon fiber/poly(etherketoneketone) (PEKK) and amorphous E-glass fiber/PEKK systems [124]. Even transcrystallization is reported in polymers that are crystallized on planar substrates (rubber and calcite), suggesting that the crystallization behavior near an incoherent (or dissimilar) interface is different from that occurring in the bulk and is independent of the nature of the substrate, be that amorphous or crystalline [125].

Now returning to the influence of preferentially organized lamellae on the toughening processes, based on TEM observations of tensile stretched samples of both polyamide 6/rubber blend and polyamide 6/clay nanocomposite, it was concluded that in the same way as rubber cavitation is a necessary condition for matrix shear yielding to impart high toughness to polymer/rubber blends, full debonding at the polymer/clay interfaces is an essential factor for effective toughening in polymer/clay nanocomposites so that shear yielding of large volumes of matrix material may be promoted. However, owing to the strong tethering junctions (formed by ionic interaction between the ends of the polymer chains and surface of the negatively charged clay layers) between individual clay layers (if they are well-dispersed) or the outermost surfaces of clay layers (if intercalated) and matrix, full-scale debonding at the clay-matrix interface is rarely seen in plastically deformed samples indicating that the constraint on the polymer has not been relaxed (Figure 11.17b). Some delamination of clay layers was observed that increased the interlayer distance and dissipating energy in deforming the polymer inside the intra-gallery and the transcrystalline lamellae that are firmly bonded to the outermost surfaces of the clay particle. But as the extent of delamination was limited and no debonding of clay layers was observed, which were necessary for creating free volume in the matrix, the material failed in a brittle manner.

Further, it is very important to note that the preferred oriented layers were only found in the core region of the injection-molded bar; near the surface regions, where the influence of shear exists, no preferentially oriented crystalline lamellae were found around the particles, suggesting that in these regions the lamellae organization is not controlled by the presence of fillers but is primarily determined by the flow. This in turn suggests that these observations are particularly important in samples where the shear influence is negligible.

11.4.3. *Approaches to improve fracture toughness of polymer/clay nanocomposites*

As evident from Section 11.4.2, even after two decades of research and development work on polymer silicate nanocomposites, substantial understanding of the parameters that result in a good balance of the elastic stiffness and fracture toughness of these materials is still needed. Many approaches have been adopted to achieve this goal, the most widely accepted of which is the addition of dispersed elastomeric particles to polymer/clay nanocomposites and more recently using "tunable" block copolymers instead of a pure rubber phase. Aside from these, some isolated studies also reported achieving good stiffness and toughness using different methodologies, although the reasons for the improvement were unknown. These include using organoclay as a compatibilizer for immiscible polymer blends [126] and further modification of organically treated clay with epoxy monomer before blending to produce an intercalated nanocomposite [101]. We have also used a different approach to improve the toughness of nanocomposites by introducing well-distributed nano- to sub-micron sized voids in the matrix during processing [127]. Below are some typical examples based on these various approaches signifying their positive and/or negative aspects on the mechanical properties of polymer nanocomposites.

11.4.3.1. *Approach 1 – ternary nanocomposites*

The best known and most used approach to improve the toughness of polymer/clay nanocomposites is to incorporate a third component, an elastomeric phase. Similar to binary polymer/rubber blends, it is expected that cavitation of the dispersed soft elastomeric particles would occur under loading conditions that could activate plastic deformation of the surrounding matrix material and improve toughness; while the nanoclay layers would reinforce the matrix, hence enhancing the stiffness/strength of the ternary nanocomposites. When three phases are present, however, several issues will arise, cuch as compatibility among the three components that affect the dispersed particle/domain sizes, microstructural features and even the fracture processes that ultimately result in complexity to understand the mechanisms responsible for improvements or deteriorations in properties. In addition, there are several ways to blend these three phases, which brings in many external parameters that in turn control their fracture processes and thus their toughening efficiency. But, as will be shown later, this toughening technique often requires a substantial elastomer concentration of the order of 15–20 wt% which implies a large loss in stiffness and strength.

Kelnar *et al.* [128] used 5 wt% of maleic anhydride-grafted ethylene-propylene rubber (EPR-g-MA) to toughen polyamide 6/organoclay nanocomposites and reported noticeable toughness enhancements (represented by tensile impact strength), particularly at lower clay contents (with 1.5 wt% organoclay, toughness ~84 kJ/m^2); but the nanocomposite containing 5 wt% clay still showed a 3-fold increase (~46 kJ/m^2) in toughness compared to the neat matrix (~16 kJ/m^2) despite a 50% drop when compared to the binary blend (~87 kJ/m^2). The significant toughening was attributed to the cavitation of the elastomer phase and activation of yielding in the matrix, thereby absorbing a large amount of plastic deformation work. While Wang et al. [129] incorporated maleic anhydride grafted ethylene-propylene-diene copolymer (EPDM-g-MA) in a polyamide 6/organoclay nanocomposite in expectation of balanced properties; though they did not provide any evidence, they reported that the clay layers resided in the polyamide matrix and EPDM-g-MA particles were completely free of clay layers. They observed a shift towards higher rubber content for the brittle-ductile transition with the addition of organoclay (Figure 11.18a) and proposed a mechanism based on Wu's interparticle distance theory [130] to explain this effect of clay on the

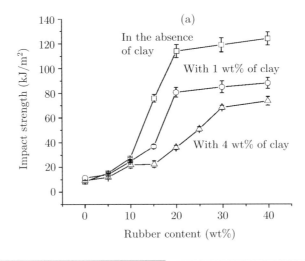

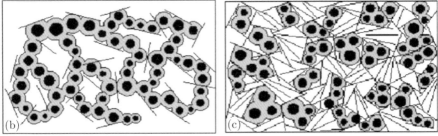

Figure 11.18. (a) Organoclay and rubber loading dependence on impact strength in polyamide 6/organoclay/EPDM-g-MA ternary nanocomposites [119]; (b) and (c) schematic representations of the "blocking" effect of clay layers (with 1 and 4 wt%, respectively) on the stress overlaps between rubber particles [129]

toughness of polyamide 6/EPDM-g-MA blends. They called this mechanism the "blocking effect" of clay layers on the overlap of the stress volume between EPDM-g-MA particles, since the key for brittle-ductile transition according to [130] is the overlap of stress volume between rubber particles. That is, they suggested that the clay layers in the polymer matrix act as physical barriers obstructing the overlap of stress volume between particles, which is obviously more effective at higher loadings of clay (Figure 11.18c). So, at a low content of clay, (~1 wt%), as the physical barrier effect is weak (Figure 11.18b), toughness and the transition were similar to that of the blends (Figure 11.18a); while at higher clay content, (~4 wt%), the barrier effect is strong due to the presence of numerous layers and therefore resulting in transition broadening and reduced toughness. However, the real situation is more complex than just physically blocking the stress volume. Indeed, the effect of clay layers on the adjacent matrix chains, their contributions to the toughening processes by delamination or their influence on the compatibility among the three phases, *etc.*, must be considered.

González *et al.* [131] in their polyamide 6/maleanized styrene-ethylene/butadiene-styrene copolymer (mSEBS)/organoclay ternary nanocomposites claimed similar morphology where the clay layers reside entirely in the polyamide matrix. The cited authors found that the brittle-ductile transition is also delayed in the presence of clay. They noted an increase in mSEBS size in the additional presence of organoclay (with 15 wt% rubber, weight-average rubber particle size in binary blend is ~0.23 μm; while it is ~0.71 μm in the additional presence of 3 wt% organoclay) and suggested it to be a result of interaction between the organic modification of clay dissolved in the matrix and the maleic anhydride modification of mSEBS that hindered the compatibilizing effect of the latter. At this point, it is interesting to note that in a topical study on the effect of clay platelets on the morphology of polyamide 6 and poly(ethylene-ran-propylene) rubber (EPR) blends where the rubber is not maleated, it was shown that the exfoliated clay layers in the polyamide matrix prevented the coalescence of the dispersed domains, resulting in decreased dispersed particle size compared to the binary polyamide 6/EPR blend [132]. Dong *et al.* [133] also observed similar behavior in their polyamide 6/clay/acrylate rubber ternary nanocomposites. They used a two-step process to synthesize ternary nanocomposite where acrylate rubber/clay composite was prepared by spray-drying a mixture of clay slurry and irradiated acrylate rubber latex; and, finally, by blending this mixture with polyamide 6. In the binary blend (with 30 wt% rubber), there were many large aggregates of rubber particles and their dispersion was poor (Figure 11.19a) due to poor compatibility between polyamide 6 matrix and acrylate rubber; but in the presence of clay layers (9 wt%), rubber particles (21 wt%) were well dispersed with a uniform diameter of ~150 nm (Figure 11.19b) and it was attributed to the clay layers which prevented the coalescence of rubber particles during blending. This suggests that while using maleic anhydride modification is important for

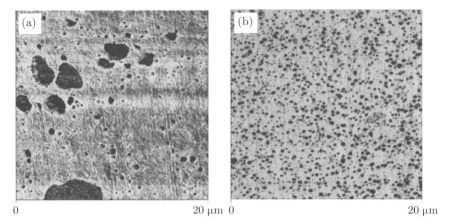

Figure 11.19. Atomic force microscope phase images of (a) binary polyamide 6/acrylate rubber blend and (b) ternary polyamide 6/clay/acrylate rubber nanocomposite showing the differences in the dispersion quality of rubber particles [133]

compatibilization with the polyamide matrix, it also has a negative influence when blended along with organoclay resulting in an increased rubber particle size. On the other hand, if the compatibility between the matrix and rubber particles is poor, this may lead to poor interface and so, interfacial debonding of the rubber particle from the matrix under loading conditions rather than cavitation, which will affect the toughening mechanisms and the fracture toughness value. In general, it is well-established that when the adhesion between rubber and matrix is strong, the cavitation phenomenon takes place inside a rubber particle due to the difference in Young's modulus and Poisson's ratio between rubber particles and matrix. In contrast, when adhesion between rubber and matrix is relatively poor, interfacial debonding may also occur.

Nevertheless, in the study of González et al. [131], as seen in Figure 11.20, super-tough ternary nanocomposites are obtained. But the amount of rubber necessary to attain super-toughness was higher in the presence of clay layers (∼30 wt%) than in the blends (∼20 wt%). But their results also revealed that super-toughness is only achieved with 30 wt% rubber and only when organoclay content is less than 3 wt%. Even with 4 wt% organoclay, notched impact strength dropped tremendously (from 730 J/m at 3 wt% to 113 J/m at 4 wt%). Again, the reasons or the mechanisms for this are not studied. Also, as expected, Young's modulus in both [131] and [129] decreased considerably in the presence of rubber.

Chiu et al. [134] also prepared polyamide 6 ternary nanocomposites with organoclay (5 wt%) and maleated polyolefin elastomer (10 wt%) as fillers. Izod impact strength of this ternary nanocomposite was 51.6 J/m, which was much higher than the binary clay nanocomposite (22.8 J/m) and neat polyamide (37.7 J/m), but significantly lower than the tough binary blend (104.8 J/m). This again ascertains the necessity to have a higher rubber content of >15 wt%

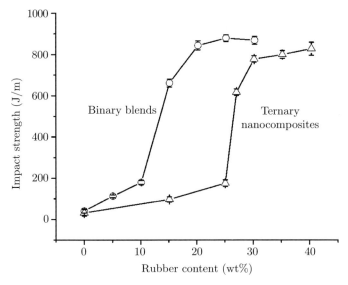

Figure 11.20. Variation of notched impact strength with mSEBS loading in binary polyamide 6/mSEBS blends and ternary polyamide 6/organoclay (3wt%)/mSEBS nanocomposites [131]

to reach the super-tough state. Similar results were obtained in other ternary nanocomposites, including polypropylene/clay/styrene-butadiene-styrene rubber [135], epoxy/carboxyl-terminated butadiene acrylonitrile (CTBN) rubber/organoclay [136], epoxy/core shell rubber (CSR)/clay [137], polypropylene/SEBS-g-MA/vermiculite [138], polypropylene/SEBS-g-MA/organoclay [139], and poly(butylene terephthalate) (PBT)/maleated poly(ethylene-co-vinylacetate) (EVA-g-MA)/organoclay [140]. In most cases, toughness was improved at the expense of stiffness/strength owing to the presence of soft elastomer particles which offset the stiffening/strengthening effects of the stiff clay layers on the resultant nanocomposites.

Sue et al. [135,141] used a double notch-4-point bend technique to understand the toughening mechanisms of their ternary nanocomposites based on epoxy and polypropylene as matrices, core shell rubber and styrene-butadiene-styrene (SBS) as toughening agents and α-zirconium phosphate (layered structure like clay) and organoclay as reinforcing agents, respectively. The fracture toughness results of these two systems are given in Table 11.2 along with their elastic moduli. Clearly, toughness drastically improved in the ternary nanocomposites compared to binary nanocomposites and surprisingly, even binary blends. The crack-tip damage zones of different samples were then investigated with TEM. In the binary nanocomposites, very a small damage zone was seen explaining the brittleness of these materials (Figures 11.21a and 11.21b); but crack deflection/bifurcation was observed. Close-up of the binary α-zirconium phosphate/epoxy nanocomposite suggested that the deformation of the matrix preferentially advanced along the intercalated α-zirconium phosphate layers (by de-

Table 11.2. Elastic modulus and fracture toughness of polypropylene/organoclay/SBS rubber and epoxy/zirconium phosphate/CSR ternary nanocomposites [135,141]

	Fracture toughness, J_C (kJ/m^2)	Flexural modulus (GPa)
Neat polypropylene	3.0	1.76±0.02
Polypropylene with 3 wt% organoclay	1.5	2.12±0.05
Polypropylene with 5 wt% SBS	9.0	1.47±0.01
Polypropylene with 3 wt% organoclay and 5 wt% SBS	10.5	1.48±0.03
	Fracture toughness, K_{IC} (MPa.m$^{1/2}$)	Tensile modulus (GPa)
Neat epoxy	0.69±0.05	3.10±0.05
Epoxy with 2 vol% zirconium phosphate	0.70±0.04	3.97±0.15
Epoxy with 3.5 vol% CSR	0.92±0.08	2.56±0.06
Epoxy with 2 vol% zirconium phosphate and 3.5 vol% CSR	1.64±0.04	3.77±0.20

laminating them) due to their weak interlayer strength (Figure 11.21b). With further addition of rubber particles, plastic deformation, as illustrated by the highly deformed and elongated rubber particles around the crack tip, was observed in both systems (Figures 11.21c and 11.21d). But contrary to their binary nanocomposite fracture mechanism, no delamination of intercalated α-zirconium phosphate layers was observed in the ternary system, which is surprising. As high magnification images are not shown for polypropylene/SBS/clay system, it is hard to identify if there is any delamination of clay layers in the crack-tip damage zone. Further, and more importantly, following the fracture mechanisms, the toughness data of the ternary nanocomposites (which is higher than the binary blends) cannot be explained if the rigid particles do not play a role during the fracture process. This is particularly true for the epoxy/α-zirconium phosphate/CSR system, where ∼45% increase in toughness is noticed when compared to binary epoxy/CSR blend. For PP/clay system, a reduction of 50% in the fracture toughness is seen compared to the neat material. In the additional presence of SBS particles, the toughness is even higher than the binary blend (though only by 10%); but this is significant as a reduction is normally expected in line with the toughness of binary nanocomposites.

Therefore, in both studies, the role of rigid particles towards the toughening processes in the additional presence of soft rubber particles is unclear. Moreover, questions like what if the layered mineral is exfoliated in the matrix or if it is present completely inside the rubber phase, or distributed between the matrix and rubber particles require clarifications. Also, in most of the above studies,

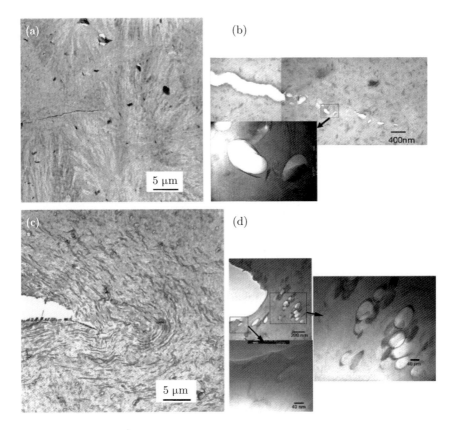

Figure 11.21. TEM micrographs of crack-tip damage zone of (a) polypropylene/organoclay, (b) epoxy/zirconium phosphate, (c) polypropylene/organoclay/SBS, and (d) epoxy/zirconium phosphate/CSR nanocomposites [135,141]

only one mixing protocol was employed to prepare the ternary nanocomposites, which obviously reduces the complications on one hand, but on the other, it suppresses a number of parameters that help in further understanding of the toughening mechanisms, such as, what happens to the rubber phase if it is extruded twice or what is the difference between the extent of clay dispersion from a one-step to a two-step extrusion process, location of clay (in the matrix or in the dispersed rubber phase), influence of exfoliation extent and aspect ratio of clay if extruded twice, and so forth.

In an attempt to clarify these questions, we have prepared polyamide 66-based ternary nanocomposites using four different blending protocols [45,94,142, 143] and the state of exfoliation of organoclay in the polyamide 66 matrix and SEBS-g-MA phase was quantified. SEBS-g-MA was used as a toughening agent (15 wt%) and a high polar organoclay as a reinforcing agent (5 wt%) to obtain balanced mechanical properties of polyamide 66. The four blending protocols adopted were: (a) N1 – polyamide 66, SEBS-g-MA and organoclay were blended

Table 11.3. Mechanical properties of neat polyamide 66 (B0), polyamide 66/organoclay nanocomposite (95/5) (B1), polyamide 66/SEBS-g-MA binary blend (85/15) (B2), and ternary nanocomposites, N1 to N4 [143]

Blending sequence	Notched impact energy (kJ/m^2)	Static fracture toughness (MPa.m$^{1/2}$)	Flexural modulus (GPa)
B0	6.3±0.3	2.6±0.35	2.95±0.05
B1	2.4±0.6	–	3.79±0.10
B2	30.3±5.6	–	2.06±0.15
N1	10.3±1.1	3.5±0.05	2.54±0.11
N2	7.9±0.4	3.0±0.12	2.53±0.07
N3	11.8±1.6	3.9±0.07	2.65±0.03
N4	6.6±0.4	3.3±0.14	2.63±0.02

simultaneously; (b) N2 – polyamide 66 was blended with SEBS-g-MA first and the polyamide 66/SEBS-g-MA blend was mixed with organoclay later; (c) N3 – polyamide 66 was reinforced with organoclay first and the polyamide 66/organoclay nanocomposite was blended with SEBS-g-MA later; and (d) N4 – SEBS-g-MA was mixed with organoclay and then the SEBS-g-MA/organoclay masterbatch was blended with polyamide 66 later. Distinct differences in microstructures were observed depending on the processing sequence, which in turn controlled their mechanical properties, particularly, modulus and toughness (Table 11.3). TEM observations suggested that in N1 and N2, SEBS-g-MA particles were finely dispersed in polyamide 66 matrix and the presence of clay did not have a significant influence on the dispersion quality of the SEBS-g-MA phase. However, depending on the interaction of clay with polyamide 66 and SEBS-g-MA, the dispersion quality of clay varied. As polyamide 66 is more polar than SEBS-g-MA, silicate layers were well dispersed in the former, whereas thick platelets of clay were evident in the latter. N3 and N4 have contrasting structures. In N3, most of the clay was present in the polyamide 66 matrix with good dispersion and distribution, suggesting that this blending sequence, preparing a binary polyamide/clay nanocomposite first, and then blending with rubber, is the best route regarding the dispersion quality of clay. While in N4 most of the clay was present in rubber particles as clay was blended with rubber first and the relatively low polar nature of SEBS-g-MA could not exfoliate the clay layers and resulted in an intercalated structure within the SEBS-g-MA phase. Based on these TEM observations of N1 to N4, a schematic is prepared and shown in Figure 11.22.

Since it is well known that the mechanical properties (Table 11.3) of the nanocomposites are affected by the state of exfoliation and dispersion of clay in polyamide 66 matrix and SEBS-g-MA phases, these parameters were quan-

Fracture of polyamide/clay nanocomposites

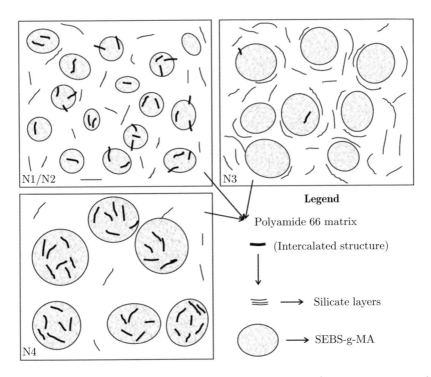

Figure 11.22. Schematic of distinct differences in microstructures of ternary nanocomposites (N1/N2, N3 and N4) prepared by different blending protocols [94]

tified to obtain a more detailed understanding on their structure-property relationships. The results obtained from the quantification process of clay particles (Table 11.4) for the binary nanocomposite, B1 and all the ternary nanocomposites, N1 to N4 clearly reveal the significance of the presence of a third component (SEBS-g-MA) and the influence of blending protocol employed on the average length, thickness, and aspect ratio of clay particles, along with the

Table 11.4. Quantification data of rubber and clay particles in B1, B2, and N1 to N4 [45]

	SEBS-g-MA size, nm	Clay in polyamide 66 matrix		Clay in SEBS-g-MA particles	
		Aspect ratio of clay particles	Apparent % of clay	Aspect ratio of clay particles	Apparent % of clay
B1	–	50	100	–	–
B2	90	–	–	–	–
N1	120	38	50	22	50
N2	137	37	46	20	54
N3	202	32	80	19	20
N4	281	22	18	15	82

apparent content of clay in polyamide 66 matrix and dispersed SEBS-g-MA particles. The increased particle size of SEBS-g-MA in N1 to N4 follows the previous discussion on the usage of maleic anhydride grafting and the organoclay affecting the compatibility between SEBS-g-MA and polyamide 66. However, depending on the blending sequence, the size varied. In N4, as SEBS-g-MA and organoclay were blended first, the interaction between the maleic anhydride groups of SEBS-g-MA and the hydroxyl-ethyl groups of organoclay suppressed the dispersion of SEBS-g-MA in polyamide 66 matrix.

More significantly, even in the binary clay nanocomposite, the aspect ratio of clay particles (~50) is not very high and is different from that under ideal conditions (several hundreds to thousand), which points to the practical difficulty of clay reinforcement. Compared to binary nanocomposite, the aspect ratio of clay particles is even worse for ternary nanocomposites. Further, the apparent percentage and aspect ratio of clay in polyamide matrix and SEBS-g-MA phase are similar in N1 and N2. Thus, notched impact strength and fracture toughness differences between them were a direct influence of the blending sequence. In N3 and N4, the maximum amount of clay particles was present in the matrix and dispersed rubber particles, respectively, and so in these two cases, this parameter was of primary importance in determining the properties and fracture mechanisms.

Post-mortem TEM analysis was conducted to study the fracture mechanisms in a plane perpendicular to the fracture surface near the notch tip on both notched impact and three-point bend fractured specimens to identify the deformation history that finally led to failure. As expected, in the binary blend, the rubber particles were cavitated and acted as triggers for large scale deformation of the matrix by relieving the state of triaxial stress at the tip of the growing crack. In the ternary nanocomposites, in brief, the presence of organoclay in SEBS-g-MA phase (in N1, N2 and N4) and the two-time extrusion of SEBS-g-MA (N2 and N4) made the latter more rigid and reduced its ability to cavitate and ultimately minimizing the stretching of the voided matrix material. Apart from this, partial (interfacial) debonding of some of the SEBS-g-MA particles and debonding of the intercalated clay layers from the less polar rubber particles was seen. The debonding of intercalated clay layers is most excessive in N4 where the majority of clay layers resided in but weakly bonded to the rubber particles; an example of this behavior is shown in Figure 11.23a. The interfacial debonding of rubber also suggests that the presence of clay has a negative effect on the interfacial interaction of SEBS-g-MA particles to polyamide 66. Despite all these plastic damage mechanisms, reduced or nil plastic stretch of the matrix was observed, even close to the notch tip in N1, N2 and N4, ultimately lowering the toughening efficiency. The maximum percentage of exfoliated clay in the continuous matrix improved the stiffness of the nanocomposite (particularly, N3). But it had a negative effect in significantly improving the fracture toughness (even though extensive

Fracture of polyamide/clay nanocomposites

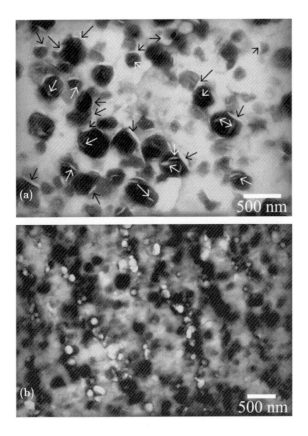

Figure 11.23. TEM micrographs taken within the sub-critically deformed zone, very close to the fracture surface for (a) N4 and (b) N3. Large scale debonding of clay from rubber particles (indicated with white arrows) along with interfacial debonding of the latter from the matrix (pointed with black arrows) are seen in N4; while a large extent of cavitation of SEBS-g-MA particles is evident in N3 [45]

cavitation of rubber particles was seen, Figure 11.23b) as it constrained the matrix mobility and limited the yielding.

This signifies the effect of dispersed fine clay layers in the matrix in restricting yielding. To address this issue and obtain superior toughness and minimum reduction in stiffness/strength, polyamide 6/organoclay nanocomposites were prepared by incorporating POE-g-MA as a toughening agent [111]. When organoclay and POE-g-MA particles are added to polyamide matrix by simultaneous (or two-step) blending, both organoclay layers (intercalated) and POE-g-MA particles are dispersed separately in the matrix and there is no evidence of organoclay layers present in the POE-g-MA particles (Figure 11.24a). A remarkable toughening effect was achieved at 20 wt% POE-g-MA in the binary blends; more interestingly, its effectiveness in toughening the polyamide 6 matrix

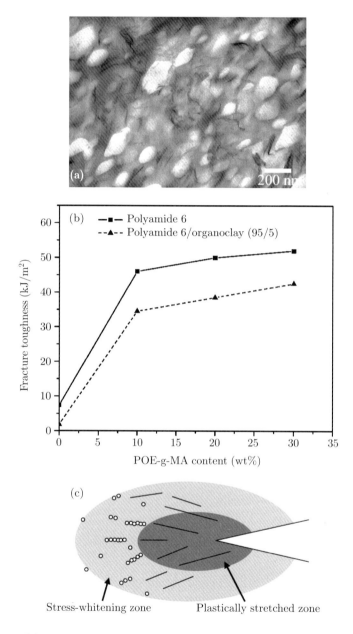

Figure 11.24. (a) Representative TEM micrograph of polyamide 6/organoclay/POE-g-MA (76/4/20) ternary nanocomposite showing that both organoclay and POE-g-MA dispersed separately in the matrix; (b) effect of POE-g-MA loading on quasi-static fracture toughness of polyamide 6 blends and nanocomposites; and (c) schematic of the deformation processes in a ternary nanocomposite showing arrays of voids (croids) and plastic zone around the crack-tip region [111]

was maintained even in the presence of 5 wt% organoclay (Figure 11.24b). The presence of intercalated clay does not seem to restrict the yielding of the matrix material, indicating that this structure may be advantageous to the toughness of the ternary nanocomposites. TEM analyses of the deformation features from single-edge-double-notch four-point-bend (SEDN–4PB) specimens revealed the internal cavitation of POE-g-MA particles along with craze-like damages consisting of line arrays of expanded voids in both un-reinforced and organoclay reinforced polyamide 6/POE-g-MA blends, followed by severe matrix plastic deformation at the crack-tip region. A schematic of these processes in the sub-critical damage zone around the crack-tip is shown in Figure 11.24c. It is this plastic work that has mainly contributed to the drastic enhancement of toughness of these ternary nanocomposites, besides energies absorbed in cavitation and stretching of rubber particles, and delamination of clay layers.

However, Young's modulus of polyamide 6/organoclay/POE-g-MA ternary nanocomposites was reduced compared to polyamide 6/organoclay binary nanocomposite at similar contents of organoclay due to the presence of the soft POE-g-MA phase.

Based on all aforementioned studies on the ternary nanocomposites, it is suggested that the nature and control of the structures of different additives along with their location determines the final material properties. Firstly, the compatibility of different additives with the matrix and amongst each other affects the dispersed particle size and their interfacial interaction; secondly, the location and role played by the different additives in triggering or resisting different plastic deformation processes in the matrix; and thirdly, restriction of the extrusion process on rubber to one-step in order to promote toughness. But the main drawback of this approach includes the requirement of a substantial elastomer concentration (usually >15 wt%) which obviously affects stiffness and strength and in improving the ductile-brittle transition temperature.

11.4.3.2. *Approach 2 – "tunable" block copolymers*

Another approach that was introduced very recently is to use "tunable" block copolymers instead of pure rubber. For example, Oshinski *et al.* [144] showed that dispersing 20% of maleated polystyrene-*block*-poly(ethylene-butene)-*block*-polystyrene triblock copolymers yields a similar toughening efficiency as 20% of pure rubber. However, most block copolymers used were rather soft with about 60–70 wt% rubber, which again raises questions on maintaining high modulus/strength. To address this issue, Corte *et al.* [145] added polystyrene-*block*-polybutadiene-*block*-poly[(methyl methacrylate)-*stat*-(methacrylic acid)] (SBM) in polyamide 12 and showed remarkable improvements in impact strength (even at low temperatures, Figure 11.25) without reducing the elastic modulus of the polyamide matrix. This was possible as with such block copolymers, rubber content was controlled by varying the size of the polybutadiene

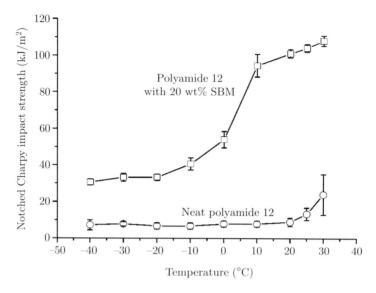

Figure 11.25. Notched Charpy impact strength as a function of temperature for neat polyamide 12 and polyamide 12 with 20 wt% SBM particles [145]

(B) middle block relative to that of the glassy polystyrene (S) and poly(methyl methacrylate) (M) end blocks; that is, by introducing glassy blocks, it was possible to reduce the overall rubber content and thereby control the loss in elastic modulus. In addition, to significantly reduce the amount of rubber, a symmetric composition was chosen, which means in the 20 wt% of SBM filler, only 7 wt% corresponds to overall rubber content. Though detailed failure mechanisms have not been studied, they attributed the toughness improvements to the cavitation of SBM particles and activation of plastic deformation of the matrix similar to conventional rubber toughened polymer blends. Nevertheless, this approach has not been used to prepare ternary nanocomposites, that is, in the additional presence of clay layers, which could be interesting.

11.4.3.3. *Approach 3 – pre-existent submicron to nano-sized voids*

We adopted another approach to improve the toughness of polymer nanocomposites without reductions in their stiffness and strength due to the presence of soft elastomer particles. Before discussing this approach, it should be noted that while the addition of rigid particles themselves, though in some cases, was shown to enhance toughness along with stiffness/strength [87,146,147], the extents of enhancement (particularly toughness) are not always significant and consistent. In any case, both soft and rigid particle toughening are based on the idea that the particles must cavitate (or debond at the interface), altering the stress state in the material around the particles and inducing extensive plastic deformation of the matrix. Thus, this leads to an idea of the presence of pre-existing submicron voids in polymers and their role on the toughening process. This kind of idea has been used to elucidate the importance of cavitation of rubber particles in toughening polymers by Bagheri and Pearson [148–150], who did a comparative examination of epoxies modified with various rubber particles having different cavitation resistance and pre-existing micro-voids (hollow spheres) (up to 40 μm). They showed that the toughening efficiency was similar with different cavitation resistance rubber particles and pre-existing micro-voids, inducing shear yielding at the crack tip. Guild and Young [151] also performed finite element analysis (FEA) of the stress distributions in the matrix for epoxy containing rubber spheres and compared to epoxy containing micro-voids. They found that the stress distributions were similar in both cases. Huang and Kinloch [152] used urea-terminated polyether amine as a micro-voiding agent during the preparation of epoxy and created voids of ∼0.70 μm; this material exhibited improved fracture toughness (K_{IC} ∼1.96 MPa.m$^{1/2}$) compared to neat epoxy (K_{IC} ∼1.03 MPa.m$^{1/2}$) without any voids, but the toughening mechanisms were not studied.

In a recent study, we adopted this idea and used polyoxyethylene nonylphenol (PN) to obtain submicron to nano-sized voids in polypropylene and polypropylene/CaCO$_3$ nanocomposites to understand their role in initiating tough plastic response in the matrix [127]. The key to the generation of voids is the low flash point of PN (∼90°C), which is below the processing temperature (200°C) of polypropylene, hence resulting in vapor or bubble formation during processing, though the size distribution and density of voids vary across the thickness of the sample. Nonetheless, the presence of only 1.5 phr PN improved greatly notched Izod impact strengths of both blend and nanocomposite (Figure 11.26a), and slightly affected their Young's modulus and yield strength. Investigations of the plastic damage occurring during the tensile and double notch–4-point bending (DN–4PB) tests revealed that submicron to nano-sized voids acted similar to cavitated rubber particles in rubber-toughened polymer systems where the plastic void growth (of the pre-existent voids) in the polypropylene matrix occurred upon deformation and subsequently resulting in the formation of isolated and domain-like structure of crazes (Figures 11.26b and 26c). In the presence of CaCO$_3$ particles, submicron voids that are associated with the debonding of these loosely bound CaCO$_3$ particles were noticed. The debonding of the particles also enables plastic void growth of polypropylene matrix. This has been reported in other polymer composite systems in the literature [153–156]. Therefore, the significant increase in notched impact strength of the ternary material system was attributed to a combination of the processes that happened in binary blend and binary nanocomposite; that is, the energy dissipated by debonding of CaCO$_3$ particles and, much more significantly, the energies associated with plastic void growth in matrix based on both pre-existent voids and those created from particle debonding, as well as void coalescence.

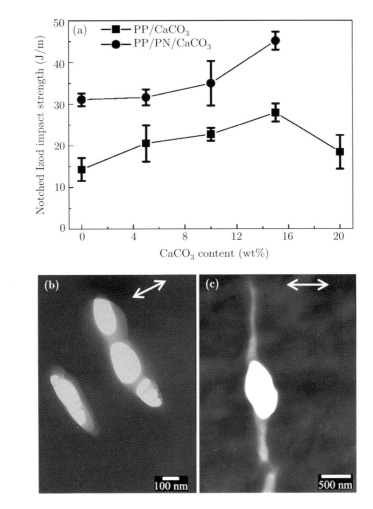

Figure 11.26. (a) The influence of the addition of $CaCO_3$ particles to polypropylene with and without PN (1.50 phr) on notched Izod impact strength of the composites; TEM micrographs of PP/PN blend at 1.50 phr PN taken along a plane parallel to the tensile direction unloaded at different extensions showing (b) the formation of domain-like structure of crazes and (c) break-down of the craze structure and coalescence of the voids. Tensile direction is shown with double-headed arrows

11.5. Conclusions and outlook

The tremendous interfacial area of nanoclay layers along with their good aspect ratios and substantial cation exchange capacities (that facilitate various modifications to promote the compatibility with a polymer or improve thermal stability or some other properties depending on the end application) made them a popular choice of "rigid" fillers in polymers to enhance their stiffness, strength, dimensional stability, barrier resistance, thermal, optical and physico-chemical properties. But their presence in the polymers affects several other facets of these materials compared to their neat counterparts. For example, the crystallization kinetics of polymer/clay nanocomposites differ from the neat polymers as the clay layers can act as heterogeneous nucleation agents and are able to change the crystalline form of the polymer, rate of crystallization, and percentage of crystallinity. These changes in the crystallization kinetics can have a significant influence on the desired ultimate properties. Therefore, to better understand the influence of clay on the polymer performance, many of these indirect effects should be accurately described and/or identified.

One major disadvantage of polymer/clay nanocomposites is their embrittlement (reduced fracture toughness) behavior, severely limiting the range of potential applications of these materials. The approaches and research efforts that were adopted to resolve and understand this behavior were discussed in this chapter. It was concluded that the best known approach to-date is the addition of a third component (soft elastomer) to the polymer/clay nanocomposites. However, it is necessary to thoroughly understand the pros and cons of this approach for the final nanostructure and in turn their effects on the toughness and other mechanical properties, along with the fracture mechanisms so as to develop better polymer nanocomposites. Also, the usage of this approach has a compromising effect on the elastic modulus. To redress this problem, "tunable" block copolymers and pre-existent submicron voids were successfully used as shown by a few isolated studies. Further thorough work from these perspectives is, however, required. Much more work is also needed to develop analytical/numerical models to predict and verify the deformation, strength and toughness of these nanocomposites, whether based on continuum mechanics models or molecular dynamic simulations. Very often, poor characterizations of polymer nanocomposites and inadequate quantitative descriptions of observed phenomena have also led to apparent contradictions and misleading impressions on these materials.

Acknowledgements

The authors thank the Australian Research Council (ARC) for the continuing support of the project on "Polymer Nanocomposites". We also acknowledge the important contributions of various members of the Polymer Nanocomposites Group in the CAMT for their original data, useful discussions and constructive comments. Finally, permissions from various publishers to reproduce some figures in this chapter are much appreciated.

References

1. Kietzke T, Neher D, Landfester K, Montenegro R, Guntner R and Scherf U (2003) Novel approaches to polymer blends based on polymer nanoparticles, *Nat Mater* **2**:408–412.
2. Alexandre M and Dubois P (2000) Polymer-layered silicate nanocomposites: preparation, properties and uses of a new class of materials, *Mater Sci Eng R* **28**:1–63.

3. Dasari A, Lim S H, Yu Z-Z and Mai Y-W (2007) Toughening, thermal stability, flame retardancy, and scratch-wear resistance of polymer-clay nanocomposites, *Aust J Chem* **60**:496–518.
4. Giannelis E P (1996) Polymer layered silicate nanocomposites, *Adv Mater* **8**:29–35.
5. Giannelis E P, Krishnamoorti R and Manias E (1999) Polymer-silicate nanocomposites – model systems for confined polymers and polymer brushes, in *Advances in Polymer Science: Polymers in Confined Environments* (Ed. Granick S) Springer, Berlin, Vol. 138, pp. 107–147.
6. Krishnamoorti R, Vaia R A and Giannelis E P (1996) Structure and dynamics of polymer-layered silicate nanocomposites, *Chem Mater* **8**:1728–1734.
7. LeBaron P C, Wang Z and Pinnavaia T J (1999) Polymer-layered silicate nanocomposites: an overview, *Appl Clay Sci* **15**:11–29.
8. Messersmith P B and Giannelis E P (1994) Synthesis and characterization of layered silicate-epoxy nanocomposites, *Chem Mater* **6**:1719–1725.
9. Ray S S and Okamoto M (2003) Polymer/layered silicate nanocomposites: a review from preparation to processing, *Prog Polym Sci* **28**:1539–1641.
10. Tjong S C (2006) Structural and mechanical properties of polymer nanocomposites, *Mater Sci Eng R* **53**:73–197.
11. Yu Z-Z, Dasari A and Mai Y-W (2006) Polymer-clay nanocomposites – a review of their mechanical and physical properties, in *Processing and Properties of Polymer Nanocomposites* (Ed. Advani S G) World Scientific Publishers, Singapore, pp. 307–358.
12. Kojima Y, Usuki A, Kawasumi M, Okada A, Fukushima Y, Kurauchi T and Kamigaito O (1993) Mechanical properties of nylon 6-clay hybrid, *J Mater Res* **8**:1185–1189.
13. Aoike T, Uehara H, Yamanobe T and Komoto T (2001) Comparison of macro- and nanotribological behavior with surface plastic deformation of polystyrene, *Langmuir* **17**:2153–2159.
14. Bhushan B, Israelachvili J N and Landman U (1995) Nanotribology – friction, wear and lubrication at the atomic-scale, *Nature* **374**:607–616.
15. Fukushima Y and Inagaki S (1987) Synthesis of an intercalated compound of montmorillonite and polyamide–6, *J Inclusion Phenom* **5**:473–482.
16. Kojima Y, Usuki A, Kawasumi M, Okada A, Kurauchi T and Kamigaito O (1993) One-pot synthesis of nylon–6 clay hybrid, *J Polym Sci Polym Chem* **31**:1755–1758.
17. Kojima Y, Usuki A, Kawasumi M, Okada A, Kurauchi T and Kamigaito O (1993) Synthesis of nylon-6-clay hybrid by montmorillonite intercalated with ε-caprolactam, *J Polym Sci Polym Chem* **31**:983–986.
18. Usuki A, Kawasumi M, Kojima Y, Okada A, Kurauchi T and Kamigaito O (1993) Swelling behavior of montmorillonite cation exchanged for omega-amino acids by ε-caprolactam, *J Mater Res* **8**:1174–1178.
19. Balazs A C, Singh C and Zhulina E (1998) Modeling the interactions between polymers and clay surfaces through self-consistent field theory, *Macromolecules* **31**:8370–8381.
20. Vaia R A and Giannelis E P (1997) Lattice model of polymer melt intercalation in organically-modified layered silicates, *Macromolecules* **30**:7990–7999.
21. Vaia R A and Giannelis E P (1997) Polymer melt intercalation in organically-modified layered silicates: model predictions and experiment, *Macromolecules* **30**:8000–8009.
22. Hasegawa N, Okamoto H, Kawasumi M, Kato M, Tsukigase A and Usuki A (2000) Polyolefin-clay hybrids based on modified polyolefins and organophilic clay, *Macromol Mater Eng* **280**:76–79.

23. Hasegawa N and Usuki A (2004) Silicate layer exfoliation in polyolefin/clay nanocomposites based on maleic anhydride modified polyolefins and organophilic clay, *J Appl Polym Sci* **93**:464–470.
24. Kato M, Usuki A and Okada A (1997) Synthesis of polypropylene oligomer-clay intercalation compounds, *J Appl Polym Sci* **66**:1781–1785.
25. Kawasumi M, Hasegawa N, Kato M, Usuki A and Okada A (1997) Preparation and mechanical properties of polypropylene-clay hybrids, *Macromolecules* **30**:6333–6338.
26. Maiti P, Nam P H, Okamoto M, Hasegawa N and Usuki A (2002) Influence of crystallization on intercalation, morphology, and mechanical properties of polypropylene/clay nanocomposites, *Macromolecules* **35**:2042–2049.
27. Nam P H, Maiti P, Okamoto M, Kotaka T, Hasegawa N and Usuki A (2001) A hierarchical structure and properties of intercalated polypropylene/clay nanocomposites, *Polymer* **42**:9633–9640.
28. Tjong S C (2006) Synthesis and structure-property-characteristics of clay-polymer nanocomposites, in *Nanocrystalline Materials: Their Synthesis-Structure-Property Relationships and Applications* (Ed. Tjong S C) Elsevier, United Kingdom, pp. 311–348.
29. Usuki A, Hasegawa N and Kato M (2005) Polymer-clay nanocomposites, in *Advances in Polymer Science*: *Inorganic Polymeric Nanocomposites and Membranes*, Springer, Berlin, Vol. 179, pp. 135–195.
30. Fornes T D, Hunter D L and Paul D R (2004) Effect of sodium montmorillonite source on nylon 6/clay nanocomposites, *Polymer* **45**:2321–2331.
31. Lan T, Kaviratna P D and Pinnavaia T J (1995) Mechanism of clay tactoid exfoliation in epoxy-clay nanocomposites, *Chem Mater* **7**:2144–2150.
32. Usuki A, Koiwai A, Kojima Y, Kawasumi M, Okada A, Kurauchi T and Kamigaito O (1995) Interaction of nylon-6 clay surface and mechanical properties of nylon-6 clay hybrid, *J Appl Polym Sci* **55**:119–123.
33. Fornes T D, Yoon P J, Hunter D L, Keskkula H and Paul D R (2002) Effect of organoclay structure on nylon 6 nanocomposite morphology and properties, *Polymer* **43**:5915–5933.
34. Fornes T D, Yoon P J, Keskkula H and Paul D R (2001) Nylon 6 nanocomposites: the effect of matrix molecular weight, *Polymer* **42**:9929–9940.
35. Fu X and Qutubuddin S (2001) Polymer-clay nanocomposites: exfoliation of organophilic montmorillonite nanolayers in polystyrene, *Polymer* **42**:807–813.
36. Lee K M and Han C D (2003) Rheology of organoclay nanocomposites: effects of polymer matrix/organoclay compatibility and the gallery distance of organoclay, *Macromolecules* **36**:7165–7178.
37. Chavarria F and Paul D R (2006) Morphology and properties of thermoplastic polyurethane nanocomposites: effect of organoclay structure, *Polymer* **47**:7760–7773.
38. Fornes T D, Hunter D L and Paul D R (2004) Nylon–6 nanocomposites from alkylammonium-modified clay: the role of alkyl tails on exfoliation, *Macromolecules* **37**:1793–1798.
39. Varlot K, Reynaud E, Kloppfer M H, Vigier G and Varlet J (2001) Clay-reinforced polyamide: preferential orientation of the montmorillonite sheets and the polyamide crystalline lamellae, *J Polym Sci Polym Phys* **39**:1360–1370.
40. Cho J W and Paul D R (2001) Nylon 6 nanocomposites by melt compounding, *Polymer* **42**:1083–1094.
41. Dennis H R, Hunter D L, Chang D, Kim S, White J L, Cho J W and Paul D R (2001) Effect of melt processing conditions on the extent of exfoliation in organoclay-based nanocomposites, *Polymer* **42**:9513–9522.

42. Reichert P, Nitz H, Klinke S, Brandsch R, Thomann R and Mulhaupt R (2000) Poly(propylene)/organoclay nanocomposite formation: influence of compatibilizer functionality and organoclay modification, *Macromol Mater Eng* **275**:8–17.
43. Kim S W, Jo W H, Lee M S, Ko M B and Jho J Y (2002) Effects of shear on melt exfoliation of clay in preparation of nylon 6/organoclay nanocomposites, *Polym J* **34**:103–111.
44. Weon J I and Sue H J (2005) Effects of clay orientation and aspect ratio on mechanical behavior of nylon-6 nanocomposite, *Polymer* **46**:6325–6334.
45. Dasari A, Yu Z-Z, Mai Y-W and Yang M S (2008) The location and extent of exfoliation of clay on the fracture mechanisms in nylon 66-based ternary nanocomposites, *J Nanosci Nanotechnol* **8**:1901–1912.
46. Yoon P J, Fornes T D and Paul D R (2002) Thermal expansion behavior of nylon 6 nanocomposites, *Polymer* **43**:6727–6741.
47. Fornes T D and Paul D R (2003) Modeling properties of nylon 6/clay nanocomposites using composite theories, *Polymer* **44**:4993–5013.
48. Morgan A B and Gilman J W (2003) Characterization of polymer-layered silicate (clay) nanocomposites by transmission electron microscopy and x-ray diffraction – a comparative study, *J Appl Polym Sci* **87**:1329–1338.
49. Eckel D F, Balogh M P, Fasulo P D and Rodgers W R (2004) Assessing organo-clay dispersion in polymer nanocomposites, *J Appl Polym Sci* **93**:1110–1117.
50. Ohmae N, Martin J M and Mori S (2005) *Micro and Nanotribology*, ASME Press, New York.
51. Morgan A B and Harris J D (2003) Effects of organoclay Soxhlet extraction on mechanical properties, flammability properties and organoclay dispersion of polypropylene nanocomposites, *Polymer* **44**:2313–2320.
52. Xie W, Gao Z M, Pan W P, Hunter D, Singh A and Vaia R (2001) Thermal degradation chemistry of alkyl quaternary ammonium montmorillonite, *Chem Mater* **13**:2979–2990.
53. Xie W, Xie R C, Pan W P, Hunter D, Koene B, Tan L S and Vaia R (2002) Thermal stability of quaternary phosphonium modified montmorillonites, *Chem Mater* **14**:4837–4845.
54. Yu Z-Z, Guo-Hua H U, Varlet J, Dasari A and Mai Y-W (2005) Water-assisted melt compounding of nylon–6/pristine montmorillonite nanocomposites, *J Polym Sci Polym Phys* **43**:1100–1112.
55. Hasegawa N, Okamoto H, Kato M, Usuki A and Sato N (2003) Nylon 6/Na-montmorillonite nanocomposites prepared by compounding Nylon 6 with Na-montmorillonite slurry, *Polymer* **44**:2933–2937.
56. Chen B, Liu J, Chen H B and Wu J S (2004) Synthesis of disordered and highly exfoliated epoxy/clay nanocomposites using organoclay with catalytic function via acetone-clay slurry method, *Chem Mater* **16**:4864–4866.
57. Wang K, Chen L, Wu J S, Toh M L, He C B and Yee A F (2005) Epoxy nanocomposites with highly exfoliated clay: mechanical properties and fracture mechanisms, *Macromolecules* **38**:788–800.
58. Kato M, Matsushita M and Fukumori K (2004) Development of a new production method for a polypropylene-clay nanocomposite, *Polym Eng Sci* **44**:1205–1211.
59. Strawhecker K E and Manias E (2000) Structure and properties of poly(vinyl alcohol)/Na$^+$ montmorillonite nanocomposites, *Chem Mater* **12**:2943–2949.
60. Fornes T D and Paul D R (2003) Crystallization behavior of nylon 6 nanocomposites, *Polymer* **44**:3945–3961.

61. Aharoni S M (1997) *n-Nylons, their synthesis, structure, and properties*, John Wiley & Sons Ltd, Chichester.
62. Fornes T D, Yoon P J and Paul D R (2003) Polymer matrix degradation and color formation in melt processed nylon 6/clay nanocomposites, *Polymer* **44**:7545–7556.
63. Lincoln D M, Vaia R A and Krishnamoorti R (2004) Isothermal crystallization of nylon-6/montmorillonite nanocomposites, *Macromolecules* **37**:4554–4561.
64. Tseng C R, Wu S C, Wu J J and Chang F C (2002) Crystallization behavior of syndiotactic polystyrene nanocomposites for melt- and cold-crystallizations, *J Appl Polym Sci* **86**:2492–2501.
65. Wu T M and Chen E C (2002) Polymorphic behavior of nylon 6/saponite and nylon 6/montmorillonite nanocomposites, *Polym Eng Sci* **42**:1141–1150.
66. Wu T M and Wu J Y (2002) Structural analysis of polyamide/clay nanocomposites, *J Macromol Sci B* **41**:17–31.
67. Devaux E, Bourbigot S and El Achari A (2002) Crystallization behavior of PA-6 clay nanocomposite hybrid, *J Appl Polym Sci* **86**:2416–2423.
68. Zapata-Espinosa A, Medellin-Rodriguez F J, Stribeck N, Almendarez-Camarillo A, Vega-Diaz S, Hsiao B S and Chu B (2005) Complex isothermal crystallization and melting behavior of nylon 6 nanoclay hybrids, *Macromolecules* **38**:4246–4253.
69. Ma C C M, Kuo C T, Kuan H C and Chiang C L (2003) Effects of swelling agents on the crystallization behavior and mechanical properties of polyamide 6/clay nanocomposites, *J Appl Polym Sci* **88**:1686–1693.
70. Maiti P and Okamoto M (2003) Crystallization controlled by silicate surfaces in nylon 6-clay nanocomposites, *Macromol Mater Eng* **288**:440–445.
71. Zhang Q-X, Yu Z-Z, Yang M S, Ma J and Mai Y-W (2003) Multiple melting and crystallization of nylon–66/montmorillonite nanocomposites, *J Polym Sci Polym Phys* **41**:2861–2869.
72. Fujimoto K, Yoshikawa M, Katahira S and Yasue K (2000) Crystal structure of nylon 6/inorganic layered silicate nanocomposite film, *Kobunshi Ronbunshu* **57**:433–439.
73. Lincoln D M, Vaia R A, Wang Z G, Hsiao B S and Krishnamoorti R (2001) Temperature dependence of polymer crystalline morphology in nylon 6/montmorillonite nanocomposites, *Polymer* **42**:9975–9985.
74. Wu H D, Tseng C R and Chang F C (2001) Chain conformation and crystallization behavior of the syndiotactic polystyrene nanocomposites studied using Fourier transform infrared analysis, *Macromolecules* **34**:2992–2999.
75. Carter C M (2001) Crystallization kinetics of poly(ethylene-terephthalate) based ionomer nanocomposite materials, in *Annual Technical Conference*, Dallas (The Society of Plastics Engineers), pp. 3210–3213.
76. Liu Z J, Zhou P L and Yan D Y (2004) Preparation and properties of nylon-1010/montmorillonite nanocomposites by melt intercalation, *J Appl Polym Sci* **91**:1834–1841.
77. Krikorian V, Kurian M, Galvin M E, Nowak A P, Deming T J and Pochan D J (2002) Polypeptide-based nanocomposite: structure and properties of poly(L-lysine)/Na+-montmorillonite, *J Polym Sci Polym Phys* **40**:2579–2586.
78. Han B, Ji G D, Wu S S and Shen J (2003) Preparation and characterization of nylon 66/montmorillonite nanocomposites with co-treated montmorillonites, *Eur Polym J* **39**:1641–1646.
79. Priya L and Jog J P (2002) Poly(vinylidene fluoride)/clay nanocomposites prepared by melt intercalation: crystallization and dynamic mechanical behavior studies, *J Polym Sci Polym Phys* **40**:1682–1689.

80. Shah D, Maiti P, Gunn E, Schmidt D F, Jiang D D, Batt C A and Giannelis E R (2004) Dramatic enhancements in toughness of polyvinylidene fluoride nanocomposites via nanoclay-directed crystal structure and morphology, *Adv Mater* **16**:1173–1177.
81. Kyotani M and Mitsuhas S (1972) Studies on crystalline forms of nylon-6 II – crystallization from melt, *J Polym Sci A2* **10**:1497–1508.
82. Murthy N S, Aharoni S M and Szollosi A B (1985) Stability of the gamma-form and the development of the alpha-form in nylon-6, *J Polym Sci Polym Phys* **23**:2549–2565.
83. Salem D R, Moore R A F and Weigmann H D (1987) Macromolecular order in spin-oriented nylon-6 (polycaproamide) fibers, *J Polym Sci Polym Phys* **25**:567–589.
84. Gurato G, Fichera A, Grandi F Z, Zannetti R and Canal P (1974) Crystallinity and polymorphism of polyamide–6, *Makromol Chem* **175**:953–975.
85. Krishnamoorti R and Giannelis E P (1997) Rheology of end-tethered polymer layered silicate nanocomposites, *Macromolecules* **30**:4097–4102.
86. Yee A F, Li D M and Li X W (1993) The importance of constraint relief caused by rubber cavitation in the toughening of epoxy, *J Mater Sci* **28**:6392–6398.
87. Chan C M, Wu J S, Li J X and Cheung Y K (2002) Polypropylene/calcium carbonate nanocomposites, *Polymer* **43**:2981–2992.
88. Bagheri R and Pearson R A (1996) Role of particle cavitation in rubber-toughened epoxies I – microvoid toughening, *Polymer* **37**:4529–4538.
89. Chen X H and Mai Y-W (1998) Micromechanics of rubber-toughened polymers, *J Mater Sci* **33**:3529–3539.
90. Bartczak Z, Argon A S, Cohen R E and Weinberg M (1999) Toughness mechanism in semi-crystalline polymer blends I – high-density polyethylene toughened with rubbers, *Polymer* **40**:2331–2346.
91. He C, Liu T, Tjiu W C, Sue H J and Yee A F (2008) Microdeformation and fracture mechanisms in polyamide-6/organoclay nanocomposites, *Macromolecules* **41**:193–202.
92. Dasari A, Yu Z-Z, Mai Y-W, Hu G H and Varlet J L (2005) Clay exfoliation and organic modification on wear of nylon 6 nanocomposites processed by different routes, *Compos Sci Technol* **65**:2314–2328.
93. Chen L, Wong S C and Pisharath S (2003) Fracture properties of nanoclay-filled polypropylene, *J Appl Polym Sci* **88**:3298–3305.
94. Dasari A, Yu Z-Z, Yang M S, Zhang Q-X, Xie X L and Mai Y-W (2006) Micro- and nano-scale deformation behavior of nylon 66-based binary and ternary nanocomposites, *Compos Sci Technol* **66**:3097–3114.
95. Yu Z-Z, Yang M S, Zhang Q-X, Zhao C G and Mai Y-W (2003) Dispersion and distribution of organically modified montmorillonite in nylon–66 matrix, *J Polym Sci Polym Phys* **41**:1234–1243.
96. Chen L, Phang I Y, Wong S C, Lv P F and Liu T X (2006) Embrittlement mechanisms of nylon 66/organoclay nanocomposites prepared by melt-compounding process, *Mater Manuf Process* **21**:153–158.
97. Zerda A S and Lesser A J (2001) Intercalated clay nanocomposites: morphology, mechanics, and fracture behavior, *J Polym Sci Polym Phys* **39**:1137–1146.
98. Fröhlich J, Golombowski D, Thomann R and Mulhaupt R (2004) Synthesis and characterisation of anhydride-cured epoxy nanocomposites containing layered silicates modified with phenolic alkylimidazolineamide cations, *Macromol Mater Eng* **289**:13–19.
99. Zilg C, Mulhaupt R and Finter J (1999) Morphology and toughness/stiffness balance of nanocomposites based upon anhydride-cured epoxy resins and layered silicates, *Macromol Chem Phys* **200**:661–670.

100. Becker O, Cheng Y B, Varley R J and Simon G P (2003) Layered silicate nanocomposites based on various high-functionality epoxy resins: the influence of cure temperature on morphology, mechanical properties, and free volume, *Macromolecules* **36**:1616–1625.
101. Liu X H and Wu Q J (2002) Polyamide 66/clay nanocomposites via melt intercalation, *Macromol Mater Eng* **287**:180–186.
102. Kim G M, Lee D H, Hoffmann B, Kressler J and Stoppelmann G (2001) Influence of nanofillers on the deformation process in layered silicate/polyamide–12 nanocomposites, *Polymer* **42**:1095–1100.
103. Kim G M, Goerlitz S and Michler G H (2007) Deformation mechanism of nylon 6/layered silicate nanocomposites: role of the layered silicate, *J Appl Polym Sci* **105**:38–48.
104. Shah D, Maiti P, Jiang D D, Batt C A and Giannelis E P (2005) Effect of nanoparticle mobility on toughness of polymer nanocomposites, *Adv Mater* **17**:525–528.
105. Gersappe D (2002) Molecular mechanisms of failure in polymer nanocomposites, *Phys Rev Lett* **89**:058301-058304.
106. Zhou T H, Ruan W H, Rong M Z, Zhang M Q and Mai Y L (2007) Keys to toughening of non-layered nanoparticles/polymer composites, *Adv Mater* **19**:2667–2671.
107. Dasari A, Yu Z-Z, Mai Y-W and Kim J-K (2008) Orientation and the extent of exfoliation of clay on scratch damage in polyamide 6 nanocomposites, *Nanotechnology* **19**:055708-055721.
108. Goda T, Váradi K, Friedrich K and Giertzsch H (2002) Finite element analysis of a polymer composite subjected to a sliding steel asperity I – normal fibre orientation, *J Mater Sci* **37**:1575–1583.
109. Yu Z-Z, Yan C, Yang M S and Mai Y-W (2004) Mechanical and dynamic mechanical properties of nylon 66/montmorillonite nanocomposites fabricated by melt compounding, *Polym Int* **53**:1093–1098.
110. Nair S V, Goettler L A and Lysek B A (2002) Toughness of nanoscale and multiscale polyamide-6,6 composites, *Polym Eng Sci* **42**:1872–1882.
111. Lim S-H, Dasari A, Yu Z-Z, Mai Y-W, Liu S L and Yong M S (2007) Fracture toughness of nylon 6/organoclay/elastomer nanocomposites, *Compos Sci Technol* **67**:2914–2923.
112. Li Y J and Shimizu H (2006) Effects of spacing between the exfoliated clay platelets on the crystallization behavior of polyamide–6 in polyamide–6/clay nanocomposites, *J Polym Sci Polym Phys* **44**:284–290.
113. Sheng N, Boyce M C, Parks D M, Rutledge G C, Abes J I and Cohen R E (2004) Multiscale micromechanical modeling of polymer/clay nanocomposites and the effective clay particle, *Polymer* **45**:487–506.
114. Muratoglu O K, Argon A S, Cohen R E and Weinberg M (1995) Toughening mechanism of rubber-modified polyamides, *Polymer* **36**:921–930.
115. Corte L, Beaume F and Leibler L (2005) Crystalline organization and toughening: example of polyamide-12, *Polymer* **46**:2748–2757.
116. Dasari A, Yu Z-Z and Mai Y-W (2007) Transcrystalline regions in the vicinity of nanofillers in polyamide-6, *Macromolecules* **40**:123–130.
117. Campbell D and Qayyum M M (1980) Melt crystallization of polypropylene – effect of contact with fiber substrates, *J Polym Sci Polym Phys* **18**:83–93.
118. Varga J and Karger-Kocsis J (1994) The difference between transcrystallization and shear-induced cylindritic crystallization in fiber-reinforced polypropylene composites, *J Mater Sci Lett* **13**:1069–1071.

119. Varga J and Karger-Kocsis J (1996) Rules of supermolecular structure formation in sheared isotactic polypropylene melts, *J Polym Sci Polym Phys* **34**:657–670.
120. Chatterjee A M, Price F P and Newman S (1975) Heterogeneous nucleation of crystallization of high polymers from melt – II – Aspects of transcrystallinity and nucleation density, *J Polym Sci Polym Phys* **13**:2385–2390.
121. Chatterjee A M, Price F P and Newman S (1975) Heterogeneous nucleation of polymer crystallization from melt – I – Substrate induced morphologies, *Bull Amer Phys Soc* **20**:341–341.
122. Goldfarb L (1980) Transcrystallization of isotactic polypropylene, *Makromol Chem* **181**:1757–1762.
123. Cho K W, Kim D W and Yoon S (2003) Effect of substrate surface energy on transcrystalline growth and its effect on interfacial adhesion of semicrystalline polymers, *Macromolecules* **36**:7652–7660.
124. Chen E J H and Hsiao B S (1992) The effects of transcrystalline interphase in advanced polymer composites, *Polym Eng Sci* **32**:280–286.
125. Bartczak Z, Argon A S, Cohen R E and Kowalewski T (1999) The morphology and orientation of polyethylene in films of sub-micron thickness crystallized in contact with calcite and rubber substrates, *Polymer* **40**:2367–2380.
126. Wang Y, Zhang Q and Fu Q (2003) Compatibilization of immiscible poly(propylene)/polystyrene blends using clay, *Macromol Rapid Commun* **24**:231–235.
127. Dasari A, Zhang Q-X, Yu Z-Z and Mai Y-W (2008) Significance of submicron to nano-sized voids in toughening polypropylene, *in preparation*.
128. Kelnar I, Kotek J, Kaprálkova L and Munteanu B S (2005) Polyamide nanocomposites with improved toughness, *J Appl Polym Sci* **96**:288–293.
129. Wang K, Wang C, Li J, Su J X, Zhang Q, Du R N and Fu Q (2007) Effects of clay on phase morphology and mechanical properties in polyamide 6/EPDM-g-MA/organoclay ternary nanocomposites, *Polymer* **48**:2144–2154.
130. Wu S H (1988) A generalized criterion for rubber toughening – the critical matrix ligament thickness, *J Appl Polym Sci* **35**:549–561.
131. González I, Eguiazábal J I and Nazábal J (2006) Rubber-toughened polyamide 6/clay nanocomposites, *Compos Sci Technol* **66**:1833–1843.
132. Khatua B B, Lee D J, Kim H Y and Kim J K (2004) Effect of organoclay platelets on morphology of nylon-6 and poly(ethylene-*ran*-propylene) rubber blends, *Macromolecules* **37**:2454–2459.
133. Dong W F, Zhang X H, Liu Y I, Gui H, Wang Q G, Gao J M, Song Z H, Lai J M, Huang F and Qiao J L (2007) Process for preparing a nylon-6/clay/acrylate rubber nanocomposite with high toughness and stiffness, *Polym Int* **56**:870–874.
134. Chiu F C, Lai S M, Chen Y L and Lee T H (2005) Investigation on the polyamide 6/organoclay nanocomposites with or without a maleated polyolefin elastomer as a toughener, *Polymer* **46**:11600–11609.
135. Li Y M, Wei G X and Sue H J (2002) Morphology and toughening mechanisms in clay-modified styrene-butadiene-styrene rubber-toughened polypropylene, *J Mater Sci* **37**:2447–2459.
136. Liu W P, Hoa S V and Pugh M (2004) Morphology and performance of epoxy nanocomposites modified with organoclay and rubber, *Polym Eng Sci* **44**:1178–1186.
137. Gam K T, Miyamoto M, Nishimura R and Sue H J (2003) Fracture behavior of core-shell rubber-modified clay-epoxy nanocomposites, *Polym Eng Sci* **43**:1635–1645.

138. Tjong S C and Meng Y Z (2003) Impact-modified polypropylene/vermiculite nanocomposites, *J Polym Sci Polym Phys* **41**:2332–2341.
139. Contreras V, Cafiero M, Da Silva S, Rosales C, Perera R and Matos M (2006) Characterization and tensile properties of ternary blends with PA-6 nanocomposites, *Polym Eng Sci* **46**:1111–1120.
140. Li X C, Park H M, Lee J O and Ha C S (2002) Effect of blending sequence on the microstructure and properties of PBT/EVA-g-MAH/organoclay ternary nanocomposites, *Polym Eng Sci* **42**:2156–2164.
141. Sue H J, Gam K T, Bestaoui N, Clearfield A, Miyamoto M and Miyatake N (2004) Fracture behavior of alpha-zirconium phosphate-based epoxy nanocomposites, *Acta Mater* **52**:2239–2250.
142. Dasari A, Yu Z-Z and Mai Y-W (2007) Nanoscratching of nylon 66-based ternary nanocomposites, *Acta Mater* **55**:635–646.
143. Dasari A, Yu Z-Z and Mai Y-W (2005) Effect of blending sequence on microstructure of ternary nanocomposites, *Polymer* **46**:5986–5991.
144. Oshinski A J, Keskkula H and Paul D R (1996) The role of matrix molecular weight in rubber toughened nylon 6 blends III – Ductile-brittle transition temperature, *Polymer* **37**:4919–4928.
145. Corte L, Rebizant V, Hochstetter G, Tournilhac F and Leibler L (2006) Toughening with little stiffness loss: polyamide filled with ABC triblock copolymers, *Macromolecules* **39**:9365–9374.
146. Bartczak Z, Argon A S, Cohen R E and Weinberg M (1999) Toughness mechanism in semi-crystalline polymer blends II – High-density polyethylene toughened with calcium carbonate filler particles, *Polymer* **40**:2347–2365.
147. Zuiderduin W C J, Westzaan C, Huetink J and Gaymans R J (2003) Toughening of polypropylene with calcium carbonate particles, *Polymer* **44**:261–275.
148. Bagheri R and Pearson R A (1995) The use of microvoids to toughen polymers, *Polymer* **36**:4883–4885.
149. Bagheri R and Pearson R A (1996) Role of particle cavitation in rubber-toughened epoxies I – Microvoid toughening, *Polymer* **37**:4529–4538.
150. Bagheri R and Pearson R A (2000) Role of particle cavitation in rubber-toughened epoxies II – Inter-particle distance, *Polymer* **41**:269–276.
151. Guild F J and Young R J (1989) A predictive model for particulate filled composite materials II – Soft particles, *J Mater Sci* **24**:2454–2460.
152. Huang Y and Kinloch A J (1992) The toughness of epoxy polymers containing microvoids, *Polymer* **33**:1330–1332.
153. Johnsen B B, Kinloch A J, Mohammed R D, Taylor A C and Sprenger S (2007) Toughening mechanisms of nanoparticle-modified epoxy polymers, *Polymer* **48**:530–541.
154. Johnsen B B, Kinloch A J and Taylor A C (2005) Toughness of syndiotactic polystyrene/epoxy polymer blends: microstructure and toughening mechanisms, *Polymer* **46**:7352–7369.
155. Tzika P A, Boyce M C and Parks D M (2000) Micromechanics of deformation in particle-toughened polyamides, *J Mech Phys Solids* **48**:1893–1929.
156. Steenbrink A C, Van der Giessen E and Wu P D (1998) Studies on the growth of voids in amorphous glassy polymers, *J Mater Sci* **33**:3163–3175.

Chapter 12

On the Toughness of "Nanomodified" Polymers and Their Traditional Polymer Composites

J. Karger-Kocsis

12.1. Introduction

Modification of polymers and traditional polymer composites by various nanoparticles has attracted considerable interest in both academia and industry. This is due to the outstanding property profile achieved at low "nanofiller" content. Property improvements may cover both structural (mechanical performance) and functional properties (*e.g.*, thermal and electric conductivities). The related "nanocomposites" contain inorganic or organic fillers the size of which, at least in one dimension, is in the nanometer range. Nanofillers can be produced *in situ* or incorporated in polymers as preformed particles. Nanofiller *in situ* may be generated by reactions of organic or inorganic precursor molecules or by the dispersion of fillers the mean particle size of which is in the micron range initially (*e.g.*, exfoliation of layered silicates). Like traditional fillers and reinforcements, the nanoparticles can be grouped with respect to their shape and thus aspect ratio (ratio of length to thickness). Accordingly, one can distinguish *spherical* (spheroidal), *platy* (disc-shaped) and *fibrous* (tube, needle-like) *fillers*.

The unexpected performance of "nanomodified" polymers is traced to the filler size, filler dispersion, and filler/matrix interactions. The effects of filler size can be highlighted by two aspects: (i) the probability of defects, causing premature failure, decreases with decreasing size of the solids, and (ii) learning from nature that in biomaterials nanometer-sized minerals (below *ca.* 30 nm) behave as perfect ones, *i.e.*, they are able to sustain the theoretical strength in spite of inherent flaws [1]. In accordance with a large body of experimental results, modeling studies (*e.g.*, [2–3]) also confirmed that the dispersion of nanofillers (arrangement, orientation, layering, packing, *etc.*) affects the mechanical response. Similarly to traditional microcomposites, the interfacial stress transfer from the "weak" polymer matrix to the "strong" particle is a key issue. Owing to the huge specific surface of well dispersed nanoparticles, the properties of the related nanocomposites are obviously controlled by interphase characteristics. This was pinpointed by Krishnamoorti and Vaia [4] calling attention to the fact that the characteristic dimensions of the polymer chain (radius of gyration), reinforcing filler (diameter or thickness), and interparticle distance of the latter are closely matched under "ideal" conditions.

The structure-property relationships in nanocomposites were already the topic of excellent monographs (*e.g.*, [5–9]). To describe the stiffness and strength as a function of the nanofiller dispersion several analytical techniques and models have been proposed and their validity checked. Many of them, well established for traditional microcomposites, worked well also for "real" nanocomposites. The attribute "real" means that instead of ideal dispersion the particles are strongly agglomerated, or stacked. The related aspects are disclosed in some chapters of this book, as well (see Chapters 3, 10 and 11). By contrast, there is still very little understanding of the dependence of toughness on characteristics of the nanofillers. Note that cover "characteristics" here the modulus, aspect ratio, dispersion quality, interphase properties, *etc.* Usually, an increase in stiffness and strength is accompanied by a prominent decrease in the ductility (*e.g.*, elongation at break), which means a reduced toughness of the related system at the same time. In the literature, as disclosed later, there are many reports quoting the opposite, *i.e.*, increase of toughness simultaneously with stiffness and strength. Nature shows us that this may happen, in fact. Albeit the stiffness of some biocomposites, like hard tissues (tooth enamel, nacre of shell) is close to their mineral constituents, their fracture energy may be orders of magnitude larger than that of the corresponding monolithic mineral. Gao *et al.* [1,10] quoted that the concept of stress concentration at flaws (recall that this is the base of fracture mechanics) is not valid in such biological nanomaterials. These biocomposites were modeled as perfectly aligned, staggered mineral platelets embedded in a protein matrix. The unexpected high fracture toughness was explained by a "tension-shear chain" model. According to this model, the tensile strength of the mineral is optimized by its size (reduced to the nanoscale) and the stress is perfectly transferred from the confined protein layers between the platelets *via* efficient shearing.

The toughness of nanocomposites deserves a separate treatise, owing to highly contradictory results in the open literature. As mentioned before, there is trade-off between stiffness, strength and toughness. Nanoparticles with reinforcing effect should result in toughness reduction. In many cases, however, the opposite tendency was found for which no straightforward explanation exists. It is most likely that the toughness enhancing mechanisms are different for thermoplastics and thermosets, and they may also depend on characteristics of the fillers. That was the driving force to prepare this chapter. The toughness response of nanomodified systems will be presented in the following sequence: spherical, platy, and fibrous fillers. It should be emphasized here that fracture mechanical studies on polymer nanocomposites are rare. On the other hand, only fracture mechanical methods provide toughness values which can be collated with each other, being material parameters. Accordingly, the fracture mechanical approach, coupled with extensive studies on the (nano)structure and morphology of the nanomodified systems, are the right tools to shed light on the structure-toughness relationship. However, due to the lack of fracture mechanical results, works using other toughness assessing standardized (*e.g.*, Charpy and Izod impact) and non-standardized techniques have also been considered. Note that the interested reader may get further valuable information on this topic when reading Chapters 10 and 11 in this book.

12.2. Toughness assessment

Nowadays, various nanomodified systems are available and used in different structural applications. *Toughness* is a key property in the related application fields that can be determined by various standardized test methods. However, the toughness results can only be compared if the specimen preparation and testing conditions were exactly the same. By contrast, when adopting the fracture mechanical concept the inherent toughness of a given material can be determined. This seems to be essential to elucidate the structure-toughness relationships.

Fracture mechanics aims at determining the response of a cracked material to applied load, and at offering methods to measure the toughness. The related approaches are grouped in linear elastic (plane strain conditions), elastic-plastic, and post-yield fracture mechanics (plane stress conditions). Linear elastic fracture mechanics works for brittle systems that fail by catastrophic crack growth after reaching a threshold load (stress) value. The related criteria rely either on the stress field ahead of the crack tip (stress intensity factor or fracture toughness, K_c) or the energy release during crack extension (strain energy release rate or fracture energy, G_c). K_c and G_c find application to polymers undergoing brittle fracture in the related test. This prerequisite usually holds for thermosets and thermoset-based composites but is not always applicable to thermoplastics. For the toughness assessment of tough polymers the J-integral, the crack tip opening displacement, and the *essential work of fracture* (EWF) methods are mostly used. The *J-integral* represents a path-independent integral around the crack tip and thus considers also the plastic deformation at the crack tip. The critical value of the J-integral (J_c) is accompanied by full crack tip blunting prior to crack growth. J_c is usually deduced from the J resistance (J_R) curve when adopting the multiple specimen technique. This approach works well for not too ductile polymeric systems. The crack tip opening displacement criterion is linked to the crack opening prior to its advance. However, this is seldom followed for polymers. The EWF approach is gaining acceptance for the determination of the toughness response of highly ductile polymers. The greatest advantage of EWF over the J-integral is that a clear distinction between surface (essential part) and volume-related (non-essential part) works is made. For a detailed description of the various fracture mechanical methods, testing standards, their applicability to polymers and polymer composites, the interested reader is addressed to valuable books [11–17].

In order to structure the knowledge on the structure-toughness relationships the literature and our own results will be grouped according to the filler type, *viz.* spherical, platy, and fibrillar. Before starting with the treatise, it has to be underlined that studies addressing "nanoeffects" by comparing the properties of micro- and nanocomposites, composed of the same constituents, are very rare.

12.3. Nanomodified thermoplastics

12.3.1. *Amorphous polymers*

Surveying the literature one recons that nanofillers are less used for amorphous glassy compared to semicrystalline thermoplastic polymers. Moreover, only few reports deal with effects of nanofillers on the toughness of amorphous thermoplastics. The possible reason for this will be disclosed later.

Spherical fillers

As spherical or quasispherical particles usually silica (SiO_2), chalk ($CaCO_3$), TiO_2, alumina (Al_2O_3 or $AlO(OH)$) and ZnO have been used to modify thermoplastics. Wu et al. [18] prepared poly(vinyl chloride) (PVC) nanocomposites by melt blending with nanosized $CaCO_3$ (40 nm) in the presence and absence of chlorinated polyethylene (CPE) as impact modifier. The Izod impact strength increased monotonously with increasing $CaCO_3$ content up to 30 weight % (wt%) for both binary (PVC/chalk) and ternary (PVC/CPE/chalk) blends. A similar tendency was observed for the elongation at break data determined in static tensile tests. This is in line with the usual belief that toughness and ductility values correlate with each other. The toughness enhancement was attributed to the stress concentration effect of the nanoparticles promoting cavitation at the interface between the particle and matrix (debonding) that is followed by plastic deformation of the PVC between the particles (matrix ligament). Lach et al. [19] determined the fracture toughness (K_c) of poly(methyl methacrylate) (PMMA) nanocomposites by indentation fracture mechanics (a rarely used technique for polymers). The related systems contained SiO_2 nanoparticles which were incorporated up to 20 wt% through solution blending. K_c increased from the initial 1.25 $MPa.m^{1/2}$ to 2.12 and 1.52 $MPa.m^{1/2}$ for the nanocomposites with 10 and 20 wt% silica, respectively. The authors interpreted this toughness response by considering the interparticle model of Wu [20]. They estimated the thickness of the interphase, which was "softer" than the matrix, based on differential scanning calorimetry (DSC) results and found it for *ca.* 9 nm. The

K_c decrease at 20 wt% SiO_2 was assigned to the formation of a percolation network, i.e., the SiO_2 particles with their bound layers (interphase) touched each other which excluded the deformation of the matrix ligaments. This interpretation, if correct, calls the attention to the key role of the matrix in respect to the toughness improvement. On the other hand, it is unclear why the interphase, containing more mobile molecules than the bulk, does not promote a plastic type deformation. Ash et al. [21] studied the mechanical behavior of PMMA/alumina micro- and nanocomposites. The related materials were produced by bulk polymerization with and without a copolymerizable monomeric dispersant. Unlike the micro- and nanocomposites with 17 nm mean particle size, the nanocomposite containing alumina with 38 nm particle size underwent a brittle-to-ductile transition in uniaxial tension. The cited authors concluded that both enhanced chain mobility and the existence of larger particles with weak interface are needed to cause the required change in the strain state, i.e., from plane strain-to-plane stress. This transition favors the shear yielding of the matrix instead of its normal yielding associated with cavitation and crazing. The high chain mobility is delivered by the added surface area of the nanoparticle causing a glass transition temperature (T_g) suppression. Note that this experimental fact contradicts the expectation. Strong interaction between the matrix molecules and particles should cause an adverse change, namely T_g increase due to hampered mobility of the molecules in the interphase. So, the most interesting finding of the above studies [20–21] is that even in the case of amorphous thermoplastics, nanofillers may enhance the molecular mobility in the interphase. Recall that ductile PMMA in the work by Ash et al. [21] was produced by using a monomeric dispersant, which likely promoted the formation of a soft interphase. As a consequence, the particle and the interphase together may be considered as a core(hard)-shell(soft) structure. The group of Müllen [22–23] elaborated a method for the in situ compatibilization of nanoparticles (ZnO, $MgCO_3$) in inverse emulsion (co)polymerization. The particles, covered by copolymers with an average molecular mass below 15 kDa, were fully redispersible in organic fluids. Incorporation of $MgCO_3$ nanoparticles (70 nm), covered by a shell composed of poly(2-ethylhexyl-stat-ethylenoxide)methacrylate, was incorporated in polystyrene (PS) in 2.5 wt%. This did not yield the expected toughness enhancement as the cavitation was followed by crazing instead of shear yielding (cf. Figure 12.1).

The patchy surface pattern in Figure 12.1 is due to the onset of crazes which broke up in their early development stage. It is also noteworthy that the incorporation of this compatibilized $MgCO_3$ (which is again a core-shell type particle from which flame retardant activity was also expected) was accompanied with a T_g increase. Similar toughness reduction was observed when synthetic boehmite alumina (AlO(OH)) in 220 nm particle size was incorporated via its water slurry predispersion [24–25] in the PS melt. The scanning electron microscopic (SEM) picture in Figure 12.2 demonstrates that cavitation, i.e., debonding of the alumina particles from the matrix, was not efficient enough to evoke the change from plane strain (triaxial stress state) to plane stress

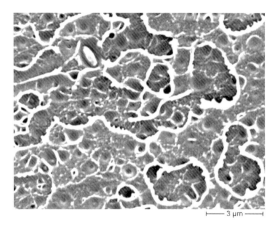

Figure 12.1. Scanning electron microscopic picture taken from the fracture surface of PS containing 2.5 wt% $MgCO_3$ coated with a statistical methacrylate copolymer ($MgCO_3$ content of these core-shell particles is 22 wt%)

(biaxial stress state). Though the water-mediated predispersion of the boehmite alumina resulted in a better dispersion of the related particles, this did not help to reach the above target.

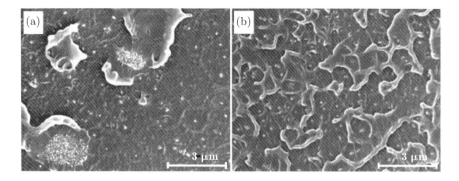

Figure 12.2. SEM pictures taken from the fracture surfaces of PS containing 3 wt% boehmite alumina. The filler (mean size in aqueous dispersion: 220 nm) was incorporated by direct melt blending (a) and by predispersion in water that was introduced in the PS melt during extrusion compounding (b)

Platy fillers

Toughness reduction was also noticed when layered silicate was incorporated in PS. In the related study PS/layered silicate micro- and nanocomposites were produced. Microcomposite was generated by direct melt blending of PS with the silicate. Intercalated nanocomposites were, however, achieved when the pristine layered silicate was swollen in water before adding the related slurry in

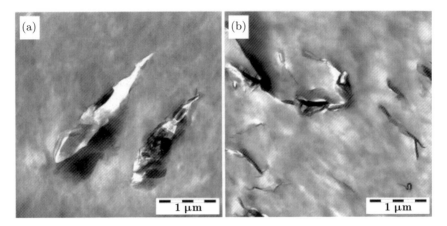

Figure 12.3. Transmission electron microscopic (TEM) frames taken from PS/pristine layered silicate (sodium fluorohectorite) composites. The microcomposite (a) was produced by direct melt blending, whereas the nanocomposite (b) was achieved by preswelling the layered silicate in water prior to dosing the corresponding slurry in the PS melt

the PS melt during melt compounding [26]. The difference in the dispersion of the layered silicate is well reflected by the TEM pictures in Figure 12.3. The only difference in the toughness characteristics was that the reduction, compared to that of the PS matrix, was considerably smaller for the nano- than for the microcomposite.

This is the right place to make the reader acquainted with the conclusion of Corté and Leibler [27]. They claimed that the modifiers should bring the matrix confinement (let us interpret it as free ligament between the particles) below the characteristic length of the matrix heterogeneity. Note that the inherent heterogeneities of amorphous glassy polymers are on the nanoscale. Further, the shear yield stresses of these polymers are quite large. Recall that the PS/boehmite and PS/layered silicate nanocomposites failed by normal fracture. So, the toughness enhancement of amorphous glassy polymers by nanofillers is a problematic issue.

Fibrous fillers

Carbon nanotubes (CNT) and carbon nanofibers (CNF) belong to the fibrous reinforcements. They possess outstanding stiffness and strength along with a very high aspect ratio. Multiwall carbon nanotubes (MWCNT), for example, have stiffness in TPa range, strength <60 GPa, and their aspect ratio may reach 1000. Based on clear analogies with traditional fibrous reinforcements, one would expect that incorporation of CNF and CNT yields improved stiffness and strength and reduced toughness. This is more or less in line with experimental results. The toughness of MWCNT filled (up to 6 wt%) polycarbonate (PC) was assessed by the J-integral and EWF approaches recently [28]. The authors found that the resistance against crack initiation (J_c, essential work of fracture) goes through a minimum (before reaching or slightly surpassing the related value of the matrix), whereas the resistance against crack propagation (tearing modulus, non-essential work of fracture term) passes a maximum (before dropping below the related value of the matrix) as a function of the MWCNT content. The related extremes (maximum, minimum), associated with a tough-to-brittle transition, were found at *ca.* 2 wt% MWCNT. This transition was matched with the percolation transition whereby an electrically conductive network forms from the attaching MWCNTs. It is noteworthy that the adverse run of the essential and non-essential work of fracture terms was already proposed by the author based on results on neat and traditionally filled polymers [29–30].

The present knowledge on the toughness of nanomodified amorphous glassy thermoplastics can roughly be summarized as follows:

– Filler type: the toughness enhancing potential is spherical > platy > fibrous;
– Interphase: the molecules involved should have a higher mobility than the bulk (*i.e.*, lower modulus than the matrix, T_g reduction). In addition, the interface should be "weak" enough to cause stress relief *via* debonding (cavitation);
– Matrix: the ligament between the particles (including interphase) should stay locally under plane stress state in order to favor shear yielding, plastic deformation.

The author believes that the toughness enhancement of amorphous polymers by nanofillers is less straightforward.

12.3.2. *Semicrystalline polymers*

Many semicrystalline polymers, such as polyethylenes (PE), polypropylenes (PP), polyamides (PA) and linear polyesters are tough when tested above ambient temperature and at low deformation rates. However, they embrittle at low temperatures and under high deformation rate loading (impact). Embrittlement may be even more pronounced when notched specimens are tested (notch sensitivity). The above polymers have been successfully toughened by incorporation of suitable rubber particles (soft inclusions). On the other hand, rubber toughening is accompanied by a considerable stiffness loss, which is undesirable. That is the reason why the research interest turned to using rigid particulates for toughening in the 1980s.

12.3.2.1. *Homopolymers*

Spherical fillers

In the case of particulate composites, the toughness is governed by debonding of the particles from the matrix and follow-up matrix deformation processes [31]. A further early wisdom is that inorganic fillers should be treated by surfactants (physically absorbed on the particle surface) or coupling agents (chemically absorbed) to guarantee their homogenous dispersion in the polymer matrix. On the other hand, contradictory results are available on how the above treatments (termed interfacial interaction) affect the toughness response of the composites. This is not unexpected, considering the interplay between interfacial adhesion and susceptibility to debonding (cavitation). Bezerédi *et al.* [32] investigated the effect of surface treatment of microsized $CaCO_3$ (surfactant and reactive silane coupling agent) on the dynamic fracture properties of PP. The *fracture energy* (G_c) proved to be more sensitive to surface treatment of the filler than the *fracture toughness* (K_c). G_c was reduced by increasing "interfacial interaction". By contrast, Levita *et al.* [33] found that the dynamic K_c of PP went through a maximum and G_c was reduced as a function of nanoscaled $CaCO_3$ (70 nm) loading (up to 40 vol%). The cited authors also claimed that the crack pinning contribution of the nanoparticles to the toughness was negligible. Wang *et al.* [34] reported that the surface treatment of nano $CaCO_3$ with stearic acid in an ultra-high sped mixer, prior to melt compounding with PP, improved the notched Izod impact strength remarkably. The maximum of the latter was found at *ca.* 12 vol%. Chan *et al.* [35] incorporated nano $CaCO_3$ (44 nm) in PP (in 4.8 and 9.2 vol%, respectively) and measured a *ca.* 300% increase in the notched Izod and *ca.* 500% improvement in the static J-integral test (J_c). The authors concluded that the particles act as stress concentration sites prior to cavitation. The nanofiller worked also as nucleant and reduced the spherulite size of the PP matrix markedly. Note that the toughness of PP with finer spherulites is greater than with coarser ones (*e.g.*, [36]). The results read from the J_R curves, constructed according to the multiple specimen technique, showed that increase in J_c was accompanied by a reduction of the tearing modulus (*i.e.*, improved resistance to crack initiation yielded reduced resistance to crack propagation) [35]. Somewhat different results were reported by Thio *et al.* [37]. They have incorporated $CaCO_3$ particles with and without surface treatment up to 30 vol% in PP. The mean sizes of the $CaCO_3$ were varied (3.5, 0.7 and 0.07 µm). The 0.7 µm particles improved the Izod impact strength markedly compared to the others. Under static loading, adopting the J-integral technique, a different scenario was found. The 0.07 µm particles were inefficient in contrast to the 0.7 and 3.5 µm ones, which enhanced the J_c prominently. On the other hand, the J-integral values decreased with increasing filler content and became less than that of the plain matrix at 30 vol% $CaCO_3$ content. The toughening mechanism disclosed was plastic deformation of the interparticle ligaments which took place after the particle-matrix debonding process. An additional energy absorption mechanism, *viz.* crack deflection, was also identified. In a follow-up paper by the authors [38] the debonding/dilatational process in the same $PP/CaCO_3$ systems was assessed by volume strain measurements. It was established that debonding starts before macroscopic yielding. This is similar to the results achieved on fiber-reinforced composites in which crack growth starts always before the maximum load is reached [39]. Lazzeri *et al.* [38] also demonstrated that after yielding, apart from plastic deformation, considerable volume increase happened. As plastic deformation is not associated with volume change, microvoiding and crazing should be responsible for that. A further conclusion was that debonding is likely not to occur at the surface of the particle but at the boundary between the stearate rich interphase (5–20 nm thickness) and the bulk PP. The effect of the interphase was earlier pinpointed by Bartczak *et al.* [40]. They studied the Izod impact response of high density polyethylene (HDPE) filled with different types of $CaCO_3$ (ground and precipitated) having different particle sizes (3.50, 0.70 and 0.44 µm) and treatments (with and without calcium stearate). The big changes found in the Izod impact strength were attributed to two effects: (i) critical threshold of the interparticle ligament (*ca.* 0.6 µm below which a great jump in the toughness occurs), and (ii) preferential crystallization of HDPE (formation of edge-on lamellae) in the interphase, whose thickness may reach 0.4 µm. Both above effects support the shear deformability of the matrix after the particle debonding process. Weon and Sue [41] studied the fracture and failure behavior of talc and nano $CaCO_3$ (44 nm, stearic acid treated, added in *ca.* 12 vol%) modified high-crystallinity PP grades. According to their results, no toughness improvement was detectable under dynamic (Izod) in contrast to the static conditions (J-integral). The authors attributed the J_c enhancement to the beta nucleation activity of the nanofiller. Thus a shell, composed of the beta polymorph, formed on the surface of the nanoparticles. Attention should be paid to the fact that the beta polymorph of PP is more prone to voiding, crazing and subsequent plastic deformation than the alpha polymorph. Accordingly, the toughness of beta PP (undergoing a beta-to-alpha transition upon mechanical loading) is higher than that of the alpha counterpart ([42–43] and references therein).

The group of Zhang (see also the related Chapter 3 in this book) [44] tailored the interphase between nano SiO_2 (<12 nm) and PP by grafting. X-ray irradiation induced grafting of various monomers on the SiO_2 surface was very efficient to achieve a homogenous dispersion of the particles during melt compounding. The better dispersion resulted in enhanced toughness. In a follow-up study [45] the authors used such monomers for grafting which contained a blowing agent (sulfonyl hydrazide) as side groups. This was expected to contribute to the break-up of the SiO_2 agglomerates in a "bubble stretch" process *via* the N_2 gas evolved by thermal decomposition of the blowing agent. The outcome was doubling of the notched Charpy impact energy.

Incorporation of nano SiO_2 (10 nm), treated with coupling agents (amine- and epoxy-functionalized silanes), in PA-6 resulted in some toughness improvement (<25%). By contrast, the unmodified SiO_2 reduced the toughness of the PA-6 by half [46]. Zhang *et al.* [47–48] studied the toughness response of nanoparticle filled PA-66 plaques using the EWF concept. They followed the energy partitioning method proposed by Karger-Kocsis [49]. The resistance to crack initiation in the yielding stage increased with increasing TiO_2 (21 nm) content (up to 3 vol%). In another work [48], apart from TiO_2 (21 nm) also SiO_2 (13 nm) and Al_2O_3 (13 nm) were involved, however, their amount was fixed at 1 vol%. The major outcome of these studies was that the specific essential work of fracture (initiation) was enhanced at the cost of the non-essential work of fracture term (propagation). Based on fractographic inspection, the nanoparticles worked as stress concentrators and triggered cavitation (also in the form of secondary cracking [50]) which was followed by plastic deformation of the matrix. It is noteworthy that the plastic zone became more and more constrained with the increasing amount of the nanoparticles. The presence of agglomerated particles, acting as stress concentrators, was made responsible for the reduced resistance to crack propagation (non-essential work of fracture term).

Platy fillers

Chen et al. [51] investigated the static fracture behavior of maleated PP (PP-g-MA) modified by organoclay (octadecylamine modified montmorillonite, MMT). The organoclay content was varied between 0 and 50 wt%. It was found that J_c decreased markedly with the clay content and this was attributed to the strongly reduced matrix deformation. The tearing modulus, i.e., the slope of the J_R curves, representing the resistance to crack propagation, decreased also with increasing organoclay content, however, not monotonously. The major conclusion of the authors was that the organoclay content should be kept below 10 wt% in order to maintain high toughness. Bureau et al. [52–54] devoted a series of papers to the EWF properties of PP with organoclay (MMT modified with various quaternary amines) in the presence and absence of PP-g-MA as compatibilizer. The contents of organoclay and PP-g-MA were kept constant at 2 and 4 wt%, respectively. The effects of processing conditions (direct, masterbatch) on the toughness of the corresponding PP nanocomposites were investigated, as well. The specific essential work of fracture was substantially reduced by incorporating organoclay and/or PP-g-MA. This reduction was accompanied with an increase in the specific non-essential work of fracture (i.e., slope of the work of fracture vs. ligament length curves). Predispersion of the organoclay in PP-g-MA (masterbatch) enhanced the specific essential work of fracture, the value of which reached that of the neat PP. Interestingly, this change was not necessarily linked with a reverse tendency in the non-essential work of fracture term, as proposed by the author of this chapter [29–30]. Low molecular weight (MW) PP-g-MA supported a homogenous, whereas the high MW version yielded a non-homogenous intercalation of the organoclay. By analyzing the fracture surfaces it was concluded that the fracture started with void formation at larger organoclay aggregates, followed by void growth and coalescence when the matrix ligaments between the particles were stretched. The toughness improvement was attributed to a change in the voiding stress as a function of the clay particle dispersion (the finer the dispersion, the higher the stress causing the voiding). It is noteworthy that due to instabilities during the crack growth, the authors followed the energy partition method of Karger-Kocsis [49] and concentrated on the initial (yielding) section. In a companion paper [55], effects of compounding methods (direct, masterbatch) and equipments (single screw extruder with and without extensional flow mixer, twin screw extruder) were investigated. The masterbatch technique yielded better clay dispersion than direct compounding. On the other hand, the processing equipments, used to dilute the masterbatch, had a marginal effect on the clay dispersion and thus also on the impact response (notched Izod) of the nanocomposites. The loading rate dependence of the EWF response (following the partitioning according to Ref. [49]) of organoclay-modified PP-g-MA compatibilized PP was the topic of a recent paper [56]. With increasing loading rate (that was in a small range, *viz.* 1 to 20 mm/min) both specific essential and non-essential work of fracture data decreased. The yielding-related specific essential work of fracture (supposedly linked with the plane strain fracture toughness [57]) increased by adding 5 wt% organoclay and 5 wt% PP-g-MA to the PP compared to that of the neat matrix. In a recent work, Yuan *et al.* [58] studied the effects of organoclay (4 wt%) on the microstructure and properties of non-compatibilized PP homopolymer after crystallization at elevated pressures. The notched Izod impact strength increased with increasing crystallization pressure for both neat and organoclay modified PP. The crystallization pressure reduced the spherulitic size and resulted in mixed polymorphs (alpha and gamma phases; the major part composed the alpha modification). This microstructure favored the shear deformability of the matrix.

Yoo *et al.* [59] studied the dynamic EWF behavior of an ionomer as a function of the organoclay content (0 to 10 wt%). The organoclay was exfoliated in this case due to the high polarity of the ionomer matrix. The limiting specific fracture energy (similar to that of the essential work of fracture) went through a maximum (at 2–3 wt%) as a function of the organoclay content. The latter became negative, indicating a ductile-to-brittle transition, above *ca.* 7 wt% organoclay contents.

McNally *et al.* [60] studied the structure-property relationships in PA-12 filled in 4 wt% with synthetic layered silicate with and without organophilic modification. The former became intercalated, whereas the latter mostly exfoliated. The impact strength of this highly plasticized PA-12 remained practically unchanged after silicate filling, but dropped steeply when tested at $T = -40°C$. The authors claimed that the poor impact resistance of the PA-12 with exfoliated organoclay is likely linked with the "freezing in" of the wax-like organophilic modifier. This explanation was in harmony with the fact that the cold impact resistance of the PA-12 nanocomposite with pristine layered silicate was higher than that of the nanocomposite with organophilic layered silicate.

Low density PE (LDPE) was filled with sodium dodecylbenzene sulfonate modified layered double hydroxide (LDH, incorporated up to 15 wt%) in the presence of PE-g-MA [61]. Note that the LDH belongs to the group of the anionic clays. The nanofiller/compatibilizer ratio was fixed at 1:2. The organophilic modified LDH was slightly intercalated in the LDPE matrix considering the change in the basal spacing. The toughness of the nanocomposites was studied by the EWF approach. Both essential and non-essential work of fracture parameters diminished with increasing LDH content. Up to 3 wt% LDH, however, the specific essential work of fracture remained constant. The large scale plastic deformation, observed in LDPE and at low LDH contents, was greatly reduced by increasing LDH content. This was attributed to the presence of agglomerates/clusters which facilitated voiding, however, without supporting the fibrillation of the matrix later on.

The nucleation activity of some nanofillers, raising the formation of a given crystalline polymorph, forced the researchers to consider and even to exploit this effect. Note that a semicrystalline polymer containing a given phase modification may have inherently a higher toughness than another modification. On the other hand, it is very challenging to generate a given polymorph selectively because usually mixed polymorphs appear. It turned out that PA-6 crystallizes instead of alpha in its gamma form in the presence of organoclays, and the PA-6 in its gamma form is tougher than the alpha version. The appearance of polymorphs is often governed by the cooling rate of the sample. Avlar and Qiao [62] showed that the work of fracture of PA-6/organoclay systems decreases with increasing organoclay content (< 4 wt%). The reduction is more pronounced for slow (air cooled) than for fast (water quenched) cooled nanocomposites.

The deformation and fracture mechanisms in PA-based nanocomposites were studied by different research groups, whereby the clay-induced nucleation and the resulting lamellar ordering have also been considered. *In situ* deformation of PA-based nanocomposite specimens under high voltage electron microscope showed that the major deformation mechanism is microvoid formation that appears first inside of clay stacks (tactoids) [63]. Afterwards the stacks, depending on their orientation, underwent splitting, opening or sliding upon further loading. Note that this scenario is fully analogous with that of the deformation of crystalline lamellae in polymers. Gamma phase of PA-6 formed at the clay particles, whereas far away from their surfaces the alpha form dominated. The orientation of the clay layers was strongly influenced by the molecular weight (MW) of the PA-6 (the higher the MW, the better was the orientation of the clay layers) [64]. He *et al.* [65] studied the embrittlement of PA-6/organoclay composites. The fracture toughness, K_c, dropped by 75% of the related value of the neat polymer when the clay content increased from 0 to 10 wt%. According to the authors, crazing was suppressed and cracking became dominant with increasing clay content. This happened because crazes were nucleated in the interphase between the clay particles and the matrix. As the craze nucleation sites increase with the clay content, many crazes are no longer able to fully develop, maturate and thus they break up by cracking at an early stage. Yalcin *et al.* [66] pointed out that toughness improvement in PA-6/organoclay (<5 wt%) nanocomposites may occur owing to the following: (i) hydrogen bonding between the PA-6 molecules is disturbed which enhances the ductility, and (ii) preferential crystallization in favor of a suitable polymorph and orientation of the crystallites which are susceptible to shear deformation. Needless to say that the above conditions can be reached only below a critical organoclay loading.

Fibrous fillers

Incorporation of MWCNT (1 vol%) in PP yielded no improvement in the impact resistance (notched Charpy) below ambient temperature, which corresponds to the glass transition temperature of the PP [67]. On the other hand, a pronounced toughness improvement was noticed above room temperature. This was assigned to nanotube fracture and pull-out events and to the presence of a fine spherulitic morphology owing to the nucleation of the MWCNT. MWCNT with grafted alkyl chains were used to reinforce PP that was compatibilized with PP-g-MA (10 wt%) [68]. The MWCNT content in the related nanocomposites was 0, 0.1 and 0.3 parts per hundred polymer (phr). Specimens were injection molded conventionally (static) and making use of the dynamic packing technique (dynamic). The latter produces oscillating shear flow due to which the MWCNT becomes aligned and the PP undergoes shear-induced crystallization. Moreover, this processing method contributes to the formation of the gamma phase [69]. The specimen produced by dynamic packing showed tenfold ductility compared to usual injection molding at 0.3 phr MWCNT content [68]. It is noteworthy that the difference was twofold in respect to the neat PP. The high ductility found was attributed to the reinforcing efficiency of the aligned MWCNTs. They are able to relieve stress concentrations and locally evolved critical stresses (by whatever means caused) and thus the probability that a crack will turn into a catastrophic one is highly reduced. The reinforcing efficiency may be supported by transcrystallization phenomena, the onset of which was demonstrated by Zhang *et al.* [70]. It should be mentioned that the appearance of a transcrystalline layer does not necessarily contribute to a better stress transfer from the matrix toward the CNT as disclosed on the example of carbon fiber-reinforced semicrystalline thermoplastics [70–73].

Satapathy *et al.* [74] studied the EWF behavior of MWCNT filled (up to 5 wt%) PP. Based on the load-displacement curves of the notched specimens used, the authors should have considered only their initial (yielding) section. As this was not followed, both their specific essential and non-essential work of fracture data are questionable. On the other hand, alignment of MWCNTs in the loading direction enhanced the resistance to crack propagation (non-essential work of fracture term) which is in concert with experimental results in Ref. [68]. Kobayashi *et al.* [75] studied the EWF behavior of amorphous polyethylene terephthalate (PET) containing CNT up to 5 wt%. The specific work of fracture was reduced (however, not monotonously), whereas the specific non-essential work of fracture went through a maximum as a function of the CNT content (*ca.* 2 wt%). Based on small-angle X-ray scattering (SAXS) the authors concluded that this increase was due to the fact that CNT supported the craze maturation process whereby a considerable amount of energy was dissipated. It is noteworthy that amorphous PET fails by shear deformation [76], which is here obviously changed for crazing due to the presence of CNT.

Interestingly, the effect of functionalized MWCNT on the toughness properties of semicrystalline polymers has not been clarified yet. This fact is in

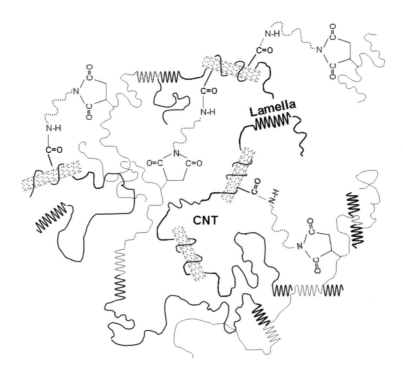

Figure 12.4. Scheme of the target morphology in semicrystalline polymers toughened by functionalized MWCNTs (–COOH), coupling agent (diamine) and compatibilizer (maleated version of the matrix polymer with similar characteristics)

strong contrast to thermosets. In the author's opinion, the use of functionalized MWCNTs is a very promising way, especially if the morphological sketch in Figure 12.4 is targeted. The functionalization of MWCNT may contain the following steps: (i) carboxyl functionalization in HNO_3/H_2SO_4, (ii) reaction of the –COOH groups by suitable diamines. As compatibilizer the same polymer as the matrix should be preferred in its maleated form. Its molecular weight should be high enough to guarantee cocrystallization with the matrix polymer. The outcome is a double networking from the viewpoint of the amorphous phase. Apart from the usual entanglement, the amorphous phase is reinforced by the MWCNTs which are "anchored" in the crystalline phase. In this way, the stress relief/redistribution upon external mechanical loading may occur in a large volume element. The cocrystallization of the compatibilizer with the matrix is important. Here the crystalline lamellae became involved in the stress redistribution process and *via* their deformation (orientation, widening, fragmentation, defolding) a large damage zone may develop. Needless to say that other chemical pathways for the MWCNT functionalization and "bonding" to the matrix can also be followed. The presented scheme, most suited for low and medium crystalline polymers (like polyethylenes and PP), considered, however, the commercial availability of maleated polyolefins as compatibilizers.

The related knowledge on the nanofiller-toughening of semicrystalline polymers, considering the conclusions of Corté and Leibler [27] and Cotterell et al. [77], can be summed up as follows. Substantial toughness improvement can be achieved if the nanofiller dispersion is fine and homogeneous (interparticle distance < 500 nm) and both the "interphase" and "bulk" morphologies of the semicrystalline matrix support dilatational failure events (multiple crazing) as non-dilatational ones (i.e., shear deformation) are usually less relevant. As voiding in the matrix is the first necessary step to relax the initial triaxial stress state (causing brittle fracture), the interphase morphology may play an important role within. Note that the term "morphology" covers both molecular (e.g., entanglement in the amorphous phase, tie molecules, lamellar build-up and arrangement) and supermolecular structures (polymorphs, spherulite type and size). Recall that these structures are strongly influenced by both nanofillers and processing conditions. Voiding is induced by cavitation/debonding in the interphase which have to be facilitated accordingly. Spherical nanofillers are more suited for this task than platy and fibrous ones. Incorporation of fibrous nanofillers may be advantageous in polymers of low crystallinity. Their effect may involve an additional strengthening of the amorphous phase as proposed by the sketch in Figure 12.4.

12.3.2.2. Copolymers and blends

Spherical fillers

PP homopolymer possessing low toughness especially below room temperature is toughened by copolymerization or melt blending [78]. In the past, for melt blending with PP usually ethylene/propylene co- and ethylene/propylene/diene terpolymers (EPM and EPDM, respectively) were used [79–80]. Nowadays, research focuses on novel elastomeric polyolefins, such as ethylene/octene copolymers (POE). PP/elastomer blends have been filled with various microscaled particulates. On the other hand, the potential of nanoscaled fillers has not yet been fully explored. Sometimes the target is to replace a PP/POE blend by a nano $CaCO_3$ (stearic acid coated, 50 nm) filled PP homopolymer [81]. The authors found that the toughening mechanisms of PP/POE and PP/nano $CaCO_3$ obey the percolation theory of Wu [20,82] and the critical ligament length is at about 0.1 µm. The major goal with the incorporation of nanoparticles, however, is to compromise the stiffness and toughness for a given system. Yang et al. [83–84] studied the effects of phase structure on the mechanical properties of PP/EPDM/nano SiO_2 ternary systems. Adding nano SiO_2 (up to 5 wt%) into PP reduced the Izod impact strength of the ternary compared to the binary blend (i.e., PP/EPDM; EPDM content up to 40 wt%). The Izod impact strength could be improved for the PP/EPDM (80/20 parts) blend when a masterbatch, composed of EPDM and hydrophilic silica (ca. 20 nm), was used for the modification. SEM pictures indicated that the nano SiO_2 particles are dispersed in the PP phase but near to the EPDM inclusions. This peculiar morphology (soft rubber surrounded by SiO_2 particles) supports the formation of a percolation network in the sense of the prevailing stress field (Wu's theory [20,82]) and thus the matrix ligaments can fully yield upon external loading. The large improvement in toughness was attributed to a craze stop mechanism achieved by the "SiO_2 shell" around the EPDM particles. The phase structure of the PP/POE/nano SiO_2 ternary system of Ma et al. [85] differed completely from the above described one. In this case, the stearic acid treated nano $CaCO_3$ (100–200 nm) were encapsulated by the POE. The Izod impact strength could be enhanced by a factor of 5 by setting this morphology – which is analogous to the one often found for microscaled fillers [78]. An even higher toughness improvement (notched Izod) was reported for PP/POE/nano $CaCO_3$ by Hanim et al. [86]. Based on SEM pictures taken from the etched fracture surface one concluded that the nanofiller was embedded into the POE phase. Note that in these works the nanofiller acted likely as a viscosity modifier for the elastomer. In this way, the viscosity ratio between the elastomer and the matrix has been influenced, which has a strong impact on the mean size of the dispersed elastomer phase [80]. The latter – also via the change in the interparticle distance or matrix ligament – is a key factor for the toughness. Note that the above findings are opposed to the opinion that the preferential location of the nanofiller in the rubber phase does not contribute to toughness improvement. This may hold for systems in which rubber cavitation is the requirement and the embedded nanofiller restricts it. The author's group found on the example of polyoxymethylene (POM) toughened by polyurethane (PU) that additional incorporation of nanofiller, though embedded in the PU phase, does not reduce the Charpy impact energy. To prepare the POM/PU/boehmite alumina (220 nm) nanocomposites the latex route [25] was followed. The alumina was dispersed in the PU latex and the related slurry injected directly in the POM melt during extrusion compounding. Separate injection of the PU latex and the alumina slurry resulted in the same morphology, i.e., alumina embedded in the PU which fulfilled the role of toughener, being well dispersed (cf. Figure 12.5).

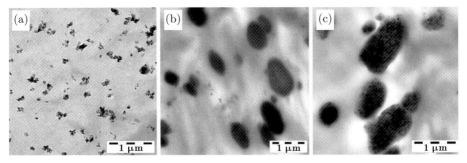

Figure 12.5. Transmission electron microscopic pictures taken from the POM/alumina (100/3 parts) (a), POM/PU (90/10 parts) (b), and POM/PU/alumina (90/10/3 parts) (c) systems produced by water-mediated extrusion melt compounding

Platy fillers

PP copolymer was filled with organoclay (up to 6 wt%) using different compatibilizers (PP-g-MA and PE-g-MA, up to 24 wt%) in an attempt to obtain balanced stiffness, strength and toughness [87]. However, the unnotched Izod impact strength was usually reduced by adding organoclay. Deshmane et al. [88] selected an elastomer toughened PP to fill with organoclay. The notched Izod impact strength could be slightly increased, however, only above room temperature. The toughness difference between the clay modified and neat PP was negligible in the subambient temperature range. The interesting finding was that the organoclay hampered the coalescence of the elastomeric particles to form clusters. The maintained or improved impact strength suggested that the organoclay did not restrict the ability of the matrix for yielding and plastic deformation.

Tjong and Bao [89] studied the fracture toughness of HDPE/SEBS-g-MA/ organoclay nanocomposites. SEBS denotes styrene/ethylene-butylene/styrene block copolymer, which is often used for toughening of thermoplastics. The toughness was assessed by the EWF method both under static and dynamic conditions. The traditional EWF data reduction method could be used only at higher temperatures, where stable tearing/necking occurred (the cited authors did not make use of the energy partitioning methods). As expected, adding organoclay (in 2 and 4 wt%) reduced both essential and non-essential work of fracture terms, whereas the opposite was found for the SEBS-g-MA (added in 5 and 10 wt%, respectively). Filling the HDPE/SEBS-g-MA blends with organoclay resulted in a slight decrease in the EWF parameters. Under impact conditions a similar tendency was noticed; however, the EWF parameters were markedly reduced in the presence of organoclay compared to the static tests. The main energy absorption mechanism of the ternary composites changed from fibrillation/plastic yielding towards restricted crazing with increasing loading rate.

Toughened PAs were also extensively filled with organoclays. Maleated polyolefin elastomer [90–91], maleated EPM [92–93], ethylene/methyl acrylate copolymer [94], hydrogenated nitrile rubber [95], SEBS-g-MA [96], EPDM-g-MA [97], ethylene/butyl acrylate/maleic anhydride terpolymer [98], POE-g-MA [99], acrylate rubber [100], maleated ethylene/acrylate co- and terpolymers [101] were tried as tougheners. Based on the above list, one can recognize that the above polar rubbers were foreseen to fulfill the role as tougher as well as the role of the compatibilizer toward the organoclay. In the cited works often a masterbatch was produced which contained the rubber and the clay. Afterwards, it was diluted by the corresponding PA in different melt blending procedures. The Charpy or Izod impact data were usually reduced by introducing organoclay in the PA/rubber blends (e.g., [91,98]). In some systems the toughness passed a maximum as a function of the clay content [93]. A clear toughness improvement was found when the additives were incorporated by the "latex route". The preirradiated latex along with the pristine clay slurry was first spray dried prior to its blending with the PA melt. Note that preirradiation served to avoid the coalescence of the latex particles which is usually the case [100]. It was reported by many groups that the organoclay affected both the size and structure of the dispersed rubber phase [93,96]. Dasari *et al.* [96] have shown that by changing the blending sequence, the location of the organoclay can be influenced. The highest toughness was provided by those systems in which the organoclay was embedded in the PA-phase. The enrichment of the clay particles near to the rubber particles' surface proved to be beneficial [93,96]. Selective incorporation of the organoclay into the rubber may have a negative effect on the toughness response as the rubber cavitation becomes restricted. Note that this is the first step in the failure mode of toughened polyamides. According to the concept of Wang *et al.* [97], the dispersion of the organoclay in the matrix may block the stress field overlapping (Wu's percolation theory is behind [20,82]) and thus yields a lower toughness. Adopting fracture mechanical methods, Baldi *et al.* [92] concluded a toughness decrease with increasing organoclay content (up to 6 wt%). Further, the fracture mechanical performance was strongly affected by the relative water content of the specimens. The dynamic J_c was slightly enhanced; however, only at very low organoclay content (*ca.* 1 wt%), according to Kelnar *et al.* [94]. By contrast, a monotonous decrease in the fracture toughness (assessed by linear elastic fracture mechanics and the J-integral) was found for a binary blend that was filled with increasing amounts of organoclay [99]. This reduction was due to a similar change in the plastic zone ahead of the crack tip. Obviously, the clay hampered the follow-up mechanisms after cavitation of the rubber particles, *viz.* crazing and yielding of the matrix and eventual stretching of the rubber particles.

PAs are blended with other thermoplastics to combine their beneficial properties. This happens usually in the presence of suitable (reactive) compatibilizers. The major goal with PA/PP combinations is to enhance the stiffness and heat deflection temperature and reduce the water uptake of the related blends [102–103]. It was demonstrated for PA-6/PP/organoclay (70/30/4 parts) systems that the organoclay is located in the PA-6 phase. By adding compatibilizer, *viz.* PP-g-MA, the organoclay moved into the interphase in the PA-6/PP/PP-g-MA/organoclay (70/30/5/4 parts) composite. Adding organoclay to the binary and ternary blends usually reduced the toughness [103].

Fibrous fillers

Interestingly, only few reports addressed the toughness of polymers copolymers and blends containing various nanofibers. Karger-Kocsis *et al.* [104] reported that the toughness of thermoplastic dynamic vulcanizates is markedly reduced when carbon nanofiber (CNF) has been introduced. Recall that the matrix phase in this kind of thermoplastic rubber is given by PP in which submicron sized crosslinked EPDM particles are dispersed. TEM work evidenced that the CNF was embedded in the PP phase. It was also surmised that CNF affected the distribution of the processing oil in this system, as well. This negatively affected the ductility and thus the toughness of the related systems.

Based on the above surveyed studies, we feel that toughened and blended semicrystalline polymers are well suited for "nanomodification". Nanofillers may contribute to the stabilization of the phase morphology of such systems and compromise their stiffness/toughness responses. The morphology of the dispersed phase may change in the presence of the nanofillers and vice versa, the nanofiller may adopt a peculiar structuring in the interphase, eventually also in the bulk. However, the toughness outcome depends on the characteristics of the actual semicrystalline polymer matrix. In toughened polymers and polymer blends, all kinds of nanofillers have the potential to improve the toughness. Researchers should consider, however, that "traditional" nucleants may be as efficient to tailor the matrix morphology upon toughness as nanofillers [105–106].

12.4. Nanomodified thermosets

Crosslinked (thermoset) resins are commonly modified by the inclusion of inorganic particulates such as chalk, silica, alumina, mica, talc. These fillers are incorporated to improve the mechanical properties (stiffness, fracture toughness), wear resistance, flame retardance, to tailor their electrical and heat conductivity, and to reduce the coefficient of thermal expansion. The modification of thermosets with microscaled fillers has been investigated in depth from the early 1980s [107]. However, inorganic fillers usually do not provide as large toughness improvement than organic ones, like functionalized liquid rubbers.

12.4.1. (Neat) Resins

Spherical fillers

A good summary on the state of wisdom of the fracture properties of epoxy (EP)/microscaled alumina is given by McGrath *et al.* [108]. The authors determined the static K_c as a function of average filler size, size distribution, particle shape, particle loading and crosslink density of the EP. The latter has been varied *via* the MW of the diamine hardener. It was established that with increasing intrinsic toughness of the unfilled matrix the toughness of the composite increased at each filler content, as well. So, improvements in the energy absorption and resistance to crack growth of the matrix are translated in the related response of the composites. The normalized toughness ($K_{c,composite}/K_{c,matrix}$) diminishes with increasing $K_{c,matrix}$– this is a well known rule of thumb for all kinds of composites. The studied systems proved to be "robust", meaning that large changes in the particle shape, size and size distribution did not seriously affect the fracture toughness. This was assigned to weak interactions between alumina and EP. The fracture mechanisms in thermosets with microscaled fillers are described by the crack front bowing theory (crack pinning, crack trapping). Its appearance is usually well detected on the fracture surface of the specimens showing peculiar "tails", or ridges. They are caused by the reunification of the crack front after passing the filler particle as an obstacle.

The tails or ridges are formed due to the fact that the crack front moving around each side of the particle never ends in the same plane. It was presumed that submicron sized particles should improve the toughness when this mechanism prevails as more sites will act for crack pinning. This is, however, not the case as shown later. The effect of the mean particle size on the dynamic fracture was moderate in an EP with 55 wt% filler. On the other hand, the impact energy, measured on blunt notched specimens, was remarkably reduced with increasing particle size [109].

Several works were devoted to compare the fracture behavior of micro- and nanofilled resins. The group of Singh [110–111] incorporated micron- and nanoscaled aluminum particles in unsaturated polyester (UP) and EP resins, respectively. The static fracture toughness, K_c, increased mostly with the particle volume fraction (up to 4.5 vol%) for the microscaled aluminum (20 and 3.5 μm) modified UP. On the other hand, it went through a maximum at ca. 2 vol% content for the 100 nm particles containing composite. This was attributed to an incomplete wetting of the aluminum agglomerates formed at higher filler content. As the dominant fracture mechanism crack trapping was quoted. In another work [111], spherical aluminum particles (in 20–100 nm and 3–4.5 μm ranges, respectively) with and without epoxy functionalized silane treatment were incorporated in 2 vol% in EP. The silane treatment was achieved differently. Direct silane addition in the mixture was less efficient than coating of the particle by deposition from aqueous solution. Both nano- and microfillers increased K_c. For the untreated filler the nano-, whereas for the silane treated versions the microfiller yielded the highest toughness. The roughness characteristics of the fracture surface correlated with the toughness (the rougher the surface, the higher the K_c value was). Adachi et al. [112–114] investigated the effect of particle size, micro/nano composition ratio and volume fraction (up to 35 vol%) on the fracture toughness of silica filled EPs. The static K_c depended strongly on the ratio of micro/nanofiller, and the highest K_c value was reached by the EP which contained solely nanofiller [112]. The particle size (240 nm, 560 nm and 1.56 μm) had little influence on the K_c at low (< 10 vol%) but a more pronounced one at higher volume fractions. K_c was the higher, the lower the median size of the particles. It was shown by Raman spectroscopy that the crack propagated by avoiding the particles, i.e., through the matrix. Moreover, the author claimed that K_c changes linearly with the reciprocal product of the square root of the mean distance between the particles' surfaces and the normalized mean stress in the matrix. The latter was considered according to the equivalent inclusion method [113]. In the vicinity of the α- (T_g) and β-relaxation transitions more or less well developed peaks could be recognized in the K_c versus T traces. Note that such correlations were reported for various thermoplastic systems in the early 1980s ([80,115] and references therein).

Jin and Park [116] found that the static K_c of a trifunctional amine hardened EP can be improved by adding nano $CaCO_3$ (40–70 nm). K_c peaked at ca. 6 wt% under static, whereas it was shifted to lower values under dynamic (impact) conditions. The toughness improvement was attributed to debonding, albeit the fracture surface displayed by the authors features crack pinning phenomena. Nanosized untreated Al_2O_3 (15 nm), when added to UP [117], did not yield enhanced toughness. Instead, the static K_c decreased by 15% as the filler volume fraction increased from 0 to 4.5 vol%. Similar K_c degradation was found for microscaled alumina (1 and 35 µm). This was attributed to the poor bonding of the matrix to the filler. Prominent reinforcement was, however, observed when nanocomposites were prepared by vinylsilane treated particles. Again, as failure mechanism crack trapping was indicated. Marur *et al.* [118] compared the static and dynamic K_c values for EP composites with 5 µm and 500 nm alumina particles. The filler content was varied between 0 and max. 40 vol%. In the presence of microparticles the static K_c did not change with the filler content, in contrast to the dynamic K_c that went through a maximum at *ca.* 10 vol%. For the nanocomposites, both static and dynamic K_c values decreased as a function of the filler content. The dynamic fracture toughness data were always below those of the static ones.

Bugnicourt *et al.* [119] studied the influence of various submicron pyrogenic silica fillers with and without silane treatment on the fracture properties of EP having either a high or a low crosslink density. The static K_c could be slightly increased (*ca.* 20% by incorporating 5 wt% treated silica). In some composites both K_c and G_c were enhanced simultaneously. Note that commonly K_c increase is accompanied with a reduction in G_c, and vice versa. The authors did not find signs of crack tip blunting (small scale damage zone comprising localized shear yielding, crack bifurcation, particle/matrix debonding) and concluded that crack pinning is likely the governing mechanism. One of the most detailed works on the toughness of EP nanocomposites containing surface coated TiO_2 (200–500 nm) and untreated Al_2O_3 (13 nm) was conducted by Wetzel *et al.* [120]. It was found that TiO_2 slightly and Al_2O_3 prominently enhanced the T_g of the modified resins. This already suggests that the nanoparticles, due to their huge interfacial area, influence locally the resin/hardener ratio, the outcome of which is a gradient-like change in the crosslink density from the surface of the particles towards the bulk matrix (if the latter still exists). So, the mixing sequence, *i.e.*, whether the preformed nanoparticles are dispersed first in the EP or in the hardener, during production of the nanocomposites may have a strong effect on the resulting morphology [121] and thus on the mechanical performance, including fracture behavior. The K_c of the nanocomposites increased with increasing nanofiller content [120]. The finer the particles, the greater the K_c improvement was (at 10 vol% alumina the K_c of the matrix has more than doubled). The cited authors attempted to disclose the governing fracture mechanism. The experimental G_c data were underestimated when considering the crack deflection theory. Recall that this considers a change in the crack plane *via* tilting (mixed mode I+II) and twisting (mixed mode I+III) caused by the particles as obstacles. No evidence was found for a prominent effect of crack tip blunting which was, however, detected on the fracture surfaces of the specimens. The fracture surfaces showed the onset of crack pinning. However, by adopting the crack pinning theory the G_c data were overestimated. Therefore, crack pinning can only partly account for the toughness increase. It is highly probable that structural heterogeneities, generated by the particle through physical and chemical absorption processes, affect the crosslink density locally, and this has significant influence on the failure mode.

The above treatise indicated the crucial effect of surface treatment of preformed nanoparticles on the mechanical properties. The surface functionalization can be tailored on request by making use of the sol-gel technique. As precursors ("sol"), usually organosilane compounds are used. They are transformed into silicates ("gel") by hydrolysis and polycondenzation processes. As the related particles are mostly produced *in situ*, this sol-gel route is restricted to some polymers. The major benefit of this technique is that the particles are individually (completely) dispersed, *i.e.*, no agglomeration occurs (*e.g.*, [122]). Organically modified silicas (ormosils), prepared by the sol-gel route, were incorporated in UP [123]. The surface of the particles was functionalized by methyl, ethyl, vinyl and phenyl groups, respectively, and the size of the particles ranged from 80 to 160 nm. J_c was enhanced by surface modification compared to the untreated ormosil. Surprisingly, the highest toughness improvement was achieved by phenyl modified ormosils. The ormosil particles triggered crack pinning; however, in the case of the phenyl modified version additionally debonding with limited plastic deformation of the matrix occurred [123].

The sol-gel route was followed also for EPs (*e.g.*, [124–125]) using tetraethoxysilane as precursor whereby the particles were produced *in situ* and *ex situ*. The impact strength slightly increased at 2 wt% silica content both at ambient and cryogenic temperatures [124]. The EP modification by nanoparticles produced by sol-gel chemistry is only of academic interest, being a complicated and "solventrich" process. The sol-gel route is far more suited for thermosets, the polymeric precursors of which are available in suitable solvent, as is the case with polyimides. Here, the precursor is polyamic acid that is converted with suitable polyamines into polyimides. In such systems, silica particles can be produced from organosilanes easily. Musto *et al.* [126] have shown that the *in situ* production of silica particles (300–500 nm; in 10 wt%) in polyimides enhanced the fracture toughness, measured by the EWF method, markedly in a broad temperature range (RT to 250°C). This was explained by the onset of interfacial cavitation (favored by the weak bonding between the particles and the matrix) and shear yielding of the matrix.

There are some options to overcome the problems related to the *in situ* sol-gel production of nanoparticles in thermosets, and especially in EPs. One of them is to prepare separately a molecule which contains an inorganic structural unit that can be dispersed or built into the polymer *via* suitable functional groups. A promising candidate for this task is polyhedral oligomeric silsesquioxane (POSS), the "core" of which is a Si_8O_{12} cage that bears eight hydrocarbon groups, one or more of which may be reactive towards the actual matrix. POSS versions can be incorporated in EP by solution blending. The POSS acts as toughener when built in (causing network heterogeneities through which the shear deformability is improved) and/or dispersed in separate phase (stress concentrator sites in submicron range) [127].

The other way is even more straightforward: the production of nanoparticles which are fully dispersed in the resin. As the related product is in close analogy to a masterbatch whose carrier is the resin itself, this term will be used for its designation. Such masterbatches can be produced basically in two ways: (i) simultaneously (the particle producing sol-gel chemistry is implemented in the production route of the corresponding resin), or (ii) consecutively (*e.g.*, first the particles are produced in a suitable solution that is replaced by the preformed resin). EP/nano SiO_2 masterbatches under the trade name Nanopox® from Nanoresins AG (Geesthacht, Germany) became the focus of extensive research interest of several groups [128–133]. Note that the precursor of the silica particles of these systems is not of organic but of inorganic nature (*viz.* water glass). Sreekala *et al.* [128] reported that the static K_c increased by *ca.* 70% when *ca.* 14 vol% silica was dispersed in a cycloaliphatic, anhydride cured EP using an EP/nano SiO_2 (*ca.* 10 nm) masterbatch. This finding has been confirmed by Zhang *et al.* [129] and Blackman *et al.* [130] for similar systems. On the other hand, the relative toughness improvement was markedly lower for an amine hardened EP containing Nanopox® masterbatch. In this case, the K_c did not change with silica content (20 nm) in the range of 0 to 8 wt% [132]. Zhang *et al.* [129] identified several toughness enhancing mechanisms: crack tip blunting (appearance of a plastic zone), shear deformation of the matrix (river lines due to crack deflection), and secondary cracking (dimple pattern). Plotting the relative toughness improvement as a function of the ratio of the interparticle distance to particle diameter, a clear change was observed close to the ratio of 1. Recall that this is in analogy with the percolation model of Wu [20,82]. Based on SEM pictures, the authors recognized that the particles are covered by an EP layer and concluded that this kind of core-shell structure is due to the EP masterbatch used. At sufficiently high content the individual core-shell particles touch each other *via* their shells and consequently the interphase material will control the fracture performance [129]. Rosso and Ye [131] demonstrated later the formation of a soft interphase around the particle, the thickness of which was between 2–3 nm. This soft interphase was argued to develop due to enrichment of piperidine in the interphase. In a follow-up study the cure kinetics and its effects on the morphology have been considered [134]. The latter was studied by atomic force microscopy (AFM) after etching the surface of the specimens by Ar^+ ion bombardment. Figure 12.6 clearly shows that a relatively small change in the nano SiO_2 content (set by using Nanopox®) had a significant effect on the resin morphology, and thus on the properties.

The group of Kinloch [130,133] investigated the fracture behavior of similar nano SiO_2 (20 nm; Nanopox®) modified EP resins, too. For an anhydride cured system the K_c increased from 0.59 to 1.42 $MPa.m^{1/2}$, whereas G_c increased from 103 to 461 J/m^2 when the silica content rose from 0 to *ca.* 14 vol%. As toughening mechanisms crack pinning, crack deflection and immobilized layer

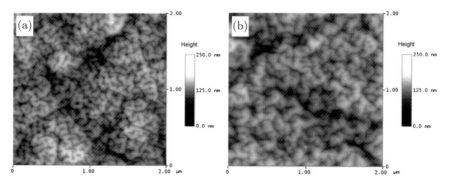

Figure 12.6. AFM scans (height images) taken from the ion eroded polished surface of "homopolymerized" EP systems containing 1 (a) and 3 vol% (b) nano SiO$_2$ (derived from Nanopox®) [134]

(shell) were discounted. The authors discarded all the above mechanisms and, based on fractographic work, concluded that plastic void growth, initiated by debonding of the particles, is the key factor of toughness improvement [133]. Moreover, the cited authors estimated the contribution of plastic void growth and found a fair agreement between experiments and theory. Nevertheless, the author of this chapter believes that network heterogeneities in both interphase and bulk, by whatever mechanisms induced, should play a dominant role in respect to the toughness response. It has to be mentioned that nanosized rubber particles containing EP masterbatches are now also commercially available (Geniomer® from Wacker, Munich, Germany).

Platy fillers

A large body of works addressed the modification of thermosetting resins with organophilic layered silicates of both natural and synthetic origin. Fundamental studies on their structure-property relationships should be credited to the group of Mülhaupt (*e.g.*, [135–137]). Effects of the type and amount of layered silicates on the structure (morphology) and properties (thermo-mechanical, mechanical and fracture mechanical) of a bifunctional anhydride cured EP was the topic of their pioneering work [135]. The authors reported that T_g mostly decreased with increasing silicate content (up to 10 wt%). The related reduction was most prominent when protonated primary amines were used as organophilic intercalants. By contrast, the T_g reduction was marginal in the case of non-functionalized quaternary amine as clay intercalant. This finding can be explained by the plasticizing effect of the organophilic modifier (which may even leave the intergallery space owing to thermal dissociation) and/or by a locally clay-induced change in the crosslinked network density. With increasing silicate content both Young's modulus and K_c increased whereas an opposite tendency was observed for the tensile strength. Interestingly, the intercalated clay morphologies resulted in higher K_c values than exfoliated ones. It was suggested that the stacks act as stress concentrators, produce nanovoids and initiate shear yielding of the matrix, the final outcome of which is enhanced toughness. In companion works [136–137] the influence of the silicate surface modification on the resulting morphology and toughness response of the EPs was studied. All nanocomposites produced possessed higher K_c and G_c data than the neat EP matrix.

Zerda and Lesser [138] reported about 100% improvement in both K_c and G_c at an organoclay content of ca. 5 wt% for amine hardened EPs. At higher clay content, however, both K_c and G_c decreased but did not drop below the corresponding values of the matrix. Crack branching, bifurcation, which create additional surfaces (of considerable roughness), were pinpointed as toughening mechanisms. Note that this failure scenario is analogous to that of discontinuous fiber-reinforced polymers. Becker et al. [139] studied the fracture toughness of organoclay modified amine hardened EPs having different functionalities. The normalized fracture toughness ($K_{c,composite}/K_{c,matrix}$) increased linearly with the clay content and was doubled for two- and tetrafunctional EPs. By contrast, a marginal increase, though linear, was found for trifunctional EP version. The authors also concluded that the clay arrangement can not be identified as either intercalated or exfoliated, as both of them were always present. Similarly to K_c, the perforation impact energy raised also linearly with the clay content for an amine hardened bifunctional EP [140]. Liu et al. [141] called attention to the fact that apart from intercalated and exfoliated structures, macroscopic particles (2–5 µm) may also be present at the same time. The K_c of a bifunctional aromatic amine hardened EP increased from the initial 0.5 to 0.9 MPa.m$^{1/2}$ when the clay content increased from 0 to 4 wt%. This enhancement was traced to the clay particle induced crack deflection associated with massive plastic deformation of the matrix. The proposed failure mode comprised the following steps: stress concentration at clay particles/tactoids. This was followed by clay/matrix debonding and cleavage of tactoids (interlaminar splitting) creating nano- and microvoids. These voids may initiate shear yielding of the matrix and undergo plastic deformation themselves. Wang et al. [142], by studying the failure mode, discarded the shear yielding (no sign of that under polarized light) and proposed the following sequence: stress concentration around the clay tactoids, interlaminar debonding (cleavage between the layers as both the EP/clay interfacial bonding and EP cohesive strength are higher than the clay interlayer strength), coalescence of the neighboring microcracks prior to the formation of catastrophic crack. On the contrary, Akbari and Bagheri [143] emphasized the onset and role of shear yielding of the matrix. The cited authors argued that the shear deformability of the matrix is supported by its heterogeneous network. A heterogeneous network may develop when EP homopolymerization occurs in the intergallery space due to the autocatalytic effect of the onium ions (e.g., [144]). The resulting heterogeneous crosslinked network is prone to shearing, in which slippage of the clay layers is also involved, upon external loading. This hypothesis was supported by showing the presence of two T_gs (bulk and intergallery EP fractions) and the onset of shear banding, however, under compressive loading. On the other hand, the author failed to demonstrate that shear yielding instead of microcracking occurred under mode I type loading. It is the right place to mention that the clay morphology depends on the interplay between cure kinetics, diffusion kinetics, and interaction between resin components and organophilic clay [144–145].

Fibrous fillers

CNTs, CNFs may suffer large strains because their walls undergo periodic buckling. This phenomenon was expected to improve the toughness of nanocomposites containing such nanotubes. As underlined earlier, CNFs and CNTs are usually highly entangled and thus their dispersion in a high viscosity polymer matrix is a challenging task. In order to support their dispersion, various techniques have been attempted, such as suitable solvents (being cosolvent for the corresponding resin), use of various surfactants, sonication, and chemical functionalization. Note that some of the abovementioned techniques already contribute to solve the other major problem with carbon nanotubes, viz. interfacial bonding. Li et al. reported [146] that the impact toughness of EP increased by ca. 70 and 90% by introducing CNT without and with functionalization (–COOH) in 0.5 wt% on a three-roll mill (calender). Hsiao et al. [147] used solvent (acetone), surfactants and in addition, sonication to produce toughened EPs containing up to 5 wt% MWCNT. Miyagawa et al. [148] introduced fluorinated single wall CNT (SWCNT) via a solution/sonication predispersion of SWCNT in an anhydride curable EP. The T_g of the nanocomposites with maximum 0.3 wt% SWCNT dropped from 140 to ca. 70°C. This was attributed to the fact that the functionalized SWCNT absorbed the EP that became "immobilized" and thus inaccessible to the hardener. So, similarly to organoclays, the stoichiometry has been changed locally due to the presence of the SWCNT. The notched Izod impact strength was slightly reduced with increasing SWCNT content. This happened also when the stoichiometry was "adjusted". On the other hand, the Young's modulus increased strongly with SWCNT content. The group of Schulte et al. (e.g., [149–151]) investigated the toughness response of EPs modified with various CNTs. Double wall CNT (DWCNT) with and without amine functionalization was added to EP on a three-roll mill. For mixing the DWCNT/EP with the hardener, sonication was also adopted. K_c of the neat EP increased from the initial 0.65 MPa.m$^{1/2}$ to 0.76 and 0.82 MPa.m$^{1/2}$ when DWCNT without and with amine functionality was incorporated at 0.1 wt% [149]. As a further reference material the authors considered an electric conductive carbon black (CB). The particle size of the CB was matched with the diameter of the DWCNT. Based on the fact that practically no difference was observed in the K_c data between the CB and DWCNT containing systems, it was concluded that instead of crack bridging, likely crack deflection is the key mechanism. Note that in this work potential stoichiometric changes in the interphase were not considered. CNT breakage, pull-out and bridging were identified by fractographic inspection. In a follow-up paper it has been quoted [150] that the toughness enhancing mechanisms are localized inelastic matrix deformation, void nucleation and crack deflection at the agglomerates. For functionalized CNTs, however, fiber-related events (fracture, pull-out) were also triggered. The pull-out contribution (occurring *via* debonding or partial debonding with simultaneous bridging) to the toughness improvement in CNT-modified thermoset was emphasized recently [151]. Xu *et al.* [152] studied the effect of amine functionalized CNF (added up to 5 wt%) and sonication parameters of the fracture toughness of "homopolymerized" EP and found practically no improvement. Sun *et al.* [153] prepared polyamidoamine dendrimer functionalized SWCNT and introduced it in EP at 1 wt% amount *via* the solvent route (acetone). As reference materials the authors prepared apart from neat also a dendrimer modified EP (the dendrimer content corresponded to that in the functionalized SWCNT). The K_c values of the neat EP, EP/SWCNT, EP/dendrimer-functionalized SWCNT and EP/dendrimer were as follows: 0.56, 0.58, 0.66 and 0.75 MPa.m$^{1/2}$. These data clearly show that the toughness improvement was caused by the dendrimer and not by the (functionalized) SWCNT. It is worth noting that dendrimers as potential impact modifiers for thermosets are preferred topics of current R&D works (*e.g.*, [154]). On the other hand, Ye *et al.* [155] found that the notched Charpy impact energy increased by almost four times, without sacrificing strength and thermal stability when 2.4 wt% natural nanotubes (*viz.* halloysite) was introduced in EP. The energy dissipation occurred by formation of damage areas ahead of the crack tip. In these areas, being halloysite rich, microcracking took place. This microcracking was stabilized by nanotube bridging. The crack traveled by nanotube fracture and pull-out events. The group of the author dispersed MWCNT in vinylester resins [156]. The strategy behind was very straightforward: Why not exploit the large amount of styrene (present as crosslinking monomer) for the "solubilization" of MWCNT? Both K_c and G_c passed a maximum (at *ca.* 1 wt%) as a function of the MWCNT content varied between 0 and 2 wt%. SEM pictures taken from the fracture surfaces suggested the onset of the following mechanisms: crack pinning and deflection at large MWCNT agglomerates and shear deformation of the matrix (river line pattern) – *cf.* Figure 12.7a. With increasing MWCNT concentration the matrix deformation became restricted and the roughness of the fracture surface reduced (*cf.* Figure 12.7b). This was associated with some toughness decrease.

Summing up the above results the author would claim that microparticles are far more efficient "toughness modifiers" than nanoparticles. The "nanoeffects" reported in numerous works should be linked with changes of the crosslink density in the interphase. The related changes are likely caused by the selective absorption of a given component of the resin by the nanoparticles. Note that this happens also when the nanoparticles are available in "masterbatch" form. Unfortunately, the related aspects (*e.g.*, cure kinetics, morphology development) have not yet been addressed by systematic studies. Nevertheless, platy fillers, present in both micro- and nanoscale at the same time, may be better "toughener" than spherical or fibrous nanofillers. Func-

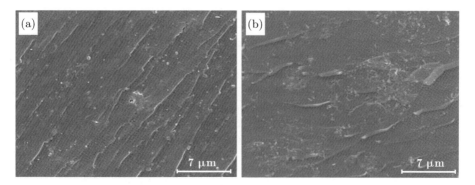

Figure 12.7. SEM pictures taken from the fracture surfaces of vinylester resins containing 0.5 (a) and 2 wt% (b) MWCNTs

tionalization of carbon nanofibers and nanotubes is only then beneficial when a molecular "spacer" is built in between their functional group and those of the resin components.

12.4.2. Toughened and hybrid resins

As nanoparticle modification did not always yield the expected toughness enhancement, researchers tried to combine the "nanomodification" strategy with that of the state-of-the-art toughening.

Spherical fillers

Kinloch et al. [157] formulated toughened EP resins which contained nano SiO_2 (20 nm) from Nanopox® masterbatch. A reactive, liquid, carboxyl terminated butadiene/acrylonitrile rubber (CTBN) served as toughener, which was pre-reacted with a bifunctional EP. In the recipes the nano SiO_2 content was varied (0 to *ca.* 15 wt%) in the presence (9 wt%) and absence of CTBN. During curing the CTBN phase was segregated in microscaled spherical particles which are known to cavitate easily [158]. The Young's modulus of the anhydride cured EP increased by nano SiO_2 filling and decreased with the additions of CTBN. The G_c of the neat EP increased from the initial 103 J/m² to over 400 J/m² by adding CTBN. This G_c value was reached by the binary EP/nano SiO_2 systems at high content of the latter (*ca.* 20 wt%). The G_c of the ternary (hybrid) system increased linearly with the nano SiO_2 content and at 15 wt% reached a value of *ca.* 1400 J/m². It is noteworthy that the Young's modulus of this hybrid still remained below that of the neat EP. Unfortunately, the failure mechanisms have not been disclosed. Nevertheless, it is intuitive that CTBN induces cavitation and strongly favors the shear deformability of the matrix. Walter et al. [159] studied the fracture mechanical behavior of EPs with preformed silicon rubber (up to 8 vol%) and nano SiO_2 (up to 8 vol%; from Nanopox® masterbatch). The static K_c increased linearly with the nano SiO_2 content at each silicon rubber content of the formulations.

Platy fillers

Gam et al. [160] studied the effect of organoclay on the toughness of core-shell rubber (CSR) particulate modified EPs cured either by amine or anhydride compounds. The K_c of the amine-hardened neat, pristine clay-containing (5 wt%), organoclay-containing (5.4 wt%) and organoclay (5 wt%) + CSR (3 wt%) containing EP formulations were as follows: 1.12, 1.51, 0.91 and 3.05 MPa.m$^{1/2}$. The above data range shows that clay microparticles (see pristine clay modified EP) are better toughness modifiers than exfoliated organoclay nanoparticles. The large difference in the K_c data was markedly reduced at $T = -20°C$. For the anhydride cured versions the following K_c data were established at room temperature: neat EP resin – 0.53, EP + 3 wt% CSR – 0.83, and EP + organoclay (2 wt%) + CSR (3 wt%) – 0.81 MPam$^{1/2}$. In this system the organoclay marginally affected the toughness of the CSR-modified EP in contrast to the Young's modulus that was prominently increased. Isik et al. [161] produced polyether polyol toughened (0 to 7 wt%) and organoclay modified (0 to 5 wt%) bifunctional amine cured EP systems. The unnotched Charpy impact energy increased with the polyol content and decreased with the organoclay loading for the related binary systems. For the ternary hybrid the impact strength diminished with increasing organoclay content. Fröhlich et al. [162] studied the fracture behavior of ternary systems composed of anhydride cured bifunctional EP, hydroxyl terminated special liquid rubber and organophilic layered silicate. The liquid rubber was rendered also stearate functionalized in order to enhance the compatibility with the organoclay. The organoclay was predispersed in the liquid rubber and the related masterbatch introduced in the EP. The organophilic layered silicate was always intercalated. Prominent fracture toughness increase (ca. 300%) along with high stiffness (ca. 10% loss compared to the neat resin) was only achieved with the stearate functionalized rubber. The latter underwent phase separation and appeared in spheres in contrast to the unmodified rubber (a specific triol compound) which was "dissolved" in the EP. Liu et al. [163] studied a ternary system composed of "homopolymerized" EP, CTBN (0 or 20 phr) and organophilic clay (0 to 6 phr). The K_c of the EP/organoclay binary systems increased monotonously, whereas G_c leveled off with the clay content. In the presence of CTBN a prominent further improvement has been found in respect of both K_c and G_c. The toughness improvement was attributed to an interplay between rubber- (cavitation, shear yielding of the EP) and organoclay-induced (crack bifurcation) events. With increasing clay content the fracture surface became more and more rough. Ratna et al. [164] reported on the perforation impact behavior of an amine cured EP toughened with epoxy-terminated hyperbranched polyester (HBP in 0 or 15 phr) and modified with organoclay (0 or 5 phr). The impact strength of the EP was increased by either adding HBP or clay. Additional incorporation of organoclay in the EP + HBP system reduced the impact resistance. Balakrishnan et al. [165–166] investigated the morphology dependence and fracture performance of piperidine cured EP containing acrylic elastomer (16 phr) and organoclay. The authors noticed a *ca.* 30% increase in the fracture energy for the ternary system with 5 wt% organoclay compared to the matrix. Subramaniyan and Sun [167] studied the toughening and organoclay modification in vinylester resin. As toughener served a CSR in a broad size range (0.5–500 μm) in 5 wt%. The binary and ternary systems were composed of: vinylester+CSR (5 wt%), vinylester+organoclay (5 wt%), and vinylester+CSR (2 wt%) + organoclay (3 wt%). The fracture toughness was assessed on both blunt and sharp notched specimens. K_c, determined on sharply notched specimens increased by 77, 36 and 25% for the modification with CSR, organoclay and combination of CSR and organoclay, respectively. Corobea *et al.* [168] recently studied the fracture behavior of organophilic layered silicate modified vinylester-urethane hybrid resins (VEUH). Note that the urethane linkage in VEUH reduces the toughness as it creates a very tight network with enhanced T_g. The organoclay contained quaternary amine compounds with and without additional vinyl functionality. The latter was expected to support the exfoliation in the VEUH resin, which however did not take place. Instead, the layered silicate was intercalated. The K_c and G_c improvements were in accordance with the silicate dispersion: the better the silicate dispersion (*cf.* Figure 12.8), the higher the K_c and G_c data were. The SEM pictures in Figure 12.8 show that the better dispersion of the silica triggers crack pinning and deflection as well as massive shear deformation of the matrix ligaments between the particles.

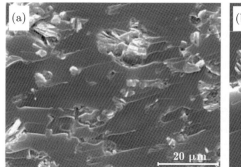 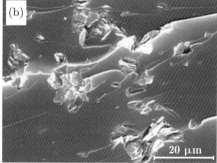

Figure 12.8. SEM pictures taken from the fracture surface of VEUHs containing 5 wt% organohilic layered silicate with various intercalants. Designation: (a) non-functional intercalant, and (b) intercalant with vinyl functionality [168]

The group of the author incorporated organophilic MMT in a vinylester/EP hybrid resin of interpenetrating network (IPN) structure. The EP component was a cycloaliphatic version hardened by an aliphatic amine ($T_{postcure}$ = 200°C), disclosed in an earlier paper [169]. AFM pictures were taken from this hybrid resin after Ar$^+$ ion erosion. Figure 12.9a shows the IPN structure of this hybrid resin in the bulk, whereas Figure 12.9b informs us about the morphology alteration in the vicinity of an intercalated organoclay microparticle.

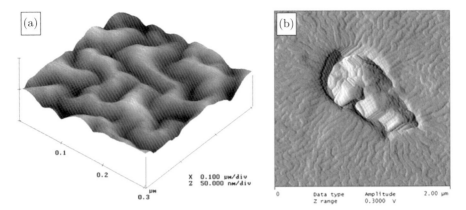

Figure 12.9. AFM scans taken from the ion eroded polished surface of a vinylester/EP system of interpenetrating network (IPN) structure containing 5 wt% organoclay. Designations: (a) IPN structure of the bulk resin (three-dimensional height image), and (b) organoclay particle-induced change in the resin morphology (amplitude image)

Fibrous fillers

Only few reports are available on fibrous nanofillers embedded in toughened or hybrid resins. For VEUH it was reported that both K_c and G_c decrease in the presence of 0.5 wt% MWCNT. Interestingly, the opposite tendency was observed for interpenetrating vinylester/EP systems as a function of MWCNT content. It has to be mentioned that the cited authors did not adopt the right fracture mechanical method when studying this very ductile hybrid resin of interpenetrating network structure [156].

Similarly to toughened thermoplastics, toughened and hybrid thermosets are more suitable matrices for nanomodification than neat resins. This conclusion is based on the fact that traditional rubber inclusions (at micron scale) and nanofillers often show synergistic effect in respect to toughness. The nanofillers along with their interphase regions "distort" the crosslinked network, but in a homogeneous way. This supports the shear deformability of the matrix, which is the major energy absorbing process. Recall that the triaxial stress state is relieved by cavitation of the rubber particles (including the formation of dilatational bands) prior to shear deformation of the matrix.

12.5. Nanomodified traditional composites

The major drawback of traditional composite laminates, composed of laminae (plies) with unidirectional (UD) endless fibers (glass, carbon, aramid), is their low damage tolerance. Composites experience massive delamination due to out-of-plane type subcritical impact loading. Delamination cracks initiate from edges of composite structure under other type of loading than impact, as well. Delamination reduces the load-carrying capacity and durability of composite structures to a great extent. Since the compression test is most sensitive to interlaminar delamination, usually the compression after impact test is adopted for composite development and benchmarking for damage tolerance. Delamination is favored by the low interlaminar shear strength (ILSS) of composite laminates. The poor interlaminar properties can be explained by the fact that the possible damage zone at the crack tip under any kind of loading situation is highly constrained owing to the neighboring, densely packed UD fibers. To overcome this problem either the space between the plies should be increased (this was followed in the 90s by the interleaf concept) or the constituent laminae should be strengthened in the z-direction (this has been started a decade ago using various concepts, like stitching, z-pinning). Recall that the many failure events caused by nanofillers, such as bifurcation, deflection, bridging, may be very useful to improve the interlaminar properties as due to their size they may be at work also in the confined space available. Bringing the nanofillers in between the laminae improves the interlaminar toughness without sacrificing other direction-dependent mechanical properties. Some nanofillers can even be aligned in the z-direction (thickness) by suitable means. They serve for crack bifurcation, deflection and thus enhance the toughness. The interested reader can find good reviews on this topic already [170–171].

12.5.1. *Thermoplastic matrices*

Their additional nanomodification was less pursued by systematic studies so far. The possible reason is that thermoplastics are inherently tough and this property is well translated to composites. Some development works were in progress with *in situ* polymerizable systems resulting in PA-6, PA-11 and polybutylene terephthalate (PBT). PBT is produced by ring-opening polymerization of cyclic butylene terephthalate oligomers (CBT®).

12.5.2. *Thermoset matrices*

The overwhelming majority of R&D works dealt with the nanomodification of thermoset resins which are prone to brittle fracture. Note that the nanofillers can be incorporated either in the bulk resin or in the interlaminar layers during composite manufacturing.

Spherical fillers

Hussein et al. [172] dispersed micro- (1 µm) and nano Al_2O_3 (25 nm) in 10 vol% in an anhydride curable EP. UD carbon fiber (CF) reinforced composite plates were produced by wet filament winding followed by curing in a hot press. The ILSS increased with increasing CF volume content and reached 80 MPa at 55 vol%. This value, though as maximum, was found with the nano alumina modified EP containing 35 vol% UD CF. The fracture toughness was increased by both alumina particles. Functionalized boehmite particles (up to 16 wt%) were introduced in EP with which CF fabrics were impregnated and cured. Unfortunately, all properties assessed were reduced [173]. Tsantzalis et al. [174] produced UD CF-reinforced EP composites, the matrix of which contained piezoelectric particles (4 vol%), CNF (1 vol%) and their mixture. The mode I interlaminar fracture energy (G_{Ic}) decreased by the piezoelectric particles, but increased by the CNF (+100%) and CNF+particles (+50%). Rosso et al. [134] incorporated nano SiO_2 from Nanopox® in a homopolymerized EP (up to 5 wt%). Using model composites (single fiber or fiber bundle) both interfacial shear and transverse tensile strengths were improved. Kinloch et al. [175] used hybrid toughened EP with glass fiber (GF) fabric reinforcement in their work. The non-crimp UD GF fabric with polyester weft stitching was impregnated by a version of resin transfer molding. The mode I and mode II initiation fracture energy values (G_{Ic} and G_{IIc}) were very impressive and at the same time dependent on the loading mode (cf. Table 12.1).

Table 12.1. Mode I (G_{IC}) and II (G_{IIC}) initiation fracture energy values of UD GF-reinforced (~57 vol%) EP composites with various toughening of the EP [175]

	Composition		G_{IC} (J/m^2)	G_{IIC} (J/m^2)
	Nano SiO_2 (wt%)	CTBN (wt%)		
GF-EP	0	0	330	1300
GF-EP + CTBN	0	9	885	1460
GF-EP + nano SiO_2	10	0	1015	1380
GF-EP + nano SiO_2 + CTBN	10	9	860	1895

Platy fillers

Organoclays were incorporated aiming at the upgrade of the interlaminar fracture toughness of composite laminates. Becker et al. [176] produced UD CF composites containing 0 to 7.5 wt% organoclay by hand lay-up followed by hot pressing. G_{Ic} of the composites was increased by 20–25% when the organoclay content in the EP matrix ranged between 2.5 and 7.5 wt%. Interestingly, this was not associated with a similar trend in the ILSS data. Anhydride cured EP with 10 wt% layered organophilic silicate yielded a better mechanical performance than without this nanomodifier also for glass mat-reinforced composites [177]. Tsai and Wu [178] prepared EP resins containing various amounts of organoclay (up to 7.5 wt%) via mechanical mixing and sonication. Composite laminates of various lay-ups were produced by a vacuum hand lay-up process. Unexpectedly, G_{Ic} decreased with increasing organoclay content from the initial 1.89 kJ/m^2 to 1.08 kJ/m^2 in the presence of 7.5 wt% organoclay. Quaresimin and Varley [179] compared the effects of milled CNF (7.5 wt%) and organoclay (5 wt%) on the interlaminar properties of UD CF/EP composites produced by hand lay-up and vacuum bag molding. G_{Ic} decreased for both organoclay and CNF, opposed to G_{IIc} where a slight and prominent improvement was found for the organoclay and CNF, respectively. CNF incorporation enhanced the

ILSS opposed to the organoclay. Siddiqui et al. [180] studied the performance of CF woven fabric reinforced amine cured EP with (up to 7 wt%) and without organoclay. The laminates were prepared by hand lay-up and cured in vacuum hot press. The fracture energy of the matrix (showing some increase with the organoclay content) was directly translated to the interlaminar fracture energy of the composite. Note that this behavior is rare and usually occurs in resins showing very low G_c initially. Incorporation of platy fillers in resins forming the matrix of other types of composites, such as syntactic foams (e.g., [181–182]), proved to be beneficial, too.

Fibrous fillers

Gojny et al. [183] used DWCNT with and without amine functionality and CB (up to 0.3 wt%) for the modification of an EP that served as matrix in a GF/EP composite. The composites, containing non-crimp GF fabric plies, were produced by resin transfer molding whereby no filtering of the particles was observed. The ILSS was improved by 20 and 9% in the case of DWCNT and CB, respectively, at 0.3 wt%. In a further paper by the same group [184], the orientation of CNT variants in the z-direction (thickness) was attempted applying an external electric field during curing. The nanoparticles (fumed silica, CB, various CNTs) improved the ILSS, which however decreased with increasing GF content. The application of an electric field in the z-direction, while curing the composite, resulted only in a slight increase in ILSS. This was assigned to the very weak fiber/matrix interface, controlling the interfacial shear. Unexpectedly, DWCNT reduced the G_{Ic} at 37 vol% GF content markedly, but only slightly at 50 vol% GF, when present in 0.3 wt%. By contrast, G_{IIc} values were slightly increased and remained at the same level for the composites with GF content 37 and 50 vol%, respectively. Koo and Pilato [185] surveyed some benefits of nanomodification of traditional composites. The ILSS of UD CF/EP laminate could be improved by CNF and some nanosilica variants when added in 2 wt%. Organoclay showed an adverse effect. CNF modification did not affect the G_{Ic} but slightly improved the G_{IIc} value. Zhu et al. [186] reported on the effect of nanotubes with and without sidewall functionalization on the mechanical properties of GF fabric reinforced vinylester composites. The nanotubes, dispersed in a solvent, were sprayed on the GF fabric (0.1 wt%). The ILSS was not improved with incorporation of nanotubes, just the opposite occurred. This was traced to resin undercure due to the free radical trapping of the nanotubes. By adjusting the cure recipe and cure cycle (postcure) the ILSS data were dramatically improved (20–45%). Functionalized nanotubes did not show additional improvement in respect to the pristine one. Fan et al. [187] examined the ILSS of GF fabric reinforced EP composites using MWCNT. For the composite preparation vacuum-assisted resin transfer molding was used. During infiltration of the GF fabrics attempts were made to orient the MWCNT in the z-direction via different techniques. The enhancement in the ILSS was between 3 and 10% at 0.5 wt% and 18–20% at 1 wt% MWCNT content. The additional improvement due to preferential alignment of the MWCNT was marginal. This finding is adverse to the theoretical model of Tong *et al.* [188] predicting a steep linear increase with the MWCNT density aligned in the z-direction. Arai *et al.* [189] dispersed CNF in between UD CF/EP prepreg layers. In this way, an interlayer, having a thickness from 50 to 200 μm, was created. The initial G_{Ic} went through a maximum as a function of the area density of CNF, exhibiting an improvement of *ca.* 50%. A similar improvement (50–80%) was found under mode II type loading in the G_{IIc} values. CNT hybridization of the matrix influenced positively also the resistance to fatigue of traditional composite laminates [190].

Though the use of nanomodified resins for traditional composites has several benefits, it would be more straightforward to place the nanofillers in between the laminae selectively. The interlaminar properties can be improved in this way, as well. The dispersion of nanofillers in the bulk resin prior to producing composite laminates may cause undesired filtering effects (*e.g.*, in the case of resin transfer molding). However, bulk dispersion of selected nanofillers in traditional composites may have additional benefits. For example, CNTs dispersed in GF-reinforced composites may work for sensing of the structural integrity of the composite. Structural failure is namely accompanied with a change in the electrical conductivity provided by the percolation network of the CNTs in the related composite [191].

12.6. Conclusions and outlook

The "nanomodification" strategy to improve the toughness of polymers will likely follow the synthesis route in the future. Apart from direct synthesis, different nanostructuring concepts through additives will be explored. Nanomodification will be a key topic for natural polymers and polymers prepared from renewable resources.

Preformed nanofillers (and impact modifiers, as well) will be offered in masterbatch forms and diluted with suitable resins upon request. Further on, great efforts will be devoted to generate novel nanofillers, like graphite sheets, CNT-clay hybrids, coiled CNTs. The potential of nanofillers of natural origin, like cellulose whiskers, will be checked in various polymers, including rubbers. In respect of nanofillers, the R&D work will focus on their possible functional properties ("smart" nanoparticles).

To improve the interlaminar properties of traditional composites the interleaving concept will be revisited. However, this time electrospun nanofiber mats, various nanomembranes will be preferentially used as interlayers.

Acknowledgements

The author thanks the German Science Foundation for the support of works, the results of which are included in this chapter. Thanks are also due to Dr. R. Thomann (Freiburg, Germany) for the TEM work done.

References

1. Gao H, Ji B, Jäger I L, Arzt E and Fratzl P (2003) Materials become insensitive to flaws at nanoscale: Lessons from nature, *Appl Phys Sci* **100**:5597–5600.
2. Termonia Y (2007) Structure-property relationships in nanocomposites, *Polymer* **48**:6948–6954.
3. Liu H and Brinson L C (2008) Reinforcing efficiency of nanoparticles: A simple comparison for polymer nanocomposites, *Compos Sci Technol* **68**:1502–1512.
4. Krishnamoorti R and Vaia R A (2007) Polymer nanocomposites, *J Polym Sci Part B: Phys* **45**:3252–3256.
5. Sinha Ray S and Okamoto M (2003) Polymer/layered silicate nanocomposites: A review from preparation to processing, *Prog Polym Sci* **28**:1539–1641.
6. Chazeau L, Gauthier C, Vigier G and Cavaillé J Y (2003) Relationships between microstructural aspects and mechanical properties of polymer-based nanocomposites, in *Handbook of Organic-Inorganic Hybrid Materials and Nanocomposites* (Ed. Nalwa H S) Amer Sci Publ, Los Angeles, Vol. 2, Ch. 2, pp. 63–111.
7. Zhang M Q, Rong M Z and Friedrich K (2003) Processing and properties of nonlayered nanoparticle reinforced thermoplastic composites, in *Handbook of Organic-Inorganic Hybrid Materials and Nanocomposites* (Ed. Nalwa H S) Amer Sci Publ, Los Angeles, Vol. 2, Ch. 3, pp. 113–150.
8. Karger-Kocsis J and Zhang Z (2005) Structure-property relationships in nanoparticle/semicrystalline thermoplastic composites, in *Mechanical Properties of Polymers Based on Nanostructure and Morphology* (Eds. Michler G H and Baltá Calleja F J) CRC, Boca Raton, FL, Ch. 13, pp. 553–602.
9. Utracki L A (2004) *Clay-Containing Polymeric Nanocomposites*, Rapra Technology, Shawbury.
10. Ji B and Gao H (2004) Mechanical properties of nanostructure of biological materials, *J Mech Phys Solids* **52**:1963–1990.
11. Williams J G (1987) *Fracture Mechanics of Polymers*, Ellis Horwood, Chichester.
12. Kausch H-H (1987) *Polymer Fracture*, Springer, Berlin.
13. Atkins A G and Mai Y-W (1988) *Elastic and Plastic Fracture*, Ellis Horwood, Chichester.
14. Friedrich K (Ed.) (1989) *Application of Fracture Mechanics to Composite Materials*, Elsevier, Amsterdam.
15. Mai Y-W, Wong S-C and Chen X-H (2000) Application of fracture mechanics for characterization of toughness of polymer blends, in *Polymer Blends Vol 2: Performance* (Eds. Paul D R and Bucknall C B) Wiley, N.Y., Ch. 20, pp. 17–58.
16. Wiliams J G and Pavan A (Eds.) (2000) *Fracture of Polymers, Composites and Adhesives* (ESIS Vol. 27) Elsevier, Oxford.
17. Moore D R (Ed.) (2004) *The Application of Fracture Mechanics to Polymers, Adhesives and Composites* (ESIS Vol. 33) Elsevier, Oxford.
18. Wu D, Wang X, Song Y and Jin R (2004) Nanocomposites of poly(vinyl chloride) and nanometric calcium carbonate particles: Effects of chlorinated polyethylene on mechanical properties, morphology, and rheology, *J Appl Polym Sci* **92**:2714–2723.
19. Lach R, Kim G-M, Michler G H, Grellmann W and Albrecht K (2006) Indentation fracture mechanics for toughness assessment of PMMA/SiO_2 nanocomposites, *Macromol Mater Eng* **291**:263–271.
20. Wu S (1985) Phase structure and adhesion in polymer blends: A criterion for rubber toughening, *Polymer* **26**:1855–1863.
21. Ash B J, Siegel R W and Schadler L S (2004) Mechanical behaviour of alumina/poly(methyl methacrylate) nanocomposites, *Macromolecules* **37**:1358–1369.

22. Khrenov V, Schwager F, Klapper M, Koch M and Müllen K (2007) Compatibilization of inorganic particles for polymeric nanocomposites. Optimization of the size and the compatibility of ZnO particles, *Polym Bull* **58**:799–807.
23. Klapper M, Clark Jr G C and Müllen K (2008) Application-directed syntheses of surface-functionalized organic and inorganic nanoparticles, *Polym Int* **57**:181–202.
24. Siengchin S (2008) *Water-mediated melt compounding to produce thermoplastic polymer-based nanocomposites: Structure-property relationships*, PhD Thesis, TU Kaiserslautern.
25. Karger-Kocsis J (2008) Water-mediated dispersion of "nanofillers" in thermoplastics: Is it the right way?, *eXPRESS Polym Lett* **2**:312.
26. Siengchin S, Karger-Kocsis J, Apostolov A A and Thomann R (2007) Polystyrene-fluorohectorite nanocomposites prepared by melt mixing with and without latex pre-compounding: Structure and mechanical properties, *J Appl Polym Sci* **106**:248–254.
27. Corté L and Leibler L (2007) A model for toughening of semicrystalline polymers, *Macromolecules* **40**:5605–5611.
28. Satapathy B K, Weidisch R, Pötschke P and Janke A (2007) Tough-to-brittle transition in multiwalled carbon nanotube (MWNT)/polycarbonate nanocomposites, *Compos Sci Technol* **67**:867–879.
29. Karger-Kocsis J (1999) Dependence of the fracture and fatigue performance of polyolefins and related blends and composites on microstructural and molecular characteristics, *Macromol Symp* **143**:185–205.
30. Karger-Kocsis J (2000) Microstructural and molecular dependence of the work of fracture parameters in semicrystalline and amorphous polymer systems, in *Fracture of Polymers, Composites and Adhesives* (ESIS Vol. 27) (Eds. Williams J G and Pavan A) Elsevier, Oxford, pp. 213–230.
31. Argon A S and Cohen R E (2003) Toughenability of polymers, *Polymer* **44**:6013–6032.
32. Bezerédi Á, Demjén Z and Pukánszky B (1998) Fracture resistance of particulate filled polypropene, *Angew Makromol Chem* **256**:61–68.
33. Levita G, Marchetti A and Lazzeri A (1989) Fracture of ultrafine calcium carbonate/polypropylene composites, *Polym Compos* **10**:39–43.
34. Wang G, Chen X Y, Huang R and Zhang L (2002) Nano-$CaCO_3$/polypropylene composites made with ultra-high-speed mixer, *J Mater Sci Lett* **21**:985–986.
35. Chan C-M, Wu J, Li J-X and Cheung Y-K (2002) Polypropylene/calcium carbonate nanocomposites, *Polymer* **43**:2981–2992.
36. Karger-Kocsis J (1995) Microstructural aspects of fracture in polypropylene and in its filled, chopped fiber and fiber mat reinforced composites, in *Polypropylene: Structure, Blends and Composites* (Ed. Karger-Kocsis J) Chapman and Hall, London, Vol. 3: Composites, Ch. 4, pp. 142–201.
37. Thio Y S, Argon A S, Cohen R E and Weinberg M (2002) Toughening of isotactic polypropylene with CaCO3 particles, *Polymer* **43**:3661–3674.
38. Lazzeri A, Thio Y S and Cohen R E (2004) Volume strain measurements on $CaCO_3$/polypropylene particulate composites: The effect of particle size, *J Appl Polym Sci* **91**:925–935.
39. Karger-Kocsis J, Harmia T and Czigány T (1995) Comparison of the fracture and failure behavior of polypropylene composites reinforced by long glass fibers and by glass mats, *Compos Sci Technol* **54**:287–298.
40. Bartczak Z, Argon A S, Cohen R E and Weinberg M (1999) Toughness mechanism in semi-crystalline polymer blends: II. High-density polyethylene toughened with calcium carbonate filler particles, *Polymer* **40**:2347–2365.
41. Weon J-I and Sue H-J (2006) Mechanical properties of talc- and $CaCO_3$-reinforced high-crystallinity polypropylene composites, *J Mater Sci* **41**:2291–2300.

42. Karger-Kocsis J (1996) How does "phase transformation toughening" work in semi-crystalline polymers, *Polym Eng Sci* **36**:203–210.
43. Chen H B, Karger-Kocsis J, Wu J S and Varga J (2002) Fracture toughness of α- and β-phase polypropylene homopolymers and random- and block-copolymers, *Polymer* **43**:6505–6514.
44. Rong M Z, Zhang M Q, Zheng Y X, Zeng H M, Walter R and Friedrich K (2001) Structure-property relationships of irradiation grafted nano-inorganic particle filled polypropylene composites, *Polymer* **42**:167–183.
45. Cai L F, Mai Y L, Rong M Z, Ruan W H and Zhang M Q (2007) Interfacial effects in nano-silica/polypropylene composites fabricated by *in-situ* chemical blowing, *eXPRESS Polym Lett* **1**:2–7.
46. Li Y, Yu J and Guo Z-X (2002) The influence of silane treatment on nylon 6/nano-SiO$_2$ *in situ* polymerization, *J Appl Polym Sci* **84**:827–834.
47. Yang J-L, Zhang Z and Zhang H (2005) The essential work of fracture of polyamide 66 filled with TiO$_2$ nanoparticles, *Compos Sci Technol* **65**:2374–2379.
48. Zhang H, Zhang Z, Yang J-L and Friedrich K (2006) Temperature dependence of crack initiation fracture toughness of various nanoparticles filled polyamide 66, *Polymer* **47**:679–689.
49. Karger-Kocsis J (1996) For what kind of polymer is the toughness assessment by the essential work concept straightforward?, *Polym Bull* **37**:119–126.
50. Karger-Kocsis J, Walter R and Friedrich K (1988) Annealing effects on the fatigue crack propagation of injection-molded PEEK and its short fiber composites, *J Polym Eng* **8**:221–255.
51. Chen L, Wong S-C and Pisharath S (2003) Fracture properties of nanoclay-filled polypropylene, *J Appl Polym Sci* **88**:3298–3305.
52. Bureau M N, Perrin-Sarazin F and Ton-That M-T (2004) Polyolefin nanocomposites: Essential work of fracture analysis, *Polym Eng Sci* **44**:1142–1151.
53. Perrin-Sarazin F, Ton-That M-T, Bureau M N and Denault J (2005) Micro- and nano-structure in polypropylene/clay nanocomposites, *Polymer* **46**:11624–11634.
54. Bureau M N, Ton-That M-T and Perrin-Sarazin F (2006) Essential work of fracture and failure mechanisms of polypropylene-clay nanocomposites, *Eng Fract Mech* **73**:2360–2374.
55. Li J, Ton-That M-T, Leelapornpisit W and Utracki L A (2007) Melt compounding of polypropylene-based clay nanocomposites, *Polym Eng Sci* **47**:1447–1458.
56. Saminathan K, Selvakumar P and Bhatnagar N (2008) Fracture studies of polypropylene/nanoclay composite. Part I: Effect of loading rates on essential work of fracture, *Polym Test* **27**:296–307.
57. Karger-Kocsis J and Ferrer-Balas D (2001) On the plane-strain essential work of fracture of polymer sheets, *Polym Bull* **46**:507–512.
58. Yuan Q, Deshmane C, Pesacreta T C and Misra R D K (2008) Nanoparticle effects on spherulitic structure and phase formation in polypropylene crystallized at moderately elevated pressures: The influence on fracture resistance, *Mater Sci Eng* **A480**:181–188.
59. Yoo Y, Shah R K and Paul D R (2007) Fracture behaviour of nanocomposites based on poly(ethylene-*co*-methacrylic acid) ionomers, *Polymer* **48**:4867–4873.
60. McNally T, Murphy W R, Lew C Y, Turner R J and Brennan G P (2003) Polyamide-12 layered silicate nanocomposites by melt blending, *Polymer* **44**:2761–2772.
61. Costa F R, Satapathy B K, Wagenknecht U, Weidisch R and Heinrich G (2006) Morphology and fracture behavior of polyethylene/Mg-Al layered double hydroxide (LDH) nanocomposites, *Eur Polym J* **42**:2140–2152.

62. Avlar S and Qiao Y (2005) Effects of cooling rate on fracture resistance of nylon 6-silicate nanocomposites, *Composites: Part A* **36**:624–630.
63. Kim G-M, Lee D-H, Hoffmann B, Kressler J and Stöppelmann G (2001) Influence of nanofillers on the deformation process in layered silicate/polyamide-12 nanocomposites, *Polymer* **42**:1095–1100.
64. Kim G-M, Goerlitz S and Michler G H (2007) Deformation mechanism of nylon 6/layered silicate nanocomposites: Role of the layered silicate, *J Appl Polym Sci* **105**:38–48.
65. He C, Liu T, Tjiu W C, Sue H-J and Yee A F (2008) Microdeformation and fracture mechanisms in polyamide-6/organoclay nanocomposites, *Macromolecules* **41**:193–202.
66. Yalcin B, Ergungor Z, Konishi Y, Cakmak M and Batur C (2008) Molecular origins of toughening mechanism in uniaxially stretched nylon 6 films with clay nanoparticles, *Polymer* **49**:1635–1650.
67. Zhang H and Zhang Z (2007) Impact behaviour of polypropylene filled with multi-walled carbon nanotubes, *Eur Polym J* **43**:3197–3207.
68. Zhao P, Wang K, Yang H, Zhang Q, Du R and Fu Q (2007) Excellent tensile ductility in highly oriented injection-molded bars of polypropylene/carbon nanotubes composites, *Polymer* **48**:5688–5695.
69. Kalay G and Bevis M J (1999) Application of shear-controlled orientation in injection molding of isotactic polypropylene, in *Polypropylene: An A-Z Reference* (Ed. Karger-Kocsis J) Kluwer, Dordrecht, pp. 38–46.
70. Zhang S, Minus M L, Zhu L, Wong C-P and Kumar S (2008) Polymer transcrystallinity induced by carbon nanotubes, *Polymer* **49**:1356–1364.
71. Wu C-M, Chen M and Karger-Kocsis J (2001) Interfacial shear strength and failure modes in sPP/CF and iPP/CF microcomposites by fragmentation, *Polymer* **42**:129–135.
72. Karger-Kocsis J and Varga J (1999) Interfacial morphology and its effects in polypropylene composites, in *Polypropylene: An A-Z Reference* (Ed. Karger-Kocsis J) Kluwer, Dordrecht, pp. 348–356.
73. Karger-Kocsis J (2000) Interphase with lamellar interlocking and amorphous adherent: A model to explain effects of transcrystallinity, *Adv Compos Letters* **9**:225–227.
74. Satapathy B K, Ganß M, Weidisch R, Pötschke P, Jehnichen D, Keller T and Jandt K D (2007) Ductile-to-semiductile transition in PP-MWNT nanocomposites, *Macromol Rapid Commun* **28**:834–841.
75. Kobayashi H, Shioya M, Tanaka T and Irisawa T (2007) Synchrotron radiation small-angle X-ray scattering study on fracture process of carbon nanotube/poly(ethylene terephthalate) composite films, *Compos Sci Technol* **67**:3209–3218.
76. Karger-Kocsis J (2002) Fracture and fatigue behavior of amorphous (co)polyesters as a function of molecular and network variables, in *Handbook of Thermoplastic Polyesters* (Ed. Fakirov S) Wiley-VCH, Weinheim, Vol. 1, Ch. 16, pp. 717–753.
77. Cotterell B, Chia J Y H and Hbaieb K (2007) Fracture mechanisms and fracture toughness in semicrystalline polymer nanocomposites, *Eng Fract Mech* **74**:1054–1078.
78. Karger-Kocsis J (Ed.) (1995) *Polypropylene: Structure, Blends and Composites*, Chapman and Hall, London, Vol. 2: Copolymers and Blends.
79. Karger-Kocsis J, Kalló A, Szafner A, Bodor G and Sényei Zs (1979) Morphological study on the effects of elastomeric impact modifiers in polypropylene systems, *Polymer* **20**:37–43.
80. Karger-Kocsis J and Kuleznev V N (1982) Dynamic mechanical and impact properties of polypropylene/EPDM blends, *Polymer* **23**:699–705.
81. Zhang L, Li C and Huang R (2004) Toughness mechanism in polypropylene composites: Polypropylene toughened with elastomer and calcium carbonate, *J Polym Sci Part B:Phys* **42**:1656–1662.

82. Wu S (1988) A generalized criterion for rubber toughening: The critical ligament thickness, *J Appl Polym Sci* **35**:549–561.
83. Yang H, Zhang Q, Guo M, Wang C, Du R and Fu Q (2006) Study on the phase structures and toughening mechanism in PP/EPDM/SiO$_2$ ternary composites, *Polymer* **47**:2106–2115.
84. Yang H, Zhang X, Qu C, Li B, Zhang L, Zhang Q and Fu Q (2007) Largely improved toughness of PP/EPDM blends by adding nano-SiO$_2$ particles, *Polymer* **48**:860–869.
85. Ma C G, Mai Y L, Rong M Z, Ruan W H and Zhang M Q (2007) Phase structure and mechanical properties of ternary polypropylene/elastomer/nano-CaCO$_3$ composites, *Compos Sci Technol* **67**:2997–3005.
86. Hanim H, Ahmad Fuad M Y, Zarina R, Mohd Ishak Z A and Hassan A (2008) Properties and structure of polypropylene/polyethylene-octene elastomer/nano CaCO$_3$ composites, *J Thermoplast Compos Mater* **21**:123–140.
87. Deenadayalan E, Vidhate S and Lele A (2006) Nanocomposites of polypropylene impact copolymer and organoclays: Role of compatibilizers, *Polym Intern* **55**:1270–1276.
88. Deshmane C, Yuan Q and Misra R D K (2007) High strength-toughness combination of melt intercalated nanoclay-reinforced thermoplastic olefins, *Mater Sci Eng* **A460–461**:277–287.
89. Tjong C S and Bao S P (2007) Fracture toughness of high density polyethylene/SEBS-g-MA/montmorillonite nanocomposites, *Compos Sci Technol* **67**:314–323.
90. Chiu F-C, Lai S-M, Chen Y-L and Lee T-H (2005) Investigation on the polyamide 6/organoclay nanocomposites with and without a maleated polyolefin elastomer as a toughener, *Polymer* **46**:11600–11609.
91. Chiu F-C, Fu S-W, Chuang W-T and Sheu H-S (2008) Fabrication and characterization of polyamide 6,6/organo-montmorillonite nanocomposites with and without a maleated polyolefin elastomer as toughener, *Polymer* **49**:1015–1026.
92. Baldi F, Bignotti F, Tieghi G and Riccò T (2006) Rubber toughening of polyamide 6/organoclay nanocomposites obtained by melt blending, *J Appl Polym Sci* **99**:3406–3416.
93. Kelnar I, Khunová V, Kotek J and Kaprálková L (2007) Effect of clay treatment on structure and mechanical behaviour of elastomer-containing polyamide 6 nanocomposite, *Polymer* **48**:5332–5339.
94. Kelnar Ščudla J, Kotek J, Kretzschmar B and Leuteritz A (2006) J-integral evaluation of PA6 nanocomposite with improved toughness, *Polym Test* **25**:697700.
95. Siengchin S and Karger-Kocsis J (2008) unpublished results.
96. Dasari A, Yu Z-Z, Yang M, Zhang Q-X, Xie X-L and Mai Y-W (2006) Micro- and nano-scale deformation behavior of nylon 66-based binary and ternary nanocomposites, *Compos Sci Technol* **66**:3097–3114.
97. Wang K, Wang C, Li J, Su J, Zhang Q, Du R and Fu Q (2007) Effects of clay on phase morphology and mechanical properties in polyamide 6/EPDM-g-MA/organoclay ternary nanocomposites, *Polymer* **48**:2144–2154.
98. Isik I, Yilmazer U and Bayram G (2008) Impact modified polyamide-6/organoclay nanocomposites: Processing and characterization, *Polym Compos* **29**:133–141.
99. Lim S-H, Dasari A, Yu Z-Z, Mai Y-W, Liu S and Yong M S (2007) Fracture toughness of nylon 6/organoclay/elastomer nanocomposites, *Compos Sci Technol* **67**:2914–2923.
100. Dong W, Zhang X, Liu Y, Gui H, Wang Q, Gao J, Song Z, Lai J, Huang F and Qiao J (2007) Process for preparing a nylon-6/clay/acrylate rubber nanocomposite with high toughness and stiffness, *Polym Intern* **56**:870–874.
101. Kelnar J, Kotek J, Kaprálková L, Hromádková J and Kratochvíl J (2006) Effect of elastomer type and functionality on the behavior of toughened polyamide nanocomposites, *J Appl Polym Sci* **100**:1571–1576.

102. Chow W S, Mohd Ishak Z A and Karger-Kocsis J (2005) Atomic force microscopy study on blend morphology and clay dispersion in polyamide-6/polypropylene/organoclay systems, *J Polym Sci Part B: Phys* **43**:1198–1204.
103. Gahleitner M, Kretzschmar B, Pospiech D, Ingolic E, Reichelt N and Bernreitner K (2006) Morphology and mechanical properties of polypropylene/polyamide 6 nanocomposites prepared by a two-step melt-compounding process, *J Appl Polym Sci* **100**:283–291.
104. Karger-Kocsis J, Felhös D and Thomann R (2008) Tribological behavior of a carbon-nanofiber-modified Santoprene thermoplastic elastomer under dry sliding and fretting conditions against steel, *J Appl Polym Sci* **108**:724–730.
105. Bai H, Wang Y, Song B, Li Y and Liu L (2008) Effect of nucleating agent on the brittle-ductile transition behaviour of polypropylene/ethylene-octene copolymer blends, *J Polym Sci Part B: Phys* **46**:577–588.
106. Grein C and Gahleitner M (2008) On the influence of nucleation on the toughness of iPP/EPR blends with different rubber molecular architectures, *eXPRESS Polym Lett* **2**:392–397.
107. Roulin-Moloney A C (Ed.) (1989) *Fractography and Failure Mechanisms of Polymers and Composites*, Elsevier, London.
108. McGrath L M, Parnas R S, King S H, Schroeder J L, Fischer D A and Lenhart J L (2008) Investigation of the thermal, mechanical, and fracture properties of alumina-epoxy composites, *Polymer* **49**:999–1014.
109. Nakamura Y, Yamaguchi M and Okubo M (1993) Instrumented Charpy impact test of epoxy resin filled with irregular-shaped silica particles, *Polym Eng Sci* **33**:279–284.
110. Singh R P, Zhang M and Chan D (2002) Toughening of a brittle thermosetting polymer: Effects of reinforcement particle size and volume fraction, *J Mater Sci* **37**:781–788.
111. Zunjarrao S C and Singh R P (2006) Characterization of the fracture behaviour of epoxy reinforced with nanometer and micrometer sized aluminium particles, *Compos Sci Technol* **66**:2296–2305.
112. Kwon S-C and Adachi T (2007) Strength and fracture toughness of nano and micron-silica particles bidispersed epoxy composites: Evaluated by fragility parameter, *J Mater Sci* **42**:5516–5523.
113. Adachi T, Osaki M, Araki W and Kwon S-C (2008) Fracture toughness of nano- and micro-spherical silica-particle-filled epoxy composites, *Acta Mater* **56**:2101–2109.
114. Kwon S-C, Adachi T and Araki W (2008) Temperature dependence of fracture toughness of silica/epoxy composites: Related to microstructure of nano- and micro-particles packing, *Composites: Part B* **39**:773–781.
115. Karger-Kocsis J and Kiss L (1987) Dynamic mechanical properties and morphology of polypropylene block copolymers and polypropylene/elastomer blends, *Polym Eng Sci* **27**:254–262.
116. Jin F-L and Park S-J (2008) Interfacial toughness properties of trifunctional epoxy resins/calcium carbonate nanocomposites, *Mater Sci Eng* **A475**:190–193.
117. Zhang M and Singh R P (2004) Mechanical reinforcement of unsaturated polyester by Al_2O_3 nanoparticles, *Mater Letters* **58**:408–412.
118. Marur P R, Batra R C, Garcia G and Loos A C (2004) Static and dynamic fracture toughness of epoxy/alumina composite with submicron inclusions, *J Mater Sci* **39**:1437–1440.
119. Bugnicourt E, Galy J, Gérard J-F and Barthel H (2007) Effect of sub-micron silica fillers on the mechanical performances of epoxy-based composites, *Polymer* **48**:1596–1605.

120. Wetzel B, Rosso P, Haupert F and Friedrich K (2006) Epoxy nanocomposites – fracture and toughening mechanisms, *Eng Fract Mech* **73**:2375–2398.
121. Philipp M, Gervais P C, Sanctuary R, Müller U, Baller J, Wetzel B and Krüger J K (2008) Effect of mixing sequence on the curing of amine-hardened epoxy/alumina nanocomposites as assessed by optical refractometry, *eXPRESS Polym Lett* **2**:546–552.
122. Hartwig A, Sebald M, Pütz D and Aberle L (2005) Preparation, characterisation and properties of nanocomposites based on epoxy resins: An overview, *Macromol Symp* **221**:127–135.
123. Jesson D A, Smith P A, Hay J N and Watts J F (2007) The effect of ormosil nanoparticles on the toughness of a polyester resin, *J Mater Sci* **42**:3230–3237.
124. Huang C J, Fu S Y, Zhang Y H and Li L F (2004) Mechanical properties of SiO_2/epoxy nanocomposites at cryogenic temperature, in *Composite Technologies for 2020* (Eds. Ye L, Mai Y-W and Su Z) Woodhead Publ., Abington, pp. 707–712.
125. Bondioli F, Cannillo V, Fabbri E and Messori M (2005) Epoxy-silica nanocomposites: Preparation, experimental characterization, and modeling, *J Appl Polym Sci* **97**:2382–2386.
126. Musto P, Ragosta G, Scarinzi G and Mascia L (2004) Toughness enhancement of polyimides by in situ generation of silica particles, *Polymer* **45**:4265–4274.
127. Fu J, Shi L, Chen Y, Yuan S, Wu J, Liang X and Zhong Q (2008) Epoxy nanocomposites containing marcaptopropyl polyhedral oligomeric silsesquioxane: Morphology, thermal properties, and toughening mechanism, *J Appl Polym Sci* **109**:340–349.
128. Sreekala M S and Eger C (2005) Property improvements of an epoxy resin by nanosilica particle reinforcement, in *Polymer Composites from Nano- to Macroscale* (Eds. Friedrich K, Fakirov S and Zhang Z) Springer, Berlin, Ch. 6, pp. 91–105.
129. Zhang H, Zhang Z, Friedrich K and Eger C (2006) Property improvements of *in situ* epoxy nanocomposites with reduced interparticle distance at high nanosilica content, *Acta Mater* **54**:1833–1842.
130. Blackman B R K, Kinloch A J, Sohn Lee J, Taylor A C, Agarwal R, Schueneman G and Sprenger S (2007) The fracture and fatigue behaviour of nano-modified epoxy polymers, *J Mater Sci* **42**:7049–7051.
131. Rosso P and Ye L (2007) Epoxy/silica nanocomposites: Nanoparticle-induced cure kinetics and microstructure, *Macromol Rapid Commun* **28**:121–126.
132. Deng S, Ye L and Friedrich K (2007) Fracture behaviours of epoxy nanocomposites with nano-silica at low and elevated temperatures, *J Mater Sci* **42**:2766–2774.
133. Johnsen B B, Kinloch A J, Mohammed R D, Taylor A C and Sprenger S (2007) Toughening mechanism of nanoparticle-modified epoxy polymers, *Polymer* **48**:530–541.
134. Rosso P, Ye L and Karger-Kocsis J (2007) unpublished results.
135. Zilg C, Mülhaupt R and Finter J (1999) Morphology and toughness/stiffness balance of nanocomposites based upon anhydride-cured epoxy resins and layered silicates, *Macromol Chem Phys* **200**:661–670.
136. Kornmann X, Thomann R, Mülhaupt R, Finter J and Berglund L (2002) Synthesis of amine-cured, epoxy-layered silicate nanocomposites: The influence of the silicate surface modification on the properties, *J Appl Polym Sci* **86**:2643–2652.
137. Fröhlich J, Golombowski D, Thomann R and Mülhaupt R (2004) Synthesis and characterisation of anhydride-cured epoxy nanocomposites containing layered silicates modified with phenolic alkylimidazolineamide cations, *Macromol Mater Eng* **289**:13–19.
138. Zerda A S and Lesser A J (2001) Intercalated clay nanocomposites: Morphology, mechanics, and fracture behaviour, *J Polym Sci Part B:Phys* **39**:1137–1146.

139. Becker O, Varley R and Simon G (2002) Morphology, thermal relaxations and mechanical properties of layered silicate nanocomposites based upon high-functionality epoxy resins, *Polymer* **43**:4365–4373.
140. Ratna D, Manoj N R, Varley R, Singh Raman R K and Simon G P (2003) Clay-reinforced epoxy nanocomposites, *Polym Intern* **52**:1403–1407.
141. Liu T, Tjiu W C, Tong Y, He C, Goh S S and Chung T-S (2004) Morphology and fracture behaviour of intercalated epoxy/clay nanocomposites, *J Appl Polym Sci* **94**:1236–1244.
142. Wang K, Chen L, Wu J, Toh M L, He C and Yee A F (2005) Epoxy nanocomposites with highly exfoliated clay: Mechanical properties and fracture mechanisms, *Macromolecules* **38**:788–800.
143. Akbari B and Bagheri R (2007) Deformation mechanism of epoxy/clay nanocomposite, *Eur Polym J* **43**:782–788.
144. Park J and Jana S C (2003) Effect of plasticization of epoxy networks by organic modifier on exfoliation of nanoclay, *Macromolecules* **36**:8391–8397.
145. Pluart L L, Duchet J and Sautereau H (2005) Epoxy/montmorillonite nanocomposites: Influence of organophilic treatment on reactivity, morphology and fracture properties, *Polymer* **46**:12267–12278.
146. Li D, Zhang X, Sui G, Wu D, Liang J and Yi X-S (2003) Toughness improvement of epoxy by incorporating carbon nanotubes into the resin, *J Mater Sci Lett* **22**:791–793.
147. Hsiao K-T, Alms J and Advani S G (2003) Use of epoxy/multiwalled carbon nanotubes as adhesives to join graphite fibre reinforced polymer composites, *Nanotechnology* **14**:791–793.
148. Miyagawa H and Drzal L T (2004) Thermo-physical and impact properties of epoxy nanocomposites reinforced by single-wall carbon nanotubes, *Polymer* **45**:5163–5170.
149. Gojny F H, Wichmann M H G, Köpke U, Fiedler B and Schulte K (2004) Carbon nanotube-reinforced epoxy-composites: Enhanced stiffness and fracture toughness at low nanotube content, *Compos Sci Technol* **64**:2363–2371.
150. Fiedler B, Gojny F H, Wichmann M H G, Nolte M C M and Schulte K (2006) Fundamental aspects of nano-reinforced composites, *Compos Sci Technol* **66**:3115–3125.
151. Wichmann M H G, Schulte K and Wagner H D (2008) On nanocomposite toughness, *Compos Sci Technol* **68**:329–331.
152. Xu L R, Bhamidipati V, Zhong W-H, Li J, Lukehart C M, Lara-Curzio E, Liu K C and Lance M J (2004) Mechanical property characterization of a polymeric nanocomposite reinforced by graphitic nanofibers with reactive linkers, *J Compos Mater* **38**:1563–1582.
153. Sun L, Warren G L, O'Reilly J Y, Everett W N, Lee S M, Davis D, Lagoudas D and Sue H-J (2008) Mechanical properties of surface-functionalized SWCNT/epoxy composites, *Carbon* **46**:320–328.
154. Karger-Kocsis J, Fröhlich J, Gryshchuk O, Kautz H, Frey H and Mülhaupt R (2004) Synthesis of reactive hyperbranched and star-like polyethers and their use for toughening of vinylester-urethane hybrid resins, *Polymer* **45**:1185–1195.
155. Ye Y, Chen H, Wu J and Ye L (2007) High impact strength epoxy nanocomposites with natural nanotubes, *Polymer* **48**:6426–6433.
156. Gryshchuk O, Karger-Kocsis J, Thomann R, Kónya Z and Kiricsi I (2006) Multiwall carbon nanotube modified vinylester and vinylester-based hybrid resins, *Composites: Part A* **37**:1252–1259.
157. Kinloch A J, Mohammed R D, Taylor A C, Eger C, Sprenger S and Egan D (2005) The effect of silica nano particles and rubber particles on the toughness of multiphase thermosetting epoxy polymers, *J Mater Sci* **40**:5083–5086.

158. Karger-Kocsis J and Friedrich K (1993) Microstructure-related fracture toughness and fatigue crack growth behavior in toughened, anhydride-cured epoxy resins, *Compos Sci Technol* **48**:263–272.
159. Walter R, Medina R, Haupert F and Schlarb A K (2007) Improved toughness and stiffness of epoxy resins modified by preformed rubber microparticles and SiO_2 nanoparticles for resin transfer molding applications, *Viennano '07 Proceedings*, (March 14–16, 2007), Öster. Tribol. Gesellschaft, Vienna.
160. Gam K T, Miyamoto M, Nishimura R and Sue H-J (2003) Fracture behavior of core-shell rubber-modified clay-epoxy nanocomposites, *Polym Eng Sci* **43**:1635–1645.
161. Isik I, Yilmazer U and Bayram G (2003) Impact modified epoxy/montmorillonite nanocomposites: Synthesis and characterization, *Polymer* **44**:6371–6377.
162. Fröhlich J, Thomann R and Mülhaupt R (2003) Toughened epoxy hybrid nanocomposites containing both an organophilic layered silicate filler and a compatibilized liquid rubber, *Macromolecules* **36**:7205–7211.
163. Liu W, Hoa S V and Pugh M (2004) Morphology and performance of epoxy nanocomposites modified with organoclay and rubber, *Polym Eng Sci* **44**:1178–1186.
164. Ratna D, Becker O, Krishnamurthy R, Simon G P and Varley R J (2003) Nanocomposites based on a combination of epoxy resin, hyperbranched epoxy and a layered silicate, *Polymer* **44**:7449–7457.
165. Balakrishnan S and Raghavan D (2004) Acrylic, elastomeric, particle-dispersed epoxy-clay hybrid nanocomposites: Mechanical properties, *Macromol Rapid Commun* **25**:481–485.
166. Balakrishnan S, Start P R, Raghavan D and Hudson S D (2005) The influence of clay and elastomer concentration on the morphology and fracture energy of preformed acrylic rubber dispersed clay filled epoxy nanocomposites, *Polymer* **46**:11255–11262.
167. Subramaniyan A K and Sun C T (2007) Toughening polymeric composites using nanoclay: Crack tip scale effects on fracture toughness, *Composites: Part A* **38**:34–43.
168. Corobea M-C, Donescu D, Grishchuk S, Castellà N, Apostolov A A and Karger-Kocsis J (2008) Organophilic layered silicate modified vinylester-urethane hybrid resins: Structure and properties, *Polym Polym Compos* (in press).
169. Karger-Kocsis J, Gryshchuk O and Schmitt S (2003) Vinylester/epoxy-based thermosets of interpenetrating network structure: An atomic force microscopic study, *J Mater Sci* **38**:413–420.
170. Njuguna J, Pielichowski K and Alcock J R (2007) Epoxy-based fibre reinforced nanocomposites, *Adv Eng Mater* **9**:835–847.
171. Dzenis Y (2008) Structural nanocomposites, *Science* **319**:419–420.
172. Hussain M, Nakahira A and Niihara K (1996) Mechanical property improvement of carbon fiber reinforced epoxy composites by Al_2O_3 filler dispersion, *Mater Letters* **26**:185–191.
173. Shahid N, Villate R G and Barron A R (2005) Chemically functionalized alumina nanoparticle effect on carbon fiber/epoxy composites, *Compos Sci Technol* **65**:2250–2258.
174. Tsantzalis S, Karapappas P, Vavouliotis A, Tsotra P, Kostopoulos V, Tanimoto T and Friedrich K (2007) On the improvement of toughness of CFRPs with resin doped with CNF and PZT particles, *Composites: Part A* **38**:1159–1162.
175. Kinloch A J, Masania K, Taylor A C, Sprenger S and Egan D (2008) The fracture of glass-fibre-reinforced epoxy composites using nanoparticle-modified matrices, *J Mater Sci* **43**:1151–1154.
176. Becker O, Varley R J and Simon G P (2003) Use of layered silicates to supplementarily toughen high performance epoxy-carbon fiber composites, *J Mater Sci Lett* **22**:1411–1414.

177. Kornmann X, Rees M, Thomann Y, Necola A, Barbezat M and Thomann R (2005) Epoxy-layered silicate nanocomposites as matrix in glass fibre-reinforced composites, *Compos Sci Technol* **65**:2259–2268.
178. Tsai J-L and Wu M-D (2008) Organoclay effect on mechanical responses of glass/epoxy nanocomposites, *J Compos Mater* **42**:553–568.
179. Quaresimin M and Varley R J (2008) Understanding the effect of nano-modifier addition upon the properties of fibre reinforced laminates, *Compos Sci Technol* **68**:718–726.
180. Siddiqui N A, Woo R S C, Kim J-K, Leung C C K and Munir A (2007) Mode I interlaminar fracture behavior and mechanical properties of CFRPs with nanoclay-filled epoxy matrix, *Composites: Part A* **38**:449–460.
181. Gupta N and Maharsia R (2005) Enhancement of energy absorption in syntactic foams by nanoclay incorporation for sandwich core applications, *Appl Compos Mater* **12**:247–261.
182. Wouterson E M, Boey F Y C, Wong S-C, Chen L and Hu X (2007) Nano-toughening versus micro-toughening of polymer syntactic foams, *Compos Sci Technol* **67**:2924–2933.
183. Gojny F H, Wichmann M H G, Fiedler B, Bauhofer W and Schulte K (2005) Influence of nano-modification on the mechanical and electrical properties of conventional fibre-reinforced composites, *Composites: Part A* **36**:1525–1535.
184. Wichmann M H G, Sumfleth J, Gojny F H, Quaresimin M, Fiedler B and Schulte K (2006) Glass-fibre-reinforced composites with enhanced mechanical and electrical properties: Benefits and limitations of a nanoparticle modified matrix, *Eng Fract Mech* **73**:2346–2359.
185. Koo J H and Pilato L A (2005) Polymer nanostructured materials for high temperature applications, *SAMPE J* **41**(No.2):7–19.
186. Zhu J, Imam A, Crane R, Lozano K, Khabashesku V N and Barrera E V (2007) Processing of glass fiber reinforced vinyl ester composite with nanotube enhancement of interlaminar shear strength, *Compos Sci Technol* **67**:1509–1517.
187. Fan Z, Santare M H and Advani S G (2008) Interlaminar shear strength of glass fiber reinforced epoxy composites enhanced with multi-walled carbon nanotubes, *Composites: Part A* **39**:540–554.
188. Tong L, Sun X and Tan P (2008) Effect of long multi-walled carbon nanotubes on delamination toughness of laminated composites, *J Compos Mater* **42**:5–23.
189. Arai A, Noro Y, Sugimoto K-I and Endo M (2008) Mode I and mode II interlaminar fracture toughness of CFRP laminates toughened by carbon nanofiber interlayer, *Compos Sci Technol* **68**:516–525.
190. Grimmer C S and Dharan C K H (2008) High-cycle fatigue of hybrid carbon nanotube/glass fiber/polymer composites, *J Mater Sci* **43**:4487–4492.
191. Li C, Thostenson E T and Chou T-W (2008) Sensors and actuators based on carbon nanotubes and their composites: A review, *Compos Sci Technol* **68**:1227–1249.

Chapter 13

Micromechanics of Polymer Blends: Microhardness of Polymer Systems Containing a Soft Component and/or Phase

S. Fakirov

13.1. Introduction

Recent years have seen the microindentation hardness technique gaining increasing application in the characterization of the structure and morphology of polymers [1–5]. The method uses a sharp indenter that penetrates the surface of the specimen upon application of a given load at a known rate. Pyramid indenters are best suited for indentation tests. Here the hardness, in principle, does not depend on the size of the indentation and the elastic recovery is minimized in comparison to other indenters. During an indentation test the response of a polymeric material is initially elastic. When the stresses exceed the elastic limit, plastic flow occurs and a permanent deformation arises. At this stage, the plastic yield stress and the elastic modulus govern the elasto-plastic response to indentation [6]. When the load is removed, the indentation depth recovers elastically while the diagonal of the impression remains nearly unaltered [7] as schematically shown in Figure 13.1.

Because of its simplicity, microindentation has become a common technique to measure the micromechanical behavior of polymers and its relation with microstructure. *Microhardness, H,* is obtained by dividing the peak contact load by the projected area of impression, which is measured under a light microscope (imaging method). Typical loads of 10^2 mN, when applied to the surface of a conventional polymer, such as poly(ethylene terephthalate) (PET), produce penetration depths of about 2–3 µm.

The complicating effects of viscoelastic relaxation are usually minimized by measuring the indentation diagonal immediately after the load release [8]. The size of the permanent area of impression has been shown to depend on the arrangement and structure of the microcrystals and the specific morphology of the polymeric material [8]. From a mechanical point of view, the polymer may be regarded as a composite consisting of alternating crystalline and disordered elements [8]. An earlier study of the hardness dependence on the density, ρ, of melt crystallized polyethylene (PE) revealed that for degrees of crystallinities larger than 50%, the plastic strain is dominated by the deformation modes of the crystals [8].

It is important to note in these introductory remarks that, like other mechanical properties of solids, microhardness obeys the *"rule of mixture"*, frequently also called the *"additivity law"* (further referred to only as additivity law):

$$H = \Sigma H_i \varphi_i \tag{13.1}$$

where H_i and φ_i are the microhardness and mass fraction, respectively, of each component and/or phase. This law can be applied to multicomponent and/or multiphase systems provided each component and/or phase is characterized by its own H. Equation (13.1) is frequently used in semicrystalline polymers for one purpose or other, operating with the microhardness values of the crystalline, H_c, and amorphous, H_a, phases, respectively. This relationship is of great value because it offers the opportunity to characterize components and/or phases of a system micromechanically, which are not accessible to direct measurement.

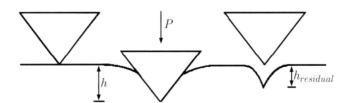

Figure 13.1. Contact geometry for a pyramid indenter at zero load P (left), maximal load (center), and complete unload (right). The residual penetration depth after load removal is given by $h_{residual}$ [1]

13.2. The peculiarity of polymer systems containing a soft component and/or phase

Application of the additivity law (Eq. (13.1)) presumes a very important requirement – each component and/or phase of the complex system should have a glass transition (T_g) or melting (T_m) temperature value above room temperature (at which the indentation is performed) and thus be capable of developing an indentation impression after the removal of the indenter. If this is not the case, the assumption $H_a = 0$ for the soft component and/or phase does not seem to be the best solution, although it is frequently made.

The assumption $H_a = 0$ would mean that the soft component and/or phase (with a T_g or T_m temperatures below room temperature and not displaying its own indentation impression) has only a *"diluting effect"*. Usually, the role of such a component and/or phase in the formation of the overall H of the complex system is not only in the "diluting effect". It creates a completely different *deformation mechanism* (in addition to the plastic deformation of the solid component and/or phase) as compared to the deformation behavior of complex systems non-comprising a soft component and/or phase. The deformation mechanism of the entire system is changed in such a way that the system does not obey anymore the additivity law (Eq. (13.1)) [9].

It has been demonstrated that many semicrystalline polymers, copolymers and blends obey the additivity law [1]. Exceptions, such as blends of high density polyethylene (HDPE) with polypropylene (PP) have been explained by a peculiarity in the morphology and characteristics of the crystallites formed (mostly related to the crystal sizes, surface free energy, and others) [1,10]. An example of such an exception is illustrated in Figure 13.2, where the experimentally obtained variation of H as a function of the weight fraction of the PE component, φ, (curve (3)) is plotted.

It is seen that the H values for the initial PE and isotactic polypropylene (iPP) gel films ($H^{PE} = 105$ MPa and $H^{PP} = 116$ MPa) do not differ substantially from each other. At the same time, one can see a very clear deviation

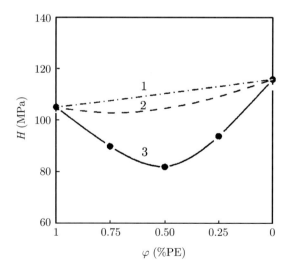

Figure 13.2. Microhardness, H, of PE/iPP blended gel films as a function of PE concentration, φ: additivity behavior from Eq. (13.1) using the w_c values of the individual homopolymers (1); H values using w_c^{PE} and w_c^{PP} data (Table 13.1) (2); experimental data (3) [1,10]

(Figure 13.2, straight line (1)) with increasing PE concentration, φ, from the additivity law:

$$H = \varphi H^{PE} + (1 - \varphi) H^{PP} \qquad (13.2)$$

Since the glass transition temperature of PE is much lower than room temperature, the microhardness contribution of the amorphous phase has been accepted to be $H_a^{PE} \sim 0$ [10]. Hence, for PE, with a degree of crystallinity, w_c^{PE}, using the additivity model of Eq. (13.1) one may write:

$$H^{PE} = w_c^{PE} H_c^{PE} \qquad (13.3)$$

On the other hand, since for iPP $H_a^{PP} \neq 0$, for iPP, with a degree of crystallinity, w_c^{PP}:

$$H^{PP} = w_c^{PP} H_c^{PP} + (1 - w_c^{PP}) H_a^{PP} \qquad (13.4)$$

By combining Eqs. (13.3) and (13.4) one obtains for the overall microhardness of the blend, H, the expression:

$$H = \varphi w_c^{PE} H_c^{PE} + (1 - \varphi) w_c^{PP} H_c^{PP} + (1 - \varphi)(1 - w_c^{PP}) H_a^{PP} \qquad (13.5)$$

which describes the microhardness of the blended gel films in terms of the microhardnesses of the independent crystalline and amorphous components assuming $H_a^{PE} \sim 0$. If one takes into account the crystallinity depression measured for the PE and iPP components in Eq. (13.5), use $H_c^{PE} = 130$ MPa and $H_c^{PP} = 145$ MPa, and let $H_a^{PP} = 90$ MPa [10] then one obtains curve (2) in Figure 13.2, which is still far from the experimental values (Figure 13.2, curve (3)).

It should be noted here that in the same paper [10] it is demonstrated that the differences between the experimentally measured and calculated H values disappear if the thermodynamically derived parameter b (accounting for the crystal surface free energy and the energy required to plastically deform the crystal [1]) is used for the calculations. However, it seems important to mention also that the values of the b parameter have been derived from the same H_{exp} values and later used for the calculation of the H values (Eq. (13.1)).

The same approach (assuming $H_a = 0$ for the soft component and/or phase) applied to thermoplastic elastomers of poly(ether ester) (PEE) type fails also to explain the large discrepancy (up to 100 MPa when the measured H values are in the range of 20–40 MPa) between the experimental values and those calculated according to Eq. (13.1) [11,12]. For this reason, one has to look for other factors, which may be responsible for such a discrepancy. Before discussing them, let us recall briefly some of the characteristic features of the structure and morphology of thermoplastic elastomers of PEE-type, which illustrate in the best way the concept under discussion.

Thermoplastic elastomers of PEE type represent polyblock copolymers comprising usually poly(buthylene terephthalate) (PBT) as the "hard" segments and poly(glycols) as "soft" segments, both of them forming "hard" and "soft" domains, respectively. Since the soft domains are characterized by T_g values around $-50°C$, they are in a liquid state at room temperature and are distinguished by a viscosity being much closer to those of low-molecular-weight liquids rather than to that of a solid amorphous polymer. In this respect it seems useful to recall that the molecular weights of the poly(tetramethylene glycol) (PTMG) and poly(ethylene glycol) (PEG) used are around 1000, *i.e.*, one deals with typical oligomeric materials. For this reason, it seems reasonable to accept that such a liquid (soft phase) will be characterized by a negligibly small microhardness, H^s, in the equation for the overall microhardness of such a copolymer, H:

$$H = \varphi[w_c H_c^h + (1 - w_c) H_a^h] + (1 - \varphi) H^s \qquad (13.6)$$

where φ is the mass fraction of hard segments (PBT in the present case), H_c^h and H_a^h are the microhardnesses of the crystalline and amorphous phases of the same hard domains, respectively, and w_c is the degree of crystallinity of PBT.

Assuming, as in the case of the PE/*i*PP blend, the microhardness of the soft domains being $H^s = 0$, the calculations of H according to Eq. (13.1) for a series of PEEs lead to a discrepancy between the measured, H_{exp}, and calculated (using again Eq. (13.1)) H'_{cal} amounting to 40–64 MPa, depending on the soft-segment composition, as will be discussed below.

A question arises about the reason for the failure of the additivity law in the above-mentioned systems, both the polyolefins blends [10] and the multiblock copolymer. Obviously, one has to assume that, for multicomponent and/or multiphase systems, when one of the components (phases) is characterized by a viscosity at room temperature close to those of the low-molecular-weight liquids, the mechanism of the response to the applied external mechanical field is different from that when all the components (phases) have T_g and T_m higher than room temperature. In the latter case all the components (phases) plastically deform as a result of the applied external force. In the former case, in addition to the plastic deformation of the harder components (phases), they are also displaced within the soft (liquid) matrix in which they are "floating". The extent of this displacement depends on the viscosity of the matrix (the softer component and/or phase). This is the reason why the harder components cannot display their inherent microhardness. The microhardness is reduced by the ability of the harder components to move. This situation is illustrated in Figure 13.3.

How can one account for this microhardness depression, *i.e.* for the "*floating effect*"? As demonstrated above, the simple assumption that the soft phases have $H^s \sim 0$, *i.e.*, if one accounts only for its "diluting effect", does not solve the problem. It is necessary to characterize the ability of the harder phase to move about within the soft matrix, and this will depend on the viscosity of the matrix, *i.e.*, the soft phase. Since T_g and viscosity are closely related to each other, it is possible to look for an analytical relationship between microhardness of the amorphous polymers and their T_g.

It is a main target of this chapter, by means of larger number of studied systems, to show that the assumption $H^s = 0$ for the soft component and/or phase being dominating in complex polymer systems leads to drastic deviations from the additivity law (Eq. (13.1)). The contribution of the soft component

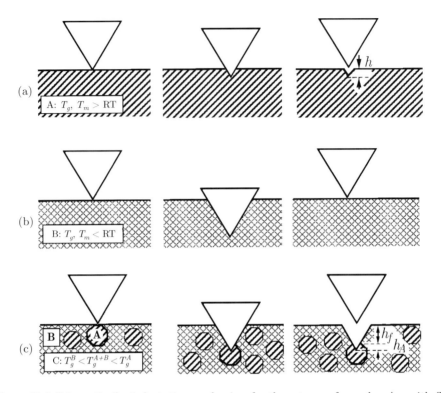

Figure 13.3. Schematic the indentation mechanism for three types of samples: A – with T_g and T_m above the test temperature (room temperatute, RT), microhardness $H \approx 1/h$, (a); B – with T_g and T_m below RT, $H \approx 0$, (b); C – complex system with matrix B and a "floating" dispersed phase A, $H \approx 1/(h_A + h_f)$, (c). In all the cases the elastic recovery of the samples is not taken into account [9]

and/or phase to the overall microhardness can be much more reliably accounted for using the relationship between T_g and the microhardness of the amorphous component and/or phase, particularly for systems characterized by dominating amorphous component and/or phase. A secondary goal is to demonstrate that the experimentally derived relationship between H and T_g can be applied for evaluation of the T_g value of a practically non-accessible component and/or phase. A further target of the chapter is to demonstrate that the microhardness analysis is a quite useful technique for characterization of polymer properties having a crucial contribution to the mechanical behavior of polymeric materials. This is of particular importance in dealing with polymer blends and composites where the interphase adhesion between their constituents plays a fundamental role in their mechanical properties on nano- and microlevels. Microhardness measurements allow one to get an idea about the quality of this adhesion also in cases when a compatibilizer or a coupling agent is used as well as a weld line appears as a result of doubly injection molding or using a mold with an obstacle.

No experimental work was undertaken for the study related to the additivity law since a good deal of data have already been reported in the literature in this respect. One needs only to evaluate the contribution of the soft component and/or phase through their glass transition temperature to the overall microhardness. Usually, samples differing in their crystallinity are used, which are prepared by means of various techniques. What deserves to be mentioned here is that in all cases the microhardness measurements have been carried out at room temperature employing a Vickers pyramid diamond. The microhardness value, H, has been always derived from:

$$H = 1.854\, P/d^2 \qquad (13.7)$$

where P is the applied load and d the diagonal of the residual impression. The permanent deformation has been measured immediately after load release to avoid long delayed recovery. Loads of a couple of hundreds of mN have been used, for an indentation time of 0.1 min, in order to correct for the instant elastic recovery. This correction was then applied to derive the H values using indentation times typically in the range of 0.1–21 min and loads of around 150 mN.

13.3. Comparison between measured and computed microhardness values for various systems

In accordance with the main goal of this study, *i.e.*, recalculations of the overall microhardness by means of additivity law (Eq. (13.1)), however, assuming (in contrast to the approach used for the already published data) that the microhardness of the soft component and/or phase is not zero. Their contribution to the overall microhardness of the complex system is evaluated by means of the recently derived relationship between H and T_g for a series of amorphous homo- and copolymers [13]:

$$H = 1.97\, T_g - 571 \quad (H \text{ in MPa}, \ T_g \text{ in K}) \qquad (13.8)$$

Subsequently, examples of comparison of the measured, H_{exp}, and calculated (with $H^s \neq 0$) microhardness values, H''_{cal} will be considered below for two-component multiphase systems comprising soft phases (blends of polyolefins), multiblock copolymers of thermoplastic elastomers (TPE) of condensation type, blends of amorphous miscible polymers, blends of copolymers with different molecular architecture, and finally, homopolymers comprising amorphous phase with very low T_g (PE).

13.3.1. *Two-component multiphase systems comprising soft phase(s) (blends of semicrystalline homopolymers)*

The best studied system in this respect is the afore-mentioned blend PE/iPP [10]. As already commented (Figure 13.2), a significant discrepancy between H_{exp} and H'_{cal} (assuming $H_a^{PE} = 0$) can be observed (compare the values of H_{exp} with H'_{cal} in Table 13.1). If one tries to account for the contribution of the PE amorphous phase, (which amounts to up to 30%), to the overall microhardness

Table 13.1. Composition, crystallinity, w_c^{PE}, and w_c^{PP}, measured microhardness, H_{exp}, calculated microhardness (according to Eq. (13.1), with $H_a = 0$), H'_{cal}, and according to Eq. (13.8), (with $H^s \neq 0$), H''_{cal}, and the differences between the measured and calculated values, $\Delta H'$ ($\Delta H' = H'_{cal} - H_{exp}$) and $\Delta H''$ ($\Delta H'' = H''_{cal} - H_{exp}$), respectively, for PE/$i$PP blends

PE/iPP (by wt)	w_c^{PE} [10]	w_c^{PP} [10]	H_{exp} [10] (MPa)	H'_{cal} [10] (MPa)	$\Delta H'$ [9] (MPa)	H''_{cal} [9] (MPa)	$\Delta H''$ [9] (MPa)
100/0	0.80	–	105	104	–1	–	–
75/25	0.77	0.39	90	108	18	72	–18
50/50	0.74	0.43	82	110	28	83	1
25/75	0.71	0.45	94	113	19	96	2
0/100		0.49	116	117	1	–	–

of the blend by means of its glass transition temperature T_g^{PE} using Eq. (13.8), for some of the samples, characterized by lower content in the blend (50 and 25 wt%) of the high crystalline PE (w_c between 70 and 80%), one obtains H''_{cal} values rather close to the measured ones (differences of 2–3%), as can be concluded from Table 13.1. For these calculations a value of $T_g^{PE} = -80°C$ is assumed being the most frequently used for the amorphous unbranched PE [14–19]. It is worth noticing here that much higher values for T_g of PE (around –30°C) are also reported [20–25], however, they are derived from branched PE [20–24] or samples characterized by highly strained amorphous chains [25].

Basically, the unsatisfactory agreement between H''_{cal} (assuming $H^s \neq 0$) and H_{exp} for the blend with the highest content of PE (75 wt%) could have another, additional origin. In order to apply Eq. (13.8) correctly, one needs to know the T_g values of the amorphous phase of the particular sample under investigation. This was not the case in the discussed system PE/iPP for which the common value of $T_g = -80°C$ was used, although the real value can be higher, particularly in the cases with a high degree of crystallinity. What is more, a higher T_g value would lead to a smaller difference $\Delta H''$ for the same sample. Therefore, for the next systems only the experimentally measured T_g values of the respective soft component and/or phase will be used for similar calculations.

13.3.2. *One-component multiphase systems containing soft phase(s) (polyblock copolymers)*

Micromechanical studies have also been carried out on thermoplastic elastomers. The latter, as briefly mentioned above, represent a special class of multiphase systems (block copolymers) exhibiting an unusual combination of properties: they are elastic and at the same time tough and they show low-temperature flexibility and also strength at relatively high temperatures (frequently *ca.* 150°C) [26,27]. In addition to the thermoplastic elastomers of PEE type, also a series of new poly(ester ether carbonate) (PEEC) multiblock terpolymers with varying amount of ether and carbonate soft-segment content will be considered. Dielectric relaxation experiments on the same PEEC revealed the existence of two relaxation processes [28]. The dielectric loss values show a relaxation maximum appearing at about 0°C for 10 kHz (β-relaxation) accompanied by a lower temperature relaxation (γ-relaxation) that appears at about –50°C. This peculiarity of thermoplastic elastomers, namely the fact that they comprise domains with T_g values far below room temperature (typically between –50 and 0°C) make them extremely attractive for the purposes of the present study because they always contain a soft, liquid-like phase.

The microhardness of films of thermoplastic elastomers based on PBT-cycloaliphatic carbonate (PBT-PCc) block copolymers has also been studied [29]. The microhardness of their amorphous films has been discussed in terms of a model given by the additivity values of the single components H_a^{PBT} and H_a^{PCc}. In the case of semicrystalline copolymers, the authors related the observed deviation from the additivity law as mainly due to the depression of the crystal microhardness of the PBT crystals and partly due to a decrease in crystallinity of the PBT phase [29]. The measured H values for the terpolymers studied [29] are very low in comparison to the values known for common synthetic polymers, even those in the amorphous state [30]. What is more, the H_{exp} values are more than three times smaller than the calculated ones. For the purpose of such calculations, the authors [29] present the additivity law (Eq. (13.1)) in the following form:

$$H = \varphi H^{PBT} + (1 - \varphi) H^s \qquad (13.9)$$

where φ is the weight fraction of the hard PBT segments and H^s the microhardness of the soft domains. Further, taking into account the fact that the T_g of the soft-segment amorphous phase lies between –50 and 0°C, (depending on the PCc content [29]), the authors assumed again $H^s \sim 0$. As a result, H will be depressed with decreasing values of φ according to the simple expression [29]:

$$H = \varphi H^{PBT} = \varphi [w_c H_c^{PBT} + (1 - w_c) H_a^{PBT}] \qquad (13.10)$$

Here, the assumption $H^s = 0$ means also to take into account only the "diluting effect" of the soft phase and no attempt to be made for considering the possibility of another deformation mechanism. So, by applying the numerical values $\varphi = 0.6$, $H_c^{PBT} = 287$ MPa, and $H_a^{PBT} = 54$ MPa [29] one can derive the calculated values H'_{cal} for the terpolymers depending on their crystallinity w_c (Table 13.2).

The differences between H_{exp} and H'_{cal} are quite obvious – the calculated values are two to three times higher than the measured ones (Table 13.2).

For a quantitative evaluation of the microhardness depression effect of the soft phase, one has to replace H^s in Eq. (13.9) with Eq. (13.8) using for T_g the experimentally measured values of the soft-segment phase T_g^s. This leads to the expression:

$$H = \varphi [w_c H_c^h + (1 - w_c) H_a^h] + (1 - \varphi)(1.97 T_g^s - 571) \qquad (13.11)$$

Calculation of H for PEE and PEEC by means of Eq. (13.11) offers data that are in very good agreement with the measured H_{exp} values as shown in Table 13.2 (for PEEC) and in Table 13.3 for PEE and PEEC (samples 1–6).

Another system of interest demonstrating the limits of the additivity law (Eq. (13.1)) (with the assumption $H^s = 0$) is again a thermoplastic elastomer of novel type – multiblock polyester-amide copolymers, synthesized recently

Table 13.2. Composition, degree of crystallinity of PBT, w_c, glass transition temperatures of the soft, T_g^s, and the hard, T_g^h, domains, measured microhardness, H_{exp}, calculated microhardness (according to Eq. (13.1) with $H^s = 0$), H'_{cal}, and according to Eq. (13.8) (with $H^s \neq 0$), H''_{cal}, and the differences between measured and calculated values, $\Delta H'$ ($\Delta H' = H'_{cal} - H_{exp}$) and $\Delta H''$, respectively for PEEC block terpolymers

PBT/ PTMG/ PCc	w_c	T_g [29]			Microhardness				
		T_g^s from DSC	T_g^s from DMTA	T_g^h from DSC	H_{exp} [29]	H'_{cal} [29]	$\Delta H'$ [9]	H''_{cal} [9]	$\Delta H''$ [9]
(by wt)	(%)	(°C)	(°C)	(°C)	(MPa)	(MPa)	(MPa)	(MPa)	(MPa)
100/0/0	59.0	55	–	–	–	–	–	–	–
60/40/0	37.7	−56	−50	57	29.6	85.1	55.5	28	−1.6
60/32/8	25.8	−45	−34	55	22.8	68.5	45.7	20	−2.8
60/20/20	24.3	−32	−17	54	18.6	66.4	47.8	28	9.4
60/12/28	21.2	−10	−3	53	17.5	62.0	44.5	41	23.5
60/0/40	13.0	1	19	52	15.5	50.6	35.1	38	22.5

Table 13.3. Composition, annealing temperature T of 5× drawn samples, degree of crystallinity of PBT, w_c, glass transition temperature of the soft domains, T_g^s, measured microhardness, H_{exp}, calculated microhardness, H'_{cal}, (according to Eq. (13.1), with $H^s = 0$) and H''_{cal} (according to Eq. (13.8), with $H^s \neq 0$), and the difference between measured and calculated values, $\Delta H'$ ($\Delta H' = H'_{cal} - H_{exp}$) and $\Delta H''$ ($\Delta H'' = H''_{cal} - H_{exp}$), respectively, for thermoplastic elastomers of PEE- or PEEC-type [1]

Copolymer	Composition	T	w_c	T_g^s	Microhardness				
					H_{exp} [1]	H'_{cal} [1]	$\Delta H'$ [9]	H''_{cal} [9]	$\Delta H''$ [9]
	(by wt)	(°C)	(%)	(°C)	(MPa)	(MPa)	(MPa)	(MPa)	(MPa)
PBT/PTMG	60/40	70	37.7	−56	29.6	87.5	56.1	29.7	0.1
PBT/PTMG/PCc	60/32/8	70	25.8	−45	22.8	69.5	46.7	21.8	1.0
PBT/PTMG/PCc	60/20/20	70	24.3	−32	18.68	59.9	41.3	22.0	3.4
PBT/PEG	57/43	170	41	−44.5	34.2	85.2	51	34.4	0.2
PBT/PEG	57/43	25	37	−37	30.7	79.9	49.2	31.6	0.9
PBT/PEG	57/43	170	41	−41	32.9	85.2	52.3	35.4	2.3
PBT/PEG	75/25	25	35	−35	47.3	101.7	54.4	85.1	37.8
PBT/PEG	75/25	150	39	−39	44.2	108.5	64.3	83.7	39.5

[31]. These materials, similarly to PEE, possess a hetero-phase structure, with two T_g values and only one melting temperature above room temperature, which corresponds to the fusion of PBT crystals. The diamide segments are chosen to mainly contribute to the amorphous domains and confer to the material an elastomeric character. Data on the molecular weight of the used oligotetrahydrofuran, blocks fractions (in mol %), temperature transitions, degree of crystallinity, and density, together with the measured and calculated H_{exp} and H'_{cal} (with $H^s = 0$) values are reported [32]. Hardness is shown to drastically decrease with increasing etherdiamide content. The experimentally measured hardness values of the copolymers clearly deviate from the linear additivity law, where the authors [32] assume $H^s \sim 0$ for the poly(etherdiamide) homopolymer, as its T_g is far below room temperature. Following the same logic as in the case of PEE, PEEC and PE, they have used the following equation, which formally describes the hardness of a two-component system in terms of the H values of the individual constituents [32]:

$$H = (1 - \varphi^{DA})H^{PBT} + \varphi^{DA}H^{DA} \tag{13.12}$$

Here, H^{PBT} and H^{DA} are the microhardness values of the PBT and etherdiamide domains, respectively, ($H^{DA} \sim 0$) and φ^{DA} is, in this particular case, the mol fraction of the soft segments (etherdiamide component). By analogy with the previous cases, the value of H^{PBT} has been expressed in terms of the crystal hardness, H_c^{PBT}, the hardness of the PBT amorphous regions, H_a^{PBT}, and as a function of the volume degree of crystallinity of PBT referred to the volume fraction of the PBT "component" in the sample, v_c^{PBT}. Therefore, one can rewrite Eq. (13.12) to yield [32]:

$$H = (1 - \varphi^{DA})[v_c^{PBT}H_c^{PBT} + (1 - v_c^{PBT})H_a^{PBT}] \tag{13.13}$$

Using the volume fraction crystallinity, $v_c = v_c^{PBT}(1 - \varphi^{DA})$, Eq. (13.13) finally reads [32]:

$$H = v_c^{PBT}H_c^{PBT} + (1 - v_c^{PBT} - \varphi^{DA})H_a^{PBT} \tag{13.14}$$

H_a^{PBT} has been reported to be of 54 MPa [29,33].

It has been found [32] that the values calculated by means of Eq. (13.14) of H'_{cal}, i.e., assuming again $H^s = 0$, are 8 to 10 times larger than the measured H_{exp} values for two-thirds of the samples under investigation. Therefore, one has to look again for another reason for the observed discrepancy. In order to apply again the "floating effect" concept, one has to replace in Eq. (13.12) the H^{DA} with $H^{DA} = 1.97 T_g^{DA} - 571$ and combining further with Eq. (13.13) to obtain for the overall microhardness:

$$H = (1 - \varphi^{DA})[w_c^{PBT}H_c^{PBT} + (1 - w_c^{PBT})H_a^{PBT}] + \varphi^{DA}(1.97 T_g^{DA} - 571) \tag{13.15}$$

where w_c is the weight fraction crystallinity of PBT.

The calculated values for H''_{cal} according to Eq. (13.15) also differ from the measured ones, and what is more, they are not consistent – only one third of them show small differences (around 3–8%) from the H_{exp} values, while the rest scatter in a large interval contrasting the results from the thermoplastic elastomers of PEE type (Tables 13.2 and 13.3). An attempt to explain these differences will be undertaken later, when discussing the obtained results.

13.3.3. Two-component one-phase systems (miscible blends of amorphous polymers)

In addition to the studies on blends of polyolefins, as well as on multiblock copolymers described above, in which some of the components and/or phases are crystallizable, investigations have also been carried out on blends of non-crystallizable components.

Amorphous films of poly(methylmethacrylate)/poly(vinylidenefluoride) (PMMA/PVDF) blends have been prepared by initial precipitation from a solvent and rapid solidification at ~15°C from the molten state [34]. Moreover, these two constituents are considered as thermodynamically miscible [35,36]. The PMMA/PVDF compositions studied have 25/75, 45/55, 50/50, 55/45, 60/40 and 75/25 ratios (by weight). The presence of a single X-ray halo as well as a single T_g value for all the blends, in the above composition range, favored the view that these materials are composed of homogeneous mixtures at molecular level [34]. For this reason, the authors assumed [34] that the parallel decrease of the microhardness obeys a simple expression for the overall microhardness of the blend, H:

$$H = H^{PMMA}(1 - \varphi^{PVDF}) \qquad (13.16)$$

where φ^{PVDF} is the mass fraction of PVDF. Since the T_g value of PVDF is known to be –40°C [35], the authors have applied the common approach at the time, assuming that PVDF molecules do not offer any mechanical contribution to the yield behavior of the blend [34].

If one follows this logic and calculates the H values from Eq. (13.16) taking into account only the H_a^{PMMA} value and PMMA mass fraction, one obtains values being quite different from the experimental ones, as can be concluded from Table 13.4, where the measured values of T_g and those of the density for the blends are given, as reported in [34].

If one applies the other approach for accounting the contribution of the soft component, as in the two cases described in the previous sections, i.e. via the T_g value, the results look quite differently, as demonstrated below.

Formally, the blend PMMA/PVDF can be considered as a three-phase one, because the reported density, ρ, values (Table 13.4, [34]) differ from the commonly accepted for ρ_a one. For the neat PVDF a density value of 1740 kg.m^{-3} is reported [34], while in the literature a value for the fully amorphous PVDF (ρ_a) of 1680 kg.m^{-3} can be found [14,36]. This difference in ρ_a suggests that some "ordering" in the system may have taken place during the sample preparation. Using the value of $\rho_c = 1930$ kg.m^{-3} for the completely crystalline PVDF [36] (which corresponds to the α-, also called type I-modification, i.e. crystallization from melt), one could estimate an apparent "degree of crystallinity" $w_c = \rho_c/\rho[(\rho - \rho_a)/(\rho_c - \rho_a)]$ for the sample with $\rho = 1740$ kg.m^{-3}, leading to $w_c = 0.25$.

Table 13.4. Composition of PMMA/PVDF blends, their glass transition temperatures, T_g^B, density, ρ, the experimentally measured microhardness values, H_{exp}, calculated by means of Eq. (13.1) (with $H^s = 0$), H'_{cal}, or calculated with $H^s \neq 0$ values, H''_{cal} and H'''_{cal}, as well as the differences between the calculated and the measured values, $\Delta H'$ ($\Delta H' = H'_{cal} - H_{exp}$), $\Delta H''$ ($\Delta H'' = H''_{cal} - H_{exp}$) and $\Delta H'''$ ($\Delta H''' = H'''_{cal} - H_{exp}$), respectively [37]

PMMA/PVDF	$T^B_{g,exp}$	$T^B_{g,cal}$	ρ	H_{exp}	H'_{cal}	$\Delta H'$	H''_{cal}	$\Delta H''$	H'''_{cal}	$\Delta H'''$
	[34]	[37]	[34]	[34]	(Eq. 13.1)		(Eq. 13.8)		(Eq. 13.19)	
(by wt)	(K)	(K)	(kg.m⁻³)	(MPa)	(MPa)	(MPa)	(MPa)	(MPa)	(MPa)	(MPa)
100/0	393	–	1183	212	201	–11	203	–9	203	–9
75/25	359	336	1276	160	91	–69	152	–8	106 (148)	–54 (–12)
55/45	337	340	1370	125	100	–25	117	–12	125	0
50/50	331	333	1390	107	84	–32	102	–5	113	6
45/55	325	325	1419	103	69	–34	92	–11	106	3
40/60	319	317	1444	70	53	–17	81	11	93	13
25/75	293	291	1546	53	2,3	–50	52	–1	65	12
0/100	233	–	1740	0 (extr.)	–112	–112	1	1	0	0

Based on the fact that all the samples (Table 13.4) have been prepared in the same manner [34], one can assume that the corresponding PVDF fraction in each blend is characterized by the same "degree of crystallinity" (25%). This finding allows us to consider formally the blend samples under investigation (Table 13.4) as two-phase systems. In the case of such blends the microhardness can be calculated by means of the additivity law as:

$$H = \varphi[w_c H_c^{PVDF} + (1 - w_c)H_a^{PDVF}] + (1 - \varphi)H_a^{PMMA} \qquad (13.17)$$

where φ is the mass fraction of PVDF in the blend, and H_c and H_a – the microhardness values for the completely crystalline and fully amorphous samples, respectively.

By combination of Eqs. (13.17) and (13.8) one obtains [37]:

$$H = \varphi[w_c H_c^{PVDF} + (1-w_c)(1.97 T_g^{PVDF} - 571)] + (1-\varphi)(1.97 T_g^{PMMA} - 571) \quad (13.18)$$

H_c^{PVDF} can be easily evaluated using the extrapolated microhardness value for the neat PVDF (with $w_c = 25\%$) of $H = 0$ MPa and the value of $H_c^{PVDF} = 336$ MPa is obtained. Using this value of H_c^{PVDF} and letting $T_g^{PVDF} = 233$ K and $T_g^{PMMA} = 393$ K (Table 13.4), one can calculate by means of Eq. (13.18) the microhardness of the studied blends. The values obtained are summarized in Table 13.4 as "H''_{cal} Eq. (13.8)".

After taking into account the fraction of the densified PVDF one observes now a better agreement between the experimental and calculated results.

The suggested treatment (Eq. (13.17)) considers formally the PMMA/PVDF blends as a three-phase system – two amorphous and one "crystalline" (PVDF).

Now, lets try to calculate the H values taking into account the real situation, i.e., (i) the two components are completely miscible in the amorphous state, and (ii) this amorphous phase displays only one T_g, that of the blend, T_g^B. Then H will be [37]:

$$H = \varphi_c H_c^{PVDF} + (1 - \varphi_c)(1.97 T_g^B - 571) \qquad (13.19)$$

where φ_c is the mass fraction of the "crystalline" PVDF phase and $(1 - \varphi_c)$ of the amorphous two-component blend. Further, using the values for T_g^B as derived from Gordon and Taylor equation [38] one obtains by means of Eq. (13.19) the values of H listed also in Table 13.4 in its last column as "H_{cal}'''". A relatively good agreement with the reported [34] H_{exp} values can be observed (Table 13.4). In specific cases the agreement is even better than in the previous three-phase model treatment (according to Eq. (13.17)). The most serious exception in this respect is the blend 75/25, for which a much better value (see the value in parentheses, Table 13.4) is obtained if one uses the measured $T_g^B = 359$ K (Table 13.4) instead of the value calculated by means of Gordon and Taylor equation. This difference can be explained by a possible higher "densification" of this particular sample, which results in a T_g- increase as reported for many polymers [39,40].

13.3.4. Two-component two-phase amorphous systems containing a soft phase

In recent publications on blends of novel copolymers of polystyrene (PS) with polybutadiene (PB), with well-defined linear block- or star block architecture, detailed morphological and mechanical investigations have been performed [41–43]. Studying the microhardness behavior of these copolymer blends, the authors, by analogy with previous cases discussed above, have drawn the following conclusion regarding the contribution of the soft phase to the overall microhardness: "since the T_g of the phase in the present case is always well below the test temperature (i.e., the room temperature, 23°C), which may be regarded as being at liquid-like state, it does not affect the measured H values. Therefore, there is no correlation between the soft phase glass transition temperature and the microhardness of the polymer blends discussed in the present study" [41]. At the same time, for the blends of the star block and a linear triblock copolymers, both consisting of PS and PB, it has been found that the experimental hardness of the blends show much lower H values than those predicted by the additivity law (again assuming $H^s = 0$), as shown in Figure 13.4. This finding is similar to the results obtained for microphase separated blends styrene/butadiene block copolymers [42].

To what extent is the cited assumption of the authors [41] regarding the lack of influence of the liquid-like phase on the H values correct and is the observed deviation from the additivity law not due to the neglecting of this influence? For the studied two-component system [41] consisting of two copolymers (linear and star-like) based on the same two monomers (styrene and butadiene), the additivity law (Eq. (13.1)) can be presented in the following way, accounting also for the contribution of the soft phases to the overall microhardness, H:

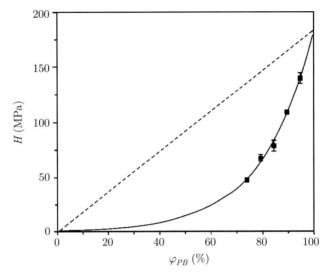

Figure 13.4. Microhardness of the blend of linear and star-like copolymers of PS and PB, H, as a function of the total PS content (assuming volume fraction ≈ weight fraction), φ_{PB}; dashed line represents the additivity law (Eq. (13.1)) [41]

$$H = \varphi_{block}[w_{PS}H^{PS} + (1-w_{PS})H^{PB}] + (1-\varphi_{block})[w'_{PS}H^{PS} + (1-w'_{PS})H^{PB}] \quad (13.20)$$

where H^{PS} and H^{PB} are the microhardness of the respective homopolymers, w_{PS}, w'_{PS} and $(1-w_{PS})$, $(1-w'_{PS})$, are the mass fractions of PS and PB in each copolymer, respectively, and φ_{block} and $(1-\varphi_{block})$ are the mass fraction of the copolymers in the blend. Taking into account the important fact that the two copolymers (as well as their blocks) are completely amorphous, one can express their microhardness by means of Eq. (13.8) and obtain for the overall H of the blend the following expression:

$$H = \varphi_{block}[w_{PS}(1.97T_g^{PS} - 571) + (1-w_{PS})(1.97T_g^{PB} - 571)] + \\ + (1-\varphi_{block})[w'_{PS}(1.97T_g^{PS} - 571) + (1-w'_{PS})(1.97T_g^{PB} - 571)] \quad (13.21)$$

Equation (13.21) reflects analytically the contribution to the overall microhardness H of each constituent of the blends (copolymers and their blocks). At the same time, one should not forget a very important fact – experimentally are observed only two glass transition temperatures (that of PS and that of PB), regardless of the fact that their blocks are incorporated in the two types of copolymers [41]. For this reason, it seems justified to use for the calculation of the overall microhardness of the copolymer blend H (assuming again $H^s \neq 0$) a more simple equation accounting only for the total mass fraction in the blend of the two species, PS and PB, and their T_g values:

$$H = \varphi_{PS}(1.97T_g^{PS} - 571) + (1-\varphi_{PS})(1.97T_g^{PB} - 571) \quad (13.22)$$

where φ_{PS} and $(1-\varphi_{PS})$ are the total mass fractions in the blends of PS and PB, respectively.

Using the reported data [41] about the total mass fractions of PS and PB in the blends (φ_{PS} and $(1 - \varphi_{PS})$, respectively), and the values of T_g for both types of phases (as summarized in Table 13.5), one can calculate the respective H values, H''_{cal}, of the studied blends by means of Eq. (13.22), and compare them with the experimentally measured ones, H_{exp}, as well as with the calculated values using Eq. (13.1) and assuming $H^s = 0$, H'_{cal}, (given also in Table 13.5).

Table 13.5. Content of the linear block copolymer, LN4, in the copolymer blends, total polybutadiene content, φ_{PB}, (assuming volume fraction $\sim$ weight fraction), glass transition temperatures of PS, T_g^{PS}, and of PB, T_g^{PB}, phases, measured microhardness of the blend, H_{exp}, calculated microhardness (according to Eq. (13.1), with $H^s = 0$), H'_{cal}, and according to Eq. (13.8) (with $H^s \neq 0$), H''_{cal}, as well as the differences between the measured and the calculated values, $\Delta H'$ ($\Delta H' = H'_{cal} - H_{exp}$) and $\Delta H''$ ($\Delta H'' = H''_{cal} - H_{exp}$), respectively

LN4 content	φ_{PB}	T_g^{PS}	T_g^{PB}	$T_{g\ cal}^{PB}$	H_{exp}	H'_{cal}	$\Delta H'$	H''_{cal}	$\Delta H''$
	[41]	[41]	[41]	[41]	[41]	[41]	[9]	[9]	[9]
(wt %)	(%)	(°C)	(°C)	(°C)	(°C)	(MPa)	(MPa)	(MPa)	(MPa)
0	26	104	−74	–	44	130	86	78	34
20	34	104	−68	−64	28	116	88	56	28
40	42	103	−53	−54	24	102	78	40	16
60	49	100	−31	−43	16	89	73	37	21
80	57	87	−25	−30	12	75	63	12	0
100	65	70	−15	–	8	61	53	−4	4

One can see that the difference between the measured H values and the calculated ones, however neglecting the contribution of the soft phase ($H^s = 0$, i.e., using a relationship similar to Eq. (13.16)), is quite large (H_{exp} values are between 2 and 5 times smaller that those of H'_{cal}). Even so, when accounting for the contribution of the soft phase ($H^s \neq 0$, i.e., by means of Eq. (13.22)), the two types of values are much closer to each other (Table 13.5). What is more, since the system under investigation is completely amorphous, it is possible to predict its microhardness for various compositions having only the T_g and weight fraction of each phase, as described in Eq. (13.21). The good agreement between calculated in this way and the measured H values demonstrate again the validity of the analytical relationship between H and T_g (Eq. (13.8)).

Quite similar is the situation regarding the microhardness behavior of binary blend comprising the polystyrene homopolymer (hPS) and a star block copolymer of styrene with butadiene over a wide composition range. As in the case of separated block copolymers and binary block copolymer blend, as described above [41,42], a clear deviation of the microhardness behavior from the additivity law has been also observed [43]. Using an equation similar to the last one (Eq. (13.22)), however, accounting for the fact that one of the components of the blends is a homopolymer and applying the respective values

Table 13.6. Content of the homo PS in the blends, hPS, total polystyrene content in the blends, φ_{PS}, (assuming volume fraction $\sim$ weight fraction), measured microhardness of the blends, H_{exp}, calculated microhardness (according to Eq. (13.1), with $H^s = 0$), H'_{cal}, and according to Eq. (13.8) with ($H^s \neq 0$), H''_{cal}, as well as the differences between the measured and the calculated values, $\Delta H'$ ($\Delta H' = H'_{cal} - H_{exp}$) and $\Delta H''$ ($\Delta H'' = H''_{cal} - H_{exp}$), respectively

hPS content (wt %)	φ_{PS} (wt %)	H_{exp} [42] (MPa)	H'_{cal} [42] (MPa)	$\Delta H'$ [42] (MPa)	H''_{cal} [9] (MPa)	$\Delta H''$ [9] (MPa)
0	74	44	127	83	81	37
20	79	64	136	72	98	34
40	84	75	144	70	115	40
60	90	100	155	55	137	37
80	95	138	163	25	154	16
100	100	180	172	8	–	–

for T_g, and mass fractions ([43], Table 13.6), the H values of the blends are calculated assuming that $H^s \neq 0$. The results are presented in Table 13.6.

Again, a much better agreement between H_{exp} and H''_{cal} values is found if one accounts for the contribution of the soft phase ($H^s \neq 0$), in contrast to the opposite case ($H^s = 0$), (Table 13.6), however, the differences $\Delta H''$ are still quite large, as compared with the case of PEE (Tables 13.2 and 13.3).

13.3.5. One-component two-phase systems (semicrystalline polymers with T_g below room temperature)

As demonstrated above, there is hardly any doubt regarding the existence of a linear relationship between T_g and the microhardness H of amorphous polymers characterized by dominating single, mostly C–C bonds in the main chain. This empirically derived analytical relationship (Eq. (13.8)) makes it possible to account quantitatively for the contribution of the soft component and/or phase to the overall microhardness of multicomponent and/or multiphase systems as demonstrated above.

The same relationship offers another challenging opportunity – to evaluate the T_g of amorphous phase(s), which are not commonly accessible, as for example, the wholly amorphous PE. Basically, this can be done starting again from the additivity law, and more specifically, by the fact that the overall microhardness depends linearly on those of its constituent components and/or phases and their respective weight fractions. For a semicrystalline polymer one can extrapolate the dependence H_c vs. degree of crystallinity, w_c, (or density) to $w_c = 0$ (or to the ρ_a value) and thus evaluate H_a for this polymer. Further, exploring the relationship between H and T_g (Eq. (13.8)) it is possible to get an idea about T_g of an inaccessible practically amorphous phase. This approach will be illustrated below using as example PE.

Polyethylene is still nowadays one of the most common and most studied polymers. However, there is not yet full consensus among researchers about such a basic property as the value of its glass transition temperature, T_g. Values as different as around –25°C and –120°C are reported [14]. These discrepancies can be found also in more recent publications: $T_g = -35$°C [44], and $T_g = -125$°C [45]. The lack of agreement is related to the fact that PE is not commonly accessible in amorphous state (below its melting temperature) due to its extremely high crystallization rate originating from the perfect chain structure. Even the preparation of samples with different degrees of crystallinity is not a routine task. The frequently used approach by varying the crystallization temperature and/or crystallization time is not applicable contrasting to many other polymers. Better results can be obtained by using PE samples with different degree of branching. By introducing various amounts of chain defects in the main chain, it is possible to control the degree of crystallinity. In this way, even at constant crystallization conditions (temperature and time), one can prepare a series of samples with a systematic variation in the structural parameters such as degree of crystallinity, crystal size, long spacing, density, paracrystalline lattice distortions, melting temperature, *etc.* [46–48]. Following a given property of such a series and extrapolating to the density of completely amorphous sample, one can find support in favor of one of the two rather different values of T_g for PE.

It has been found that H increases linearly with the rise of crystallinity, as reported for many polymers [1,8]. From the straight line of the plot of H *vs.* density, ρ, for differently branched PE samples, the H value for the completely amorphous PE ($\rho = 855$ kg.m^{-3} [14]) has been evaluated [49]. The obtained H value has been further used for the evaluation of T_g of PE by means of the linear relationship between H and T_g for completely amorphous polymers (Eq. (13.8)), and a value of $T_g = -23$°C has been obtained [50].

Taking into account the findings of Geil *et al.* [15–19] based on the direct study of wholly amorphous linear PE, that the T_g of PE is –80°C on the one hand, and on the other, that branched PE is supposed to have a much higher T_g value (corresponding to the β-relaxation peak), one can consider the extrapolated T_g value of –23°C from branched PE samples as the T_g of completely amorphous branched PE [50].

Noteworthy in this respect is also the report of Perena *et al.* [20] who studied microhardness using dynamic mechanical (DMTA) measurements at low temperatures (between –60°C and 25°C) with five commercial samples PE, two of them of high density (HD) and another three of low density (LD). The experimental data for H show clear transition around –30°C (for LD samples) and around –10°C (for HD) samples. The data from DMTA show this transition only for the LD samples in agreement with the observation [21] that β-relaxation is clearly detected by DMTA only in branched PE and has been not detected at all in linear PE of medium molecular weight. Having in mind the fact that the PE used for the application of Eq. (13.8) is also a branched one, it should be emphasized that there is very good agreement between the experimentally observed transition temperatures (between −20°C and −30°C) [20] and those predicted by means of Eq. (13.8) −23°C and −25°C) [24].

The same approach was recently applied to PE samples characterized by chain-folded and chain-extended crystals [25]. Considering the fact that in these samples the amorphous phase amounts to a low percentage (in the chain-extended samples around 5 wt%) and because of the highly strained chains in the amorphous areas, one can expect relatively high values for T_g of these amorphous phases. The calculations lead to $T_g = -1°C$ (for the chain-folded samples) and $T_g = 10°C$ (for the chain-extended samples) [25].

Summarizing this section, we have to note that thanks to the empirical linear relationship between H and T_g in a rather broad range of T_g, i.e., from −50°C up to 250°C), which covers most commonly used polymers of the polyolefin type and also polyesters and polyamides [13], it is possible to calculate not only the microhardness value of many amorphous polymer provided its T_g is known ($H = 1.97 T_g - 571$) or to account for the contribution of soft components and/or phases to the microhardness of the entire system, but also, to evaluate the T_g values of practically inaccessible amorphous phases in semicrystalline polymers.

13.4. Main factors determining the microhardness of polymer systems containing a soft component and/or phase

13.4.1. *Importance of the ratio hard/soft components (or phases)*

Analyzing the results summarized in Table 13.1 one can conclude that the agreement between the H_{exp} values and the calculated H_{cal} ones (assuming $H_a \neq 0$) is not as good as expected for all the samples. Quite close to each other are the values for the samples characterized by lower w_c values (below 80%) and higher amount of PP in the blends. This could mean that for the cases where the crystalline phase (or component) dominates (80% or more), the overall microhardness is determined by the "diluting effect" of the amorphous phase, i.e., the amount of the amorphous phase is not enough in order to be considered as a matrix in which the crystallites are immersed. As a matter of fact, in such cases the matrix is represented by the crystalline phase in which the much smaller amount (30% or less) of the amorphous phase is dispersed. It is quite obvious that in such a situation one cannot apply the concept of the "floating effect" for explanation of the mechanical behavior of the complex system and one has to accept the plastic deformation mechanism of the solid component and/or phase as dominating.

The above considerations are supported by the results of the thermoplastic elastomers (PEE) with various compositions (hard/soft segments ratio). For example, in Table 13.3 data for other two PEEs (samples 7 and 8) with not such a good agreement between H_{exp} and H''_{cal} (according to Eq. (13.11)) values are presented. A possible explanation for the different behavior of these two PEE samples is their composition. Samples 1–6 are characterized by hard/soft segments ratios of roughly 60/40 while in the samples 7 and 8 this ratio is 75/25.

The fact that in the latter case the hard PBT segments dominate, suggests another response mechanism to the mechanical field – the PBT hard segments are no longer "floating" in the liquid-like matrix of soft domains.

Further support in favor of the importance of the hard/soft components ratio can be found in the blend hPS/star block copolymers of PS and PB (Table 13.6). A more precise inspection of Table 13.6 shows that the total amount of hard, glassy homopolymer PS in the blends varies between 75 and 95 wt%. In other words, we are coming to the same conclusion as for the above two systems (PE/iPP and PEE), where for the cases where the hard, (crystalline) phase dominates (for example, amounting to 80% and more) the concept of "floating effect" cannot be applied as successfully as for the cases where the hard component or phase does not dominate.

13.4.2. Crystalline or amorphous solids

In order to illustrate to what extent the presence or absence of order (crystalline) in the hard component and/or phase is important, let us come back first to the system polyester-polyamide copolymers, for which serious deviations from the additivity law (assuming $H^s=0$) have been reported [32]. Before suggesting some possible reasons for the failure of the calculations using Eq. (13.15) (assuming $H^s \neq 0$), it should be stressed that the paper under consideration [32] is distinguished by a couple of peculiarities being rather important for this discussion.

In this study [32], a relatively large number of interesting samples, differing in their hard/soft segments ratio as well as in the molecular weight of the segments, is investigated. The measured H_{exp} values differ drastically from the H'_{cal} values (assuming $H^s=0$), the difference being up to ten times (while 9 from a total of 12 samples have H_{exp} values between 10 and 30 MPa, the H'_{cal} for all the samples vary between 70 and 110 MPa). What is more striking is the fact that the degree of crystallinity of PBT domains w_c (determined by wide-angle X-ray scattering, WAXS) for all the samples studied varies between 10 and 30% (for 9 of the samples – between 10 and 20%!). Nevertheless, the drastic differences between H_{exp} and H'_{cal} values (assuming $H^s = 0$) are explained exclusively on the basis of the changes in the crystal's characteristics, such as crystal sizes, crystal surface free energy, the energy needed for plastic deformation of crystals (usually approximated to the enthalpy of fusion of crystals [1,32]).

At this stage important questions could arise: how is it possible to explain (or even to predict) the mechanical behavior of a complex system consisting of 80 wt% (and more!) soft, liquid-like (T_g between –40 and –70°C) matrix in which not more than 20 wt% crystallites are dispersed only accounting for the properties of crystallites? Does it look reasonable to neglect the contribution of the prevailing (up to 5 times) soft phase in the formation of the mechanical response of the system to the external load?

The samples studied [32] represent an excellent example for the case where the solid particles (crystallites in this case) are "floating" in the dominating soft matrix. As noted above, our attempt to recalculate the H''_{cal} values taking into account the "floating effect" ($H^s \neq 0$) failed. The data obtained are inconsistent, possibly because the molar concentration has been used [32] for the fraction of the soft segments.

It should be noted that there are no obvious reasons, except the misleading fact that the soft liquid-like substances do not produce Vickers indentation impressions, to accept that the soft component and/or phase does not contribute to the overall microhardness, *i.e.*, $H^s = 0$. Only taking into account the fact that the soft component changes dramatically the deformation mechanism of the complex system, one is able to avoid more "sophisticated" explanations for the observed discrepancies between the H_{exp} and H_{cal} values (assuming $H^s = 0$). For example, in the case of the blends of PS and PB copolymers, as well as with various partners, these differences are explained by different origins: the molecular architecture which modifies the effective phase ratio, the presence of a microphase separated morphology and some specific effects, such as yielding of thin layers, *etc.* [43] or by the assumption that the volume fraction of styrene and butadiene phases in the block copolymer blends does not reflect the effective hard/soft phase volume ratio owing to the modified copolymer architecture and microphase separated morphologies [41].

Quite similar explanations of the observed deviations from the additivity law dealing with semicrystalline polymers and their blends involving peculiarity in the crystalline characteristics (crystal sizes, crystal surface free energy, *etc.*) are offered, as mentioned in the previous sections. The most serious drawback of such explanations, even if they indicate some of the possible reasons for the observed deviations from the additivity law (Eq. (13.1)), is that they cannot account quantitatively for the observed differences as, for example, Eq. (13.8) does.

Let us consider now the blend of homo PS with star block copolymer of PS and PB, where the agreement between the values of H_{exp} and H''_{cal} (even assuming $H^s \neq 0$) was not as good as expected (Table 13.6).

It seems important to stress here on a peculiarity of this system, namely, its similarity with the blend PE/iPP, with respect to their microhardness behavior. Regardless of the seemingly important facts that the blends of hPS with copolymers are completely amorphous, non-crystallizable, and that the blends of iPP/PE consist of semicrystalline homopolymers, to neither one the concept of the "floating effect" can be applied successfully to explain the deviations from the additivity law (Eq. (13.1)) with the assumption $H^s = 0$. The concept of the "diluting effect", *i.e.*, the domination of the plastic deformation mechanism of the solid particles, seems to be more appropriate. The reason for this is, as stated in the preceding section, the fact that the dominating (in amount) solid particles (crystalline or amorphous ones) form the matrix in which the minor in amount soft component and/or phase is dispersed.

From the comparison of these two different (with respect to crystallinity) systems, one can conclude that for the explanation (and overcoming) of the frequently observed deviations from the additivity law (Eq. (13.1) assuming $H^s = 0$), the crystal characteristics (crystal sizes, crystal free surface energy, *etc.*) are not of basic importance. What counts in this case is the mechanism of deformation of the solid particles under the indenter, *i.e.*, if one deals *only* with their plastic deformation under the indenter (the case of "diluting effect") or with the same mechanism, however, *paralleled* by a displacement of the solid particles in the soft matrix (the case of "floating effect"). The domination of one or the other deformation mechanism depends exclusively on the ratio solid/soft (liquid-like) components and/or phases and not on the fact whether the solids particles are amorphous or crystalline.

Finally, the fact that the analytical relationship between T_g and H (Eq. (13.8)) helps to solve "the problem" regarding the deviations from the additivity law (Eq. (13.1)) means that the very basic starting assumption regarding the deformation mechanism, *i.e.*, the "floating effect" of the solid particles in the dominating soft component and/or phase, is quite reasonable. This fact should always be taken into account when the mechanical behavior of such systems is considered.

13.4.3. *Copolymers vs. polymer blends*

In the present section we will analyze various amorphous systems – homopolymers, their miscible blends, copolymers and their blends. This variety makes it possible to follow the effect of presence or lack of chemical linkages between the constituent monomers, *i.e.*, if one is dealing with a blend of two homopolymers or with a copolymer of the same monomers. Interesting conclusions in this respect can be drawn from the miscible blend PMMA/PVDF (Table 13.4). The results obtained allow one to conclude that the amorphous blends of miscible partners can be treated in the same way as the amorphous neat homo- and copolymers regarding the relationship between their glass transition temperature and microhardness. What is more, it is not necessary to measure not only their microhardness but also their T_g values for any blend composition because these values can be evaluated by means of the Gordon and Taylor equation [38].

Particularly striking in the present results (Table 13.4) is the observation that the calculated H data for the discussed blends do not significantly depend on the model applied, *i.e.*, if the amorphous blend is considered as a two-phase one (or even as a three-phase one) or as a one-phase two-component system. The only parameter that counts is the mass fraction of each component and/or phase and the respective T_g value. It is noteworthy that a similar behavior has been observed between blends of homopolymers and copolymers prepared from the same monomers [1,50].

Important support of this conclusion can be found in the microhardness behavior of the glassy block copolymers of PS and PB with various architectures, their blends, as well as their blends with amorphous homopolymers (hPS). As shown above, Eq. (13.21) considers the blend of star-like and linear type block copolymers of PS and PB as comprising four components, the two types of blocks in the two different with respect to the molecular architecture block copolymers. The use of this equation presumes the knowledge of four T_g values (for each of the four blocks). Experimentally have been detected only two T_g, that of PS and that of PB [41]. For this reason, Eq. (13.21) was modified into Eq. (13.22) that accounted only for two glassy phases, PS and PB. The data obtained by means of Eq. (13.22) are satisfactory, particularly for the samples with dominating soft phase (PB), (Table 13.5).

Quite similar is the situation with the amorphous blends of hPS and star block copolymers of PS and PB (Table 13.6). This two-component system consists of homo PS and a block copolymer of PS and PB, however, again only two T_g have been experimentally revealed [42].

The cases described lead to the important conclusion that the microhardness of a complex multicomponent and/or multiphase system depends only on the number of the actually observed components and/or phases (on their individual microhardness values and their respective mass fraction), regardless of whether these components and/or phases consist of homopolymer(s), their blends or parts (blocks) of copolymers. It seems that for amorphous polymer systems the microhardness depends exclusively on the chemical composition and structure of a specific monomer, rather than on the type of chemical linkages (homo- or copolymers) in agreement with our former observations [1,51].

13.4.4. New data on the relationship between H and T_g of amorphous polymers

Since the publication of the equation relating H and T_g of amorphous polymers in 1999 [13], new data about the microhardness values of interest were published. This includes results on PS materials with various molecular architectures (highly branched) displaying some higher T_g values as compared to the linear PS [52].

Samples of glassy PS with different amounts of long branches have been investigated (PS-165 is a linear PS, and the amount of branches for samples PS-174, PS-177, and PS-179 increases with the number code) [51]. The molecular weight, polydispersity, glass transition temperature, and microhardness data of these materials are listed in Table 13.7.

A fair agreement between the measured, H_{exp}, values [52] and those calculated by means of Eq. (13.8) for H_{cal} can be observed (Table 13.7). This fact

Table 13.7. Molecular weight, M_w, polydispersity, M_w/M_n, glass transition temperature, T_g, experimentally measured microhardness, H_{exp}, calculated microhardness (according to Eq. (13.8)), H_{cal}, and the difference between the calculated and the measured values, ΔH ($\Delta H = H_{cal} - H_{exp}$), of various types of PS

Material	M_w [51] (kg.mol^{-1})	M_w/M_n [51]	T_g [51] (°C)	H_{exp} [51] (MPa)	H_{cal} [9] (MPa)	ΔH [9] (MPa)
PS-165	313	2.1	104	171	172	1
PS-174	311	2.5	105	176	174	2
PS-177	317	2.6	106	178	176	2
PS-179	316	2.8	111	185	186	1

confirms the validity of Eq. (13.8) for the group of amorphous polymers (characterized by dominating single, mostly C–C bonds). In addition, one can again conclude from the results in Table 13.7 that the molecular architecture does not affect the validity of Eq. (13.8).

With regard to the future development of the present study it seems important to mention that Eq. (13.8) was recently modified in such a way that it also accounts for the temperature dependence of H for amorphous polymers [53]:

$$H^T = 1.97\,T_g - 0.6\,T - 395 \text{ (MPa)},\ (T_g \text{ and } T \text{ in K}) \quad (13.23)$$

where H^T is the microhardness value at the test temperature T. In other words, the microhardness of the same group of amorphous polymers, covered by Eq. (13.8), can be calculated for any temperature T below T_g if T_g is known.

The experimentally derived Eq. (13.23) based on results for 4 amorphous polymers, (PS, PMMA, PEN and PET) was recently verified by the reported data on the temperature dependence of H for PMMA synthesized by radiation polymerization [54]. In Table 13.8, the experimentally measured H^T_{exp} values at various temperatures (ranging between 15 and 78°C) are compared with the calculated ones, H^T_{cal}, applying Eq. (13.23).

A quite good agreement between the two types of values can be found. What is striking in this case, is the observation that the difference between H^T_{cal} and H^T_{exp} tends to zero with increasing temperature of measurements, T. This could mean that T_g, being indicative for the viscosity of the amorphous material, is getting more sensitive in this respect when the test temperature T approaches the softening point, T_g.

Table 13.8. Measured microhardness values, H^T_{exp}, and the calculated ones, H^T_{cal}, according to Eq. (13.23), as well as the difference, $\Delta H^T = H^T_{cal} - H^T_{exp}$, for various test temperatures, T, of glassy PMMA. The measured by DSC T_g value of PMMA is reported to be 354 K [54]

T (K)	288	299	312	317	324	330	335	338	340	342	344	347	350	351
H^T_{exp} (MPa)	161	141	132	127	120	117	111	109	107	105	101	97	95	91
H^T_{cal} (MPa)	132	124	117	114	110	106	103	101	100	99	98	96	95	93
ΔH^T (MPa)	−29	−17	−15	−13	−10	−11	−8	−8	−7	−6	−3	−1	0	2

In the same paper [54], careful measurements of H^T have been performed in the same temperature interval for the blend PMMA/natural rubber (NR), the latter ranging in content up to 5 wt%. The blend films have been prepared from a common solvent [54]. Table 13.9 presents the data on H^T_{exp} (as reported in [54]), and the H^T_{cal} values calculated by means of Eq. (13.23).

Surprisingly, again a fairly good agreement between H^T_{cal} and H^T_{exp} values can be seen, particularly with the progress of the test temperature (for two-thirds of a total of 12 measurements, the difference $\Delta H^T = H^T_{cal} - H^T_{exp}$ amounts to between 2 and 6 MPa), which, occasionally, correspond to the same percentage of deviations (Table 13.9).

Micromechanics of Polymer Blends

Table 13.9. Measured microhardness values, H_{exp}^T and the calculated ones, H_{cal}^T according to Eq. (13.23), as well as the difference $\Delta H^T = H_{cal}^T - H_{exp}^T$ for various test temperatures, T, of blend of glassy PMMA with natural rubber (up to 5 wt%). The measured by DSC T_g value of PMMA is reported to be 354 K [54]

T (K)	288	300	312	317	324	330	335	338	340	342	345	347
H_{exp}^T (MPa)	128	121	113	110	106	103	99	98	96	95	93	92
H_{cal}^T (MPa)	115	108	101	101	100	100	92	94	92	92	91	90
ΔH (MPa)	−13	−13	−12	−9	−6	−3	−7	−4	−4	−3	−2	−2

13.4.5. Modified additivity law for systems containing soft component and/or phase

Coming back to the main goal of the present study, the application of the additivity law (Eq. (13.1)) to complex polymer systems containing a soft component and/or phase, we would like to suggest its modification by incorporating the relationship between H and T_g (Eq. (13.8)). The advantage of this modification consists in the possibility to use it for accounting for the contribution of any amorphous phase and/or component to the overall microhardness of the system, provided the T_g value of this phase and/or component is known, regardless of their value. Hence, for systems, which contain more than one crystalline and/or amorphous phases with crystalline microhardness, glass transition temperatures and mass fractions H_{ci}, T_{gi} and φ_i, respectively, the additivity law can be presented in the following way [9]:

$$H = \Sigma \varphi_i w_{ci} H_{ci} + \Sigma \varphi_i (1 - w_{ci})(1.97 T_{gi} - 571) \quad (13.24)$$

For systems where the solid (hard) components and/or phases are not crystallizable materials, Eq. (13.24) can be simplified as [9]:

$$H = \Sigma \varphi_i (1.97 T_{gi} - 571) \quad (13.25)$$

In these two last forms, contrasting to the traditional one (Eq. (13.1)), the additivity law is applicable to multicomponent or multiphase systems comprising soft components or phases displaying a more complex deformation mechanism than the case in which all the components and/or phases have T_m and T_g above room temperature.

13.5. Microhardness on the interphase boundaries in polymer blends and composites and doubly injection molding processing

13.5.1. Microhardness on the interphase boundaries in polymer blends

Polymer blends attracted the attention of academia and industry some decades ago as an alternative to new polymers synthesis for satisfying the steadily increasing demands of polymer materials with new exploitation properties profile [55,56]. Similarly to copolymers, polymer blends offer the possibility to combine in one material the properties of the individual components. However, in contrast to copolymers where the chemically different monomer units are linked by chemical bonds, blending and particularly the achieving and preservation of a desired final degree of dispersion creates serious problems. This is valid for polymer partners that are not thermodynamically miscible. In order to enhance the dispersion process, and mostly to stabilize the reached degree of dispersion, which in turn assures the constancy of the desired blend properties, a third component (*compatibilizer*) is used [56]. This third blend component is expected to be distributed on the interphase boundary between the two components. The compatibilizer, usually in an amount of a few percents, not only enhances the blending and stabilizes the blend but also contributes to the improvement of the adhesion between the blend components, as does the coupling agent in the case of glass fiber reinforced composites, and in this way significantly improves the mechanical properties of the blend material. This last task of the compatibilizer can be fulfilled only if it is actually distributed over the interface boundary. For this reason, it is important to find appropriate techniques to introduce the compatibilizer, as well as methods for localization of its distribution. For blends of condensation polymers, in which the compatibilizer usually represents a copolymer of blended components and is commonly created during melt blending, the distribution is perfect because the copolymer is synthesized just on the interphase boundary where the two components are in contact and chemically reacting [57]. For the rest of the polymers this is not the case, and for this reason appropriate techniques for analyzing the compatibilizer distribution have to be used.

The microhardness test has become a common technique due to its simplicity. It appears to be a promising tool for the micromechanical and microstructural investigation of polymer blends and composites as outlined in the previous sections [1–8]. More specifically, it was demonstrated [58] that the microindentation hardness, H, is a very sensitive tool for testing the interphase boundary of doubly injection-molded glassy polymers. For this purpose, microindentation scanning across the contact surface has to be performed. In the case of glassy polystyrene (PS) and polycarbonate (PC), depending on the temperature of the melt fronts within the mold, a well-defined weld line can be identified [58]. This weld line is distinguished by a drop in H when scanning across. The hardness drop is followed by an immediate increase, thus defining a weld line width of 0.2–0.4 mm [58].

Microhardness indentation measurements have recently been used in the blend of glassy poly(methyl methacrylate) (PMMA) and natural rubber (NR) prepared by means of a common solvent [54]. These measurements revealed that when approaching the phase boundary between the two immiscible partners, the microhardness value drastically drops (by 30%) in a narrow phase boundary (around 0.75 mm thickness). This strong decrease in the H value very close to the rubber particle surface indicates the existence of a narrower or wider phase boundary between the two components.

From the foregoing it looks interesting to try to characterize the change in the interphase boundary between two immiscible polymers in the presence and in the absence of a compatibilizer by means of microindentation hardness and scanning electron microscopy (SEM). For this purpose, solution blended PS and NR as well as styrene-butadiene-styrene (SBS) block copolymer as compatibilizer were used.

Before completing this task an attempt was undertaken to demonstrate, mostly by means of SEM, the changes in the interphase boundary between the polymer (PS) and its compatibilizer (SBS) [59]. Blend cast films were prepared using a common solvent (toluene) in order to obtain a 5% (by wt) solution of SBS in PS.

The SEM observations on the film surface (Figure 13.5a) show that SBS is regularly dispersed in the PS matrix in the form of nearly monodispersed spheres (diameter of about 20–40 μm) that are too small for studying the interface boundary between the two components (diagonals of indentation are about 40 μm in diameter). For this reason, much larger rarely observed spheres (diameters ca. 250 μm, Figure 13.5a) were selected for this purpose. Only those spheres that "floated" on the surface of the PS matrix could be used for this purpose, i.e., those particles that are not in the bulk of the film or covered by the thin layer of PS. While Figure 13.5a illustrates such an SBS spherical particle surrounded by many indentations; in the lower half of Figure 13.5b one can only appreciate a section of such a large SBS sphere being in contact with the dominating PS matrix occupying the upper half of the electron-micrograph (Figure 13.5b).

The two different polymers can be easily identified because of the completely different nature of their surfaces. While the surface of the glassy PS is quite smooth, the SBS shows a "wavy" surface that is typical for the elastomers. Furthermore, for the correct interpretation of Figure 13.5 it is important to mention that the glassy PS is a hard material that shows well-shaped indentations (Figure 13.5), in contrast to SBS which, owing to its very low glass transition temperature, and to the fact that its elastic deformation is practically reversible, does not show any indentations on its surface. This very strong difference in the microhardness behavior of the two polymers allows us to make an estimate about the phase boundary extension between the two polymers. While indentation A in Figure 13.5b shows symmetric diagonals of the same length, the indentation B (Figure 13.5b), which is much closer to the phase boundary and possibly touches it, is distinguished by asymmetric diagonals (a shorter vertical diagonal as compared to the horizontal one). The last observation is very well expressed in the last indentation spot C (Figure 13.5b) which covers the interface boundary. One can easily observe that only the horizontal diagonal shows symmetry with respect to the center of the indentation. The vertical diagonal shows only the upper part arising from the PS matrix. The expected lower part of the diagonal from the SBS component, due to the complete elastic recovery of this polymer, is missing.

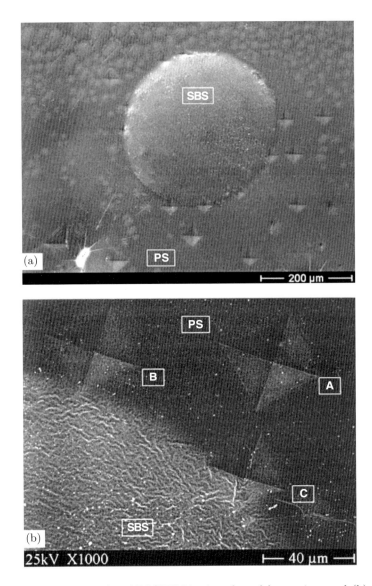

Figure 13.5. SEM micrographs of PS/SBS blend surface: (a) overview, and (b) closer to PS-SBS interphase boundary [59]

The most striking observation in the present case is the fact that no sharp edges of the two polymers on the contact region are observed. This contrasts with the finding on glassy poly(methyl methacrylate)/natural rubber (PMMA/NR) blend films [60] prepared in the same way as described above for PS/SBS blends. In Figure 13.5b one can also see that the transition zone between the two polymers is rather smooth. This observation is obviously related to the fact that SBS, being a compatibilizer for the blends of PS with natural and synthetic rubbers, contains PS (two of the three blocks are of styrene). The styrene blocks of SBS penetrate through the surface of PS and thus ensure a smooth transition from one polymer material to the other across the interphase boundary, which, by the way, is the main task of the compatibilizer.

The next step was, as mentioned above, to study the compatibilized with SBS blends of PS/NR, focusing this time on the quantitative evaluation of the transition zone between the two blend components comprising SBS or not. For this reason various blends were prepared using a common solvent, namely: PS/NR = 95/5, PS/SBS = 95/5 and PS/NR/SBS = 95/4.5/0.5 (by wt). In order to obtain a larger number of reliable measurements close to the interphase boundary, the measurements were performed not on a straight line across the interphase boundary (the radius of the spherical particles) but on a curved trajectory gradually approaching the dispersed sphere. Since the dimensions of the indentation diagonals are comparable to the thickness of the interphase boundary, the lengths of the diagonals at the interphase are slightly different. The measured H was derived from the average value of the diagonals.

SEM observations on the PS/SBS film surface demonstrated that SBS is dispersed in the PS matrix in the form of nearly mono-dispersed particles as those shown in Figure 13.5. Quantitative information about the broadening and "depth" of the interphase boundary can be gained from the H data. The variation of H as a function of the distance z from the interphase boundary is plotted in Figure 13.6 for neat PS, as well as for blends PS/NR, PS/SBS and PS/NR/SBS. One sees that neat PS shows the highest and most constant H values (around 175 MPa). The H values obtained for neat PS are in accordance with previously

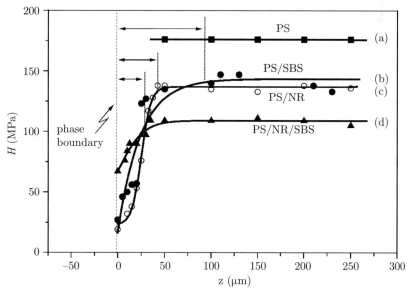

Figure 13.6. Variation of microhardness H as a function of the distance, z, from the phase boundary for the blends PS/NR, PS/SBS and PS/NR/SBS as well as for neat PS [61]

reported ones [1]. The blend of PS with the compatibilizer SBS (Figure 13.6b) shows a completely different behavior. Far away from the interphase particle boundary the microhardness is rather constant (around 130–140 MPa), though showing a gradual decrease when approaching the interphase boundary (ca. 100 µm from it) and reaching a final value of $H = 25$ MPa (Figure 13.6b).

Quite similar is the trend for the binary blend PS/NR (Figure 13.6c), reflecting again the observed variation of H and the distance from the interphase boundary. The only difference here is the fact that the H values far away from the phase boundary are slightly lower as compared to the previous ones (PS/SBS, Figure 13.6b). In addition, the decrease of H here is much steeper (starting at a distance of about 50 µm from the boundary instead of 100 µm as in Figure 13.6b) and reaches the same final H value of about 25 MPa.

The behavior of the ternary blend PS/NR/SBS (Figure 13.6d) is in contrast to the previous ones (Figures 13.6b and 13.6c). First of all, the H values on the plateau of the curve are the smallest (around 100 MPa in comparison with 175 MPa for neat PS, Figure 13.6a, and around 130–140 MPa for the binary blends PS/SBS and PS/NR, Figures 13.6b and 13.6c, respectively). Another striking observation for the compatibilized blend PS/NR/SBS is the fact that the H-decrease from the constant value of 100 MPa starts very near to the phase boundary between PS and NR (only 25 µm as compared to 50 and 100 µm for the binary blends). The second peculiarity of the PS/NR/SBS blend is the relatively high final H value of 75 MPa (Figure 13.6d).

What can be the reason for the observed differences in the behavior of the binary and the ternary (compatibilized) blends, and more specifically in (i) the H values of the respective matrix (far away from the interphase boundary), (ii) the distance from the latter where the H-decrease starts, and (iii) in the lowest final H values?

First of all, it is convenient to note that the H value of neat PS of 175 MPa is in agreement with the reported ones for glassy PS [62]. Taking into account the fact that the second blend component represents a typical rubber (NR) or a rubber-like polymer (SBS), both of them with very low glass transition temperatures T_g (around –70°C for NR [62] and –110°C for the soft part of SBS [62]), one can expect a depression in H for the blends far away from the interphase boundary. This depression is the largest for the compatibilized blend containing two rubbery components ($H = 100$ MPa, Figure 13.6d, against $H = 175$ MPa for neat PS, Figure 13.6a, and around 130–140 MPa for the binary blends, Figures 13.6b and 13.6c). The slightly higher H values for the PS/SBS blend as compared to the PS/NR blend are due to the fact that in the block copolymer styrene is dominant and, thus, SBS has a higher T_g value and, consequently, a higher micro-hardness as compared to NR.

Most interesting is the behavior of the blend of PS with the compatibilizer SBS (Figure 13.6b). One can see that the transition zone between the two components is very smooth, being the largest among the blends studied (compare the horizontal arrows in Figure 13.6). As mentioned above, this observation is obviously related to the fact that SBS, being rich in PS, its styrene blocks penetrate through the PS surface and thus ensure a smooth transition from one blend partner to the other across the interphase boundary.

Quite important, with respect to the aim of this study, are the results from the non-compatibilized PS/NR blend and the compatibilized one (PS/NR/SBS) (Figures 13.6c and 13.6d, respectively); the blend without compatibilizer shows a larger and "deeper" transition zone (Figure 13.6d) (compare 25 with 50 µm and final H values of 25 with 75 MPa, respectively). A possible explanation for the observed differences could be that the SBS compatibilizer is distributed just on the interphase boundary between PS and NR. In accordance with this assumption, parts of the SBS molecules admittedly penetrate in one of the components, and other parts of the compatibilizer in the second component. In this way, SBS is fulfilling its role to improve the adhesion between the two polymers through their compatibilization. When the "empty" space within the interphase boundary in the blend PS/NR is filled up with another polymer (SBS), the microhardness in this transition zone has to be higher, resulting in a higher final H value, as actually observed (compare Figure 13.6c with 13.6d). This finding suggests that the compatibilizer is displaced just on the interphase boundary between PS and NR where it is expected. It should also be noted that the phase boundary ($z = 0$ in Figure 13.6) is well defined by a few H measurements on the surface of the dispersed particles (NR or SBS, Figures 13.6b and 13.6c) showing a constant value of around 20–25 MPa. However, these latter values are somewhat questionable because of the elastic recovery of the material. The data cannot be regarded as a feature of the rubbery material but they help us define the interphase boundary extension between PS and NR.

The conclusion regarding the distribution of the compatibilizer SBS in the blend PS/NR based on the microhardness analysis is supported by other observations using transmission electron microscopy (TEM) [63] and small-angle X-ray scattering (SAXS) from a synchrotron source [64].

Studying microfibrillar composites (MFC) based on the blend isotactic polypropylene/poly(ethylene terephthalate) in respect to their ability to form transcrystalline layers of *i*PP around the PET microfibrils, it was found *via* TEM that such layers are missing if the blend contains a compatibilizer (ethylene glycidyl methacrylate (E-GMA) in this particular case) in contrast to the blends without a compatibilizer [63].

Quite similar is the conclusion regarding the distribution of the compatibilizer drawn from the SAXS analysis of the same *i*PP/PET/E-GMA blend during drawing (between 0 and 10% relative deformation) [64]. The few holes generated in the blend during cold drawing are needle-shaped for the material without compatibilizer, whereas they are shorter, wider and less pronounced for the material containing compatibilizer. This result supports the expected role of the compatibilizer, namely that it is enriched in the interphase between the two blend components where it changes the delamination behavior of the components during the cold drawing [64].

The suggested interpretation of these TEM [63] and SAXS [64] results assumes that the PET microfibrils (particularly after the isotropization of the second blend component, the PP in this case) are coated by a thin layer of the added compatibilizer. The latter prevents the direct contact between PET and PP and this suppresses the epitaxial as well as the nucleating effect of PET microfibrils on the crystallization, *i.e.*, transcrystalline layers cannot be formed. What is more, in an extended study on the same topic it was found that if the compatibilizer is added to the blend before drawing, the starting microspheres of the minor blend component coated with compatibilizer cannot coalesce in order to be transformed during the cold drawing into microfibrils [65].

In conclusion, the microindentation hardness study performed reveals conspicuous differences at the interphase boundary of blends of PS and natural rubber compatibilized with SBS. In the PS/SBS and PS/NR blends, the decrease of hardness through the phase boundary (50–100 μm distance) occurs very rapidly when approaching the soft phase, showing a drastic fall of the initial hardness value. On the other hand, the compatibilized PS/NR/SBS blend exhibits a thinner phase boundary of about 25 μm, showing a smoother decrease of the hardness. Thus, it is shown that the microhardness technique, particularly in combination with SEM, is a sensitive tool for the quantitative evaluation of the nature and quality of the interphase boundary in polymer blends.

13.5.2. *Microhardness on the interphase boundaries in polymers after double injection molding processing*

Earlier studies [66] have shown that microhardness may provide useful information about the correlation of processing parameters between, near and at the weld line or knit line, *i.e.*, the region where separated melt fronts reunite. In practice, *weld lines* occur in injection molding, *i.e.*, after a flow obstacle or, in the case of multiple gating, for melt streams of the same material or, in the case of two-component injection molding, for melt streams of different polymer materials [67]. In general, the presence of a weld line has an effect on the surface appearance and the mechanical properties of the molded parts [68].

The use of microhardness to characterize changes in the microstructure, molecular orientation, and micromechanical properties of injection-molded polymer materials has been the object of increasing interest [1,8]. In addition, it is known that process variables induce important changes in the microstructure and properties of the molded material. Hardness variations often occur in the surface and across the thickness of the molded samples. As a result, the mechanical properties can be controlled by processing variables such as melt and mold temperature, injection pressure, *etc.* [69].

Recently, a study reported [58] of the microindentation hardness measurements across the weld line arising when two opposite flow fronts are filling the cavity of a mold using two glassy polymers, polycarbonate (PC) and polystyrene. A large hardness difference between H measured away from the weld line and H measured at $z = 0$ ($H_{\min}$) was found. For PC this difference was about 20 MPa (14%), while for PS it was larger than 50 MPa (~30%) [58]. The H measurement along the injection direction z for the PC and PS samples containing a pigment for better visualization of the flow front was also performed [70], and the H_{min} value defining the weld line was identified. Asymmetry of H values, between the sample side containing the pigment and the other side without pigment, was observed. For both polymers, the higher H values measured on the side containing the pigment show the hardening effect of the pigment within the polymer. Such an asymmetry contrasts with that when no pigment is used [58].

It seemed interesting to extend the above studies [58] to the two glassy polymers (PC and PS) processed using a two-component injection molding system. Specifically, the target was to examine the effect of the processing temperature on the H-value across the weld line arising when the two opposite flow fronts are filling the cavity of the mold [60]. In addition, for the sake of comparison, the measurements were carried out close to the edge and in the middle of the sample as shown schematically in Figure 13.7b.

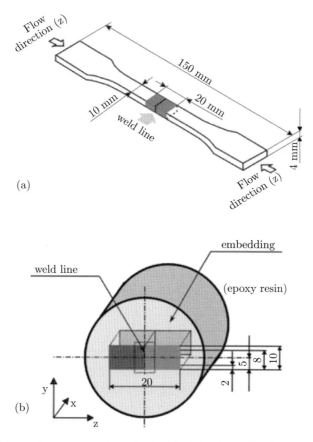

Figure 13.7. Schematic representation of a doubly injection-molded sample showing front weld line (a) and the measuring lines (at 2, 5, and 8 mm from the sample edge) parallel to the injection direction and across the weld line (b) [60]

The moldings were prepared in the form of ISO 3167 Type A tensile bars using a two-component injection-molding machine in which both melt streams (of the same material, though differently colored by 2 wt% of PS or PC color-batch, respectively) could be controlled independently (Figure 13.7a). The bars were molded using for PC melt temperatures T_m of 270°C and 300°C, respectively and a mold temperature (at the mold surfaces), $T_w = 80°C$. In the case of PS, values of $T_m = 230°C$ and 270°C and of $T_w = 50°C$ were used. From the injection-molded tensile bars, the central part containing the weld line with a total length of 20 mm was cut. The sample was then embedded in an epoxy resin in order to enhance the sample fixing during subsequent measurements (Figure 13.7b, z is the flow direction).

Microhardness was measured across the weld line at the surface of each of the moldings. Special care was taken to make indentations with the diagonals parallel and perpendicular to the injection direction (z), since the melt flow induced orientation may affect the diagonal sizes and thus the microhardness values too.

While Figure 13.7 presents the schematics of an injection-molded sample showing the shape of the melt front (containing a pigment) and the location of the measurements across the weld line, Figure 13.8 shows the optical micrograph of the surface of a PC sample after performing several indentations.

In contrast to previous measurements on the same systems [54], in which practically only one or two measurements characterized the deep decrease of

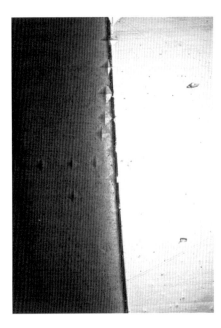

Figure 13.8. Optical micrograph of indentations at the surface of an injection-molded PC sample near the weld line area (left-hand side with pigment) [60]

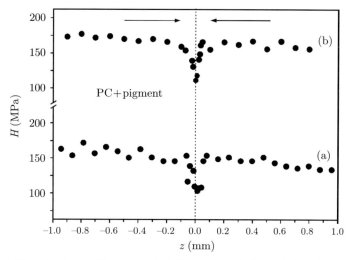

Figure 13.9. Microhardness, H, measured at the surface of the molding along the injection direction, z, for PC with melt temperature of 270°C. Arrows denote the direction of the two opposing fronts: measurements are performed along the injection direction (Figure 13.7) at a distance from the sample edge as follows: (a) – 2 mm and (b) – 5 mm. The left part of the sample contains a pigment. The right one does not contain the pigment [60]

H on the weld line, in the present case the same line is defined by 5–10 indentations that makes the reported microhardness decrease more reliable.

Figure 13.9 illustrates the H variation on the surfaces of the molding along the injection direction, z, for the PC samples with a melt temperature of 230°C taken at 2 mm (Figure 13.9a) and 5 mm (Figure 13.9b) from the edge.

Results clearly show the gradual H decrease for both 2 and 5 mm measurements along the z direction, until a minimum value at $z = 0$ is reached. Then, one observes a further increase of H when indenting away from $z = 0$. The symmetry of the hardness profile about $z = 0$ is evident in both cases (Figures 13.9a and 13.9b). The weld line zone containing the H changes is defined in the PC case within a 0.10 mm region, which is much smaller than the previously reported one (1 mm) for the same polymers [58]. The observed difference in both cases for the same material obviously originates from the more reliable definition of the weld line due to the much larger number of experimental points in the present case as compared to the previous one [58]. The H value was determined within $\Delta H/H \approx 0.5$.

Most interesting is to note the large hardness difference found between H measured away from the weld line and H measured at $z = 0$ (H_{min}). For PC this difference is about 50 MPa (~30%) for both measurements (Figures 13.9a and 13.9b). It should be noted that H measurements at a distance of 8 mm from the same sample edge (Figure 13.7) were also performed. The values obtained are the same as those measured at 2 mm distance from the edge (Figure 13.7) and therefore they are not plotted on the Figure 13.9 as well as on the subsequent one.

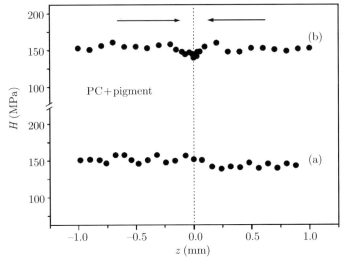

Figure 13.10. Microhardness, H, measured at the surface of the molding along the injection direction, z, for PC with melt temperature of 300°C. Arrows denote the direction of the two opposing fronts: measurements are performed along the injection direction (Figure 13.7) at a distance from the sample edge as follows: (a) – 2 mm and (b) – 5 mm. The left part of the sample contains the pigment [60]

Figure 13.10 shows the H data taken in the same way again on the PC sample processed at a melt temperature of 300°C.

Comparison of the data from Figures 13.9 and 13.10 suggests two important conclusions illustrating the difference between the two cases, i.e., the effect of melt temperature on microhardness. First of all, measurements performed at 2 mm from the edge (Figure 13.7b) do not detect any weld line (Figure 13.10a), and secondly, the measurements at 5 mm from the edge (Figure 13.10b) show a slight decrease (around 20 MPa, i.e., ~15%). What can be the reason for the different microhardness behavior as depending on the melt temperature and location of measurement?

The observation that injection-molded samples with higher melt temperature (300°C, Figure 13.10) show a much smaller microhardness depression at the weld line (Figure 13.10b) if any (Figure 13.10a) as compared to those molded with a lower melt temperature (270°C) (Figure 13.9) could be related with the diffusion conditions when the two polymer fronts meet. Taking into account the fact that the mold temperature and the T_g are the same ($T_w = 80°C$ and $T_g = 148°C$) as well as the flow front speed (200 ms^{-1}) for the two melt temperatures, it is quite clear that the temperature of the "melt," when the two fronts meet, should be lower if the starting temperature of the melt is lower. This means that the opportunities for mutual diffusion of chains resulting in their randomization and a reduction or disappearance of the weld line are not favorable. Increasing the melt temperature, such as from 270°C to 300°C in the present case, leads to a strong decrease of the weld line (Figure 13.10b) and even to its disappearance (Figure 13.10a).

Regarding the effect of the microindentation location, namely, whether it is closer or further apart from the sample edge (or, in other words, from the mold walls), it is noteworthy that the differences observed (Figures 13.9 and 13.10) are also related to the temperatures of the two molten polymer fronts at the instant they meet. Let us try to estimate the temperature difference of the melt closer or further from the mold walls during the flow. For this, we have to take into account the observation that the two flow fronts as well as the weld "line" are not planar but rather rounded. If so, the central part of the mold cavity will first contact the melt (top of the "cone") followed by the sidewalls of the cavity (base of the "cone"). In such a case, when the side parts of the mold are filled up with the melt, the sidewalls will be preheated because of the quick heat conduction in the mold. Such a difference in the temperature of the molded parts results in larger undercooling of the melt in the central part of the mold cavity as compared with both sides. The latter offers better diffusion conditions at the weld line than the central part. Obviously, this is the reason for the different H values through the weld line measured closer or further apart from the sample edge.

The above interpretation concerning the effect of melt temperature and location of the microindentations across the weld line, as derived from the experiments with PC, seems to be supported by the analogous measurements on PS. Again, two melt temperatures (230°C and 270°C) and the same mold temperature ($T_w = 50°C$) in both cases are used. Similarly to the previous case (Figure 13.9), for the lower melt temperature (230°C), a well defined strong H-decrease (from 200 MPa to 100 MPa) followed by a H increase at both measuring zones (2 and 5 mm from the edge of the sample) is observed. In addition, this drastic change in the H values takes place in a quite narrow z-range (about 0.10 mm). The width of the weld line is similar to that in the case of PC (Figures 13.9 and 13.10), while the drop in the H values is much larger.

When the melt temperature is increased up to 270°C, a small drop (about 10%) in the H value across the weld line is observed, closer to the edge such a drop is even absent. This result is similar to the case of PC at the higher melt temperature (Figure 13.10). Therefore, this finding supports the interpretation given above, i.e., again the dominating role of the melt temperature, particularly when the two opposite polymer fronts meet, is stressed. One should note here that the case under discussion is very close to the well-studied phenomenon of physical healing [71,72] where the mutual interdiffusion is determined by chain flexibility and diffusion conditions (temperature, molecular weight, etc.).

In conclusion, the broadening of the weld line (expressed in the microhardness changes across the $z = 0$) depends primarily on the mutual chain diffusion from the two polymer fronts coming from opposite sides. Further, it is shown that the main factors affecting the quality of the welding on the line $z = 0$ are, on the one hand, the T_g value of the polymer under investigation, and on the other hand, the melt and the mold temperatures.

As mentioned above, a weld line is formed not only in the cases described but also when processing by injection molding is carried out using a mold with a cylindrical obstacle. For this reason it was of interest to extend the above studies on PS using a mold with obstacle [73]. In addition, for the sake of comparison, the measurements were carried out on the surface of the molded plate as well as on the "bulk" surface after cutting and polishing at different locations with reference to the obstacle edges across the injection direction. The plates were molded using a melt temperature of 240°C and a mold temperature $T_w = 50°C$. From the central part of an injection-molded plate, containing the weld line, six test samples were cut and embedded in a thermoset resin to enhance their fixing during polishing and subsequent measurements.

Figure 13.11 illustrates the variation in H as a function of z, for various indentation series placed at various distances from the obstacle (up to 60 mm) on the surface of the molding along the injection direction. For the sake of

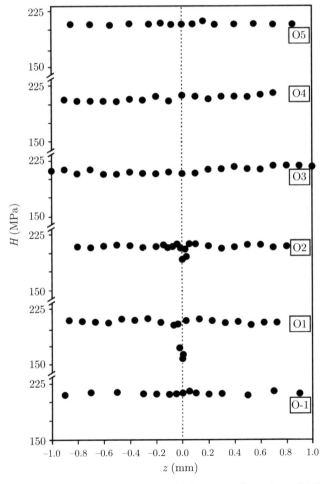

Figure 13.11. Microhardness H as a function of distance z from the weld line measured on the surface of the molded plate with obstacle for glassy PS [73]

comparison, a measurement before the obstacle (O-1) was also carried out. The indentation series were performed perpendicular to the weld line and denoted by O-1 to O5, while the (O) point denotes the center of the obstacle, and the z = 0 denotes the position of the weld line.

Results show a rather different microhardness behavior depending on the location, before or after the obstacle (O), and particularly how far the test location was from the obstacle. First of all, one has to stress the fact that, in accordance with expectation, before the obstacle there is no indication for the existence of any weld line (Figure 13.11, curve O-1), where rather constant values of H of about 210 MPa are observed. The situation changes drastically with the first measurement after the obstacle (Figure 13.11, O1) being situated closest to the obstacle, where a gradual decrease of H along the z direction starts, until a minimum value at z = 0 is reached. Then, one observes a further increase of H when indenting away from z = 0. The weld line zone containing the H changes is defined within a 0.20–0.25 mm region, which is much smaller than the previously reported one (1 mm) for the same polymer [70] and about the same in a more recent measurement (0.20–0.30 mm) [58].

Most interesting is to note the large hardness difference found between H measured away from the weld line and H measured at z = 0 (H_{min}). This difference is about 70 MPa (~30%) (Figure 13.11, curve O1). The same difference becomes smaller (35 MPa) for the next measurement series being further away from the obstacle (Figure 11, curve O2) until it disappears completely (Figure 13.11, curves O3–O5). What is more, the lines observed (Figure 13.11, curves O3–O5) as well as the H-values obtained are the same as those for the test before the obstacle (Figure 13.11, curve O-1). These results appear to indicate that after some distance from the obstacle (~5–10 mm in the present case, being of order of magnitude of the obstacle diameter), the obstacle does not have any further influence on the melt flow in the sense of formation of a weld line. Similar findings have been reported in the literature [74].

As concerns the microindentation hardness on the bulk surface, it should be noted that in contrast to the previous case (Figure 13.11), the decrease in H is just a few percent and indicates only a not well defined weld line. Measurements of H at greater distances from the obstacle show that even this small decrease in H is no longer present.

From the results of the bulk surface measurements one can conclude that very close to the obstacle edge (~5 mm) only a slight indication for the existence of a weld line can be detected. For greater distances there are no hints of the existence of such a weld line. The differences observed are mostly related to the temperature of the two fronts at the instant they meet. Let us try to get an idea about the temperature difference of the melt closer or further from the obstacle wall. When the polymer melt comes into contact with the cylindrical obstacle, it splits into two fronts. What is of particular importance about these two fronts is that being in contact with the obstacle (with a temperature $T_w \sim 50°C$), they decrease significantly their initial starting temperature of

240°C. Therefore, when the two fronts meet immediately after the obstacle there is no mutual chain diffusion resulting in randomization and homogenization. As a consequence, a more or less well developed weld line can be detected at the locations closest to the obstacle. The remainder of the two melt flows, not in contact with the obstacle, do not experience its cooling effect and for this reason, when they meet behind the obstacle, they preserve their cavity temperature typical for this part of the mold, which allows their homogenization without formation of any weld line.

This consideration is also supported by the observation that, even for measurement locations being closest to the obstacle but performed on the original sample surface (Figure 13.11) or on the "bulk" surface, the weld line is differently expressed. Because the melt in the middle part of the sample is not in direct contact with the mold cavity walls, their welding conditions are more favorable compared with those on the sample surface. The above interpretation (effect of melt temperature and location of the microindentations across the welding line, as derived from the H measurements on the surface and in the bulk of glassy PS processed using a mold with a cylindrical obstacle) supports the analogous measurements on PS and PC both processed at two melt temperatures but the same mold temperature on a two-component injection-molding machine. Again, the dominating role of the melt temperature, particularly when the two polymer fronts meet, is crucial and the mutual interdiffusion is determined by chain flexibility and diffusion conditions (temperature, molecular weight, *etc.*). For this reason, the broadening of the weld line as well as its appearance and disappearance (expressed in the microhardness changes across the $z = 0$) depend primarily on the possibility for mutual chain diffusion from the two polymer fronts meeting behind the obstacle. More specifically, it is shown that welding behind the obstacle is more effective in the bulk than on the surface of the injection-molded sample.

13.6. Conclusions and outlook

Using the reported data on the experimentally derived values of glass transition temperature, T_g, degree of crystallinity, w_c, Vickers indentation microhardness, H, and blend compositions for homopolymers, block copolymers, blends of polyolefins, or of polycondensates, blends of miscible amorphous polymers and copolymers (some of them with rather complex molecular architecture), all of them containing a soft component and/or phase at room temperature, an attempt is undertaken to look for the reasons for the frequently reported drastic deviations of the experimentally derived H values from the calculated ones by means of the additivity law assuming that the contribution of the soft component and/or phase is negligibly small.

In contrast to this commonly used approach, it is suggested in the present study that the soft component and/or phase can dramatically change the deformation mechanism under the indenter, and thus contribute significantly to the formation of the overall H value. It is demonstrated that this contribution can be quantitatively accounted for *via* the empirically derived relationship between H and T_g. The above disclosed results allow one to derive the following conclusions:

- The additivity law can be successfully applied also to complex polymer systems comprising soft component(s) and/or phase(s) if one accounts for their contribution to the overall microhardness of the system *via* their T_g and applying the linear relationship between H and T_g derived for solid amorphous polymers [13]. In this way one takes into account the fact that the deformation mechanism under the indenter is rather different compared to the systems distinguished by T_g and T_m values being above the ambient temperature.

- This approach allows to overcome to some extent the main disadvantage of the indentation technique for measuring H, the necessity to obtain observable, well defined indentation impressions on the sample surface.

- The microhardness behavior of blends of completely amorphous homo- and copolymers supports our previous conclusion [1] that the contribution of a component and/or phase to the overall microhardness depends mostly on the chemical nature of the respective monomers. Whether the monomers are chemically linked giving a homopolymer, a copolymer (even with a complex molecular architecture) or one deals with blends of them does not play any significant role.

- In contrast to the "floating effect" concept, the application of the thermodynamic approach (accounting for the crystal sizes, crystal surface free energy, *etc.*) for the explanation of the deviations from the additivity law is possible only for systems comprising crystalline component and/or phase; what is more, even for the last systems, the calculation of H requires the knowledge of parameters, which are not easily accessible.

- A modified additivity law is suggested, which contains a term accounting for the contribution of the soft component and/or phase to the overall microhardness *via* the relationship between H and T_g; its application results in much smaller differences between the measured and calculated H values.

Further, the microindentation hardness studies performed around the interphase boundary of blends and the weld line reveal conspicuous differences. The decrease of hardness through the phase boundary (50–100 μm distance) occurs very rapidly when approaching the soft blend component, showing a drastic fall of the initial hardness value. On the other hand, the compatibilized blend exhibits a thinner phase boundary of about 25 μm, showing a much smaller decrease of hardness. Thus, it is shown that the microhardness technique, particularly in combination with SEM, is a sensitive tool for the quantitative evaluation of the nature and quality of the interphase boundary in polymer blends existing in blends of thermodynamically non-miscible partners or created as a result of special processing conditions as double injection molding or use of molds with an obstacle.

Taking into account the results representing the experimentally measured microhardness in dependence on the temperature for polystyrene and its blends as compared with the calculated ones it seems challenging to study to what extent the equation relating H and T_g at various temperatures (Eq. (13.23)) can be applied to complex polymer systems comprising dominating soft component and/or phase, *i.e.*, if it would be possible to predict the overall microhardness of such systems for any temperature below T_g and/or T_m of the solid component and/or phase without experimental measurements. Such an expectation seems feasible, as long as the T_g value of the soft component and/or phase account for the changes in their viscosity according to Eq. (13.23), as demonstrated by the data obtained. The verification of this expectation is the target of the next step of this study.

Acknowledgements

The author acknowledges gratefully the financial support of the Foundation for Research, Science and Technology of New Zealand, making possible his sabbatical stay at the Department of Mechanical Engineering and the Centre for Advanced Composite Materials of The University of Auckland, Auckland, New Zealand, where this chapter was prepared. The hospitality of The University of Auckland is also appreciated.

References

1. Balta Calleja F J and Fakirov S (2000 and 2007) *Microhardness of Polymers*, Cambridge University Press, Cambridge.
2. Jawhai T, Merino J C, Rodriguez-Cabello J C and Pastor M (1993) Raman mapping of the microdeformed zone produced by Vickers and Knoop microindentation techniques in poly(vinylidene fluoride), *Polymer* **34**:1613–1619.
3. Osawa S and Porter M (1996) Anisotropy in mechanical properties of forged isotactic polypropylene, *Polymer* **37**:2095–2101.
4. Kiely J D, Hwang R Q and Houston J E (1998) Effect of surface steps on the plastic threshold in nanoindentation, *Phys Rev Lett* **81**:4424–4427.
5. Lee E H, Rao G R and Mansur L K (1996) Super-hard-surfaced polymers by high-energy ion-beam irradiation, *Trends Polym Sci* **4**:229–237.
6. Briscoe B J and Sebastian K S (1996) The elastoplastic response of poly(methyl methacrylate) to indentation, *Proc R Soc Lond A* **452**:439–457.
7. Eyerer P (1985) Recent developments and applications of non-destructive tests methods, *Kunststoffe* **75**:763–769.
8. Balta Calleja F J (1985) Microhardness relating to crystalline polymers, *Adv Polym Sci* **66**:117–148.
9. Fakirov S (2007) On the application of the "rule of mixture" to microhardness of complex polymer systems containing a soft component and/or phase, *J Mater Sci* **42**:1131–1148.
10. Balta Calleja F J, Santa Cruz C, Bayer R K and Kilian H G (1990) Microhardness and surface free energy in linear polyethylene: The role of entanglements, *Colloid Polym Sci* **268**:440–446.

11. Apostolov A A, Boneva D, Balta Calleja F J, Krumova M and Fakirov S (1998) Microhardness under strain. III. Microhardness behavior during stress-induced polymorphic transition in blends of poly(butylene terephthalate) and its block copolymers, *J Macromol Sci Phys* **B37**:543–555.
12. Balta Calleja F J, Fakirov S, Roslaniec Z, Krumova M, Ezquerra T A and Rueda D R (1998) Microhardness of condensation polymers and copolymers. 2. Poly(ester ether carbonate) thermoplastic terpolymers, *J Macromol Sci Phys* **B37**:219–237.
13. Fakirov S, Balta Calleja F J and Krumova M (1999) On the relationship between microhardness and glass transition temperature of some amorphous polymers, *J Polym Sci Polym Phys Ed* **37**:1413–1419.
14. Brandrup J and Immergut E H (1989) *Polymer Handbook*, John Wiley and Sons, New York.
15. Geil P H (1987) Ultraquenching, double Tg, order, and motion in amorphous polymers, in *Order in the Amorphous "State" of Polymers* (Eds. Keinath S K, Miller R L and Rieke J K) Plenum, New York, pp. 83–91.
16. Breedon Jones J, Barenberg S and Geil P H (1979) Amorphous linear polyethylene, *Polymer* **20**:903–916.
17. Lam R and Geil P H (1978) T_g of amorphous linear polyethylene – torsion braid analysis, *Polym Bull* **1**:127–131.
18. Lam R and Geil P H (1981) Amorphous linear polyethylene – annealing effects, *J Macromol Sci Phys* **B20**:37–58.
19. Miyaji H and Geil P H (1981) Annealing of nodular linear polyethylene crystallized from the glass, *Polymer* **22**:701–703.
20. Perena J M, Martin B and Pastor M (1989) Microhardness and dynamic mechanical measurements in polyethylene near the beta-relaxation, *J Mater Sci Lett* **8**:349–351.
21. Popli R, Glotin M, Mandelkern L and Benson R S (1984) Dynamic mechanical studies of alpha-relaxation and beta-relaxation of polyethylene, *J Polym Sci Polym Phys Ed* **22**:407–448.
22. Gonzales C C, Perena J M, Bello A, Martin B, Merino J C and Pastor J M (1989) Temperature-dependence of microhardness indentations and dynamic mechanical moduli of polyesters in the vicinity of the glass-transition, *J Mater Sci Lett* **8**:1418–1419.
23. Lorenzo V, Benavente R, Perez E, Bello A and Perena J M (1993) Microhardness as a thermally activated process – indentation time-dependence for an amorphous copolyester, *J Appl Polym Sci* **48**:1177–1181.
24. Fakirov S and Krasteva B (2000) On the glass transition temperature of polyethylene as revealed by microhardness measurements, *J Macromol Sci Phys* **B39**:297–301.
25. Boyanova M and Fakirov S (2004) Effect of an obstacle during processing on the weld line of injection-molded glassy polystyrene: Microhardness study, *Polymer* **45**:2093–2098.
26. Sperling L H (1986) *Introduction to Physical Polymer Science*, Wiley-Science, New York.
27. Legge N R, Holden G and Schroeder H E (Eds.) (1987) *Thermoplastic Elastomers. A Comprehensive Review*, Hanser, Munich.
28. Roslaniec Z, Ezquerra T A and Balta Calleja F J (1995) Dielectric-relaxation of poly(ester-ether-carbonate) multiblock terpolymers, *Colloid Polym Sci* **273**:58–65.
29. Giri L, Roslaniec Z, Ezquerra T A and Balta Calleja F J (1997) Microstructure and mechanical properties of PET-PCc block copolymers: Influence of composition, structure, and physical aging, *J Macromol Sci Phys* **B36**:335–343.

30. Ania F, Martinez-Salazar J and Balta Calleja F J (1989) Physical aging and glass-transition in amorphous polymers as revealed by microhardness, *J Mater Sci* **24**:2934–2338.
31. Pietkiewicz D and Roslaniec Z (1999) Synthesis of copoly(ester-amide)s with alpha, omega-di(3-aminopropyl)oligotetrahydrofuran, *Polimery* **44**:115–122.
32. Flores A, Pietkiewicz D, Stribeck N, Roslaniec Z and Balta Calleja F J (2001) Structural features of random polyester-amide copolymers as revealed by X-ray scattering and microindentation hardness, *Macromolecules* **34**:8094–8100.
33. Fakirov S, Boneva D, Balta Calleja F J, Krumova M and Apostolov A A (1998) Microhardness under strain. Part I. Effect of stress-induced polymorphic transition of poly(butylene terephthalate) on microhardness, *J Mater Sci Lett* **17**:453–457.
34. Martinez-Salazar J, Canalda Camara J C and Balta Calleja F J (1991) Mechanical studies of poly(vinylidene fluoride)/poly(methyl methacrylate) amorphous blends, *J Mater Sci* **26**:2579–2582.
35. Noland J S, Hsu N N C, Saxon R and Schmitt J M (1971) Compatible high polymers – poly(vinylidene fluoride) blends with homopolymers of methyl and ethyl methacrylate, *Adv Chem Ser* **99**:15–18.
36. Jungnickel B J (1996) Poly(vinylidene fluoride) (overview) in *Polymeric Materials Encyclopedia*, (Ed. Salamone J C) CRC Press, Boca Raton, Vol. 9, pp. 7115–7122.
37. Fakirov S, Balta Calleja F J and Boyanova M (2003) On the derivation of microhardness of amorphous blends of miscible polymers from glass transition temperature values, *J Mater Sci Lett* **22**:1011–1013.
38. Gordon M and Taylor J S (1952) Ideal copolymers and the 2nd-order transitions of synthetic rubbers. 1. Noncrystalline copolymers, *J Appl Chem* **2**:493–500.
39. Ngai K L and Plazek D J (1996) Temperature dependences of the viscoelastic response of polymer systems in *Physical Properties of Polymers Handbook* (Ed. Mark J M) American Institute of Physics, Woodbury, New York, pp. 455–478.
40. Simov D, Fakirov S and Mikhailov M (1970) On the relationship between the transition temperatures and the tension stress in oriented polymers, *Kolloid Z Z Polym* **238**:521–522.
41. Adhikari R, Michler G H, Cagiao M E and Balta Calleja F J (2003) Micromechanical studies of styrene/butadiene block copolymer blends, *J Polym Eng* **23**:177–190.
42. Michler G H, Balta Calleja F J, Puente I, Cagiao M E, Knoll K, Henning S and Adhikari R (2003) Microhardness of styrene/butadiene block copolymer systems: Influence of molecular architecture, *J Appl Polym Sci* **90**:1670–1677.
43. Balta Calleja F J, Cagiao M E, Adhikari R and Michler G H (2004) Relating microhardness to morphology in styrene/butadiene block copolymer/polystyrene blends, *Polymer* **45**:247–254.
44. Wunderlich B (1990) *Thermal Analysis*. Academic, Boston.
45. Lide D R (1994) *CRC Handbook of Chemistry and Physics*, CRC Press, Boca Raton FL.
46. Rueda D R, Balta Calleja F J and Hidalgo A (1974) IR differential method for quantitative study of unsaturations in polyethylene, *Spectrochim Acta* **30a**:1545–1549.
47. Martinez-Salazar J and Balta Calleja F J (1980) Influence of chain defects on the crystallization of polyethylene with reference to crystal size and perfection, *J Cryst Growth* **48**:283–294.
48. Gonzales-Ortega J C and Balta Calleja F J (1974) *An Fiz* **70**:92–97.
49. Martinez-Salazar J (1979) *PhD Thesis*, Universidad Autonoma de Madrid, Madrid, Spain.
50. Fakirov S, Krumova M and Rueda D R (2000) Microhardness model studies on branched polyethylene, *Polymer* **42**:3047–3056.

51. Balta Calleja F J, Giri L, Esquerra T A, Fakirov S and Roslaniec Z (1997) Microhardness of condensation polymers and copolymers. 1. Coreactive blends of poly(ethylene terephthalate) and polycarbonates, *J Macromol Sci Phys* **B36**:655–665.
52. Garcia Gutierez M C, Michler G H, Henning S and Schade C (2001) Micromechanical behavior of branched polystyrene as revealed by *in situ* transmission electron microscopy and microhardness, *J Macromol Sci Phys* **B40**:795–812.
53. Fakirov S, Krumova M and Krasteva B (2000) On the temperature dependence of microhardness of some glassy polymers, *J Mater Sci Lett* **19**:2123–2125.
54. Mina M F, Ania F, Balta Calleja F J and Asano T (2004) Microhardness studies of PMMA/natural rubber blends, *J Appl Polym Sci* **91**:205–210.
55. Paul D R and Bucknall C B, (Eds.) (2000), *Polymer Blends*, Vol. 1 and 2; John Wiley & Sons, Inc., New York.
56. Deanin R D and Manion M A (1991) Compatibilization of polymer blends; in *Polymer Blends and Alloys* (Eds. Shonaike G O and Simon G P) Marcel Dekker, New York, Ch. 1, pp. 1–13.
57. Fakirov S (Ed) (2000) *Transreactions in Condensation Polymers*, Wiley-VCH, Weinheim.
58. Garcia Gutierrez M C, Rueda D R, Balta Calleja, F J, Kuehnert I and Mennig G, (1999) Microhardness study across the weld line in doubly injection-molded glassy polymers, *J Mater Sci Lett* **18**:1237–1238.
59. Boyanova M, Balta Calleja F J and Fakirov S (2003) Study of the phase boundary in PS/SBS blends as revealed by microhardness analysis, *J Mater Sci Lett* **22**:1741–1743.
60. Boyanova M, Balta Calleja F J, Fakirov S, Kuehnert I and Mennig G (2005) Influence of processing conditions on the weld line in doubly injection-molded glassy polycarbonate and polystyrene: Microindentation hardness study, *Adv Polym Techn* **24**:14–20.
61. Boyanova M, Mina M F, Balta Calleja F J and Fakirov S (2003) Effect of SBS compatibilizer on the interphase boundary of polymer blends of polystyrene and natural rubber, *e-Polymers* no. 047.
62. Plazek D J and Hgai K L (1996) The Glass Temperature in *Physical Properties of Polymers Handbook* (Ed. Mark J E) American Institute of Physics, Woodbury New York, pp. 187–216.
63. Krumova M, Michler H G, Evstatiev M, Friedrich K. Stribeck N and Fakirov S (2005) Transcrystallization with reorientation of polypropylene in drawn PET/PP and PA66/PP blends. Part 2. Electron microscopic observations on the PET/PP blend, *Progr Colloid Polym Sci* **130**:167–173.
64. Stribeck N, Evstatiev M, Boyanova M, Almendarez Camarillo A, Fakirov S, Friedrich K, Cunis S and Gehrke R (2003) SAXS at BW4. Straining of PET/PP blends, *Hasylab Ann Rep* **1**:953–954.
65. Fakirov S, Bhattacharyya D, Lin R J T, Fuchs C and Friedrich K (2007) Contribution of coalescence to microfibril formation in polymer blends during cold drawing, *J Macromol Sci Phys* **B46**:183–193.
66. Rueda D R, Balta Calleja F J and Bayer R K (1981) Influence of processing conditions on the structure and surface microhardness of injection-molded polyethylene, *J Mater Sci* **16**:3371–3380.
67. Rueda D R, Bayer R K and Balta Calleja F J (1989) Microhardness and mechanical anisotropy of elongational flow-injection molded polyethylene, *J Macromol Sci Phys* **B28**:267–284.

68. Balta Calleja F J, Baranowska J, Rueda D R and Bayer R K (1993) Correlation of microhardness and morphology in injection-molded poly(ethylene terephthalate), *J Mater Sci* **28**:6074–6080.
69. Birley A W, Hawort B and Batchelor J (1992) *Physics of Plastics: Processing, Properties and Materials Engineering*; Hanser, Munich.
70. Mennig G (1995) Processing-induced structure formation in *Polypropylene — Structure, Blends and Composites* (Ed. Karger-Kocsis J) Chapman & Hall: London, Vol. 1, pp. 205–226.
71. Kausch H H (1987) *Polymer Fracture*, Springer, Heidelberg.
72. Wool R P (1996) *Polymer Interfaces*, Hanser, Munich.
73. Nguyen-Chung T, Mennig G, Boyanova M, Fakirov S and Balta Calleja F J (2004) Effect of an obstacle during processing on the weld line of injection-molded glassy polystyrene: Microhardness study, *J Appl Polym Sci* **92**:3362–3367.
74. Mennig G (1991) Influence of molecular-weight on macroscopic weld lines, *Angew Makromol Chem* **185/186**:179–188.

PART V
NANOCOMPOSITES: MODELING

Chapter 14

Some Monte Carlo Simulations on Nanoparticle Reinforcement of Elastomers

J. E. Mark, T. Z. Sen, A. Kloczkowski

14.1. Introduction

In spite of its great fundamental interest and commercial importance, one of the most important unsolved problems in the area of elastomers and rubberlike elasticity is the lack of a good molecular understanding of the reinforcement provided by fillers such as carbon black and silica [1–5]. More specifically, the reinforcement of elastomers is an interesting aspect in the basic research of nanocomposites in general, and is of much practical importance since the improvements in properties fillers provide are critically important with regard to the utilization of elastomers in almost all commercially significant applications. Some of the work on this problem has involved analytical theory [6–12], but most of it is based on a variety of computer simulations [13–46].

In this context, the present review describes one way in which computational modeling has been used to elucidate the structures and properties of elastomeric polymer networks. One of the main goals has been to provide guidance on how to optimize the mechanical properties of an elastomer, in the present case by the incorporation of reinforcing fillers.

In the present approach, the simulations focus on the ways the filler particles change the distribution of the end-to-end vectors of the polymer chains making up the elastomeric network, from the fact that the filler excludes the chains from the volumes it occupies. The changes in the polymer chain distributions from this filler "excluded volume effect" then cause associated changes in the mechanical properties of the elastomer host matrix. Single polymer chains are modeled, in the standard rotational isomeric state representation [47–49], and Monte Carlo techniques are used to generate their trajectories in the vicinities of collections of filler particles. A brief overview of the approach is given in the following Section.

14.2. Description of simulations

14.2.1. *Rotational isomeric state theory for conformation-dependent properties*

In rotational isomeric state models, the continuum of rotations occurring about skeletal bonds is replaced by a small number (generally three) of rotational states that are judiciously chosen. Preferences among these states are then characterized by Boltzmann factors as statistical weights, with the required energies obtained by either potential energy calculations or by interpreting available conformation-dependent properties in terms of the models. Multiplication of matrices containing these statistical weights is then used to generate the partition function and related thermodynamic quantities, and multiplications of similar matrices containing structural information are then used to predict or interpret various properties of the chains [47–49]. Examples of such properties are end-to-end distances, radii of gyration, dipole moments, optical anisotropies, *etc.* as unperturbed by intramolecular excluded volume interactions between chain segments [50].

14.2.2. *Distribution functions*

The extension of these ideas most relevant in the present context is the use of this model to generate *distributions* of end-to-end distances, instead of simply their averages [51]. The same statistical weights were used in Monte Carlo simulations to generate representative chains, and their end-to-end distances, r, were calculated. The corresponding distribution function was obtained by accumulating large numbers of these Monte Carlo chains with end-to-end vectors within various space intervals, and dividing these numbers by the total number of the chains, N. The distances were then placed into a histogram to produce the desired end-to-end vector probability distribution function, $P(r)$ or $P(r/nl_0)$, where n is the number of skeletal bonds of length l_0. The histogram generally consisted of 20 equally spaced intervals over the allowed range $0 < (r/nl_0) < 1$, since previous studies showed that this choice was the most suitable for obtaining probability distribution functions [52]. The function $P(r/nl_0)$ was smoothed using the IMSL (International Mathematics and Statistics Library) cubic spline subroutine CSINT (Cubic Spline INTerpolant). The smoothing procedure is necessary for the proper calculation of the stress-strain isotherms from the Monte Carlo histogram [52].

These distributions are very useful for chains that cannot be described by the Gaussian limit, specifically chains that are too short, too stiff, or stretched too close to the limits of their extensibility [51]. In particular, they have documented how inadequate the Gaussian distribution is for short chains of polyethylene and poly(dimethylsiloxane) (PDMS) particularly in the region of high extension that is critical to an understanding of ultimate properties.

14.2.3. *Applications to unfilled elastomers*

The present application of these calculated distribution functions is the prediction of elastomeric properties of the chains within the framework of the Mark-Curro theory [51,53] described below.

The distribution, $P(r)$, of the end-to-end vectors, r, is directly related to the Helmholtz free energy, $A(r)$, of a chain by

$$A(r) = c - kT \ln P(r), \tag{14.1}$$

where c is a constant. The resulting perturbed distributions are then used in the "three-chain" elasticity model [54] to obtain the desired stress-strain isotherms in elongation. For the specific case of this model, the general expression for ΔA takes the form

$$\Delta A = \frac{\nu}{3}\left[A(r_0\alpha) + 2A(r_0\alpha^{-\frac{1}{2}}) - 3A(r_0)\right] \tag{14.2}$$

for elongations that are "affine" (in which the molecular deformations parallel the macroscopic deformation in a linear manner). Here, α is the elongation ratio L/L_i, ν is the number of chains in the network, and r_0 is the value of root-mean-square end-to-end distance of the undeformed network chains.

One quantity of primary interest here is the nominal or engineering stress, f^*, defined as the elastic force at equilibrium per unit cross-sectional area of the sample in the undeformed state:

$$f^* = -T\left(\frac{\partial \Delta A}{\partial \alpha}\right)_T. \tag{14.3}$$

Substitution of Eq. 14.2 into Eq. 14.3 then gives

$$f^* = -\frac{\nu k T r_0}{3}\left[G'(r_0\alpha) - \alpha^{-\frac{3}{2}}G'(r_0\alpha^{-\frac{1}{2}})\right] \tag{14.4}$$

where $G(r) = \ln P(r)$, and $G'(r)$ denotes the derivative dG/dr. The modulus (or "reduced stress") is defined by $[f^*] \equiv f^*/(\alpha - \alpha^{-2})$ and is often fitted to the Mooney-Rivlin semi-empirical formula $[f^*] \equiv 2C_1 + 2C_2\alpha^{-1}$ [54–56], where C_1 and C_2 are constants independent of deformation α.

Such results are often shown as Mooney-Rivlin plots [51], in which the calculated values of the reduced stress or modulus normalized by the value given by the Gaussian limit are shown as a function of reciprocal elongation. In a test case, the value of unity was obtained for long chains, in this case those having $n = 250$ skeletal bonds, as expected. In the case of shorter chains (having $n = 20$ or 40 skeletal bonds) there were upturns in modulus with increasing elongation that were similar to those shown in bimodal networks in which short chains are introduced to give advantageous increases in ultimate strength and modulus [57–59].

14.2.4. *Applications to filled elastomers*

In this case, the same Monte Carlo simulations were carried out as was done for the unfilled networks, but now each bond of the chain was tested for overlapping with a filler particle as the chain was being generated [28]. If any bond penetrated a particle surface, the entire chain conformation was rejected and a new chain started. Some specific illustrative examples of such investigations are given below.

14.3. Spherical particles

14.3.1. *Particle sizes, shapes, concentrations, and arrangements*

The particle sizes of greatest interest are those used commercially, with small particles giving significantly better reinforcement than larger ones. The primary particles are generally assumed to be spherical. The concentrations or "loadings" in the simulations are generally relatively small, smaller than those used commercially, since larger concentrations lead to unacceptably high attritions from chains running into particles. In actual filled elastomers, the particles are dispersed at least relatively randomly, but it is of interest to do simulations on regular particle arrangements as well [28].

14.3.1.1. *Regular arrangements, on a cubic lattice*

In these simulations, a filled PDMS network was modeled as a composite of cross-linked polymer chains and spherical filler particles arranged in a regular array on a cubic lattice [14,40]. The arrangement is shown schematically in Figure 14.1. The filler particles were found to increase the non-Gaussian behavior of the chains and to increase the moduli, as expected. It is interesting to note that composites with such structural regularity have actually been produced [60,61], and some of their mechanical properties have been reported [62,63].

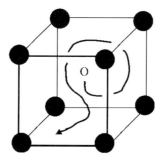

Figure 14.1. Schematic view of a polymer chain being generated within a series of filler particles in cubic arrangements

14.3.1.2. *Random arrangements, within a sphere*

In a subsequent study [16], the reinforcing particles were randomly distributed, as is illustrated in Figure 14.2. The system was taken to be a sphere having a radius equal to the end-to-end distance of the completely stretched out chain. The chain being generated was started at the center of the sphere, and this was the only place a filler particle could not be placed. Otherwise, the particles required to give the desired loading were randomly dispersed over the sample volume shown.

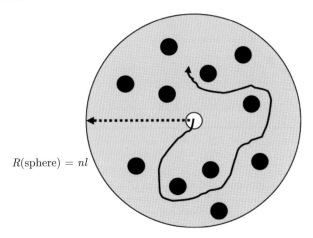

Figure 14.2. Schematic view of a polymer chain and randomly-distributed filler particles. The origin of the chain was placed at the center of the sphere of radius $R(\text{sphere}) = nl$ (maximum extension, r_{max}). All the filler particles were placed randomly in non-overlapping arrangements within the sphere, except of course at its center (where the chain started its trajectory). Chain conformations that trespassed on any particle were rejected, and statistical calculations were performed on the remaining, acceptable conformations

14.3.2. *Distributions of chain end-to-end distances*

Of greatest interest here is whether the particles cause increases or decreases in the end-to-end distances, with this expected to depend particularly on the size of the filler particles, but presumably on other variables such as their concentration in the elastomeric matrix as well.

14.3.2.1. *Typical results*

Some illustrative results for filler particles within a PDMS matrix are described in Figure 14.3 [16,40]. One effect of the particles was to increase the dimensions of the chains, in the case of filler particles that were small relative to the dimensions of the network chains. In contrast, particles that were relatively large tended to decrease the chain dimensions. Since these changes in dimensions arising from the filler excluded volume effects are of critical importance, it is necessary to put them into a larger context.

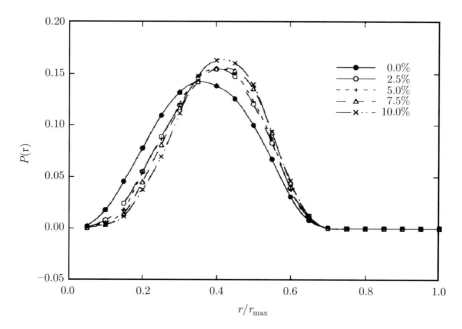

Figure 14.3. Radial distribution functions $P(r)$ at $T = 500$ K for network chain end-to-end distances obtained from the Monte Carlo simulations. The results are shown as a function of the relative extension, $r/r_{\max}$, for PDMS networks having 50 skeletal bonds between cross links [16]. The radius of the filler particles was 5 Å, and the values of the volume % of filler present are indicated in the inset

14.3.2.2. *Relevant neutron scattering results*

These simulation results on the distributions are in agreement with some subsequent neutron scattering experiments on deuterated and non-deuterated chains of PDMS [64,65]. The polymers contained silica particles that were surface treated to make them inert to the polymer chains, as was implicitly assumed in the simulations. These experimental results also indicated chain extensions when the particles were relatively small, and chain compressions when they were relatively large. Increases in chain dimensions have also been recently reported in scattering studies on heavily cross-linked polystyrene (PS) spheres introduced as filler particles into a PS matrix [66].

14.3.2.3. *Comparisons with some related simulations*

Some recent dynamic Monte Carlo simulations have also reported increases in chain dimensions in filled systems [67]. There have been several reports of simulations, however, that have yielded results in disagreement with the described simulations and the two corresponding scattering reports mentioned above. The major difference in approach was the use of dense collections of chains instead of single chains sequentially generated in the vicinities of the filler particles. Specifically, the simulations by Vacatello [18,38] and by Kumar et al. [39] find chain dimensions that are either unchanged by the filler particles or are decreased.

In a rather different type of simulation, Mattice et al. [68] generated particles within a matrix by collapsing some of the chains into domains that would act as reinforcing filler. They found that small particles did lead to significant increases in chain dimensions, while large particles led to moderate decreases, in agreement with the single-chain simulations and scattering experiments. These simulations parallel the already cited experimental scattering study of PS spheres in a PS matrix [66].

It was suggested that the increases in dimensions could have come from inadvertent increases in free volume. Specifically, collapsing some of the chains into particles could decrease their packing efficiency (thus increasing the free volume of the remaining polymer matrix). These remaining chains could then expand into the new free volume, increasing their end-to-end distances. This possibility was tested by arbitrarily increasing the free volume (at constant numbers of chains) by decreasing the density by approximately 4% [69]. The mean-square radii, $<s^2>_{matrix}$, of the matrix chains did not change at all, indicating that free volume changes were not important in this context.

14.3.2.4. *Improvements in the model*

Because of these discrepancies, the present simulations were refined in an attempt to understand the differences described [35]. This involved (i) relocating the particles periodically during a simulation, (ii) starting the chains at different locations, (iii) using Euler matrices to change the orientations of the chains being generated, and (iv) replacing the "united atom" approach by detailed atom specifications. None of these modifications significantly changed the results obtained. An additional modification, generating dense collections of chains, is in progress [70].

14.3.2.5. *Distributions of particle diameters*

Also in progress are simulations to determine any effects of having multimodal distributions of particle sizes [71]. Looking at this issue was encouraged by the improvements in properties obtained by using bimodal distributions of network chain lengths in elastomers [58] and thermosets [72], and bimodal distributions of the diameters of rubbery domains introduced into some thermoplastics [73–75].

14.3.3. *Stress-strain isotherms*

There are two items of primary interest here, specifically increases in modulus in general, and upturns in the modulus with increasing deformation. Results are typically expressed as the reduced nominal or engineering stress as a function of deformation. The area under such curves up to the rupture point of the sample then gives the energy of rupture, which is the standard measure of the toughness of a material [57].

14.3.3.1. Typical results

Figure 14.4 shows the stress-strain isotherms in elongation [16] corresponding to the distributions shown in Figure 14.3. There are substantial increases in modulus that increase with increase in filler loading, as expected. Additional increases would be expected by taking into account other mechanisms for reinforcement such as physisorption, chemisorption, *etc.*, as described below. Similar studies can be found elsewhere [41–46].

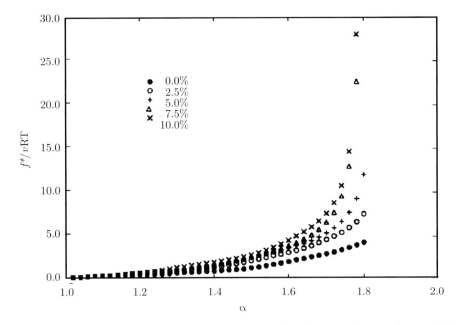

Figure 14.4. Normalized stresses calculated from the distributions shown in Figure 14.3

14.3.4. Effects of arbitrary changes in the distributions

One additional interesting result is the observation that in some cases, chain compression can also cause increases in modulus. This is being clarified by making some arbitrary changes in the distributions obtained and documenting the effects these changes have on the corresponding simulated stress-strain isotherms. For example, the curves can be shifted to lower and higher values of the chain dimensions, as is illustrated by two of the curves in Figure 14.5 [40]. The "fitted curve" is produced as follows: the distribution of end-to-end distances for the 500,000 Monte Carlo polyethylene chains of 50 bonds at 550 K is fitted to a Gaussian curve and this curve is called "fitted". Then this curve is shifted in different directions mathematically to obtain other representative curves called left-shifted, right-shifted, and up-shifted.

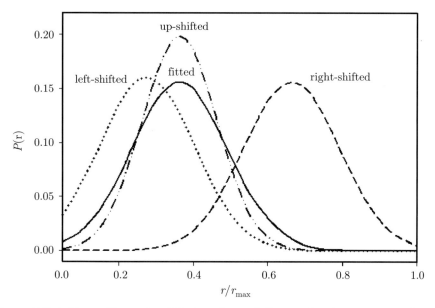

Figure 14.5. Arbitrary illustrative shifts in end-to-end distance distributions, to smaller and larger values of r. Also shown is an arbitrary illustrative narrowing of an end-to-end distance distribution, at the same most-probable value of r

This gives the isotherms shown in Figure 14.6, which show the expected increases in modulus when the chains are extended by the filler excluded volume

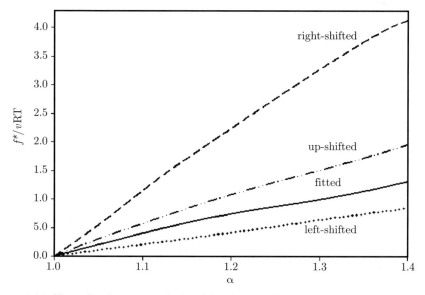

Figure 14.6. Normalized stresses calculated for the distribution shifts shown in Figure 14.5

effect, and decreases when the chains are compressed. Unexpected results are obtained, however, when the distribution is narrowed at the same most-probable value of the chain dimensions, as illustrated in Figure 14.5 [40]. The narrowing causes the peak defining the most-probable value to shift upward to keep the area under the curve the same, as is required. In this case, the change in the shape of the distribution does indeed cause an increase in the modulus, as shown in Figure 14.6. This is consistent with other simulations finding increases in modulus even when the chains are compressed, since it demonstrates that the mechanical properties can depend on subtle features of the distribution, beyond merely some average value of the chain dimensions!

14.3.5. *Some preliminary results on physisorption*

Preliminary studies have been carried out to model the effects of physical adsorption of some of the chains onto the particle surfaces [70]. The goal was to determine the relative importance of the two major effects expected. These are the increase in the effective number of chains or cross links (which would certainly increase the elastic force, stress, and modulus), and the changes in the end-to-end distances of the chains that are adsorbed (which could conceivably either increase or decrease these elastomeric properties). Specifically, amorphous PE chains having 50 skeletal bonds were Monte Carlo generated in the presence of filler particles having 20 Å diameters, with the first atom of each chain being attached to the particle surface.

The reference case of "no adsorption" was treated as follows. After the chains were generated, the conformations overlapping the filler were discarded. Over 300,000 chain conformations survived and were kept for subsequent calculations. For the "adsorption" case, any chain being generated that hit the filler surface was assumed to have been adsorbed onto the surface. If the chain could not escape the filler surface because of conformational constraints, then that chain was discarded. Also discarded were chains that did not hit the filler surface at all. Over 50,000 chains survived elimination, and were accepted as "adsorbed" chains.

Every time a chain hit the filler surface, the number x of bonds adsorbed onto the filler surface was taken to be either 1, 2, or 3 (described as 1-bond, 2-bond, and 3-bond adsorption). These bonds were assumed to be adsorbed in such a way that they formed a loop on the filler surface, so that the end-to-end vectors of the adsorbed parts of the chains on the surfaces were zero (even, tentatively, for the 1-bond adsorption case).

The results showed that if a chain had hit the filler surface, it did it 1.7 times per chain on the average. The first interesting difference between the no-adsorption and the adsorption cases is the decrease in end-to-end distances of the adsorbed chains, as is illustrated in Figure 14.7. When adsorption occurred, then the PE should be stretchable to higher elongations before its modulus increased markedly. This is shown in Figure 14.8. Similarly, as is shown in Figure 14.9, the moduli of the adsorbed chains were lower. This suggests the

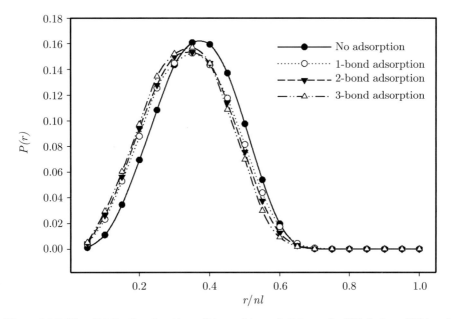

Figure 14.7. The distribution function of the end-to-end distance for PE chains of 50 bonds, with a filler radius of 20 Å at 550 K

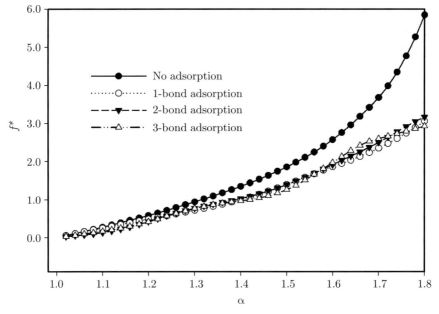

Figure 14.8. Nominal stress, f^*, for PE chains as a function of extension ratio, α, for the PE chains described in Figure 14.7

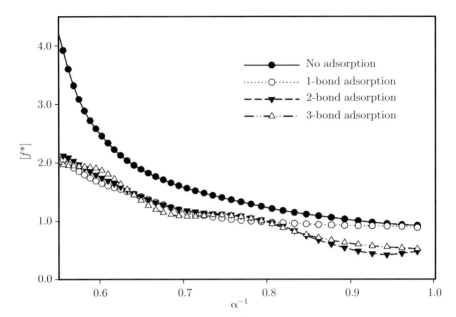

Figure 14.9. Reduced stress $[f^*]$ as a function of α^{-1} for the PE chains described in Figure 14.7

induced changes in chain conformations were less important with regard to these mechanical properties than the increases in the effective number of chains or cross links resulting from the adsorptions.

Calculations performed for a range of temperatures also suggested interesting differences. Specifically, increase in temperature also caused decreases in chain end-to-end distances, as is illustrated in Figure 14.10. This may be due at least in part to the fact that amorphous PE chains shrink upon increase in temperature. More specifically, the temperature coefficient of the unperturbed dimensions is $d\ln<r^2>_0/dT = -1.1\times 10^{-3}$ deg^{-1} [47–49]. The nominal stress and the reduced stress or modulus decrease correspondingly, as is shown in Figures 14.11 and 14.12, respectively.

14.3.6. *Relevance of cross linking in solution*

The cases where the filler causes compression of the chain are relevant to another area of rubberlike elasticity, specifically the preparation of networks by cross linking in solution followed by removal of the solvent [57]. This is shown schematically in Figure 14.13. Such experiments were initially carried out to obtain elastomers that had fewer entanglements and the success of this approach was supported by the observation that such networks came to elastic equilibrium much more rapidly. They also exhibited stress-strain isotherms in elongation that were closer in form to those expected from the simplest molecular theories of rubberlike elasticity.

In these procedures, the solvent disentangles the chains prior to their cross linking, and its subsequent removal by drying puts the chains into a "super-

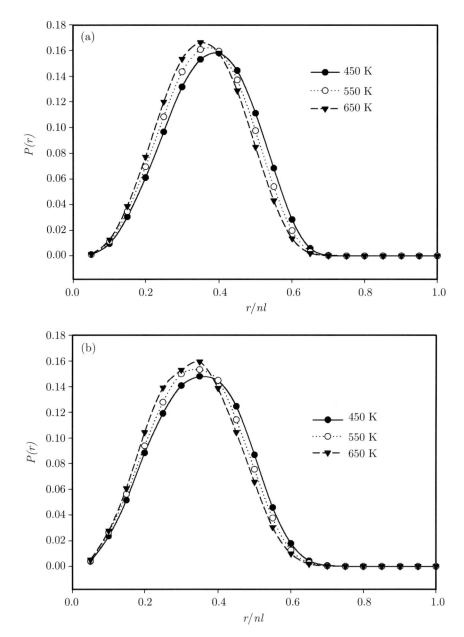

Figure 14.10. The effects of temperature on the distribution function for the end-to-end distances for the PE chains for (a) no adsorption; (b) 2-bond adsorption

contracted" state [57]. Experiments on strain-induced crystallization carried out on such solution cross-linked elastomers indicated that the decreased entangling was less important than the supercontraction of the chains, in that crystalliza-

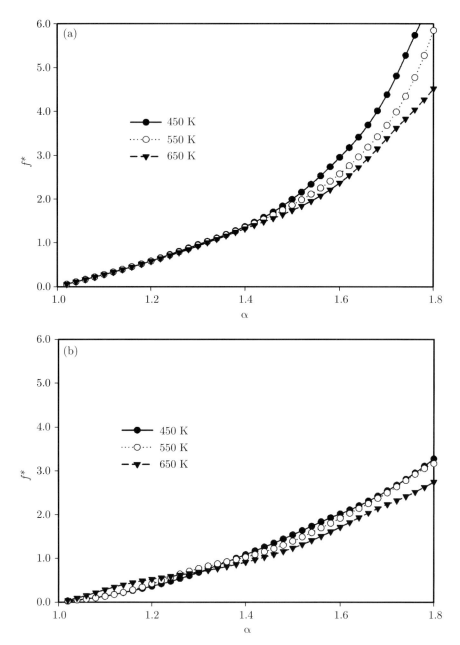

Figure 14.11. The effects of temperature on the nominal stress, f^*, for the PE chains for (a) no adsorption; (b) 2-bond adsorption

tion required larger values of the elongation than was the case for the usual elastomers cross linked in the dry state [76,77]. More recent work in this area has focused on the unusually high extensibilities of such elastomers [78–80].

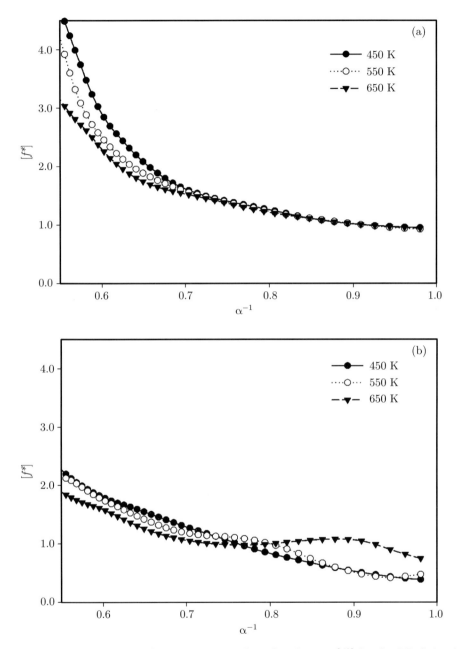

Figure 14.12. The effects of temperature on the reduced stress [f^*] for the PE chains for (a) no adsorption; (b) 2-bond adsorption

In any case, the present simulations should help elucidate molecular aspects of phenomena in this research area as well.

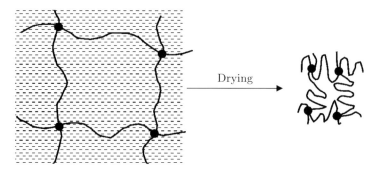

Figure 14.13. Forming a "super-compressed" network by cross linking in solution, followed by drying

14.3.7. *Detailed descriptions of conformational changes during chain extension*

An illustration of this application involves the nominal stress for syndiotactic polypropylene at T = 481 K as a function of elongation for different chain lengths, for a filler radius of 10 Å [40]. The Monte Carlo simulations were performed using recently derived conditional bond probabilities for stereoregular vinyl chains [81].

Some typical results obtained for chains having either 100 or 200 skeletal bonds are shown elsewhere [40]. At the beginning of the elongation, the chains of the two different lengths followed the same linear curve, which corresponds to the elastomeric region. This linearity is consistent with the equation for the deformation of a single chain in which the stress, f^*, is directly proportional to its end-to-end distance, r [82]. Specifically,

$$f^* = (3kT/<r^2>_0)r \qquad (5)$$

where $<r^2>_0$ represents the mean-square unperturbed dimension of the chain.

A "plastic" region (characterized by large increases in stress) appeared at lower elongations for chains having 100 skeletal bonds, as compared with those having 200. Chains of 100 bonds required greater stresses to be elongated once this critical point was reached, and this need for higher stresses can be explained in terms of its end-to-end distance distribution [40]. Since the chains of 100 bonds are already more extended than the chains of 200 bonds, the amount of additional elongation they can endure until the elastic region ends is more limited. Once the plastic region is reached, the stress development showed a non-linear character as the elongation was increased.

14.4. Ellipsoidal particles

14.4.1. *General features*

Non-spherical filler particles are also of considerable interest. Prolate (needle-shaped) particles can be thought of as a bridge between the roughly spherical particles used to reinforce elastomers [83] and the long fibers frequently used for this purpose in thermoplastics and thermosets [84]. Oblate (disc-shaped) particles can be considered as analogues of the much-studied clay platelets used to reinforce a variety of materials [85–96].

14.4.1.1. *Regular arrangements of prolate ellipsoids*

In one particularly relevant series of experiments, initially spherical particles of polystyrene were deformed into prolate ellipsoids by (i) heating the elastomeric PDMS matrix in which they resided above the glass transition temperature of the PS, (ii) stretching the matrix uniaxially, and then (iii) cooling it under the imposed deformation [97]. The technique is illustrated schematically in Figure 14.14. It is important to note that this approach also orients the axes of the non-elliptical particles, as shown in the top portion of the figure. If desired, the orientation can be removed by dissolving away the host matrix, and then re-dispersing the particles randomly within another polymer that is subsequently cross linked. This is illustrated in the bottom portion of the sketch.

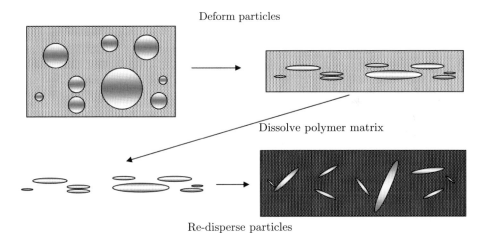

Figure 14.14. Originally spherical filler particles being deformed into prolate (needle-shaped) ellipsoids by stretching a polymer matrix in which they reside. This *in situ* approach also orients the axes of the deformed particles in the direction of the stretching. The orientation can be removed by dissolving away the host polymer matrix and then redispersing the ellisoidal particles isotropically within another polymer (giving reinforcement that is presumably isotropic)

Some relevant simulations [19,42] were presented as the moduli as a function of reciprocal elongation for particles having various values of the radius and loading volume fraction. The anisotropy in structure causes the values of the modulus in the longitudinal direction to be significantly higher than those in the transverse directions.

These simulated results are in at least qualitative agreement with the experimental differences in longitudinal and transverse moduli obtained experimentally [97]. Quantitative comparisons are difficult, in part because of the non-uniform stress fields around the particles after the deforming matrix is allowed to retract.

14.4.1.2. *Randomized arrangements of prolate ellipsoids*

In this case, isotropic behavior is expected, due to the lack of orientation dependence between the non-spherical particles and the deformation axis regardless of the shapes of the particles. The simulated results confirmed this expectation that the reinforcement from randomly-oriented non-spherical filler particles is isotropic regardless of the anisometry of their shapes. There may be difficulties on the experimental side in obtaining completely randomized orientations (and dispersions), because of the tendency of non-spherical particles to order themselves, particularly in the types of flows that accompany processing techniques or even the simple transfers of polymeric materials.

14.4.2. *Oblate ellipsoids*

In spite of their inherent interest, relatively little has been done on fillers of this shape.

14.4.2.1. *Regular arrangements*

The particles were again placed on a cubic lattice [20], and were oriented in a way consistent with their orientation in PS–PDMS composites that were the subject of an experimental investigation [98]. In general, the network chains tended to adopt more compressed configurations relative to those of prolate particles having equivalent sizes and aspect ratios. The elongation moduli were found to depend on the sizes, number, and axial ratios of the particles, as expected. In particular, the reinforcement from the oblate particles was found to be greatest in the plane of the particles, and the changes were in at least qualitative agreement with the corresponding experimental results [98]. In the experimental study, axial ratios were controllable, since they were generally found to be close to the values of the biaxial draw ratio employed in their generation. The moduli of these anisotropic composites were reported, but only in the plane of the biaxial deformation [98]. It was not possible to obtain moduli in the perpendicular direction, owing to the thinness of the films that had to be used in the experimental design.

14.4.2.2. *Randomized arrangements*

With regard to the simulations, it would be of considerable interest to investigate the reinforcing properties of such oblate particles when they are randomly oriented and also randomly dispersed. Such work is in progress [71].

Some Monte Carlo Simulations on Nanoparticle Reinforcement

14.5. Aggregated particles

14.5.1. *Real systems*

The silica or carbon black particles used to reinforce commercial materials are seldom completely dispersed [1–5], as is assumed in the simulations described. As is shown schematically in Figure 14.15, the primary particles are generally aggregated into relatively stable "aggregates" and these are frequently clustered into less-stable arrangements called "agglomerates" [40].

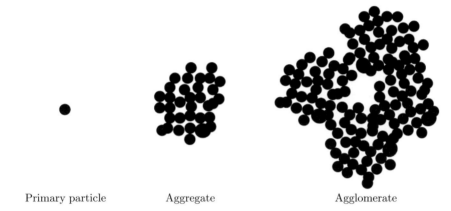

Figure 14.15. Sketches of primary particles, aggregates, and agglomerates occurring in fillers such as carbon black and silica

14.5.2. *Types of aggregates for modeling*

Simulations should be carried out on such more highly ordered structures, some limiting forms of which are sketched in Figure 14.16 [71]. It is well known in the industry that such structures are important in maximizing the reinforcement, as evidenced by the fact that being too persistent in removing such aggregates and agglomerates in blending procedures gives materials with less than optimal mechanical properties [1–5].

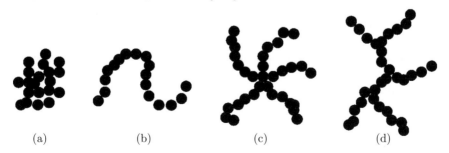

Figure 14.16. Four illustrative types of aggregates: (a) globular; (b) chainlike; (c) star-shaped; and (d) branched

14.5.3. *Deformabilities of aggregates*

Friedlander *et al.* have demonstrated that such aggregates have a remarkable deformability, by carrying out elongation experiments both reversibly, and irreversibly to their rupture points [99–104]. This is of considerable importance, since when these structures are within elastomeric matrices, their deformations upon deformation of the filled elastomer means that they must contribute to the storage of the elastic deformation energy. This would have to be taken into account both in the interpretation of experimental results and in more refined simulations of filler reinforcement.

14.6. Potential refinements

The characterized excluded volume effect is only one aspect of elastomer reinforcement [6–12], but some additional effects could be investigated by additional modeling of the adsorption of the elastomer chains onto the filler surface. The preliminary physisorption results described above should obviously be refined. Then, the calculations could be extended to include chemical adsorption by assuming that there are randomly-distributed, active particle sites interacting very strongly with the chains (by a Dirac δ-function type of potential). If the distance between the chain (generated using the Monte Carlo method) and the active site becomes less than the range of the short-range interactions, then the chain would become chemisorbed. The distribution of other active sites on the filler surface and the Lennard-Jones interactions would determine if the remaining parts of the chain are absorbed onto the surface. Simulations for chains sufficiently long to partially adsorb onto several filler particles would be especially illuminating, in that they could shed new light on the general problem of polymer adsorption. The distribution of the chain contours between the polymer bulk and various filler particles could also be of considerable importance.

14.7. Conclusions

Although there are obviously unresolved issues, the broad overview presented here should demonstrate the utility of simulations to give a better molecular understanding of how fillers reinforce elastomeric materials. It is also hoped that some of the unsolved problems described will encourage others to contribute to elucidating this important area of polymer science and engineering.

Acknowledgments

It is a pleasure to acknowledge the financial support provided J. E. Mark by the National Science Foundation through Grant DMR–0314760 (Polymers Program, Division of Materials Research). Also, A. Kloczkowski gratefully acknowledges the financial support provided by NIH grant 1R01GM072014-01.

References

1. Boonstra B B (1979) Role of Particulate Fillers in Elastomer Reinforcement: A Review, *Polymer* **20**:691–704.
2. Warrick E L, Pierce O R, Polmanteer K E and Saam J C (1979) Silicone Elastomer Developments 1967–1977, *Rubber Chem Technol* **52**:437–525.
3. Rigbi Z (1980) Reinforement of Rubber by Carbon Black, *Adv Polym Sci* **36**:21–68.
4. Donnet J-B and Vidal A (1986) Carbon Black: Surface Properties and Interactions with Elastomers, *Adv Polym Sci* **76**:103–127.
5. Donnet J and Custodero E (2005) Reinforcement of Elastomers by Particulate Fillers, in *Science and Technology of Rubber*, 3rd ed. (Eds. Mark J E and Erman B) Elsevier, Amsterdam, pp. 367–400.
6. Heinrich G and Vilgis T A (1993) Contribution of Entanglements to the Mechanical Properties of Carbon Black Filled Polymer Networks, *Macromolecules* **26**:1109–1119.
7. Witten T A, Rubinstein M and Colby R H (1993) Reinforcement of Rubber by Fractal Aggregates, *J Phys II France* **3**:367–383.
8. Kluppel M and Heinrich G (1995) Fractal Structures in Carbon Black Reinforced Rubbers, *Rubber Chem Technol* **68**:623–651.
9. Kluppel M, Schuster R H and Heinrich G (1997) Structure and Properties of Reinforcing Fractal Filler Networks in Elastomers, *Rubber Chem Technol* **70**:243–255.
10. Heinrich G, Kluppel M and Vilgis T A (2002) Reinforcement of Elastomers, *Curr Opinion Solid State Mats Sci* **6**:195–203.
11. Heinrich G and Kluppel M (2002) Recent Advances in the Theory of Filler Networking in Elastomers, *Adv Polym Sci* **160**:1–44.
12. Kluppel M (2003) The Role of Disorder in Filler Reinforcement of Elastomers on Various Length Scales, *Adv Polym Sci* **164**:1–86.
13. Kloczkowski A, Sharaf M A and Mark J E (1993) Molecular Theory for Reinforcement in Filled Elastomers, *Comput Polym Sci* **3**:39–45.
14. Sharaf M A, Kloczkowski A and Mark J E (1994) Simulations on the Reinforcement of Elastomeric Poly(Dimethylsiloxane) by Filler Particles Arranged on a Cubic Lattice, *Comput Polym Sci* **4**:29–39.
15. Kloczkowski A, Sharaf M A and Mark J E (1994) Computer Simulations on Filled Elastomeric Materials, *Chem Eng Sci* **49**:2889–2897.
16. Yuan Q W, Kloczkowski A, Mark J E and Sharaf M A (1996) Simulations on the Reinforcement of Poly(dimethylsiloxane) Elastomers by Randomly-Distributed Filler Particles, *J Polym Sci, Polym Phys Ed* **34**:1647–1657.
17. Hooper J B, McCoy J D and Curro J G (2000) Density Functional Theory of Simple Polymers in a Slit Pore. I. Theory and Efficient Algorithm, *J Chem Phys* **112**:3090–3093.
18. Vacatello M (2001) Monte Carlo Simulations of Polymer Melts Filled with Solid Nanoparticles, *Macromolecules* **34**:1946–1952.
19. Sharaf M A, Kloczkowski A and Mark J E (2001) Monte Carlo Simulations on Reinforcement of an Elastomer by Oriented Prolate Particles, *Comput Theor Polym Sci* **11**:251–262.
20. Sharaf M A and Mark J E (2002) Monte Carlo Simulations on Filler-Induced Network Chain Deformations and Elastomer Reinforcement from Oriented Oblate Particles, *Polymer* **43**:643–652.
21. Vacatello M (2002) Molecular Arrangements in Polymer-Based Nanocomposites, *Macromol Theor Sims* **11**:757–765.

22. Gersappe D (2002) Molecular Mechanisms of Failure in Polymer Nanocomposites, *Phys Rev Lett* **89**:58301–58304.
23. Fuchs M and Schweizer K S (2002) Structure of Colloid-Polymer Suspensions, *J Phys Conden Mat* **14**:R239–269.
24. Vacatello M (2002) Chain Dimensions in Filled Polymers: An Intriguing Problem, *Macromolecules* **35**:8191–8193.
25. Ozmusul M S and Picu R C (2002) Structure of Polymers in the Vicinity of Convex Impenetrable Surfaces: The Athermal Case, *Polymer* **43**:4657–4665.
26. Picu R C and Ozmusul M S (2002) Structure of Linear Polymeric Chains Confined Between Impenetrable Spherical Walls, *J Chem Phys* **118**:11239–11248.
27. Starr F W, Schroeder T B and Glotzer S C (2002) Molecular Dynamics Simulation of a Polymer Melt with a Nanoscopic Particle, *Macromolecules* **35**:4481–4492.
28. Mark J E (2002) Some Simulations on Filler Reinforcement in Elastomers, *Molec Cryst Liq Cryst* **374**:29–38.
29. Schmidt G and Malwitz M M (2003) Filler Simulations, *Curr Opinion Coll Interfac Sci* **8**:103–108.
30. Vacatello M (2003) Phantom Chain Simulations of Polymer-Nanofiller Simulations, *Macromolecules* **36**:3411–3416.
31. Vacatello M (2003) Predicting the Molecular Arrangements in Polymer-Based Nanocomposites, *Macromol Theor Sims* **12**:86–91.
32. Hooper J B, Schweizer K S, Desai T G, Koshy R and Keblinski P (2004) Structure, Surface Excess and Effective Interactions in Polymer Nanocomposite Melts and Concentrated Solutions, *J Chem Phys* **121**:6986–6997.
33. Vacatello M (2004) Monte Carlo Simulations of Polymers in Nanoslits, *Macromol Theor Sims* **13**:30–35.
34. Barbier D, Brown D, Grillet A C and Neyertz S (2004) Interface between End-Functionalized PEO Oligomers and a Silica Nanoparticle Studied by Molecular Dynamics Simulations *Macromolecules* **37**:4695–4710.
35. Sharaf M A and Mark J E (2004) Monte Carlo Simulations on the Effects of Nanoparticles on Chain Deformations and Reinforcement in Amorphous Polyethylene Networks, *Polymer* **45**:3943–3952.
36. Doxastakis M, Chen Y L, Guzman O and de Pablo J J (2004) Polymer-Particle Mixtures: Depletion and Packing Effects, *J Chem Phys* **120**:9335–9342.
37. Lin H and Mattice W L (2004) Monte Carlo Simulations of the Chain Dimensions in Filled Polyethylene Melts, Abstracts, *POLY Workshop on Molecular Modeling of Macromolecules*, Hilton Head, 2004:10.
38. Vacatello M and Vacatello M (2005) Molecular Arrangements in Polymer Nanofiller Systems, in *Computer Simulations of Liquid Crystals and Polymers* (Eds. Pasini P, Zannoni C and Zumer S) Kluwer Academic Publishers, Dordrecht, pp. 109–133.
39. Ozmusul M S, Picu R C, Sternstein S S and Kumar S K (2005) Lattice Monte Carlo Simulations of Chain Conformations in Polymer Nanocomposites, *Macromolecules* **38**:4495–4500.
40. Mark J E, Abou-Hussein R, Sen T Z and Kloczkowski A (2005) Some Simulations on Filler Reinforcement of Elastomers, *Polymer* **46**:8894–8904.
41. Sen T Z, Sharaf M A, Mark J E and Kloczkowski A (2005) Modeling the Elastomeric Properties of Stereoregular Polypropylenes in Nanocomposites with Spherical Fillers, *Polymer* **46**:7301–7308.
42. Sharaf M A, Kloczkowski A, Sen T Z, Jacob K I and Mark J E (2006) Molecular Modeling of Matrix Chain Deformation in Nanofiber Filled Composites, *Colloid Polymer Sci* **284**:700–709.

43. Sharaf M A, Kloczkowski A, Sen T Z, Jacob K I and Mark J E (2006) Filler-Induced Deformations of Amorphous Polyethylene Chains. The Effects of the Deformations on Elastomeric Properties, and Some Comparisons with Experiment, *Eur Polym J* **42**:796–806.
44. Mark J E, Abou-Hussein R, Sen T Z and Kloczkowski A (2007) Monte Carlo Simulations on Nanoparticles in Elastomers. Effects of the Particles on the Dimensions of the Polymer Chains and the Mechanical Properties of the Networks, *Macromol Symp* **256**:40–47.
45. Sen S, Thomin J D, Kumar S K and Keblinski P (2007) Molecular Underpinnings of the Mechanical Reinforcement in Polymer Nanocomposites, *Macromolecules* **40**:4059–4067.
46. Wu S and Mark J E (2007) Some Simulations and Theoretical Studies on Poly(dimethylsiloxane), *J Macrom Sci, Polym Revs* **47**:463–485.
47. Flory P J (1969) *Statistical Mechanics of Chain Molecules*, Interscience, New York.
48. Mattice W L and Suter U W (1994) *Conformational Theory of Large Molecules. The Rotational Isomeric State Model in Macromolecular Systems*, Wiley, New York.
49. Rehahn M, Mattice W L and Suter U W (1997) Rotational Isomeric State Models in Macromolecular Systems, *Adv Polym Sci* **131/132**:1–18.
50. Flory P J (1953) *Principles of Polymer Chemistry*, Cornell University Press, Ithaca NY.
51. Mark J E and Curro J G (1983) A Non-Gaussian Theory of Rubberlike Elasticity Based on Rotational Isomeric State Simulations of Network Chain Configurations. I. Polyethylene and Polydimethylsiloxane Short-Chain Unimodal Networks, *J Chem Phys* **79**:5705–5709.
52. DeBolt L C and Mark J E (1988) Theoretical Stress-Strain Isotherms for Elastin Model Networks, *J Polym Sci, Polym Phys Ed* **26**:865–874.
53. Mark J E and Curro J G (1984) A Non-Gaussian Theory of Rubberlike Elasticity Based on Rotational Isomeric State Simulations of Network Chain Configurations. III. Networks Prepared from the Extraordinarily Flexible Chains of Polymeric Sulfur and Polymeric Selenium, *J Chem Phys* **80**:5262–5265.
54. Treloar L R G (1975) *The Physics of Rubber Elasticity*, 3rd ed., Clarendon Press, Oxford.
55. Mark J E (2003) Some Recent Theory, Experiments, and Simulations on Rubberlike Elasticity, *J Phys Chem, Part B* **107**:903–913.
56. Mark J E and Erman B (2007) *Rubberlike Elasticity. A Molecular Primer*, 2nd ed., Cambridge University Press, Cambridge.
57. Erman B and Mark J E (1997) *Structures and Properties of Rubberlike Networks*, Oxford University Press, New York.
58. Mark J E (2003) Elastomers with Multimodal Distributions of Network Chain Lengths, *Macromol Symp, St. Petersburg issue* **191**:121–130.
59. Mark J E (2004) Some Interesting Things About Polysiloxanes, *Acct Chem Res* **37**:946–953.
60. Sunkara H B, Jethmalani J M and Ford W T (1994) Composite of Colloidal Crystals of Silica in Poly(methyl methacrylate), *Chem Mater* **6**:362–364.
61. Sunkara H B, Jethmalani J M and Ford W T (1995) Solidification of Colloidal Crystals of Silica, in *Hybrid Organic-Inorganic Composites* (Eds. Mark J E, Lee C Y-C and Bianconi P A) American Chemical Society, Washington, Vol. 585, pp. 181–191.
62. Pu Z, Mark J E, Jethmalani J M and Ford W T (1996) Mechanical Properties of a Poly(methyl acrylate) Nanocomposite Containing Regularly-Arranged Silica Particles, *Polym Bull* **37**:545–551.

63. Pu Z, Mark J E, Jethmalani J M and Ford W T (1997) Effects of Dispersion and Aggregation of Silica in the Reinforcement of Poly(methyl acrylate) Elastomers, *Chem Mater* **9**:2442–2447.
64. Nakatani A I, Chen W, Schmidt R G, Gordon G V and Han C C (2001) Chain Dimensions in Polysilicate-Filled Poly(Dimethylsiloxane), *Polymer* **42**:3713–3722.
65. Nakatani A I, Chen W, Schmidt R G, Gordon G V and Han C C (2002) Chain Dimensions in Polysilicate-Filled Poly(Dimethylsiloxane), *Int J Thermophys* **23**:199–209.
66. Mackay M E, Tuteja A, Duxbury P M, Hawker C J, Horn B V, Guan Z, Chen G and Krishnan R S (2006) General Strategies for Nanoparticle Dispersion, *Science* **311**:1740–1743.
67. Huang J, Mao Z and Qian C (2006) Dynamic Monte Carlo Study on the Polymer Chain in Random Media Filled with Nanoparticles, *Polymer* **47**:2928–2932.
68. Erguney F M, Lin H and Mattice W L (2006) Dimensions of Matrix Chains in Polymers Filled with Energetically Neutral Nanoparticles, *Polymer* **47**:3689–3695.
69. Erguney F M and Mattice W L, Response of Matrix Chains to Nanoscale Filler Particles, *Polymer* (submitted).
70. Sen T Z and Kloczkowski A (unpublished results).
71. Abou-Hussein R and Mark J E (unpublished results).
72. Holmes G A and Letton A (1994) Bisphenol-A Bimodal Epoxy Resins. Part I: The Dynamic Mechanical Characterization of a 6300 (340/22,500) Weight Average Molecular Weight System, *Polym Eng Sci* **34**:1635–1642.
73. Okamoto Y, Miyagi H, Kakugo M and Takahashi K (1991) Impact Improvement Mechanism of HIPS with Bimodal Distribution of Rubber Particle Size, *Macromolecules* **24**:5639–5644.
74. Chen T K and Jan Y H (1992) Fracture Mechanism of Toughened Epoxy Resin with Bimodal Rubber-Particle Size Distribution, *J Mater Sci* **27**:111–121.
75. Takahashi J, Watanabe H, Nakamoto J, Arakawa K and Todo M (2006) In-Situ Polymerization and Properties of Methyl Methacrylate-Butadiene-Styrene Resin with Bimodal Rubber Particle Size Distributions, *Polym J* **38**:835–843.
76. Premachandra J and Mark J E (2002) Effects of Dilution During Cross Linking on Strain-Induced Crystallization in *cis*-1,4-Polyisoprene Networks. 1. Experimental Results, *J Macromol Sci, Pure Appl Chem* **39**:287–300.
77. Premachandra J, Kumudinie C and Mark J E (2002) Effects of Dilution During Cross Linking on Strain-Induced Crystallization in *cis*-1,4-Polyisoprene Networks. 2. Comparison of Experimental Results with Theory, *J Macromol Sci, Pure Appl Chem* **39**:301–320.
78. Urayama K and Kohjiya S (1997) Uniaxial Elongation of Deswollen Polydimethylsiloxane Networks with Supercoiled Structure, *Polymer* **38**:955–962.
79. Kohjiya S, Urayama K and Ikeda Y (1997) Poly(Siloxane) Network of Ultra-High Elongation, *Kautschuk Gummi Kunststoffe* **50**:868–872.
80. Urayama K and Kohjiya S (1998) Extensive Stretch of Polysiloxane Network Chains with Random- and Super-Coiled Conformations, *Eur Phys J B* **2**:75–78.
81. Kloczkowski A, Sen T Z and Sharaf M A (2005) The Largest Eigenvalue Method for Stereo-Regular Vinyl Chains, *Polymer* **46**:4373–4383.
82. Erman B and Mark J E (2005) The Molecular Basis of Rubberlike Elasticity, in *Science and Technology of Rubber*, 3rd ed. (Eds. Mark J E, Erman B and Eirich F R) Academic, San Diego, pp. 157–182.
83. Medalia A I and Kraus G (1994) Reinforcement of Elastomers by Particulate Fillers, in *Science and Technology of Rubber*, 2nd ed. (Eds. Mark J E, Erman B and Eirich F R) Academic, San Diego, pp. 387–418.

84. Fried J R (2003) *Polymer Science and Technology*, 2nd ed., Prentice Hall, Englewood Cliffs, NJ.
85. Okada A, Kawasumi M, Usuki A, Kojima Y, Kurauchi T and Kamigaito O (1990) Nylon 6-Clay Hybrid, in *Polymer-Based Molecular Composites* (Eds. Schaefer D W and Mark J E) Materials Research Society, Pittsburgh, Vol. 171, pp. 45–50.
86. Pinnavaia T J, Lan T, Wang Z, Shi H and Kaviratna P D (1996) Clay-Reinforced Epoxy Nanocomposites: Synthesis, Properties, and Mechanism of Formation, in *Nanotechnology. Molecularly Designed Materials* (Eds. Chow G-M and Gonsalves K E) American Chemical Society, Washington, Vol. 622, pp. 250–261.
87. Giannelis E P (1996) Organoceramic Nanocomposites, in *Biomimetic Materials Chemistry* (Ed. Mann S) VCH Publishers, New York, pp. 337–59.
88. Vaia R A and Giannelis E P (2001) Liquid Crystal Polymer Nanocomposites: Direct Intercalation of Thermotropic Liquid Crystalline Polymers into Layered Silicates, *Polymer* **42**:1281–1285.
89. Pinnavaia T J and Beall G (Eds.) (2001) *Polymer-Clay Nanocomposites*, Wiley, New York.
90. Auerbach S M, Carrado K A and Dutta P B (Eds.) (2004) *Handbook of Layered Materials*, Marcel Dekker, New York.
91. Fischer H (2003) Polymer Nanocomposites: From Fundamental Research to Specific Applications, *Mat Sci Engin, C* **23**:763–772.
92. Ray S S and Okamoto M (2003) Polymer/Layered Silicate Nanocomposites: A Review from Preparation to Processing, *Prog Polym Sci* **28**:1539–1641.
93. Ahmadi S J, Huang Y D and Li W (2004) Synthetic Routes, Properties and Future Applications of Polymer-Layered Silicate Nanocomposites, *J Mater Sci* **39**:1919–1925.
94. Kawasumi M (2004) The Discovery of Polymer-Clay Hybrids, *J Polym Sci, Polym Chem Ed* **42**:819–824.
95. Usuki A, Hasegawa N and Kato M (2005) Polymer-Clay Nanocomposites, *Adv Polym Sci* **179**:135–195.
96. Zhu J and Wilkie C A (2007) Inercalation Compounds and Clay Nanocomposites, in *Hybrid Materials. Synthesis, Characterization, and Applications* (Ed. Kickelbick G) Wiley-VCH Verlag, Weinheim, pp. 151–173.
97. Wang S and Mark J E (1990) Generation of Glassy Ellipsoidal Particles within an Elastomer by In-situ Polymerization, Elongation at an Elevated Temperature, and Finally Cooling Under Strain, *Macromolecules* **23**:4288–4291.
98. Wang S, Xu P and Mark J E (1991) Method for Generating Oriented Oblate Ellipsoidal Particles in an Elastomer and Characterization of the Reinforcement They Provide, *Macromolecules* **24**:6037–6039.
99. Ogawa K, Vogt T, Ullmann M, Johnson S and Friedlander S K (2000) Elastic Properties of Nanoparticle Chain Aggregates of TiO_2, Al_2O_3, and Fe_2O_3 Generated by Laser Ablation, *J Appl Phys* **87**:63–73.
100. Suh Y J, Ullmann M, Friedlander S K and Park K Y (2001) Elastic Behavior of Nanoparticle Chain Aggregates (NCA): Effects of Substrate on NCA Stretching and First Observations by a High-Speed Camera, *J Phys Chem B* **105**:11796–11799.
101. Friedlander S K, Jang H D and Ryu K H (1998) Elastic Behavior of Nanoparticle Chain Aggregates, *Appl Phys Lett* **72**:11796–11799.
102. Suh Y J, Prikhodko S V and Friedlander S K (2002) Nanostructure Manipulation Device for Transmission Electron Microscopy: Application to Titania Nanoparticle Chain Aggregates, *Microsc Microanal* **8**:497–501.

103. Suh Y J and Friedlander S K (2003) Origin of the Elastic Behavior of Nanoparticle Chain Aggregates: Measurements Using Nanostructure Manipulation Device, *J Appl Phys* **93**:3515–3523.
104. Rong W, Pelling A E, Ryan A, Gimzewski J K and Friedlander S K (2004) Complementary TEM and AFM Force Spectroscopy to Characterize the Nanomechanical Properties of Nanoparticle Chain Aggregates, *Nano Lett* **4**:2287–2292.

Chapter 15

Modeling of Polymer Clay Nanocomposites for a Multiscale Approach

P. E. Spencer, J. Sweeney

15.1. Introduction

Polymer nanocomposites have been in existence for more than 20 years. Recently, Okada & Usuku [1] have presented a historical review and Hussain *et al.* [2] have reviewed the current science and technology. Filler particles, with at least one dimension at the nanometer level, are embedded within a polymer matrix, and can effect significant improvements in mechanical properties – modulus, yield stress, fracture toughness – when present at quite low levels of a few percent by weight (wt%). This is associated with particle geometries of very high aspect ratio and resultant high surface areas per unit volume. Significant improvements are obtained at filler concentrations so low that the optical properties of the polymer matrix are largely unaffected, so that a transparent matrix results in a transparent composite. After the pioneering success with the Nylon 6/clay system [1], further advances were reported using other polymer matrices, examples being polyolefins [3], epoxy, [4], polyesters [5], and polyamides [6].

The mechanisms responsible for the property enhancements are generally accepted as associated with the inhibition of polymer molecular motions near the filler surface. Macroscopic measurements of composite properties show that for the nylon/clay system, basic mechanical properties such as modulus, strength and impact strength cease to improve beyond a concentration of 5 wt% (equivalent to 2 vol%) [1]. If we accept the primary role of filler surface, and therefore filler surface area, the existence of a ceiling on property enhancement suggests that an increase in filler volume does not produce a simply related increase in polymer/clay interface area. This is to be expected in systems in which the particles tend to agglomerate, a tendency that would become more difficult to overcome at high concentrations. The importance of filler surface has been demonstrated by Boo *et al.* [7], where different levels of dispersion were found to correlate with the resultant improvements in mechanical properties. In systems where good dispersion can be maintained, such as that of Zhang *et al.* [8] consisting of silica spheres in an epoxy matrix, improvements in properties have been observed at up to 14 vol% filler.

In clay systems, platelets of high aspect ratio suitable for nanoreinforcement occur naturally in stacks (*"galleries"*), so that platelets internal to the stack are isolated from the polymer matrix and relatively little clay surface is exposed to the polymer. Spacing between platelets in this configuration can be measured by X-ray diffraction (*e.g.*, [3,6,9]). More effective reinforcement results from polymer entering the space between the platelets – *"intercalation"* – which may be detectable as increased platelet spacing. Complete separation of the platelets – *"exfoliation"* – is usually viewed as producing the most effective nanocomposite, though this is an ideal condition and well-dispersed clay nanocomposites will usually contain both exfoliated and intercalated clay [10]. The question of the effectiveness of intercalated galleries a few platelets thick in comparison with fully exfoliated systems is yet to be quantified. Typical photomicrographs of PP-clay nanocomposite are shown in Figure 15.1.

Many workers have observed the distribution and orientation of platelets using transmission electron microscopy (TEM) of thin films. In the polypropylene (PP)/clay systems that are our particular interest, these observations are a major source of information on the distribution and orientation of the platelets. The degree of exfoliation and the prevalence of intercalated platelets have been investigated, along with their shapes, which show various degrees

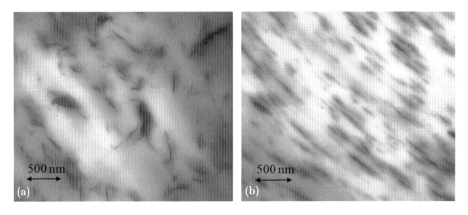

Figure 15.1. Transmission electron microscopy images of polypropylene filled with nanoclay to 5 wt%. (a) The combination of exfoliated and intercalated clay platelets is evident. (b) After stretching the platelets appear to be less curved and are aligned in the stretch-direction

of curvature [9,11–19]. On the other hand, there have been a number of theoretical studies of nanoreinforcement in which an idealized distribution of platelets is assumed. It is possible to use conventional elastic composite theory to calculate moduli, using the Halpin-Tsai or Mori-Tanaka Equations [20]. In the Halpin-Tsai model, perfectly aligned continuous reinforcement is assumed, and this theory has been exploited for clay nanocomposites by a number of workers [21–27]. In the Mori-Tanaka model, an array of ellipsoidal inclusions is assumed, and this theory has been similarly exploited [27,28]. Hbaieb et al. [28] compared simple models with two- and three-dimensional numerical continuum approaches, and found significant differences.

One of the basic issues of the functioning of nanocomposites is whether, and if so how, different nanoparticles interact. In filled rubbers, it has been suggested that the filler (carbon black or silica) particles can influence one another, so that at sufficient concentration they in effect form a network, resembling an additional molecular network. Klüppel and co-workers have developed a theory of filler networking in a number of publications [29–31], whereby particles up to 1 nm apart can interact *via* the immobilized or glassy layer of material (the "interphase") that surrounds them. More relevant to work on polymer nanocomposite is that of Shen et al. [32], who studied the rheology of polyamide nanocomposite at various filler levels. They observed a filler concentration threshold, above which their melts exhibited solid-like behavior, and introduced a "grafting-percolated" model. Monte Carlo simulations with this model suggested that the network would become effective at around 2 vol%, in line with their observations. In a study of PP/clay nanocomposite, Wang et al. [33] discussed the concept of physical "jamming" of platelets within the three-dimensional polymer network, and concluded from rheological observations that it became effective at around 1 wt%.

The issues that arise here – the distribution and orientation of platelets, their size, shape and aspect ratio, the effectiveness of simplified reinforcement models, and the filler network – can be addressed by numerical simulation. We shall use Finite Element (FE) modeling within a stochastic framework in the development below.

15.2. Sequential multiscale modeling

In the nanocomposite material, phenomena on the molecular scale (angstrom lengthscale), as well as on the scale of the platelet configuration (nano/micro lengthscale), affect the overall macroscopic material behavior. We could not hope to include *all* the underlying degrees of freedom in a macroscopic-sized numerical model, as this would be impossible to realize in practice. Rather, we adopt a multiscale modeling approach – linking a hierarchy of models on the important lengthscales, as indicated in Figure 15.2 – that reduces the overall number of degrees of freedom by an averaging (homogenization) process.

There are two basic multiscale modeling approaches: *sequential* (in which a smaller-scale model provides the relevant parameters that form the basis of a larger-scale model, conducted one after the other), and *concurrent* (in which

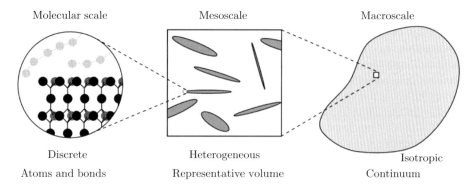

Figure 15.2. The relevant lengthscales involved in the hierarchical multiscale modeling of nanocomposite materials

information is interchanged between models on two different scales, executed in parallel). A comprehensive review of multiscale strategies for polymer nanocomposites has recently been presented by Zeng *et al.* [34]. As the length-scales involved in our multi-scale problem appear to be well separated, we have adopted a sequential ("parameter passing") hierarchical approach, in which structural information is averaged as we step up through the hierarchy of models (Figure 15.2). On each level, in the sequential approach, the averaged information is *completely* determined before attempting to simulate the level above.

On the intermediate lengthscale, the configuration of platelets is modeled using a statistical representative volume approach. As explained in detail in the following section, an representative volume element (RVE) is a "sample" of the heterogeneous material. On this meso-lengthscale, the individual components (platelet and matrix) are treated using continuum mechanics. The FE method is used to determine the "effective" material properties of the nanocomposite (which involves averaging over the sample), which may form the basis of a macroscopic model.

In this work, it is assumed that the platelets are perfectly bonded to the matrix material – a highly oversimplified view. In order to provide a more realistic description, atomistic modeling techniques (such as Monte Carlo or Molecular Dynamics, as discussed in [34]) can be used to provide important information regarding the polymer/platelet interface (*e.g.*, degree of adhesion, slip, *etc.*).

Within the framework of the sequential multiscale approach, it is necessary to determine the RVE effective material properties (*e.g.*, the Young's modulus or stiffness tensor) in advance of the macroscopic analysis. It is the determination of the effective RVE elastic properties that is the central subject of this work.

15.3. Representative volume element

The "*Representative volume element*", RVE, was originally defined by Hill [35] as a sample of model material that is "structurally entirely typical of the whole mixture on average". Since its introduction in 1963, many alternative interpretations of the RVE concept have been suggested: with different ideas about the minimum RVE size and the appropriate boundary conditions, in particular. For a recent discussion regarding RVE definition and size in relation to random heterogeneous material, see [36].

The RVE approach assumes that it is possible to replace the *heterogeneous* (polymer/platelet) material with an equivalent *homogeneous* one, which takes on the average properties of the original heterogeneous material. The volume of model material over which this averaging takes place (the size of the RVE) would ideally be infinitely large. However, in practice (for the purposes of numerical simulation), we define the RVE as a typical sample of model heterogeneous material that contains enough information to reasonably simulate an infinite medium, from which the average ("effective") material properties may be obtained in a meaningful way.

Within the context of the sequential multiscale approach discussed in the previous section, the effective RVE properties are all predetermined, and then applied at the material points of the macroscopic-scale model – e.g., FE analysis. (In the alternative concurrent multiscale approach, an RVE exists or is invoked for each material point involved in the analysis.)

Below we define the proper averaging procedure required to determine the effective properties of the nanocomposite; and its use in a statistical interpretation of the above RVE concept. We also describe the details of how to randomly generate RVE geometries that are suitable for FE analysis under periodic boundary conditions.

15.3.1. *Effective elastic material properties*

Consider the case of an RVE subject to only small strains and linear elasticity (*i.e.*, stress varies linearly with the infinitesimal strain tensor for each individual material). We assume that the continuum approximation is valid at all points within the separate materials, that the polymer and matrices are perfectly bonded, and that there is no voiding or cracking. Note that the use of these assumptions creates a somewhat idealistic material model.

Let ϑ be a general macroscopic observable quantity. The standard RVE approach defines the "*effective*" value of ϑ as the volume average of the property over the RVE, that is

$$\langle \vartheta \rangle = \frac{1}{V} \int_V \vartheta(\mathbf{x}) \, dV, \qquad (15.1)$$

where V is the total volume of the RVE, and $\mathbf{x}$ is a position vector. Given a stress field $\sigma_{ij}(x_k)$ and strain field $\varepsilon_{ij}(x_k)$ (where x_k in index notation simply lists components of position), the effective stress and strain components are thus

$$\langle \sigma_{ij} \rangle = \frac{1}{V} \int_V \sigma_{ij}(x_k) \, dV \qquad (15.2)$$

and

$$\langle \varepsilon_{ij} \rangle = \frac{1}{V} \int_V \varepsilon_{ij}(x_k) \, dV. \tag{15.3}$$

The effective stiffness and compliance tensor elements for the RVE are then defined using these averages *via*

$$\langle \sigma_{ij} \rangle = C_{ijkl} \langle \varepsilon_{kl} \rangle \tag{15.4}$$

and

$$\langle \varepsilon_{ij} \rangle = S_{ijkl} \langle \sigma_{kl} \rangle \tag{15.5}$$

respectively. As discussed above, we can take C_{ijkl} and S_{ijkl} to represent the stiffness and compliance, respectively, of an equivalent homogeneous material – having the same properties as the original heterogeneous material.

The problem of finding the effective elastic material properties, C_{ijkl} or S_{ijkl} requires the stress and strain fields for a typical volume of model composite material under small deformation to be evaluated. In this work we generate RVE geometries with randomly distributed platelets, apply small deformations by displacing the RVE boundaries, and evaluate the resulting stress $\sigma_{ij}(x_k)$ and strain $\varepsilon_{ij}(x_k)$ fields using FE analysis.

15.3.2. *Statistical ensemble*

We model the nanocomposite on the intermediate nano- and micro-lengthscale using a *statistical* interpretation of the above RVE approach. This entails the generation of a *statistical ensemble* for the RVE, in a random fashion. A statistical ensemble is a collection of many similar copies (or "realizations") of the system under consideration (a representative sample of heterogeneous nanocomposite material).

Each individual realization of the ensemble (which we shall refer to as an "*RVE realization cell*" or "*RVE cell*") must be spatially large enough to reasonably capture the essential physics responsible for the property enhancement (in terms of any long-range order in the stress/strain fields). It follows that the minimum size of the RVE realization will depend on the details of the platelet configuration: the filling fraction, dispersion, shape, size, material properties, *etc.*, as well as on the properties being measured. In practice, the minimum size of the RVE cell was determined, for a given set of parameters, by observing the convergence of measured quantities with increasing RVE cell size. In our results below, RVE cells typically involved of the order of 10–100 platelets.

The ensemble must then contain enough RVE cells to give a good statistical representation of the nanocomposite material. The traditional RVE approach involves only a single realization, which must contain enough information to reasonably represent the infinite medium. In our statistical approach, essentially that same amount of information is contained within the entire ensemble of many smaller RVE cells. Thus, in considering the RVE statistically, we have effectively replaced a *single large* volume with *many small* ones. In general, it is more computationally efficient to simulate many smaller volumes.

The effective stiffness and compliance tensors, defined for the classical RVE in Equations (15.4) and (15.5), now involve an average over the ensemble. Let $\bar{\sigma}_{ij}^{(k)}$ and $\bar{\varepsilon}_{ij}^{(k)}$ be the volume-averaged stress and strain components for the kth RVE cell in the ensemble. The *ensemble average* of the stress components is given by the arithmetic mean over all RVE cells

$$\langle \sigma_{ij} \rangle = \frac{1}{N} \sum_{k=1}^{N} \bar{\sigma}_{ij}^{(k)}, \qquad (15.6)$$

and similarly for the strain components

$$\langle \varepsilon_{ij} \rangle = \frac{1}{N} \sum_{k=1}^{N} \bar{\varepsilon}_{ij}^{(k)}, \qquad (15.7)$$

where N is the number of realizations in the ensemble (sometimes referred to as the "size" of the ensemble – not to be confused with the size of the RVE realization cell). These are essentially equivalent to Equations (15.2) and (15.3), and so the effective stiffness and compliance tensor elements, Equations (15.4) and (15.5), still hold over the ensemble.

A major advantage to the statistical approach is that it naturally provides a good indication of error for measured quantities. If $\vartheta^{(k)}$ is a general quantity measured in the kth RVE cell ($k = 1,2,...,N$), then in addition to the ensemble average of ϑ

$$\langle \vartheta \rangle = \frac{1}{N} \sum_{k=1}^{N} \vartheta^{(k)} \qquad (15.8)$$

we may also calculate the sample standard deviation in ϑ over the ensemble

$$s(\vartheta) = \left[\frac{1}{N-1} \sum_{k=1}^{N} \left(\vartheta^{(k)} - \langle \vartheta \rangle \right)^2 \right]^{\frac{1}{2}}. \qquad (15.9)$$

In practice this allows us to recognize whether the final results have been determined to a reasonable accuracy – increasing the number of realizations if not.

15.3.3. *Periodic boundary conditions*

The most appropriate set of boundary conditions (BC) to impose on an RVE realization cell is *periodic* BC, as this best simulates the infinite medium limit. With periodic BC, the RVE cell can be thought of as a unit cell, which is repeated in all directions forming an infinite continuous body – thus defining the simulated material, as illustrated in Figure 15.3. Each copy of the unit cell is referred to as an "image". During simulation, we only need to keep track of information within a single RVE unit cell.

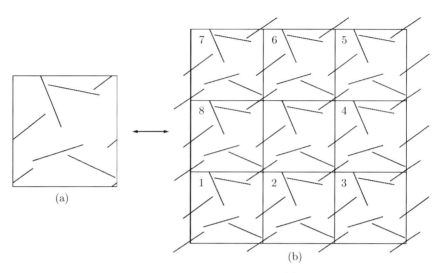

Figure 15.3. Unit cell with periodic boundary conditions. (a) Any platelet passing through an edge of the unit cell reappears on the opposite edge. Thus the unit cell may be repeated in all directions to form an infinite continuous body. In (b), we show the (central) unit cell with its nearest "images" (labeled 1–8). A platelet (in the central cell) will *only* "see" the *nearest* image of any other platelet (within the 9 cells shown)

In this work we restrict our attention to the case of a 2D nanocomposite. As can be seen in Figure 15.3a, any platelet crossing an edge must reappear at the opposite edge. Indeed, as often happens if a platelet is close to a corner, a single platelet may reappear in several sections.

In practice, periodic BC are easily implemented as follows. Given that the RVE cell (unit cell) is defined in the range $0 \leq x_i \leq L$ in the ith direction (in x and y), then a value of x_i outside this range (but within an adjacent image – $L < x_i \leq 2L$ or $-L \leq x_i < 0$) is adjusted according to

$$x_i \to \begin{cases} x_i - L, & \text{if } x_i > L \\ x_i + L, & \text{if } x_i < 0 \end{cases}. \tag{15.10}$$

Also, the periodicity must be taken into account in any operation involving relative distance across the RVE cell: such as checking whether two plate segments intersect, or whether they lie close to one another, which we shall demonstrate later in Section 15.4. It is often useful for one to imagine, for the ith direction, that the two ends of the interval $0 \leq x_i \leq L$ are joined together to form a continuous ring.

A FE mesh must be constructed in a way that allows the FE analysis to be conducted under periodic BC. We accomplish this by insisting that every node appearing on an edge has a corresponding node on the opposite edge. Within the FE calculation, relating all the pairs of opposite boundary-nodes *via* constraints enforces the periodicity. Consequently, after deformation, the profile of opposite boundaries is identical.

The alternative to the use of periodic BC would be to unrealistically assume reflective symmetry either side of the RVE cell edges, and implement uniformly prescribed BC. This approach would be much simpler to implement – but would introduce undesirable errors near the edges of the RVE cell. The classical RVE approach, which essentially implements a single larger RVE cell, *does* make use of these unrealistic BC. However, the use of reflective BC is justifiable for a *large* realization: because as the system size increases, the edge-errors incurred as a result of inappropriate BC become less and less significant. The same is not true for our statistical approach, for which the use of many smaller RVE cells makes the use of periodic BC essential.

A similar FE approach to the one used in this work was introduced in 1997 by Gusev [37] to determine the elastic properties using a periodic RVE containing randomly distributed non-overlapping spheres, and later platelets [38]. More recently, the method was applied to polymer/clay nanocomposites by Sheng and Boyce [39] within a multiscale approach.

15.4. Generating RVE geometry

The configuration of platelets observed in the real nanocomposite material very much depends on the exact nature of its preparation. As it would not be feasible to simulate the complex dynamics of the whole preparation stage, we recreate the characteristics of the platelet configuration within the RVE cell in a random fashion.

The characteristics of the platelets (their size, shape, orientation and dispersion) are randomly selected according to given predefined sets of statistical distributions. These distributions may be "extracted" from a large number of images taken of the real nanocomposite – resulting in a statistical recreation of the platelet configuration existing in the real material.

However, in the present study we assume a set of *idealized* statistical distributions (*e.g.*, all platelets having the same length and aspect ratio, having an entirely random orientation, or all being aligned, *etc.*) The distribution of platelets in space is modeled as being completely random, subject to geometrical constraints. The resulting platelet configurations are ideal for investigating the effect each platelet characteristic has on the overall properties, and thus which are important for property enhancement.

15.4.1. *Number of platelets*

The *"filling fraction"* is a measure of the amount of platelet material contained in the nanocomposite material. It is most appropriate here to measure the filling fraction by *area* (or volume) rather than by weight. Thus we define the area (volume) filling fraction f as the total area (volume) of platelet divided by the total area (volume) of the combined material.

Given a desired filling fraction, how many platelets must each RVE realization cell contain? The use of periodic BC implies that the RVE must contain a whole (integer) number of platelets. We are assuming that the platelets are positioned randomly in space (*i.e.*, given by a spatial Poisson process), subject to geometrical constraints. Accordingly, the number of platelets in an individual cell should *not* be taken as constant. Rather, it should be randomly selected from the Poisson distribution: the probability of finding exactly n ($n = 0,1,2...$) platelets within a particular RVE cell is given by

$$P_n = \frac{\bar{n}^n}{n!}\exp(-\bar{n}). \tag{15.11}$$

Here $\bar{n}$ is the average number of platelets expected in a single RVE cell, which is easily calculated from the desired filling fraction, RVE cell size, and the distributions in platelet length and aspect ratio. In order to produce random numbers drawn according to the Poisson distribution (Eq. (15.11)), we used the rejection method given in "Numerical Recipes" [40].

Thus, different RVE cells will contain different numbers of platelets – and some may contain no platelets at all – however, the ensemble average for the number of platelets $\langle n \rangle$ will tend to the desired average $\bar{n}$ as the number of realizations increases.

15.4.2. *Generation of platelet configurations*

In this work, for the sake of simplicity, we have restricted our attention to the 2D case, corresponding to a thin slice taken through the real 3D nanocomposite material. An example RVE cell is shown in Figure 15.4. The FE analysis will be performed under conditions of plane stress. Below we give details of the simple rejection-type Monte Carlo algorithm used to construct the RVE cell. Each individual platelet is randomly placed into the RVE cell *sequentially* – the position only being accepted if it "fits in" with the existing platelets (if any), otherwise a new random position is tried.

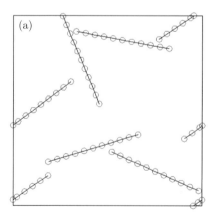
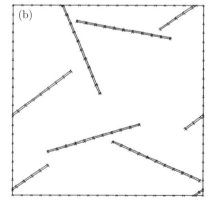

Figure 15.4. Construction of a single RVE realization cell geometry, representing a thin slice through the real 3D nanocomposite. (a) The platelets are initially represented by 1D straight line segments, and then (b) "filled out" to the desired width

The imposition of periodic BC is made much easier (and the above mentioned algorithm made more efficient) by initially representing each platelet as a series of straight line segments, as shown in Figure 15.4a. The segmentation allows the representation of curved platelets. In our implementation, the degree of segmentation ultimately controls the FE mesh density, as the seeding of the mesh depends on the segmentation. Thus, a more complex platelet geometry (which requires a greater segmentation to properly represent it) tends to produce a greater local FE mesh density.

To generate the configuration of platelets within an RVE cell, the following algorithm was used to randomly place each of its n (randomly selected according to Eq. (15.11)) platelets in sequence:

1. Select the current platelet characteristics: length, thickness, curvature and orientation. These will *not* alter.
2. Randomly generate a trial position $(x_{\text{trial}}, y_{\text{trial}})$ within the RVE cell by generating two unbiased random numbers in the range $[0,L]$.
3. Check for acceptance or rejection – in the attempt to place the current platelet in the trial position $(x_{\text{trial}}, y_{\text{trial}})$: a) Reject if any part of the trial platelet overlaps or intersects with an existing platelet; b) Reject if any part of the trial platelet is within a certain small distance from an existing platelet. This condition takes into account the thickness associated with the platelets.
4. If accepted, add the trial platelet at $(x_{\text{trial}}, y_{\text{trial}})$, otherwise return to Step 2.

Individual aspects of this algorithm and the complications incurred by the periodic BC are explained in more detail below.

The use of this algorithm *ensures* that the platelet characteristics appear according to the desired statistical distributions. For example, if the selection of platelet orientation $0 \leq \theta \leq 2\pi$ in Step 1 is entirely random, say, then the fact that the platelet is placed by "position-rejection" only, without changing θ – ensures that the distribution in θ for all platelets over a sufficiently large number of realizations will indeed be entirely random.

The periodicity of the RVE realization cell (unit cell) must be taken into account when considering any operation involving distance. This is the case in Step 3 above, where we check to see if two platelets (or rather the segments making up the platelets) intersect, overlap or are within a small distance of one another. An arbitrary point in the unit cell (x^a, y^a), should only "see" the *nearest image* of another point (x^b, y^b). That is, the "appropriate" image (of point b) may be in the same cell (as point a), or the eight neighboring cells, as depicted in Figure 15.3b, depending on which image is the nearest. This is known as the "minimum image criterion". To test distances in practice, one can either (i) explicitly reproduce the whole cell geometry in the nearest images, (ii) shift any point under consideration (point b above) to each of its nearest images positions, or (iii) make the adjustment

$$\Delta x_i \to \begin{cases} \Delta x_i - L, & \text{if } \Delta x_i > L/2 \\ \Delta x_i + L, & \text{if } \Delta x_i < -L/2 \end{cases}, \qquad (15.12)$$

where $\Delta x_i = x_i^b - x_i^a$ is the ith component (x and y) of the distance between two points.

The periodic BC also means that any platelet crossing an edge must reappear at the opposite edge. In our implementation, for simplicity, we insist that the RVE cell is initially a square, with straight rather than jagged edges. Accordingly, if a platelet segment is found to cross an edge, it is "split" at the intersection point, and the remaining length translated (in x or y) to the opposite edge (forming an extra segment). As is sometimes the case near a corner of the cell, this splitting process may need to be iterated several times until *all* the original segment length has been placed.

After all the platelets have been placed in the RVE cell, the line segments that we used to represent the platelets are "filled out" to form trapezoids, or rectangles if the platelet is not curved, as shown in Figure 15.4b. The reason we disallowed the line segments to be placed within a certain distance from one another, in Step 3(b) of the above algorithm, was to allow enough room for them to be filled out to the desired width.

Occasionally the positioning of a platelet causes a very fine mesh to be produced locally. This tends to happen, for example, if the end of the platelet is very close to the edge of the cell, particularly if the platelet is almost parallel to the edge. It is tempting to consider extending the rejection criteria in the above algorithm to disallow such placements. However, if at all possible this should be avoided. As with many Monte Carlo methods, introducing a subtle bias in this way can lead to incorrect results over the ensemble.

As the RVE is filled with more and more platelets, it becomes harder and harder to accommodate further platelets in the spaces left over. This leads to a tendency for platelets to align close to one another almost in parallel. It should be remembered that this is merely an artefact of this simplistic algorithm. Eventually, it becomes impossible to accommodate any more platelets in the RVE cell.

15.5. Periodic finite element mesh

In order to produce a mesh that is suitable for FE analysis with periodic BC, we ensure that every node on the RVE cell boundary has a corresponding node on the opposite edge. These pairs of corresponding nodes, which represent the same point in space, are subject to a "master-slave" type constraint in the FE calculation. Initially (*i.e.*, before any deformation), pairs of corresponding nodes on the left/right faces are geometrically related *via*

$$x_{\text{right}} = x_{\text{left}} + L; \; y_{\text{right}} = y_{\text{left}}, \qquad (15.13)$$

and bottom/top pairs *via*

$$x_{\text{top}} = x_{\text{bottom}}; \; y_{\text{top}} = y_{\text{bottom}} + L, \qquad (15.14)$$

where L is the initial length of each side of the square RVE cell. The edges of the RVE cell are exactly straight, rather than jagged.

The meshing was performed using the "Triangle" unstructured mesh generator of Shewchuk [41], which produces guaranteed-quality meshes by the Delaunay-based refinement algorithm of [42]. The mesh is "seeded" by supplying node and segment information defining the boundary around each platelet and the boundary around the whole RVE cell. All the supplied segments remain in the resulting mesh. The density of seeding sets a target for the local mesh density, as the algorithm attempts to generate a mesh that "fits in" with the seed segments. In order to enforce the quality of the mesh, the algorithm is ordinarily allowed to subdivide the supplied boundary segments, creating extra boundary-nodes (known as *"Steiner points"*). However, we have the ability to suppress the formation of these Steiner points if required.

In addition to the seeding, the size of each triangular element is controlled *via* a minimum angle constraint, in which nodes are added such that no angle is below a given minimum. We chose this particular mesher for its excellent ability to change from a high mesh-density (required near the platelet ends) to a much lower density (in the spaces between platelets) over a relatively short distance. This tends to reduce the number of nodes in the RVE cell, decreasing the computational effort required to perform the FE analysis.

As the mesher is not intrinsically periodic (*i.e.*, does not observe Equations (15.12–15.14)), we used the following procedure to produce a mesh that is suitable for FE analysis with period BC:

1. *Seed the boundaries.* Produce a list of nodes and segments, along the boundary of each platelet, and along the RVE cell edges, ensuring that there are matching nodes on opposite faces.
2. *Perform an initial "trial" meshing.* Here we allow the mesher to create Steiner points on the boundaries in order to enforce the mesh quality. In doing so, the nodes of opposite faces of the RVE cell may now no longer match.
3. *Re-seed the boundaries.* In addition to the original seed points from Step 1, we also include pairs of points corresponding with any Steiner points identified in the previous step – such that each and every node on the RVE cell boundary has a corresponding node on the opposite edge. This stage is shown in Figure 15.5(a).
4. *Construct the final mesh.* A fresh mesh is produced, based on the updated seeding, in which we suppress the creation of Steiner points on the RVE boundary, so that nodes on opposite faces match. This stage is shown in Figure 15.5(a).
5. *Set the material regions.* The material type (platelet or polymer) is assigned to each element of the mesh (by testing whether an element centroid lies inside a polygon formed from the platelet boundary segments).

The above procedure ensures that a high quality mesh exists over the boundaries. For example, if there is a high density of nodes locally near the edge of the RVE cell, due to the presence of a platelet, then the above procedure will cause the high density of nodes to be carried on over the boundary, to the opposite edge of the cell. Also, the condition that all the nodes appearing

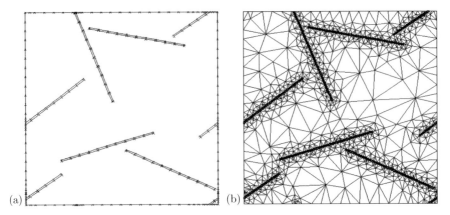

Figure 15.5. (a) The mesh is seeded such that every node on the left edge has a corresponding node on the right edge. The same is true for bottom/top nodes. An initial pass through the mesher provides appropriate seeding in regions of high mesh density. (b) The resulting mesh has highest density in and near to the platelets. The seeding ensures that regions of high mesh density pass through the boundaries to the opposite face

on an edge must have corresponding node on the opposite edge is enforced. The geometry of the problem is thus represented by: the position of each node, the element connectivity, the material type of each element, and a list of corresponding RVE cell boundary nodes pairs.

15.6. Numerical solution process

As previously described in Section 15.3.1, the effective elastic properties (stiffness matrix elements) may be deduced by measuring the volume-average stresses that result from the imposition of small applied strains on the RVE cell boundaries. Different components of the stiffness matrix may be "picked out" by individually applying a pure x-stretch, a pure y-stretch, or a simple shear strain. Below we describe the solution process, in which the resulting boundary value problem is solved using FE analysis. The measured quantities are then averaged over the whole statistical ensemble of RVE cell realizations. Our implementation of this method is fully automated.

15.6.1. *Finite element analysis of boundary value problem*

The set of BC required to apply a pure stretch ΔL in the x-direction, with zero lateral strain ("constant-width x-strain") is illustrated in Figure 15.6. The corner nodes are given prescribed displacements. If u and v are the displacements in the x- and y-directions, respectively, then

$$\begin{aligned} u_1 &= 0 & v_1 &= 0 \\ u_2 &= \Delta L & v_2 &= 0 \\ u_3 &= \Delta L & v_3 &= 0 \\ u_4 &= 0 & v_4 &= 0 \end{aligned} \quad (15.15)$$

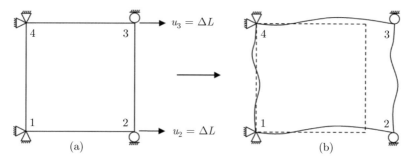

Figure 15.6. Boundary conditions imposed on the RVE cell mesh, required to produce a constant-width applied strain in the x-direction, both (a) before and (b) during deformation. The corner nodes have prescribed displacements, and the displacements of all other boundary nodes are related to those of the equivalent node on the opposite edge

where the subscripts label the corner nodes, numbered anti-clockwise starting from lower-left. The other boundary nodes are related by the constraints

$$u_{\text{right}} = u_{\text{left}} + \Delta L \quad v_{\text{right}} = v_{\text{left}}$$
$$u_{\text{top}} = u_{\text{bottom}} \quad v_{\text{top}} = v_{\text{bottom}}, \quad (15.16)$$

where u_{right}, v_{right}, u_{left} and v_{left} refer to the x- and y-displacements of a right/left corresponding pair of edge-nodes; and u_{top}, v_{top}, u_{bottom} and v_{bottom} to top/bottom edge-nodes. Thus, the profiles of the top/bottom edges, and also the right/left edges, will always be exactly the same during the deformation.

Within each material region (platelet and matrix), the elastic properties are assumed to be uniform and isotropic. The Young's modulus and Poisson's ratio for the matrix material were taken to be

$$E_{\text{poly}} = 0.9 \text{ GPa} \quad \nu_{\text{poly}} = 0.42, \quad (15.17)$$

being typical values for PP at an ambient temperature of 20°C. For the silicate clay platelet material, we used

$$E_{\text{plate}} = 160 \text{ GPa} \quad \nu_{\text{plate}} = 0.24, \quad (15.18)$$

based on the general consensus of reported values over a small subset of the literature. (A broader survey into the elastic moduli of smectite clay platelets has recently been conducted [43], which found the convergence of opinion to be in the range 178–265 GPa.) The bonding between platelet and matrix is assumed to be perfect.

On imposition of the small applied stain $\varepsilon_{ij}^{\text{applied}} \ll 1$, standard FE plane stress analysis was used to solve the boundary value problem, and thus determine the resulting stress field $\sigma_{ij}(x_k)$ within the RVE cell. The field points are in fact the integration points associated with each element. As there is only a single integration point for the 3-noded triangular elements we used, there is only a single set of stress components $\sigma_{ij}(x_k)$ associated with each element. We used the solver capabilities of the commercial FE software package Abaqus/Standard

(SIMULIA Corp.). The effective stresses (Eq. (15.2)) associated with this RVE realization is simply the average over all M elements (both matrix and platelet) contained in the RVE cell

$$\bar{\sigma}_{ij} = \frac{1}{A_{\text{tot}}} \sum_{m=1}^{M} A_m \sigma_{ij}^{(m)} \qquad (15.19)$$

where $\sigma_{ij}^{(m)}$ is the stress associated with the mth element, and

$$A_{\text{tot}} = \sum_{m=1}^{M} A_m \qquad (15.20)$$

is the total area of the RVE cell. (If we were to have used elements containing more than one integration point, we would have had to average over the integration area.) The components of $\bar{\sigma}_{ij}$ for the RVE cell are recorded (appended to a text file) in order to form the average over all RVE cell realizations that make up the statistical ensemble.

The other types of strains mentioned above (pure y-strain and simple shear) may be applied in a similar manner. However, in this work we concentrate on the elastic properties produced for the constant-width stretch in a single direction.

15.6.2. Ensemble averaged elastic properties

As previously discussed in Section 15.3.2, there are two main reasons for our use of a statistical ensemble of RVE cells, rather than the traditional single-cell approach. The first is that it is often more efficient to solve a large number of small FE problems, rather than one large one (particularly if we were to consider a non-linear problem, where the CPU time rises steeply with the number of nodes in the problem). The second main reason for choosing the statistical approach is that the measured values, such as the elements of the stiffness matrix, have an associated measure of error.

If the average stress measured and recorded for the kth member of the ensemble is $\bar{\sigma}_{ij}^{(k)}$, then the effective stress $\langle \sigma_{ij} \rangle$ is the average over the whole ensemble given by Eq. (15.6). For a constant-width x-stretch, the effective stress is given by

$$\begin{bmatrix} \langle \sigma_{11} \rangle \\ \langle \sigma_{22} \rangle \\ \langle \sigma_{12} \rangle \end{bmatrix} = \begin{bmatrix} C_{11} & C_{12} & 0 \\ C_{12} & C_{22} & 0 \\ 0 & 0 & C_{66} \end{bmatrix} \begin{bmatrix} \varepsilon_{11}^{\text{applied}} \\ 0 \\ 0 \end{bmatrix}, \qquad (15.21)$$

where $\varepsilon_{11}^{\text{applied}} = \Delta L/L$, which gives

$$C_{11} = \frac{\langle \sigma_{11} \rangle}{\varepsilon_{11}^{\text{applied}}}, \qquad (15.22)$$

and also

$$C_{12} = \frac{\langle \sigma_{22} \rangle}{\varepsilon_{11}^{\text{applied}}}. \qquad (15.23)$$

(The constant-width y-stretch would provide C_{22} and C_{12}; and the pure shear C_{66}.) We are particularly interested in the degree of property enhancement due to the presence of the platelets. Thus, rather than presenting C_{11}, it is useful to define the *stiffness enhancement* as $C_{11}/C_{11}^{\text{poly}}$, where C_{11}^{poly} is the stiffness for the pure polymer matrix material only. For the special case of the platelet orientations being completely random, the nanocomposite as a whole is isotropic, and so the results may be presented in terms of the effective Young's modulus $E = C_{11}(1-v^2)$ and Poisson's ratio $v = C_{12}/C_{11}$ (plane stress).

15.6.3. *Automation*

As an ensemble typically consists of over a hundred realizations, the RVE generation, meshing and FE solution process was automated. The main control parameters were the size of the RVE and the applied strain, together with the platelet length, aspect ratio, orientation distribution, curvature, and the number of platelets in the stack (forming an effective particle). There were also some meshing control parameters, such as the seeding density and the minimum angle for the triangulation, which affect the accuracy of the resulting FE analysis. The whole process was broken up into the following steps:

1. Randomly select the number of platelets n to be included in this RVE realization cell, according to Eq. (15.11).
2. Sequentially place each of the n platelets in the RVE cell, using the position-rejection Monte Carlo algorithm, as described in Section 15.4.2, ensuring that platelets crossing the edge of the cell reappear on the opposite side.
3. Mesh the geometry using the two-pass scheme described in Section 15.5, which is based on the Triangles mesher. For every edge node there is a corresponding node on the opposite side, geometrically related *via* Equations (15.13) or (15.14).
4. Make an input (.inp) file for the FE solver Abaqus. This is a text file that contains all the information required for the analysis: the node coordinates, element connectivity, material type of each element, a list of edge-node pairs, and the boundary strain.
5. Solve the FE problem by executing Abaqus/Standard.
6. Perform averages over the RVE cell elements using Equations (15.19) to (15.20): extracting the required data from the Abaqus output files (.dat and .fil). Append a line containing the results for this realization to an "ensemble results" text file, for later analysis.

After repeating the procedure for all N members of the ensemble, the ensemble averages were then calculated, Equations (15.6) to (15.9), using the values recorded in the final step above. If the desired accuracy in the ensemble average (as indicated by its standard deviation) was not achieved, then more realizations were performed and added to the ensemble, until the required accuracy was reached. The effective elastic properties were deduced from the average stresses as explained in Section 15.6.2.

Each process in the scheme was implemented separately: with text file input and text file output. This is extremely useful for testing and debugging purposes. There are various ways in which one could automate the process. We wrote a menu driven "controlling" program that called each program – by passing a string to the command-line of the operating system – either singly or iteratively. Thus, an entire ensemble of RVE realizations, or indeed a set of ensembles, could be "setup" and executed automatically.

15.7. Elastic RVE numerical results

Continuing with the example geometry we have been using thus far, Figure 15.7 shows the final (deformed) geometry and the stress field after undergoing a 1% constant-width x-strain. Here, the periodic boundary value problem was solved using FE analysis under the condition of plane stress, as explained in Section 15.6.1. Note that this cell is much smaller than the proper RVE realization cells used later: the size was chosen simply in order to show the various features more clearly. One can see that, due to the imposed periodic boundary constraints given by Eq. (15.16), the profiles of opposite edges are indeed identical. Also, the extent to which the deformation has affected the edge profiles, in this example, demonstrates that significant edge-errors would have been incurred had we imposed uniform displacement BC (where the edges remain straight), rather than the more appropriate periodic BC.

As can be seen in Figure 15.7a, upon deformation, the stress is mainly taken up by the platelets: particularly those that are close to being aligned parallel with the stretch-direction. In fact, the magnitude of the stress in the platelets is up to an order of magnitude greater than that in the polymer. Note that platelets orientated almost perpendicular to the stretch direction carry *much* less stress compared with those parallel to it.

For platelets that are in relative isolation, the largest stress appears to occur towards the centre of the platelet. As shown quantitatively by Sheng and Boyce *et al.* [39], this indicates that the load-transfer from matrix to platelet is mainly through interface shear stress, giving rise to the classical "shear-lag" effect [44] along the length of the platelet. One can also clearly see that the stress in the platelets is affected by the presence of other nearby platelets (platelet-platelet interaction).

In Figure 15.7b, the maximum contour value has been reduced in order to reveal the stress field that exists in the polymer matrix. One can see that the stress in the polymer tends to be lowest near the long edge of the platelets (particularly towards the middle), and relatively high close to the platelet "tips". This important feature results from the fact that a platelet is much less strained than polymer material away from its immediate vicinity. As there is perfect bonding between platelet and polymer, the presence of the platelet constrains polymer that is close to its long edge, causing large deformation to be produced in the polymer close to its tips.

The areas of high stress that appear to "emanate" from a platelet tip tend to be greater when another platelet is nearby, in fact causing the nearby platelet

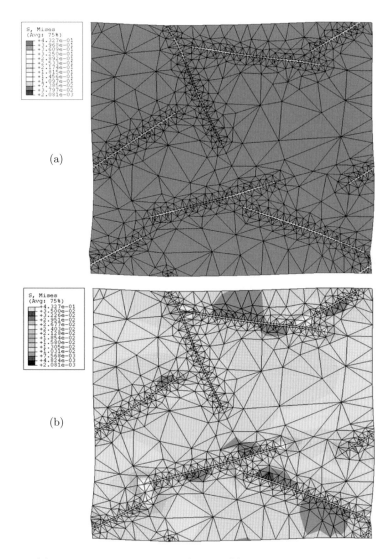

Figure 15.7. (a) The Mises stress resulting from a 1% constant-width x-stretch. The stress is taken up by the platelets. (b) The same cell geometry but with the scale altered so that the variation of the Mises stress in the polymer is visible

to bend. Note that this interaction is particularly large when the platelets are pointing directly toward one another, and when they are parallel with the stretch direction. Thus, there appears to be long-range order in the stress field that extends beyond the length of a single platelet – possibly forming some kind of network.

The cell shown in Figure 15.7 is clearly not large enough to properly represent the problem: as there may be a significant interaction between the "head" end of a platelet and its *own* "tail" end, across the periodic boundary. We must choose a cell size L that is sufficiently large, not only to render this periodic artifact negligible, but also to reasonably encompass the relevant long-range order in the stress field. The long-range order will depend on the various platelet characteristics: particularly the platelet filling fraction.

The "standard" RVE realization cell used in the work below contained platelets of aspect ratio 100:1 to a filling fraction of 4% (by area). The standard elastic constants for the matrix and platelet materials are given by Equations (15.17) and (15.18), and the applied strain was always 1%. Both straight and curved platelets were generally constructed using 10 segments (the corners of which acted as seed points), and meshed such that no angle was below 30°.

The method used to determine a reasonable value to take for L was to observe the convergence of the measured stress with increasing L, keeping the platelet characteristics fixed. Two such curves are shown in Figure 15.8, using the "standard" parameters above, for randomly aligned platelets, as in Figure 15.9, and for platelets aligned with the x-direction, as in Figure 15.15a. Each point represents the average of the stress $\bar{\sigma}_{11}$ taken over all the elements in the RVE cell, averaged over an ensemble of between 25 and 250 such realization cells. (The larger the value of L, the fewer realizations were required to produce an ensemble average to a given accuracy, as each realization contains more information.) As can be seen from Figure 15.8, the curves converge quickly to their respective infinite cell limit. We took the RVE cell size L to be 5 times the standard platelet length. Depending on the filling fraction, each RVE cell

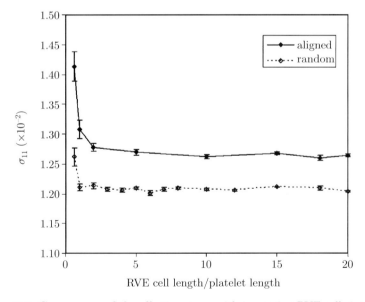

Figure 15.8. Convergence of the effective stress with increasing RVE cell size, for straight platelets of aspect ratio 100:1 and an area filling fraction of 4%. Each point is averaged over an ensemble of 25–250 realizations. Typically, we took the RVE cell size to be 5 times the platelet length

realization contained of the order of 10–100 platelets. The CPU time required to solve the problem for each realization was typically less than 1 minute on a modern PC. In this work, averages were taken over ensembles of between 25 and 250 realizations.

In reality, polymer/clay nanocomposites typically consist of a mixture of fully exfoliated single silicate layer platelets, and partially exfoliated particles consisting of a number of silicate layers together (multi-layer stacks of platelets). We begin below by considering the case where all platelets are fully exfoliated, and examine the way in which the various platelet characteristics affect the effective elastic properties. In particular, we investigate the effect of altering the curvature of the platelets and also the platelet orientation. Finally, we construct RVE geometries containing multi-layer stacks of platelets, enabling us to study how the degree of exfoliation affects the effective properties of the nanocomposite material.

15.7.1. *Fully exfoliated straight platelets*

Consider an RVE containing only randomly-orientated single-layered straight platelets. This is the "ideal" case of full exfoliation and complete dispersion, with straight platelets. An example RVE realization cell for such a configuration is shown in Figure 15.9, where the platelet aspect ratio (length/width) is 100:1, the length of the RVE cell is 5 times the platelet length, and the filling fraction

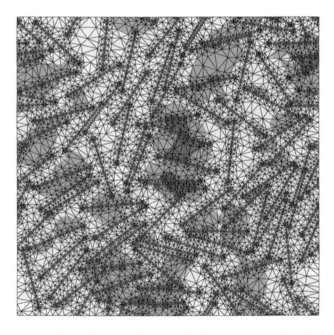

Figure 15.9. A typical RVE realization cell for randomly orientated and randomly distributed single-layer straight platelets, on application of a 1% constant-width x-strain

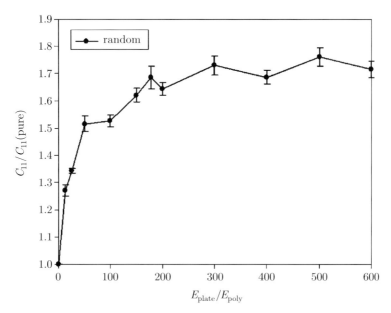

Figure 15.10. Variation of the stiffness enhancement $C_{11}/C_{11}(\text{pure})$ with the platelet stiffness for randomly orientated straight platelets. We expect the platelet stiffness to be at least $E_{\text{plate}}/E_{\text{poly}} \approx 200$ in the real material

is 4% by area. Below we will examine the way in which the various platelet characteristics affect the overall effective stiffness in the x-direction C_{11} by applying a 1% constant-width x-stretch. The results for C_{11} will be presented in units of its value for pure polymer $C_{11}(\text{pure}) = 1.09273$. Thus, a "stiffness enhancement" $C_{11}/C_{11}(\text{pure}) > 1$ indicates an improvement of mechanical properties in the x-direction due to the presence of the platelets; and $C_{11}/C_{11}(\text{pure}) < 1$ would indicate a reduction.

Firstly, let us see how the relative platelet stiffness $E_{\text{plate}}/E_{\text{poly}}$ affects the overall effective stiffness of the nanocomposite, where E_{poly} is given by Eq. (15.17). Figure 15.10 shows the stiffness enhancement $C_{11}/C_{11}(\text{pure})$ for values of platelet stiffness up to 600 times that of the polymer, whilst keeping the aspect ratio and filling fraction constant at their standard values of 100:1 and 4% respectively. Each point represents an average over an ensemble of 30–40 realizations. As one can see, the stiffness initially rises sharply, and then levels out towards the infinite platelet-stiffness limit. If the value of $E_{\text{plate}}/E_{\text{poly}} \approx 200$ for the real polymer/clay nanocomposite, then we can see the stiffness enhancement is already fairly close to the maximum possible value. That is, using a stiffer material in the platelet would not lead to a significant improvement in the mechanical properties.

Now consider the way in which the aspect ratio (defined as the platelet length divided by its width) of the platelets changes the overall RVE stiffness. The variation of stiffness enhancement with aspect ratio is shown in Figure 15.11,

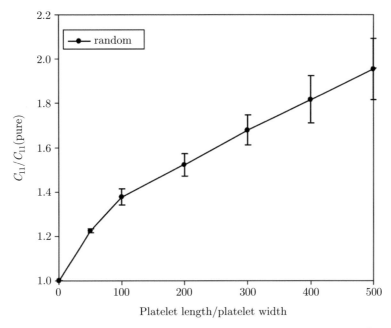

Figure 15.11. Variation of stiffness enhancement with changing aspect ratio, for randomly orientated straight platelets, with a fixed area filling fraction of 4%

keeping the filling fraction constant. Thus, as the value taken for the aspect ratio increases, the RVE contains a *larger* number of *longer/thinner* platelets. As can be seen, for aspect ratios above 100, the increase in stiffness enhancement appears to be linear up to the aspect ratio shown in the plot.

In Figure 15.12, we present the variation of stiffness enhancement with platelet area filling fraction f, holding the platelet aspect ratio constant at 100:1. Each point represents an average over 25–100 realizations, where cell configurations involving fewer platelets require a larger number of realizations (there is greater variation in the number of platelets selected from the Poisson distribution). Here we compare the case of the randomly orientated and randomly positioned platelets, with the case of platelets that are all aligned in the x-direction, and are randomly positioned. In both cases there is an increase in stiffness enhancement that is approximately linear with increasing f. (In the real nanocomposite, we would expect the increase to be less rapid with increasing f due to the effects of increased particle agglomeration.) We observe from Figure 15.12 that, for all filling fractions, the stiffness enhancement for the aligned platelets is more than *double* that for the randomly aligned platelets. This dependence on platelet orientation is investigated in greater detail below.

15.7.2. *Effect of platelet orientation*

The orientation of the platelets in the real nanocomposite appears to be approximately random. However, during processing (*e.g.*, extrusion), the platelets tend

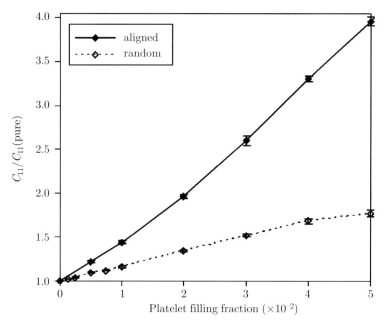

Figure 15.12. Enhancement in stiffness *vs.* the area platelet filling fraction, for both randomly aligned platelets, and platelets aligned with the *x*-direction. The platelets were straight and had a fixed aspect ratio (length/width) of 100:1

to become preferentially orientated in the processing direction, as can be seen from the images in Figure 15.1. We have already observed above, from Figure 15.12, that the stiffness enhancement for the case of perfectly aligned platelets is more than double that for the case of completely random platelets, over all filling fractions.

Staying with fully exfoliated (single-layer) straight platelets that are randomly positioned in space, we now investigate the way in which the *degree* of platelet orientation affects the overall elastic properties. Accordingly, two separate ways of gradually altering the platelet orientation distribution are considered below. All data points presented were averaged over ensembles of 25 RVE cell realizations, each containing platelets of aspect ratio 100:1 and filled to 4% by area.

Firstly, consider an RVE platelet configuration in which *every* platelet is given the *same* angle of tilt $\pm\theta$ to the *x*-direction. Figure 15.13 shows the stiffness enhancement for tilt angles ranging from $\theta = 0$ (all platelets perfectly aligned with the direction of applied strain) to $\theta = 90°$ (all platelets are aligned in the direction normal to the applied strain). Thus, this plot encompasses the two furthest possible extremes in platelet alignment. As can be seen from the figure, the stiffness enhancement has a Gaussian-like shape: with its maximum possible value for perfect alignment at $\theta = 0$, reducing to its minimum possible value for the case of perpendicular alignment (for which there is virtually no enhancement).

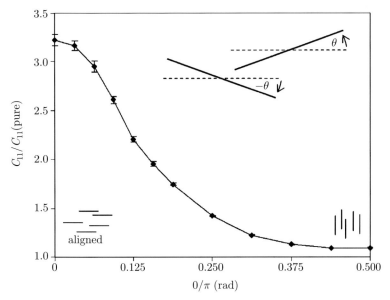

Figure 15.13. Stiffness enhancement for RVE geometries in which all platelets have a fixed tilt angle of $\pm\theta$ to the x-direction, for straight platelets with a fixed aspect ratio of 100:1 and filling fraction of 4%. On the left of the plot ($\theta=0$) the platelets are aligned perfectly with the stretch-direction, and on the right ($\theta=90°$) they are all aligned perpendicular to it

Now consider the somewhat more physically realistic RVE configuration in which the platelet orientation θ is uniformly random up to a maximum tilt angle of $\pm\theta_{max}$ from the x-direction. In Figure 15.14 we present the stiffness enhancement from $\theta_{max}=0$ (all platelets fully aligned with the direction of applied strain) to $\theta_{max}=90°$ (completely random platelet orientation). Here we traverse between the two main orientation distributions of interest. Configurations corresponding to different values of θ_{max} are similar to those one might expect for various "degrees" of processing: from the randomly orientated state of exfoliation with no processing ($\theta_{max}=90°$), through to the perfectly aligned state representing "extreme" processing ($\theta_{max}=0$). As can be seen, there is an approximately linear reduction in the stiffness enhancement with increasing θ_{max}.

Clearly we may conclude that the effective properties in the direction of any preferential platelet orientation are enhanced considerably. The load transfer from polymer to platelet is more effective for platelets that are close to being aligned with the direction of applied strain, leading to more stress being taken up by the platelets, and therefore greater reinforcement. There also appears to be a complicated platelet-platelet interaction taking place. In the transverse direction to any platelet alignment, the mechanical properties are likely to be reduced.

15.7.3. Curved platelets

On exfoliation, the platelets in the real unstretched nanocomposite material tend to be curved, rather than straight, as can be clearly seen in Figure 15.1a.

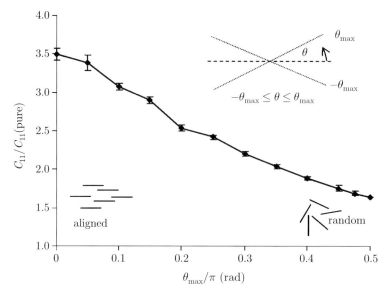

Figure 15.14. Enhancement for RVE geometries which have randomly-orientated straight platelets up to a maximum tilt angle of $\pm\theta$. On the left of the plot ($\theta = 0$) the platelets are all aligned with the x-direction, and on the right ($\theta_{\max} = 90°$) the alignment of the platelets is completely random

We now investigate how platelet curvature might affect the overall elastic properties: by considering RVE geometries containing randomly positioned platelets that *all* have the *same* constant curvature.

The same construction method was used to generate the curved platelets as for the straight platelets, discussed in Section 15.4.2. As a platelet is made up of a series of straight-line segments – causing the platelet to have constant curvature is simply a matter of applying a small offset angle between each segment.

We compared RVE configurations containing either straight or constant-curvature platelets. In both cases, the platelets consisted of 10 segments, were of aspect ratio 100:1, and were filled to an area fraction of 4%. Figure 15.15 shows a comparison of RVE cells for straight platelets aligned with the x-direction, and constant-curvature platelets aligned in the x-direction (the direction being that of the line joining its end-points). Notice that we appear to get stress "shielding" in the polymer where it is "cupped" by the curvature of the platelet (the polymer is constrained). We also considered straight and constant-curvature configurations both with completely random platelet alignment.

The stiffness enhancement for each of the 4 cases described above is plotted against filling fraction in Figure 15.16, averaged over 25–60 realizations. It shows that there is little difference in stiffness enhancement between straight and curved platelets if they are randomly oriented. However, there is a significant difference between straight and curved platelets if they are aligned with the direction in which the stain is applied.

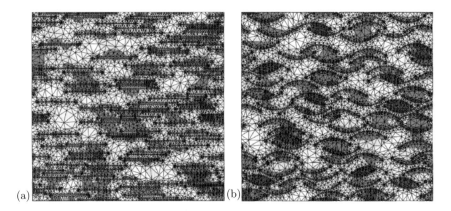

Figure 15.15. Typical RVE geometries for (a) straight platelets all aligned with the x-direction, and (b) curved platelets all aligned with the x-direction, for the same platelet filling fraction and aspect ratio

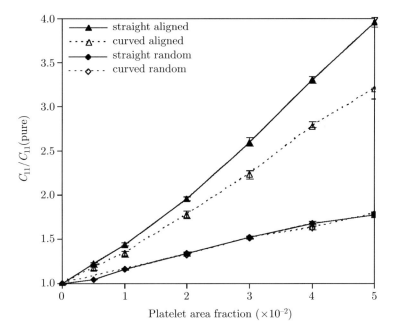

Figure 15.16. Stiffness enhancement for 4 different cases involving either all straight or all curved platelets that are either randomly orientated or aligned with the x-direction, at the same filling fraction of 4%, and aspect ratio of 100:1

The difference in overall stiffness enhancement for the aligned case is mainly due to the fact that the curved platelets cannot achieve the maximum polymer-to-platelet load-transfer which exists for the straight platelets. This is because only a small proportion of a curved platelet is in good alignment with the direction of applied strain (also one interface is in fact shielded), whereas in *whole* of a straight platelet is aligned. (We have already observed that the load transfer is highly dependant on orientation.) The curvature may also cause a difference in the platelet-platelet interaction.

15.7.4. *Multi-layer stacks of intercalated platelets*

In general, real polymer/clay nanocomposites are *not* in a state of complete exfoliation (in which *all* the clay is fully separated into single-layered platelets and dispersed in the polymer matrix). Rather, these materials typically consist of a mixture of *some* fully exfoliated platelets together with stacks of intercalated platelets a few layers thick. An intercalated stack is one in which the inter-layer gallery material has been replaced by the polymer matrix material. Below we investigate the way in which the number of layers in these stacks alters the property enhancement.

The same position-rejection method previously used for single-layer platelets, discussed in Section 15.4.2, is easily extended to generate multi-layer "effective" particles. Given an orientation θ and the randomly generated trial position $(x_{\text{trial}}, y_{\text{trial}})$ as before, then the start position of the kth layer in the stack is (x_k, y_k) where

$$\begin{aligned} x_k &= x_{\text{trial}} + d(k-1)\cos(\theta + \pi/2) \\ y_k &= y_{\text{trial}} + d(k-1)\sin(\theta + \pi/2) \end{aligned}, \quad (15.24)$$

which produces a stack of parallel platelets, separated by the inter-layer spacing d, with ends aligned perpendicular to the length. During the construction of the stack, if any part is found to overlap or intersect with existing platelets (already successfully placed in the RVE cell) then this trial is rejected, and another $(x_{\text{trial}}, y_{\text{trial}})$ tried. Thus we may construct configurations of multi-layer stacks which tend to the desired statistical distributions (in θ, for example) as the number of realizations increases. In all cases, the aspect ratio of platelets making up the stacks was 100:1, and the inter-layer spacing d was taken to be 4 times the platelet width (thus the thickness of the polymer separating the layers was taken to be 3 times the platelet width).

We constructed RVE configurations containing multi-layer stacks *all* of which had the *same* number of layers n_{layer}. The polymer material in the inter-layer regions was assumed to have the same elastic properties as the polymer surrounding the stacks. Also, perfect bonding is still assumed at the polymer/platelet interface. An example RVE geometry containing 5-layer stacks of platelets filled to an area fraction of 4% is shown in Figure 15.17. We observe that the stacks appear to act as "effective particles".

We considered the stiffness enhancement both for randomly aligned stacks (shown in Figure 15.18), and for stacks aligned with the applied-strain direction (shown in Figure 15.19), for $n_{\text{layer}} = 1,2,3,4$, and 5 at various area filling fractions f (of 1%, 2%, 3% and 4%). Each point in the plots represents an ensemble average over 250 realizations (thus the data presented in Figures 15.18 and 15.19 was

Modeling of Polymer Clay Nanocomposites for a Multiscale Approach 573

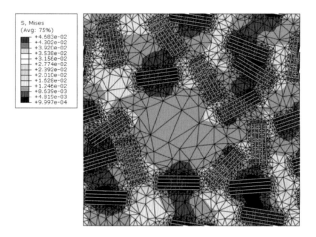

Figure 15.17. A typical RVE cell geometry containing randomly orientated and randomly distributed 5-layered particle-stacks, representing partial exfoliation

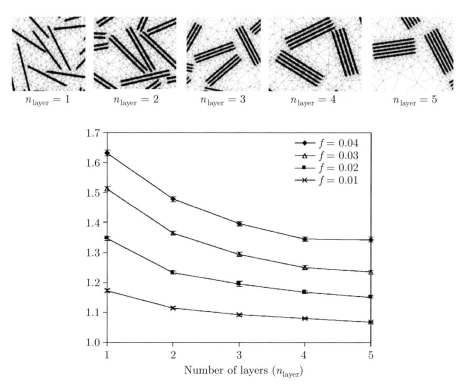

Figure 15.18. (a) Small sections of RVE cell involving stacks of 1, 2, 3, 4, and 5 layers. (b) Stiffness enhancement *vs.* the number of layers in the stack, for various area filling fractions f. The multilayer stacks are randomly aligned and randomly distributed

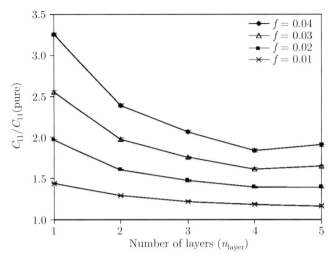

Figure 15.19. The variation in stiffness enhancement with the number of layers in the stack, at various area filling fractions, where all the particles are aligned with the x-direction

produced using 10,000 separate RVE cell realizations). One can see that the plots for random- and aligned-orientation are very similar in shape, but the magnitude of the stiffness enhancement is larger for the case of the aligned stacks. This indicates that the multi-layer stacks have a similar orientation-dependence as we found for the single-layer platelets.

It is immediately obvious that, for all the n_{layer} values studied, the stiffness enhancement increases with increasing filling fraction. However, one can also see that, for a given value of the filling fraction f, the stiffness enhancement decreases and also *levels off* as the value of n_{layer} increases. This leveling off effect is mainly due to the fact that the platelets internal to the stack tend to be shielded from the stress, making them less effective than fully exfoliated single platelets.

Clearly, it would be desirable to produce a nanocomposite containing mainly single fully-exfoliated platelets at high filling fractions. In practice, the increase in agglomeration at higher filling fractions (which we did not consider in our model) prevents this. However, from the results above, we observe that stacks of 2 and 3 layers still provide reasonable reinforcement at higher filling fractions.

15.8. Conclusions

The improved mechanical properties in polymer/clay nanocomposites are associated with particle geometries of high aspect ratio and the resulting high interfacial area per unit volume. In this work, an elastic Finite Element analysis of idealized clay platelet configurations was carried out: identifying which platelet characteristics are important in producing the property enhancement.

We used a *statistical* interpretation of the Representative Volume Element approach, which averages the material properties over a large number of relatively small periodic realization cells of the model system. The advantage in using many small realization cells rather than a single large cell (the traditional approach) is that the solution process is computationally more efficient, and it provides a good indication of error in the final results. A "position-rejection" algorithm was used to randomly place the clay platelets within each realization cell *sequentially*, such that they do not overlap or intersect. This simple algorithm guarantees that the desired statistical distributions – in platelet filling fraction, size, shape, position and orientation – are produced over the ensemble. The effective elastic material properties were deduced by measuring the response to a small applied strain for each cell – using standard Finite Element analysis to solve the periodic boundary value problem – and then averaging over the entire ensemble. In our implementation, the whole process was automated.

On application of the applied strain, the resulting stress is mainly taken up by the platelets, which thus act to reinforce the material. The axial stress in the platelets is highest towards the platelet centre, indicating that the load-transfer from polymer to platelet is primarily through interface shear stress (shear-lag effect). We observed that the presence of a platelet tends to constrain the polymer along its long edge ("stress shielding"), and produces high deformation near the platelet ends (giving the appearance of high stress "emanating" from platelet tips). The stress fields in and around the platelets were clearly affected by the presence of other nearby platelets, which may possibly give rise to some kind of network in the real material.

For the "ideal" case of full exfoliation, in which the platelets are all separate and randomly distributed, we found that the stiffness enhancement (ratio of the stiffness of the nanocomposite to that of the pure polymer) increased with increasing platelet aspect ratio, platelet stiffness, and filling fraction. The effect of platelet orientation is of particular importance (platelets tend to become preferentially aligned on processing). By allowing platelets to be randomly orientated only up to a maximum tilt angle, we found that the stiffness enhancement reduced approximately linearly as the degree of alignment was reduced: from all perfectly aligned with the stretch direction, to completely random. The load transfer from polymer to platelet is more effective for platelets that are close to being aligned with the stretch direction. We also studied configurations of platelets with constant-curvature (platelets tend to be curved on exfoliation). We found little difference in stiffness enhancement between curved and straight platelets with random orientation; however, there was a significant difference between curved and straight platelets aligned with the stretch-direction. This is because curved platelets only have a small proportion of their length orientated close to the stretch-direction.

We also considered the case of multilayer stacks of intercalated platelets: comparing configurations containing only stacks of a particular number of layers ($n_{\text{layer}} = 1,2,3,4$, and 5), at various filling fractions, for both aligned and random stack orientations. For all the cases studied, the stiffness enhancement increased with increasing filling fraction. Also, the stiffness enhancement tends to decrease and *level off* as the number of layers in the stack increases. This is because the platelets internal to the stack are shielded from the stress by the outermost platelets, making them less effective. We observe that the stiffness enhancement has a similar value for stacks containing 4 and 5 layers for a given filling fraction and alignment. The results show that the greatest enhancement occurs in the case of full exfoliation. However, we still see reasonable stiffness enhancement for stacks of 2 and 3 layers at higher filling fractions.

Acknowledgements

We acknowledge EPSRC (UK) for funding this project.

References

1. Okada A and Usuki A (2006) Twenty Years of Polymer-Clay Nanocomposites, *Macromol Mater Eng* **291**:1449–1476.
2. Hussain F, Jojjati M, Okamoto M and Gorga E R (2006) Review article: Polymer-matrix Nanocomposites, Processing, Manufacturing, and Application: An Overview, *J Compos Mater* **40**:1511–1575.
3. Gopakumar T G, Lee J A, Kontopoulou M and Parent J S (2002) Influence of clay exfoliation on the physical properties of montmorillonite/polyethylene composites, *Polymer* **43**:5483–5491.
4. Kornmann X, Thomann R, Mülhaupt R, Finter J and Berglund L A (2002) High Performance Epoxy-Layered Silicate Nanocomposites, *Polym Eng Sci* **42**:1815–1826.
5. Ke Y C, Yang Z B and Zhu C F (2002) Investigation of properties, nanostructure, and distribution in controlled polyester polymerization with layered silicate, *J Appl Polym Sci* **85**:2677–2691.
6. McNally A, Raymond Murphy W, Lew C Y, Turner R J and Brennan G P (2003) Polyamide-12 layered silicate nanocomposites by melt blending, *Polymer* **44**:2761–2772.
7. Boo W J, Sn L, Liu J, Moghbelli E, Clearfield A, Sue H-J, Pham P and Verghese N (2007) Effect of nanoplatelet dispersion on mechanical behavior of polymer nanocomposites *J Polym Sci, Part B: Polym Phys* **45**:1459–1469.
8. Zhang H, Zhang Z, Friedrich K and Eger C (2006) Property improvements of *in situ* epoxy nanocomposites with reduced interparticle distance at high nanosilica content, *Acta Mater* **54**:1833–1842.
9. Chiu F-C, Lai S-M, Chen J-W and Chu P-H (2004) Combined Effects of Clay Modifications and Compatibilizers on the Formation and Physical Properties of Melt-Mixed polypropylene/clay nanocomposites, *J Polym Sci, Part B: Polym Phys* **42**:4139–4150.
10. Morgan A B and Gilman J W (2003) Characterization of polymer-layered silicate (clay) nanocomposites by transmission electron microscopy and X-ray diffraction: A comparative study, *J Appl Polym Sci* **87**:1329–1338.
11. Liu X and Wu Q (2001) PP/clay nanocomposites prepared by grafting-melt intercalation, *Polymer* **42**:10013–10019.
12. Pereira de Abreu D A, Paseiro Losada P, Angulo I and Cruz J M (2007) Development of new polyolefin films with nanoclays for application in food packaging, *Eur Polym J* **43**:2229–2243.
13. Lee E C, Mielewski D F and Baird R J (2004) Exfoliation and dispersion enhancement in polypropylene nanocomposites by *in-situ* melt phase ultrasonication, *Polym Eng Sci* **44**: 1773–1782.

14. Naderi G, Pierre G. Lafleur P G and Dubois C (2007) Microstructure-properties correlations in dynamically vulcanized nanocomposite thermoplastic elastomers based on PP/EPDM, *Polym Eng Sci* **47**:207–217.
15. Shi D, Wei Y, Li R K Y, Ke Z and Yin J (2007) An investigation on the dispersion of montmorillonite (MMT) primary particles in PP matrix, *Eur Polym J* **43**:3250–3257.
16. Yuan Q and Misra R D K (2006) Impact fracture behavior of clay-reinforced polypropylene nanocomposites, *Polymer* **47**:4421–4433.
17. Maiti P, Nam P H, Okamoto M and Kotaka T (2002) The effect of crystallization on the structure and morphology of polypropylene/clay nanocomposites, *Polym Eng Sci* **42**:1864–1871.
18. Yuan Q, Awate S and Misra R D K (2006) Nonisothermal crystallization behavior of polypropylene-clay nanocomposites, *Eur Polym J* **42**:1994–2003.
19. Kanny K and Moodley V K (2007) Characterisation of polypropylene nanocomposite structures, *J Eng Mater Technol* **129**:105–112.
20. Ward I M and Sweeney J (2004) in *An introduction to the mechanical properties of solid polymers (2nd Ed.)* Wiley, Chichester.
21. Stretz H A, Paul D R, Li R, Keskkula H and Cassidy P E (2005) Intercalation and exfoliation relationships in melt-processed poly(styrene-co-acrylonitrile)/montmorillonite nanocomposites, *Polymer* **46**:2621–2637.
22. Vlasveld D P N, Groenewold J, Bersee H E N, Mendes E and Picken S J (2005) Analysis of the modulus of polyamide-6 silicate nanocomposites using moisture controlled variation of the matrix properties, *Polymer* **46**:6102–6113.
23. Osman M A, Rupp J E P and Suter U W (2005) Tensile properties of polyethylene-layered silicate nanocomposites, *Polymer* **46**:1653–1660.
24. Zhong Y, Janes D, Zheng Y, Hetzer M and De Kee D (2007) Mechanical and oxygen barrier properties of organoclay-polyethylene nanocomposite films, *Polym Eng Sci* **47**:1101–1107.
25. Rao Y Q (2007) Gelatin-clay nanocomposites of improved properties, *Polymer* **48**:5369–5375.
26. Kalaitzidou K, Fukushima H, Miyagawa H and Drzal L T (2007) Flexural and tensile moduli of polypropylene nanocomposites and comparison of experimental data to Halpin-Tsai and Tandon-Wang models, *Polym Eng Sci* **47**:1796–1803.
27. Fornes T D and Paul D R (2003) Modeling properties of nylon 6/clay nanocomposites using composite theories, *Polymer* **44**:4993–5013.
28. Hbaieb K, Wang Q X, Chia Y H J and Cotterell B (2007) Modeling stiffness of polymer/clay nanocomposites, *Polymer* **48**:901–909.
29. Heinrich G, Klüppel M and Vilgis T (2007) in *Physical Properties of Polymers Handbook (2nd Ed.)* (Ed. Mark J E) Springer, Ch. 36, Reinforcement theories, pp. 599–610.
30. Heinrich G and Klüppel M (2002) Recent advances in the theory of filler networking in elastomers, *Adv Polym Sci* **160**:1–44.
31. Meier J G, Mani J W and Klüppel M (2007) Analysis of carbon black networking in elastomers by dielectric spectroscopy, *Phys Rev B* **75**:054202-1-054202-10.
32. Shen L, Lin Y, Du Q, Zhong W and Yang Y (2005) Preparation and rheology of polyamide-6/attapulgite nanocomposites and studies on their percolated structure, *Polymer* **46**:5758–5766.
33. Wang K, Liang S, Deng J, Yang H, Zhang Q, Fu Q, Dong X, Wang D and Han C C (2006) The role of the clay network on molecular chain mobility and relaxation in polypropylene/organoclay nanocomposites, *Polymer* **47**:7131–7144.

34. Zeng Q H, Yu A B and Lu G Q (2008) Multiscale modeling and simulation of polymer nanocomposites, *Prog Polym Sci* **33**:191–269.
35. Hill R (1963) Elastic properties of reinforced solids: some theoretical principles, *J Mech Phys Solids* **11**:357–372.
36. Gitman I M, Askes H and Sluys L J (2007) Representative volume: Existence and size determination, *Eng Fract Mech* **74**:2518–2534.
37. Gusev A A (1997) Representative volume element size for elastic composites: A numerical study, *J Mech Phys Solids* **45**:1449–1459.
38. Gusev A A (2001) Numerical Identification of the Potential of Whisker and Platelet Filled Polymers, *Macromolecules* **34**:3081–3093.
39. Sheng N, Boyce M C, Parks D M, Rutledge G C, Abes J J and Cohen R E (2004) Multiscale Micromechanical Modeling of Polymer/Clay Nanocomposites and the Effective Clay Particle, *Polymer* **45**:487–506.
40. Press W H, Flannery B P, Teukolsky S A and Vetterling W T (1986) in *Numerical recipes, the art of scientific computing*, Cambridge University Press, London, Ch. 7, pp. 212–253.
41. Shewchuk J R (2005) Triangle, A Two-Dimensional Quality Mesh Generator and Delaunay Triangulator, version 1.6 http://www.cs.cmu.edu/~quake/triangle.html.
42. Ruppert J (1995) A Delaunay Refinement Algorithm for Quality 2-Dimensional Mesh Generation, *J Algorithms* **18**:548–585.
43. Chen B and Evans J R G (2006) Elastic moduli of clay platelets, *Scripta Mater* **54**:1581–1585.
44. Cox H L (1952) The Elasticity and Strength of Paper and Other Fibrous Materials, *Brit J Appl Phys* **3**:72–79.

List of Acknowledgements

The authors gratefully acknowledge permissions to reproduce copyrighted materials from a number of sources. Every effort has been made to trace copyright ownership and to give accurate and complete credit to copyright owners, but if, inadvertently, any mistake or omission has occurred, full apologies are herewith tendered

Chapter 1. Nano- and Micromechanics of Crystalline Polymers

1. Figure 1.1 is from *Macromolecules*, Vol. 21, 1988, Authors: Galeski A, Argon A S and Cohen R E , Title: Changes in the Morphology of Bulk Spherulitic Nylon 6 Due to Plastic Deformation, pp. 2761–2770, Copyright 1988, with permission from The American Chemical Society.
2. Figures 1.2–1.3 are from *Polymer*, Vol. 48, 2007, Author: Pawlak A, Title: Cavitation During Tensile Deformation of High Density Polyethylene, pp. 1397–1409, Copyright (2007), with permission from Elsevier Ltd., Oxford, UK.
3. Table 1.1 and Figure 1.21 are from *Macromolecules*, Vol. 38, 2005, Authors: Pawlak A and Galeski A, Title: Plastic Deformation of Crystalline Polymers: The Role of Cavitation and Crystal Plasticity, pp. 9688–9697, Copyright 2005, with permission from The American Chemical Society.
4. Figure 1.6 is from *Macromolecules*, Vol. 25, 1992, Authors: Galeski A, Bartczak Z, Argon A S and Cohen R E, Title: Morphological alterations during texture-producing plastic plane strain compression of high-density polyethylene, pp. 5705–5718, Copyright 1992, with permission from The American Chemical Society.
5. Figures 1.7–1.10 are from *Polymer*, Vol. 46, 2005, Authors: Kazmierczak T, Galeski A and Argon A S, Title: Plastic deformation of polyethylene crystals as a function of crystal thickness and compression rate, pp. 8926–8936, Copyright (2005), with permission from Elsevier Ltd., Oxford, UK.
6. Figures 1.11–1.13 are from Polymer, Vol. 46, 2005, Authors: Argon A S, Galeski A and Kazmierczak T, Title: Rate mechanisms of plasticity in semi-crystalline polyethylene, pp. 11798–11805, Copyright (2005), with permission from Elsevier Ltd., Oxford, UK.

7. Figures 1.16 and 1.20 are from *Acta Materialia*, Vol. 54, 2006, Authors: Bedoui F, Diani J, Regnier G and Seiler W, Title: Micromechanical modeling of isotropic elastic behaviour of semicrystalline polymers, pp. 1513–1523, Copyright (2006), with permission from Elsevier Ltd., Oxford, UK.

8. Figures 1.17–1.19 reprinted from *Polymer*, Vol. 45, 2004, Authors: Bedoui F, Diani J and Regnier G, Title: Micromechanical modeling of elastic properties in polyolefins, pp. 2433–2442, Copyright (2004), with permission from Elsevier Ltd., Oxford, UK.

Chapter 2. Modeling Mechanical Properties of Segmented Polyurethanes

1. Figures 2.3–2.12 and Table 2.1 are from *Journal of Polymer Science: Part B: Polymer Physics*, Vol. 45, 2007, Authors: Ginzburg V V, Bicerano J, Christenson C P, Schrock A K and Patashinski A Z, Title: Theoretical Modeling of the Relationship Between Young's Modulus and Formulation Variables for Segmented Polyurethanes, pp. 2123–2135, Copyright (2007), with permission from Wiley-VCH Verlag, Weinheim, Germany.

2. Figure 2.13 is reprinted from *Macromolecular Materials and Engineering*, Vol. 291, 2006, Authors: Laity P R, Taylor J E, Wong S S, Khunkamchoo P, Cable M, Andrews G T, Johnson A F and Cameron R E, Title: Morphological Behaviour of Thermoplastic Polyurethanes During Repeated Deformation, pp. 301–324, Copyright (2006), with permission from Wiley-VCH Verlag, Weinheim, Germany.

3. Figure 2.14 is reprinted from *Mechanics of Materials,* Vol. 37, 2005, Authors: Qi H and Boyce M C, Title: Stress-strain behavior of thermoplastic polyurethanes, pp. 817–839, Copyright (2005), with permission from Elsevier Ltd., Oxford, UK.

4. Figures 2.15 and 2.17 are reprinted from *Polymer,* Vol. 46, 2005, Authors: Christenson E M, Anderson J M, Hiltner A and Baer E, Title: Relationship between nanoscale deformation processes and elastic behavior of polyurethane elastomers, pp. 11744–11754, Copyright (2005), with permission from Elsevier Ltd., Oxford, UK.

Chapter 3. Nanoparticles/Polymer Composites: Fabrication and Mechanical Properties

1. Table 5 is reprinted from *Advanced Materials*, Vol. 19, 2007, Authors: Zhou T H, Ruan W H, Rong M Z, Zhang M Q and Mai Y L, Title: Keys to toughening of non-layered nanoparticles/polymer composites, pp. 2667–2671, Copyright (2007), with permission from Wiley-VCH Verlag GmbH & Co. KGaA.

2. Figures 3.2–3.4 and Table 1 are reprinted from *Polymer*, Vol. 42, 2001, Authors: Rong M Z, Zhang M Q, Zheng Y X, Zeng H M, Walter R and Friedrich K, Title: Structure-property relationships of irradiation grafted nano-inorganic particle filled polypropylene

composites, pp. 167–183, Copyright (2001), with permission from Elsevier.

3. Figures 3.5 and 3.8 are reprinted from *Journal of Applied Polymer Science*, Vol. 92, 2004, Authors: Rong M Z, Zhang M Q, Pan S L and Friedrich K, Title: Interfacial effects in polypropylene-silica nanocomposites, pp. 1771–1781, Copyright (2004), with permission from John Wiley & Sons, Inc.

4. Figures 3.9 and 3.10 are reprinted from *Journal of Materials Science*, Vol. 39, 2004, Authors: Ruan W H, Zhang M Q, Rong M Z and Friedrich K, Title: Polypropylene composites filled with *in-situ* grafting polymerization modified nano-silica particles, pp. 3475–3478, Copyright (2004), with permission from Springer Science and Business Media.

5. Figures 3.11 and 3.12 are reprinted from *Composites Science and Technology*, Vol. 67, 2007, Authors: Ruan W H, Mai Y L, Wang X H, Rong M Z and Zhang M Q, Title: Effects of processing conditions on properties of nano-SiO_2/polypropylene composites fabricated by pre-drawing technique, pp. 2747–2756, Copyright (2007), with permission from Elsevier.

6. Figures 3.13–3.15 and Table 2 are reprinted from *Macromolecular Rapid Communications*, Vol. 27, 2006, Authors: Ruan W H, Huang X B, Wang X H, Rong M Z and Zhang M Q, Title: Effect of drawing induced dispersion of nano-silica on performance improvement of polypropylene based nanocomposites, pp. 581–585, Copyright (2006), with permission from Wiley-VCH Verlag GmbH & Co. KGaA.

7. Figures 3.16, 3.18 and Table 3 are reprinted from *Polymer*, Vol. 47, 2006, Authors: Cai L F, Huang X B, Rong M Z, Ruan W H and Zhang M Q, Title: Effect of grafted polymeric foaming agent on the structure and properties of nano-silica/polypropylene composites, pp. 7043–7050, Copyright (2006), with permission from Elsevier.

8. Figures 3.20–3.23 are reprinted from *Polymer Engineering and Science*, Vol. 47, 2007, Authors: Zhou H J, Rong M Z, Zhang M Q, Ruan W H and Friedrich K, Title: Role of reactive compatibilization in preparation of nanosilica/polypropylene composites, pp. 499–509, Copyright (2007), with permission from John Wiley & Sons, Inc.

9. Figures 3.24 and 3.26–3.28 are reprinted from *Composites Science and Technology*, Vol. 47, 2007, Authors: Zhou T H, Ruan W H, Yang J L, Rong M Z, Zhang M Q and Zhang Z, Title: A novel route for improving creep resistance of polymers using nanoparticles, pp. 2297–2302, Copyright (2007), with permission from Elsevier.

10. Figures 3.29–3.31 and Table 5 and are reprinted from *Advanced Materials*, Vol. 19, 2007, Authors: Zhou T H, Ruan W H, Rong M Z, Zhang M Q and Mai Y L, Title: Keys to toughening of non-layered nanoparticles/polymer composites, pp. 2667–2671, Copyright (2007), with permission from Wiley-VCH Verlag GmbH & Co. KGaA.

Chapter 5. Organoclay, Particulate and Nanofibril Reinforced Polymer-Polymer Composites: Manufacturing, Modeling and Applications

1. Figure 5.1 reprinted from *Polymer Composites From Nano- to Macroscale* (Eds. Friedrich K, Fakirov S and Zhang Z), Authors: Privalko V P, Shantalii T A and Privalko E G, Title: Polyimides reinforced by a sol-gel derived organosilicon nanophase: synthesis and structure-property relationships, pp 63–76, Copyright (2008) from Springer.

2. Figures 5.2, 5.3, and 5.4 reprinted from *Composites Part A*, Vol. 39, 2008, Authors: Dong Y and Bhattacharyya D, Title: Effects of clay type, clay/compatibiliser content and matrix viscosity on the mechaical properties of polypropylene/organoclay nanocomposites, pp. 1177–1191, Copyright (2008) with permission from Elsevier.

3. Figures 5.4, 5.5, 5.6, and 5.7 reprinted from *Mechanics of Advanced Materials and Structures* (in press), Authors: Dong Y, Bhattacharyya D, and Hunter P, Title: Morphological-Image Analysis Based Numerical Modelling of Prganoclay Filled Nanocomposites, Copyright (2008) from Taylor & Francis.

4. The text on pages 185 and 186 (104 words) is reprinted from *Angewandte Chemie – International Edition*, Vol. 46, 2007, Authors: Greiner A and Wendorff J, Title: Electrospinning: A fascinating method for preparation of ultrathin fibres, pp 5670–5703, Copyright (2008) from Wiley-VCH Verlag GmbH & Co.

5. The text on page 196 (197 words) is reprinted from *Macromolecules*, Vol. 40, 2007, Authors: Schaefer D W and Justice R S, Title: How Nano Are Nanocomposites, pp 8501–8517, Copyright (2008) from American Chemical Society.

Chapter 9. Creep and Fatigue Behaviour of Polymer Nanocomposites

1. Figure 9.7 adapted from *Macromolecular Rapid Communications*, Vol. 27, 2006, Authors: Siengchin S and Karger-Kocsis J, Title: Creep behavior of polystyrene/fluorohectorite micro- and nanocomposites, pp. 2090–2094, Copyright (2008) with permission from Wiley-VCH Verlag GmbH & Co.

2. Figure 9.8 adapted from *Journal of Applied Polymer Science*, Vol. 93, 2004, Authors: Hasegawa N, Okamoto H and Usuki A, Title: Preparation and properties of ethylene propylene rubber (EPR)–clay nanocomposites based on maleic anhydride-modified EPR and organophilic clay, pp. 758–764, Copyright (2008) with permission from Wiley Periodicals, Inc.

3. Figure 9.9 data taken from *Macromolecular Materials and Engineering*, Vol. 291, 2006, Authors: Shen L, Wang L, Liu T and He C, Title: Nanoindentation and morphological studies of epoxy nanocomposites, pp. 1358–1366.

4. Figure 9.10 adapted from *Macromolecular Rapid Communications*, Vol. 28, 2007, Authors: Ganß M, Satapathy B K, Thunga M, Weidisch

R, Potschke P and Janke A, Title: Temperature dependence of creep behavior of PP–MWNT nanocomposites, pp. 1624–1633, Copyright (2008) with permission from Wiley-VCH Verlag GmbH & Co.

5. Figure 9.11 adapted from *Nanotechnology*, Vol. 18, 2007, Authors: Zhang W, Joshi A, Wang Z, Kane RS and Koratkar N, Title: Creep mitigation in composites using carbon nanotube additives, 185703 (5pp), Copyright (2008) with permission from IOP Publishing Ltd.

6. Figure 9.12 adapted from *Polymer*, Vol. 45, 2004, Authors: Zhang Z, Yang J L and Friedrich K, Title: Creep resistant polymeric nanocomposites, pp. 3481–3485, Copyright (2008) with permission from Elsevier Ltd.

7. Figure 9.13 adapted from *Composites Science and Technology*, Vol. 67, 2007, Authors: Zhou T H, Ruan W H, Yang J L, Rong M Z, Zhang M Q and Zhang Z, Title: A novel route for improving creep resistance of polymers using nanoparticles, pp. 2297–2302, Copyright (2008) with permission from Elsevier Ltd.

8. Figure 9.14 adapted from *Composite Interfaces*, Vol. 6, 1999, Authors: Yamashita A, Takahara A and Kajiyama T, Title: Aggregation structure and fatigue characteristics of (nylon 6/clay) hybrid, pp. 247–258, Copyright (2008) with permission from BRILL-VSP.

9. Figure 9.15 adapted from *Materials Science and Engineering A Structural Materials Properties Microstructure and Processing*, Vol. 402, 2005, Authors: Zhou Y X, Rangari V, Mahfuz H, Jeelani S and Mallick P K, Title: Experimental study on thermal and mechanical behavior of polypropylene, talc/polypropylene and polypropylene/clay nanocomposites, pp. 109–117, Copyright (2008) with permission from Elsevier B. V.

10. Figure 9.16 data taken from *Polymer Composites*, Vol. 25, 2004, Authors: Bellemare S C, Bureau M N, Denault J and Dickson J I, Title: Fatigue crack initiation and propagation in polyamide-6 and in polyamide-6 nanocomposites, pp. 433–441.

11. Figure 9.17 data taken from from *Journal of Applied Polymer Science*, Vol. 90, 2003, authors: Song M, Hourston D J, Yao K J, Tay J K H and Ansarifar M A, Title: High performance nanocomposites of polyurethane elastomer and organically modified layered silicate, pp. 3239–3243.

12. Figure 9.18 reprinted from *Polymer International*, Vol. 54, 2005, Authors: Song M, Wong C W, Jin J, Ansarifar A, Zhang Z Y and Richardson M, Title: Preparation and characterization of poly(styreneco-butadiene) and polybutadiene rubber/clay nanocomposites, pp. 560–568, Copyright (2008) with permission from Wiley Periodicals, Inc.

13. Figure 9.19 adapted from *Advanced Composites Letters*, Vol. 12, 2003, authors: Ren Y, Lil F, Cheng H M and Liao K, Title: Fatigue behaviour of unidirectional single-walled carbon nanotube reinforced epoxy composite under tensile load, pp. 19–24, Copyright (2008) with permission from Adcotec Ltd.

14. Figure 9.20 adapted from *Journal of Materials Science*, Vol. 42, 2007, Authors: Zhou YX, Pervin F and Jeelani S, Title: Effect vapor grown carbon nanofiber on thermal and mechanical properties of epoxy, pp. 7544–7553, Copyright (2008) with permission from Springer Science+Business Media LLC.

15. Figure 9.21 data taken from *Journal of Materials Science*, Vol. 42, 2007, Authors: Zhou Y X, Pervin F and Jeelani S, Title: Effect vapor grown carbon nanofiber on thermal and mechanical properties of epoxy, pp. 7544–7553, Copyright (2008) with permission from Springer Science+Business Media LLC.

16. Figure 9.22 data taken from *Journal of Biomedical Materials Research Part A*, Vol. 77A, 2006, Authors: Marrs B, Andrews R, Rantell T and Pienkowski D, Title: Augmentation of acrylic bone cement with multi-wall carbon nanotubes, pp. 269–276, Copyright (2008) with permission from Wiley Periodicals, Inc.

17. Figure 9.23 data taken from *Carbon*, Vol. 45, 2007, Authors: Marrs B, Andrews R and Pienkowski D, Title: Multiwall carbon nanotubes enhance the fatigue performance of physiologically maintained methyl methacrylate-styrene copolymer, pp. 2098–2104, Copyright (2008) with permission from Elsevier Ltd.

18. Figure 9.24 adapted from *Engineering Fracture Mechanics*, Vol. 73, 2006, Authors: Wetzel B, Rosso P, Haupert F and Friedrich K, Title: Epoxy nanocomposites – fracture and toughening mechanisms, pp. 2375–2398, Copyright (2008) with permission from Elsevier Ltd.

19. Figure 9.25 adapted from *Composite Structures*, Vol. 67, 2005, Authors: Chisholm N, Mahfuz H, Rangari V K, Ashfaq A and Jeelani S, Title: Fabrication and mechanical characterization of carbon/SiC-epoxy nanocomposites, pp. 115–124, Copyright (2008) with permission from Elsevier Ltd.

Chapter 10. Deformation Mechanisms of Functionalized Carbon Nanotube Reinforced Polymer Nanocomposites

1. Figure 10.1 reprinted from *Advanced Materials*, Vol. 18, 2006, Authors: Barber B A, Cohen S R, Eitan A, Schadler L S and Wagner H D, Title: Fracture transitions at a carbon nanotube/polymer interface, pp. 83–87, Copyright (2007), with permission from Wiley-VCH.

2. Figure 10.2 reprinted from *Polymer*, Vol. 44, 2003, Authors: Wong M, Paramsothy M, Xu X J, Ren Y, Li S and Liao K, Title: Physical interactions at carbon nanotube-polymer surface, pp 7757–7764, Copyright (2007), with permission from Elsevier.

3. Figures 10.3 and 10.4 reprinted from *Composite Structures*, Vol. 76, 2006, Authors: Lee W J, Lee S F and Kim C G, Title: The mechanical properties of MWNT/PMMA nanocomposites prepared by modified injection molding, pp. 406–410, Copyright (2007), with permission from Elsevier.

4. Figure 10.5 reprinted from *Journal of Polymer Science: Part B: Polymer Physics*, Vol. 42, 2004, Authors: Gorga R E and Cohen R E, Title: Toughness enhancements in poly(methyl methacrylate) by addition of oriented multiwall carbon nanotubes, pp. 2690–2702 Copyright (2007), with permission from John Wiley & Sons, Inc.

5. Figures 10.6 and 10.7 reprinted from *Chemical Materials*, Vol. 15, 2003, Authors: Velasco-Santos C, Martinez-Hernandez A L, Fisher F T, Ruoff R and Castano V M, Title: Improvement of thermal and mechanical properties of carbon nanotube composites through chemical functionalization, pp. 4470–4475, Copyright (2007), with permission from The American Chemical Society.

6. Figures 10.8, 10.9 and 10.10 reprinted from *Advanced Functional Materials*, Vol. 16, 2006, Authors: Blond D, Barron V, Ruether M, Ryan K P, Nicolosi V, Blau W J and Coleman J N, Title:. Enhancement of modulus, strength, and toughness in poly(methyl methacrylate)-based composites by the incorporation of poly(methyl methacrylate)functionalized nanotubes, pp. 1608–1614, Copyright (2007), with permission from Wiley-VCH.

7. Figure 10.11 reprinted from *Advanced Functional Materials*, Vol. 13, 2003, Authors: Singh S, Pei Y, Miller R and Sundararajan P, Title: Long-range, entangled carbon nanotubes networks in polycarbonate, pp. 868–871, Copyright (2007), with permission from Wiley-VCH.

8. Figure 10.12 reprinted from *Nano Letters*, Vol. 3, 2003, Authors: Ding W, Eitan A, Fisher F T, Chen X, Dikin D A, Andrews R, Brinson L C, Schadler L S and Ruoff R S, Direct observation of polymer sheathing in carbon nanotube-polycarbonate composites, pp. 1593–1597, Copyright (2007), with permission from The American Chemical Society.

9. Figure 10.13 and Table 10.1 reprinted from *Composites Science and Technology*, Vol. 66, 2006, Authors: Eitan A, Fischer F T, Andrews R, Brinson L C and Schadler L S, Title: Reinforcement mechanisms in MWNT-filled polycarbonate, pp 1162–1173, Copyright (2007), with permission from Elsevier.

10. Figure 10.14 and Scheme 10.2 reprinted from *Chemical Materials*, Vol. 18, 2006, Authors: Schoner M L, Khabashesku V N and Barrera E V, Title: Processing and mechanical properties of fluorinated single-walled carbon nanotube-polyethylene composites, pp. 906–913, Copyright (2007), with permission from The American Chemical Society.

11. Figure 10.15 reprinted from *Chemical Materials*, Vol. 18, 2006, Authors: Mcintosh D, Khabashesku VN and Barrera EV, Title: Nanocomposite fiber systems processed from fluorinated single-walled carbon nanotube and a polypropylene matrix, pp. 4561–4569, Copyright (2007), with permission from The American Chemical Society.

12. Figure 10.16 and Scheme 10.3 reprinted from *Journal of American Chemical Society*, Vol. 126, 2004, Authors: Blake R, Gun'ko Y K, Coleman J, Cadek M, Fonseca A, Nagy J B and Blau W J, Title: A generic organometallic approach toward ultra-strong carbon nanotube polymer composites, pp. 10226–10227, Copyright (2007), with permission from The American Chemical Society.
13. Figures 10.17. and 10.18 reprinted from *Macromolecules*, Vol. 37, 2004, Authors: Liu T X, Phang L Y, Shen L, Chow S Y and Zhang W D, Title: Morphology and mechanical properties of multiwalled carbon nanotubes reinforced nylon-6 composites, pp. 7214–7222, Copyright (2007), with permission from The American Chemical Society.
14. Figure 10.19, Schemes 10.4 and 10.5 reprinted from *Nano Letters*, Vol. 3, 2003, Authors: Zhu J, Kim J D, Peng H, Margrave J L, Khabashesku V N and Barrera E V, Title: Improving the dispersion and integration of single-walled carbon nanotubes in epoxy composites through functionalization, pp. 1107–1113, Copyright (2007), with permission from The American Chemical Society.
15. Figures 10.20, 10.21 and Scheme 10.6 reprinted from *Chemical Materials*, Vol. 19, 2007, Authors: Tseng C H, Wang C C and Chen, C Y, Title: Functionalizing carbon nanotubes by plasma modification for the preparation of covalent-integrated epoxy composites, pp. 308–315, Copyright (2007), with permission from The American Chemical Society.
16. Figure 10.22 and Table 10.2 reprinted from *Carbon*, Vol. 44, 2006, Authors: Kim J A, Song D G, Kang T J and Young J R, Title: Effects of surface modification on rheological and mechanical properties of CNT/epoxy composites, pp. 1898–1905, Copyright (2007), with permission from Elsevier.
17. Figure 10.23 reprinted from *Chemical Physics Letters*, Vol. 370, 2003, Authors: Gojny F H, Nastalczyk J, Roslaniec Z and Schulte K, Title: Surface modified multi-walled carbon nanotubes in CNT/epoxy composites, pp. 820–824, Copyright (2007), with permission from Elsevier.
18. Figure 10.24 reprinted from *Composites Science and Technology*, Vol. 65, 2005, Authors: Gojny H, Wichmann M H, Fiedler B and Schulte K, Title: Influence of different carbon nanotubes on the mechanical properties of epoxy matrix composites – A comparative study, pp. 2300–2313, Copyright (2007), with permission from Elsevier.
19. Figure 10.25 reprinted from *Journal of Polymer Science: Part B: Polymer Physics*, Vol. 45, 2007, Authors: Buffa F, Abraham G A, Grady B P and Resasco D, Title: Effect of nanotube functionalization on the properties of single-walled carbon nanotube/polyurethane composites, pp. 490–501, Copyright (2007), with permission from John Wiley & Sons, Inc.
20. Scheme 10.1 reprinted from *Chemical Materials*, Vol. 15, 2006, Authors: Eitan A, Jiang K, Dukes D, Andrews R and Schadler L S, Title:

Surface modification of multiwalled carbon nanotubes: Toward tailoring of the interface in polymer composites, pp. 198–3201, Copyright (2007), with permission from The American Chemical Society.

Chapter 11: Fracture Properties and Mechanisms of Polyamide/Clay Nanocomposites

1. Figure 11.1 reprinted from *Progress in Polymer Science*, Vol. 28, 2003, Authors: Ray S S and Okamoto M, Title: Polymer/layered silicate nanocomposites: a review from preparation to processing, pp. 1539–1641, Copyright (2003), with permission from Elsevier.

2. Figure 11.3 reprinted from *Macromolecules*, Vol. 37, 2004, Authors: Fornes T D, Hunter D L and Paul D R, Title: Nylon-6 nanocomposites from alkylammonium-modified clay: the role of alkyl tails on exfoliation, pp. 1793–1798, Copyright (2004), with permission from American Chemical Society.

3. Figure 11.6 reprinted from *Advanced Materials*, Vol. 16, 2004, Authors: Shah D, Maiti P, Gunn E, Schmidt D F, Jiang D D, Batt C A and Giannelis E R, Title: Dramatic enhancements in toughness of polyvinylidene fluoride nanocomposites via nanoclay-directed crystal structure and morphology, pp. 1173–1177, Copyright (2004), with permission from Wiley-VCH Verlag GmbH & Co. KGaA.

4. Figure 11.9 reprinted from *Journal of Applied Polymer Science*, Vol. 105, 2007, Authors: Kim G M, Goerlitz S and Michler G H, Title: Deformation mechanism of nylon 6/layered silicate nanocomposites: role of the layered silicate, pp. 38–48, Copyright (2007), with permission from John Wiley & Sons, Inc.

5. Figure 11.10 reprinted from *Advanced Materials*, Vol. 17, 2005, Authors: Shah D, Maiti P, Jiang D D, Batt C A and Giannelis E P, Title: Effect of nanoparticle mobility on toughness of polymer nanocomposites, pp. 525–528, Copyright (2005), with permission from Wiley-VCH Verlag GmbH & Co. KGaA.

6. Figure 11.11 reprinted from *Advanced Materials*, Vol. 19, 2007, Authors: Zhou T H, Ruan W H, Rong M Z, Zhang M Q and Mai Y L, Title: Keys to toughening of non-layered nanoparticles/polymer composites, pp. 2667–2671, Copyright (2007), with permission from Wiley-VCH Verlag GmbH & Co. KGaA.

7. Figure 11.12 reprinted from *Nanotechnology*, Vol. 19, 2008, Authors: Dasari A, Yu Z-Z, Mai Y-W and Kim J-K, Title: Orientation and the extent of exfoliation of clay on scratch damage in polyamide 6 nanocomposites, pp. 055708-055721, Copyright (2008), with permission from Institute of Physics.

8. Figure 11.14 reprinted from *Materials Manufacturing Processes*, Vol. 21, 2006, Authors: Chen L, Phang I Y, Wong S C, Lv P F and Liu T X, Title: Embrittlement mechanisms of nylon 66/organoclay

nanocomposites prepared by melt-compounding process, pp. 153–158, Copyright (2006), with permission from Taylor & Francis Group, LLC.

9. Figures 11.15 and 11.24a reprinted from *Composites Science and Technology*, Vol. 67, 2007, Authors: Lim S-H, Dasari A, Yu Z-Z, Mai Y-W, Liu S L and Yong M S, Title: Fracture toughness of nylon 6/organoclay/elastomer nanocomposites, pp. 2914–2923, Copyright (2007), with permission from Elsevier.

10. Figures 11.16b, 11.16c reprinted from *Macromolecules*, Vol. 41, 2008, Authors: He C, Liu T, Tjiu W C, Sue H J and Yee A F, Title: Microdeformation and fracture mechanisms in polyamide-6/organoclay nanocomposites, pp. 193–202, Copyright (2008), with permission from American Chemical Society.

11. Figure 11.17 reprinted from *Macromolecules*, Vol. 40, 2007, Authors: Dasari A, Yu Z-Z and Mai Y-W, Title: Transcrystalline regions in the vicinity of nanofillers in polyamide-6, pp. 123–130, Copyright (2007), with permission from American Chemical Society.

12. Figures 11.18b and 11.18c reprinted from *Polymer*, Vol. 48, 2007, Authors: Wang K, Wang C, Li J, Su J X, Zhang Q, Du R N and Fu Q, Title: Effects of clay on phase morphology and mechanical properties in polyamide 6/EPDM-g-MA/organoclay ternary nanocomposites, pp. 2144–2154, Copyright (2007), with permission from Elsevier.

13. Figure 11.19 reprinted from *Polymer International*, Vol. 56, 2007, Authors: Dong W F, Zhang X H, Liu Y Q, Gui H, Wang Q G, Gao J M, Song Z H, Lai J M, Huang F and Qiao J L, Title: Process for preparing a nylon-6/clay/acrylate rubber nanocomposite with high toughness and stiffness, pp. 870–874, Copyright (2007), with permission from John Wiley & Sons Limited.

14. Figures 11.21a and 11.21c reprinted from *Journal of Materials Science*, Vol. 37, 2002, Authors: Li Y M, Wei G X and Sue H J, Title: Morphology and toughening mechanisms in clay-modified styrene-butadiene-styrene rubber-toughened polypropylene, pp. 2447–2459, Copyright (2002), with permission from Kluwer Academic Publishers.

15. Figures 11.21b and 11.21d reprinted from *Acta Materialis*, Vol. 52, 2004, Authors: Sue H J, Gam K T, Bestaoui N, Clearfield A, Miyamoto M and Miyatake N, Title: Fracture behavior of alpha-zirconium phosphate-based epoxy nanocomposites, pp. 2239–2250, Copyright (2004), with permission from Elsevier.

16. Figure 11.23 reprinted from *Journal of Nanoscience and Nanotechnology*, Vol. 8, 2008, Authors: Dasari A, Yu Z-Z, Mai Y-W and Yang M, Title: The location and extent of exfoliation of clay on the fracture mechanisms in nylon 66-based ternary nanocomposites, pp. 1901–1912, Copyright (2008), with permission from American Scientific Publishers.

Chapter 12. On the Toughness of "Nanomodified" Polymers and Their Traditional Polymer Composites

1. Figure 12.3 reproduced from *Journal of Applied Polymer Science*, Vol. 106, 2007, Authors: Siengchin S, Karger-Kocsis J, Apostolov A A and Thomann R, title: Polystyrene-fluorohectorite nanocomposites prepared by melt mixing with and without latex precompounding: Structure and mechanical properties, pp. 248–254, Copyright (2008) with permission from Wiley Periodicals, Inc.

2. Table 12.1 data were taken from *Journal of Materials Science*, Vol. 43, 2008, Authors: Kinloch A J, Masania K, Taylor A C, Sprenger S and Egan D, Title: The fracture of glass-fibre-reinforced epoxy composites using nanoparticle-modified matrices, pp. 1151–1154, Copyright (2008) with permission from Springer Science+Business Media.

Chapter 13. Micromechnics of Polymer Blends: Microhardness of Polymer Systems Containing a Soft Component and/or Phase

1. Figures 13.1 and 13.2 reprinted from *Microhardness of Polymers*, 2007, Authors: Balta Calleja F J and Fakirov S, Copyright (2007), with permission from Cambridge University Press.

2. Figure 13.3 reprinted from *Journal of Material Science*, Vol. 42, 2007, Author: Fakirov S, Title: On the application of the "rule of mixture" to microhardness of complex polymer systems containing a soft component and/or phase, pp. 1131–1148, Copyright (2007), with permission from Springer-Verlag GmbH.

3. Figure 13.4 reprinted from *Journal of Polymer Engineering*, Vol. 23, 2003, Authors: Adhikari R, Michler G H, Cagiao M E and Balta Calleja F J, Title: Micromechanical studies of styrene/butadiene block copolymer blends, pp. 177–190, Copyright (2007), with permission from Freund Publishing House Ltd.

4. Figure 13.5 reprinted from *Journal of Material Science Letters*, Vol. 22, 2003, Authors: Boyanova M, Balta Calleja F J and Fakirov S, Title: Study on the phase boundary in PS/SBS blends as revealed by microhardness analysis, pp. 1741–1743, Copyright (2007), with permission from Springer-Verlag GmbH.

5. Figure 13.6 reprinted from *e-Polymer*, 2003, Authors: Boyanova M, Mina M F, Balta Calleja F J and Fakirov S, Title: Effect of the SBS compatibiliser on the interphase boundary of polymer blends of polystyrene and natural rubber, No. 047, Copyright (2007), with permission from e-Polymer.

6. Figures 13.7, 13.8, 13. 9, and 13.10 reprinted from *Advances of Polymer Technology*, Vol. 24, 2005, Authors: Boyanova M, Balta Calleja F J, Fakirov S, Kuehnert I and Mennig G, Title: Influence of processing conditions on the weld line in doubly injection-molded glassy polycarbonate and polystyrene: Microindentation hardness study, pp. 14–20, Copyright (2007), with permission from Wiley InterScience.

Chapter 14. Some Monte Carlo Simulations on Nanoparticle Reinforcement of Elastomers

1. Figures 14.3 and 14.4 reprinted from *Journal of Polymer Science: Part B: Polymer Physics*, Vol. 34, 1996, Authors: Yuan Q W, Kloczkowski A, Mark J E and Sharaf M A. Title: Simulations on the Reinforcement of Poly(dimethylsiloxane) Elastomers by Randomly Distributed Filler Particles, pp. 1647–1657, Copyright (1996), reprinted with permission of John Wiley & Sons, Inc.

2. Figures 14.5 and 14.6 reprinted from *Polymer* Vol. 46, 2005, Authors: Mark J E, Abou-Hussein R, Sen T Z and Kloczkowski A. Title: Some Simulations on Filler Reinforcement in Elastomers, pp. 8894–8904, Copyright (2005), reprinted with permission from Elsevier Ltd.

Chapter 15. Modeling of Polymer Clay Nanocomposites for a Multiscale Approach

1. Figure 15.1 reproduced with kind permission of prof. Eileen Harkin-Jones, School of Mechanical and Aerospace Engineering, Queen's University, Belfast, UK.

Author Index

Adachi T, 445
Ahzi S, 39
Akbari B, 450
Arai A, 460
Archer L A, 233, 225
Argon A S, 6, 12, 25, 27, 31
Asai S, 219, 220
Ash B J, 429
Avlar S, 437

Bagheri R, 413, 450
Baldi F, 443
Bao S P, 442
Barrera E V, 356
Bartczak Z, 18, 19
Becker O, 388, 450, 458
Bedoui F, 40, 43
Bellare A, 19
Bellemare S C, 323
Benoit H, 67
Bezerédi A, 433
Bicerano J, 66
Billon N, 8
Blackman B R K, 448
Blake R, 357
Bleda T, 42
Bokobza L, 230
Bonart R, 271, 277
Bondioli F, 320
Boo W J, 546
Boyce M C, 78, 79, 553, 562
Brooks N W J, 26, 28, 29, 32
Bucknall C B, 8
Bueche F, 214
Bufa F, 370
Bugnicourt E, 446
Bureau M N, 435
Bousmina M, 196

Cakmak M, 294
Cassagnau P, 231, 234, 235
Chakraborty A K, 219
Chan C M, 117, 433
Chandrasekhar S, 217
Chen L, 435
Chen X, 272
Chiu F C, 402
Chowdhury S C, 362
Choy C L, 41
Christenson C P, 61, 63
Christenson E M, 79
Cohen R E, 6, 12, 347
Coleman J N, 351
Cooper S L, 60
Corobea M-C, 455
Corté L, 397, 411, 431, 440
Cotterell B, 440
McCrackin F L, 217, 218
Crisholm N, 333
Crist B, 26
Critchfield F E, 75

Dahoun A, 39
Dasari A, 443
Davis D D, 226
Debye P, 274, 276
Deshmane C, 442
Diani J, 45
DiBenedetto A T, 222, 224, 225, 235
DiMarzio E A, 217, 218
Doi M, 253
Dommelen, van J A W, 29, 40, 45
Dong W F, 401
Dresselhaus M S, 344
Drozdov A A, 77
Dušek K, 83

Eckel D F, 382
Ediger M D, 226
Edward G, 325
Edwards S F, 253
Eshelby J D, 38
Estes G M, 79

Fan Z, 459
Fantner G E, 252
Ferry J D, 82
Feynman R, 209
Flory P J, 42
Frank F C, 23
Frankland S J, 362
Friedrich K, 6, 317
Fröhlich J, 388, 454

Galeski A, 6, 12, 26
Gam K T, 454
Ganss M, 315
Garrett J T, 60, 84
Gaucher-Miri V, 14
Gaylord R J, 218
Geil P H, 24
deGennes P G, 214, 253
Gersappe D, 129, 390
Giannelis E P, 129, 390
Gier D R, 63
Gilman J W, 382
Gleiter H, 29
Glotzer S C, 217
Goettler L A, 301
Gojny F H, 459
González I, 401, 402
Goodier J, 12
Gorga R E, 347
Goritz D, 150, 231
Granick S, 221, 231
Greiner A, 185
McGrath L M, 444
Gruenbauer H J M, 66
Guild F J, 413
Guiu F, 24
Gusev A A, 553
Guth E, 156

Hadley D W, 15
Hadziioannou G, 67
Halpin J C, 39, 156
Hanim H, 441
Hansma P K, 252
Harik V M, 362
Hasegawa N, 312
Hasegawa R, 104
Hbaieeb K, 547
He C, 394, 437
Henning S, 5
Hill R, 37, 548
Hine P J, 41
Hsiao K-T, 451
Huang Y, 413
Hussain F, 545
Hussein M, 457

Ikehara T, 23
Isik I, 454

Jancar J, 210, 251, 254
Jang B Z, 5, 8
Janzen J, 39
Jia Z, 349
Jin F-L, 445
Joshi M, 311
Justice R S, 196
Juwono A, 325

Kalfus J, 224, 225, 227, 230, 233–235,
    251, 254
Kardos J L, 39
Karger-Kocsis J, 311, 443, 434, 435
Kato M, 383
Kaufman S, 219, 220
Kausch H H, 6
Kazmierczak T, 29, 31
Keblinski P, 218, 231
Kelnar I, 399
Kelnar Š J, 443
Kendall K, 99
Kilian H G, 148
Kim G M, 389, 390
Kim J A, 367

Author index

Kinloch A J, 413, 448, 453, 458
Klüppel M, 547
Kobayashi H, 438
Koberstein J T, 60
Kolarřik J, 69
Koo J H, 459
Krishnamoorti R, 426
Kumar S K, 218, 525

Lach R, 428
Lakrout, H, 81
Lazzeri A, 433
Lee B J, 29, 39, 345
Lee S F, 345
Lee W J, 345
Leibler L, 431, 440
Lequeux F, 226, 225, 227
Lesser A J, 387, 388, 450
Levita G, 433
Levresse P, 233
Lezak E, 20, 21
Li D, 451
Lietz S, 313
Lin E K, 220
Lin Y H, 213 , 253
Lipatov Y S, 235
Liu T, 450
Liu T X, 361
Liu W, 454
Liu X H, 389
Liu Y, 5
Lv R, 16

Ma C C M, 386
Mackenzie J D, 146
Maier P G, 150, 231
Maiti P, 385
Mallick P K, 323
Mark J E, 145, 218
Marrs B, 330
Marur P R, 446
Matějka L, 225
Matsuo M, 41
Mattice W L, 525
Menke H, 276
Miyagawa H, 451

Montes H, 220, 225
Morgan A B, 382
Mori T, 38, 71
Mukhtar M, 29, 32
Muratoglu O K, 12, 396, 397
Musto P, 447
Mülhaupt R, 449
Müllen K, 429

Nabarro F R N, 24
Nair S V, 393
McNally T, 436
Narisawa I, 5
Nikolov S, 40
Nitta K H, 5
Nowicki W, 218, 251

Okabe T, 362
Okada A, 301, 545
Okamoto M, 385
Oshinski A J, 411
O'Sickey M J, 60

Paris P C, 308
Park S-J, 445
Parks D M, 39
Patashinski A, 81
Patel J, 39
Paul D R, 380
Pawlak A, 16
Pearson R A, 413
Pegoretti A, 310, 311
Peierls R, 24
Perena J M, 488
Peterlin A, 11
Peterman J, 29
Peterson J M, 24
Phillips P J, 39
Pilato L A, 459
Pope E J A, 146
Porod G, 271, 274

Qi H, 78,79
Qian D, 345, 362
Qiao Y, 437
Quaresimin M, 458

Ramier J, 150
Ranade A, 310
Raos G, 231
Ratna D, 454
Raviv U, 231, 233
Reichert P, 380
Ren Y, 326
Riekel C, 294
Rosso P, 448, 458
Rubinstein M, 221, 233
Ruland W, 274
Rutledge G C, 41
Ryan A J, 61, 84
Ryan A J,

Satapathy B K, 438
Sawatari C, 41
Schadler L S, 354
Schaefer D W, 196
Schneider K, 295
Schulte K, 368, 451
Schweitzer K S, 231
Seguela R, 14, 16, 26
Shadrake L G, 24
Shah D, 386
Shen L, 314, 547
Sheng N, 396, 553, 562
Sherliker F R, 99
Shewchuk J R, 557
Schneider K, 295
Siddiqui N A, 459
Siengchin S, 311
Singh R P, 445
Song M, 325, 369
Sonnenschein M F, 61
Špirkova M, 77
Spitalsky Z, 42
Sreekala M S, 448
Staniek E, 29
Stein R S, 60
Sternstein S S, 225, 226, 233, 234, 235, 251
Strobl G, 16
Subramaniyan A K, 455
Subbotin A, 254
Sue H J, 382, 403,

Sue H-J, 434
Sun C T, 455
Sun L, 452

Takayanagi M, 39
Tanaka K, 38, 71
Tashiro K, 40, 41
Thio Y S, 433
Tjong C S, 442
Tobolsky A V, 60
Toda A, 23
Tokita N, 224
Tong L, 460
Treloar L R G, 211
Tsai J-L, 458
Tsai S W, 39
Tsantzalis S, 457
Tseng C H, 363
Turner R B, 61

Usuki A, 380
Usuku A, 545
Utracki L A, 301

Vacatello M, 217, 218, 525
Vaia R A, 426
Varley R J, 458
Velasco-Santos C, 349, 350
Viswanathan V, 311
Vlasveld D P N, 311
Vonk C G, 271

Wagner H D, 344
Walter R, 100
Wang G, 433
Wang K, 388, 400, 443, 450, 547
Wang S Q, 231
Wang Y D, 294
Ward I M, 15, 40, 41
Wendorff J H, 185
Weon J I, 382
Weon J-I, 434
Wetzel B, 332, 446
Wong M, 345
Wong S C, 393
Wu D M, 113

Author index

Wu J, 187
Wu M-D, 458
Wu Q J, 389
Wu S, 428, 440, 448

Xia H, 369
Xiong J, 369
Xu G, 25, 27, 29 31
Xu L R, 452

Yalcin B, 437
Yamaguchi M, 5
Yamashita A, 322, 323
Yang H, 440
Yang J L, 315
Ye L, 448
Ye Y, 452
Yoo Y, 436

Young R J, 16, 24, 26, 29, 413
Yuan Q, 436

Zdrahala R J, 75
Zeng Q H, 548
Zerda A S, 387, 388, 450
Zhang C, 29, 31
Zhang H, 448, 546
Zhang M Q, 196
Zhang Q-X, 385
Zhang S, 438
Zhang W, 316
Zhang X C, 5, 9, 29
Zhang Y X, 434
Zhou T H, 320
Zhou Y H, 328
Zhu J, 363, 459
Zilg C, 388

Subject Index

Agent
   coupling, 248, 476, 496
   macromolecular foaming, 114
   silane coupling, 141, 320
Agglomerate, 537
Agglomeration, 168
Aggregate, 537
Alumina, 317
Analysis, 173
   Pareto ANOVA, 173, 176
Approach
   water-assisted, 383
   top-bottom, 245
   bottom-up, 245
Approximation
   Fraunhofer, 272
   Pade, 81

Ball milling, 104
Berghmans' point, 84
Berkovich, three-side pyramid, 314
Blending
   protocols, 405
   sequence, 408
Blends
   highly drawn, 188
Boltzmann constant, 213, 253
Bone cement, 330, 331
Bottlenecks, 45
Bridging, 328
Brittle, 398
   failure, 328
Brownian diffusive motion, 211
Bubble-stretching
   in situ, 113
Burgers vector, 25

Carbon black, 301
Cation exchange capacity, 378

Cavitation, 11, 428
Chain
   dynamics, 242, 253
   Gaussian, 211
   immobilization, 251
   Ising, 211
   reptation, 252
   statsitics, 242
Characteristic volume, 255
Clay, 378, 379, 380, 400
   layers, 401
Code
   two-dimensional object-oriented finite element (OOF2), 181
Coefficient
   chain friction, 253
Coil
   Gaussian, 218
   Gaussian random, 253
   Langevin, 253
Composites
   MWNT/epoxy, 368
   epoxy, 344
   fiber reinforced (FRC), 241
   microfibrillar, 188
   multi-length-scale, 245
   nanofibrillar, 187
   structural, 333
   toughness, 456
Compatibilization
   reactive, 121
Compatibilizer, 476, 496, 499
Concept
   "floating effect", 511
   Koberstein-Stein, 61
Conditions
   periodic boundary, 549
Conformation
   entropy, 251
   statistics, 251

Copolymer(s)
  and blends, 440
  block, "tunable", 411
  multiblock, 67
Crack
  bridging, 353, 362, 368, 371
  initiation, 324
  pinning, 446
  propagation, 324
Crazes, 14
Crazing, 5, 429
Creep, 302
  behavior, 127
  compliance, 309
  lifetime, 315
Criterion
  Coulomb, 34
  minimum image, 555
Crosslinking, 326
  in situ, 125
  in solution, 530
Crystal
  α-crystal, 113
  β-crystal, 113
  plasticity and cavitation, 34
Crystallinity, 314
Crystallization, 378, 396, 397
  behavior, 384
  preferential, 437
  rate, 386
  temperatures, 387
Curve(s)
  Gaussian, 526
  master, 321
  tensile stress-strain, 95
  Wohler, 307
Cyclic loads, 306

Deborah number, 215
Deformation, 353, 362
  behavior, 343, 371
  characteristics, 343
  mechanical, 362
  mechanism, 353, 371
  non-affine, 257
  sword-sheath, 362
Delamination, 390, 398, 404, 456, 457

Delaunay-based refinement algorithm, 557
Density
  crosslink, 77
Dentistry, 330
Dispersion, 94, 380
  of clay, 378, 387
  quality, 406
Distribution
  chord length (CLD), 274
  Gaussian, 521
  interface, 274
  functions, 520
DMS, 75
Doi-Edwards relation, 214
Domains
  "hard", 474
  "soft", 474
Ductility, 387
  tensile, 349

Effect
  diluting, 473, 475, 479, 489, 491, 492
  excluded volume, 519
  floating, 475, 481, 489, 490, 491, 492
  floating, concept of, 490, 511
  Fox-Flory, 75
  Langevin, 226
  Mullins, 144, 148
  Payne, 150, 153, 228, 229, 231, 233, 251
  reinforcing, of nanoparticles, 99
  toughening, 95
Ehrenfest definition, 217
Elasticity
  Bernoulli-Euler continuum, 243, 251, 258
Elastomer, 59, 343, 363, 369, 371
Elastomeric plateau region, 83
Elastomers, 519
  thermoplastic, 474
Elongation
  at-break, 389
  tensile, 361
Ensemble average, 551
Entanglement, 99
Epoxy, 363, 371, 413

Equation
    Davies', 61
    Gordon and Taylor, 484, 492
    Halpin-Tsai, 242, 547
    Mooney-Rivlin, 143
    Mori-Tanaka, 547
    Klüppel-Schramm, 77
    Kohlrausch-Watts-Williams, 305, 306
    Paris law, 308
    Vogel-Fulcher-Tamann (VFT), 213
Euclidean dimensionality, 211
Exfoliation, 168, 309
    full, 565
    organoclay, 381
Experiments
    dynamic, 286
    neutron scattering, 524
    with a macrobeam, 293, 294
    modulus recovery, 252

Fabrication, 93
Fatigue, 306
    crack propagation, 308
    lifetime, 322
    tests, 271
Fibrous reinforcements, 431
Filler
    surface, 147
Fillers, 425, 519
    fibrous, 425, 443, 451, 456, 459
    platy, 425, 435, 442, 449, 458
    micro, 311
    nano, 301
    nano, mobility of, 129
    nano, surface modification of, 93
    spherical, 425, 428, 432, 440, 444, 453, 457
Filling fraction, 553
Fourier
    backtransform, 274
    transform, 272, 273, 276
Fracture, 345, 387, 404
    energy, 433
    essential work of (EWF), 427
    mechanics, 321, 427
    mechanisms, 408

    process, 397
    properties, 378
    surface, 351
    surfaces, 363
    toughness, 330, 362, 368, 388, 393, 399, 403, 408, 413, 433
    toughness, interlaminar, 458
Frank-Read source, 25
Free energy
    Gibbs, change, 31
    Helmholtz, 521
Frequency, 308
Function
    3D chord distribution, 274
    correlation, 274
    Langevin, 81
    Patterson, 274
Functionalization, 342, 343, 353, 363, 371
    of MWCNT, 439
    chemical, 343
    reaction, 356

γ-Ray irradiation, 95
Growth
    plastic void, 413, 449
    crack, resistance to, 444
Göbel mirrors, 278

Halpin-Tsai predictions, 157
Hard phase, 64
    elements, elastically active, 70
    glass transition, 83
    percolated, 70
    percolation, 70
    vitrification, 84
Homogenization, 547
Hysteresis, 78

Improved toughness, 387
Indentation, 314
Interaction
    filler/filler, 94
    filler/matrix, 94
Intercalation, 168, 309
Interdiffusion, 99
Interfacial bonding, 153

Interparticle ligament, 434
Interphase, 241, 242, 250, 429
　boundaries, 495, 502
　nano-scale, 250, 251
　micro-scale, 246

J-integral, 427, 433, 434, 443

Kelvin unit, 320
Kiessig camera, 278

Langevin loops, 231
Laplacian operator, 274
Law
　additivity, 472
　Coulomb, 31
　modified additivity, 495
　Paris, equation, 308
Layer(s)
　clay, 401
　transcrystalline 501, 502
Layered double hydroxide, 436
Length
　scales, 241
　threshold hard segment, 61
Limit
　Gaussian, 521
　indurance, 308
Links
　adaptive, 77
Load transfer, 343, 344, 345, 354, 355, 371
Ludwig-Davidenkov-Orovan hypothesis, 15

Machine background, 284
Markov statistics, 211
Matrices
　Euler, 525
Matrix
　epoxy, 363, 367
　polymer, mobility of, 393
　polyamide 66, 408
　polyamide, 401
Matrix-crack bridging, 353, 362
Mechanics
　continuum, 243

Mechanism(s)
　nanomechanical deformation, 368
　fracture, 408
　Frank-Read, 25
　reinforcing, 243
Mechanochemical graft, 104
Melt blending, 345, 347
Method, 172
　sol-gel 144
　energy partitioning, 434
　Monte Carlo, 217, 538
　Taguchi, 172
Micro necking, 11
Microhardness, 314, 471, 495, 502
Microphase separation, 67
Mixing protocol, 405
Model
　Mori-Tanaka, 547
　Burgers, 305, 310, 311, 313, 319
　Burgers (or four elements), 304
　equivalent box, 69
　Guth, 157, 224
　Halpin-Tsai, 39, 157, 179, 180, 223, 547
　Hui-Shia, 180
　Ising, 210, 211, 212
　Kelly-Tyson, 246
　Kelvin, 304, 305, 306
　Kerner-Nielsen, 222, 223, 224, 225
　Klüppel-Schramm, 78
　micro-mechanics, 243
　Qi-Boyce, 80
　random walk, 211
　Rubinstein, 254
　of sacrificial bonds, 252
　Weibull, 331
　of Young, 32
　Young's, 26
Modeling
　computational, 519
　finite element (FE), 547
　finite element method (FEM), 170
　molecular dynamics (MD), 170
　multiscale, 547
　object-oriented finite element (OOF), 170, 181

representative volume elements
(RVE), 170
Models
  Drozdov, 81
Moduli
  elastic, 248
Modulus
  loss, 82
  storage, 73, 82
  tearing, 435
  tensile, 361, 365
  algorithm, 554
  chains, 520
  histogram, 520
  polyethylene chains, 526
  reduction factor (MRF), 180
  simulations, 519, 520, 522, 524, 534
  Young's, 59, 179, 341, 351, 356, 370
Molding
  double injection, 476, 511
  double injection processing, 502
Montmorillonite, 310, 378, 380, 386
Mooney-Rivlin semi-empirical formula, 521

Nanocomposites, 93, 167, 168, 242, 341, 342, 343, 345, 347, 349
  amino-functionalized CNT/epoxy, 368
  CNT/elastomer, 369
  CNT/polymer, 362, 371
  CNT/thermoplastic, 362
  delaminated, 170
  dispersion problem, 185
  epoxy, 362, 363, 365, 388
  exfoliated, 170
  intercalated, 170
  MWNT/epoxy, 367, 368
  MWNT/PA6, 361
  MWNT/PC, 354
  MWNT/PMMA, 349
  MWNT/PP, 357
  organoclay, 409
  PE, 356
  peculiarities, 169
  polyamide 6, 382
  polyamide/clay, 377
  polyamide 6/clay, 383
  polyamide 6/organoclay, 387
  polyamide 6/organoclay, scratching of, 391
  polyamide 66/organoclay, 389
  polymer, 377, 412, 545
  polymer/clay, 378
  polymer/clay, brittleness of, 393
  polymeric, 93
  polymeric, manufacturing of, 93
  semicrystalline thermoplastic, 356
  SWNT-g-PU/PU, 369
  SWNT/PP, 356
  tensile properties of, 365
  ternary, 399, 402, 403, 405, 408, 411, 412
  true, 242
Nanofibers
  carbon, 301
  carbon (CNF), 431
  carbon (CNFs), 341
Nanofillers, 301
  mobility of, 129
  surface modification of, 93
Nanoindentation, 314
Nanomaterials
  biological, 426
Nanoparticles, 93, 141, 301
  "mobility" of, 390
  reinforcing effect of, 99
Nanopox®, 448
Nano-SiO$_2$, 95
Nanotube
  carbon, 341
Nanotubes
  carbon, 141, 153, 301
  carbon (CNT), 167, 431
  carbon (CNTs), 341
  carbon, functionalized, 361
Network
  Gaussian, 215
  interpenetrating, 455
  percolation, 429
Networks
  bimodal, 521
  polymer, 519
Nucleants, 444

Obstacle, 508, 509, 511
Organoclay, 178, 383, 388, 406, 408
  exfoliation, 381
Orthogonal arrays, 172

Parameter(s)
  Flory-Huggins, 67, 73
  Hermans orientation, 22
  van Krevelen solubility, 67
Particle
  mobility, 391
  primary, 537
Particles
  aggregated, 537
  Brownian, 212, 214
  ellipsoidal, 534
  spherical, 522
Peierls-Nabarro force, 25
Percolation
  exponent, 70
  threshold, 70, 159, 254
Periodic finite element mesh, 556
Permanent set, 79
Physical healing, 507
Physisorption, 528
Platelet(s), 546
  intercalated, 572
  orientation, 568
Plot(s)
  Mooney-Rivlin, 144, 158, 159, 521
  Paris, 309
POE-g-MA, 409, 411
Poly(dimethylsiloxane), 142
Poly(vinylidene difluoride) (PVDF),
    386, 389, 390
Polyamide (PA)
  lamellae, 396
  PA 66, 406
  PA 6, 380
  PA 6/maleic anhydride grafted
    polyethylene-, 397
Polyamides, 384, 386
Polymer
  grafting, 99
  matrix, mobility of, 393
  nanotubes, 377, 412, 545
  networks, 519

Polymers
  amorphous, 428
  glassy, doubly injection molded, 496
  particulate filled, 241
  semicrystalline, 432
  thermodynamically immiscible, 186
Polymerization
  graft, 94
  in situ, 342, 349, 351, 371, 369
Polymorph, 434
Polymorphism, 385
Polyoxyethylene nonylphenol (PN),
    413
Polypropylene, 95, 356, 357, 413
Polyurethanes, 59
Pre-drawing, 107
Property enhancement, 561

Radius of gyration, 242, 251
Ratio(s)
  aspect, 71, 190
  poisson, 42, 43, 223
  signal-to-noise, 172
Reinforcement, 519
Relationship between H and T_g, 477
Relaxation
  process, 255
  segment, 242
  time, 82
Representative volume element, 548
Reptation, 252
Resin
  epoxy, 344, 362, 363, 368
Resins
  crosslinked (thermoset), 444
  hybrid, 453
Response
  Gaussian, 230
  mechanical, 241
  viscoelastic, 251
Rubber, 406, 408, 413
  elasticity, 77
  silicone, 143
Rule of mixture, 180, 472
  Voigt, 180
  Reuss inverse, 180
RVE realization cell, 550

Scattering
    Compton, 272
    small-angle X-ray (SAXS), 270
    small-angle X-ray (SAXS), analysis, 501
    ultra-small angle X-ray (USAXS), 272
    wide-angle X-ray (WAXS), 190
SEBS-g-MA, 405, 406, 408
Segmental immobilization, 242
Segments
    soft, 474
    soft, blocks, 59
    hard, 474
    hard, blocks, 59
    Rouse, 214
Shear yielding, 413, 450
Silica, 141, 317
Silicate(s)
    layered, 301, 378, 387, 430
Simulations
    molecular dynamics, 129
    modulus, 519, 520, 522, 524, 534
Soft phase, 64, 475
Sol-gel
    process, 141
    route, 447
Statistical ensemble, 550
Steiner points, 557
Step
    fibrillation, 186
    isotropization, 186
    mixing, 186
Stiffness
    effective, 551
    enhancement, 561
    tensile, 341, 345, 368
Strain at break, 351, 371
Strength
    interlaminar shear (ILSS), 457
    Izod impact, 442
    notched impact, 387
    tensile, 341, 345, 351, 356, 371
    yield, 357, 361
Stress
    concentrators, 435
    constant, 302
    interface shear, 562
    transfer, 241
    tensile, 369
    yield, 324
Stress-strain
    behavior, 77
    isotherms, 525
Stretch-hold technique, 269, 270
Structure
    discrete molecular, 242
    semi-IPN, 125

Tearing energy, 326
Telescopic pullout, 353, 362, 368, 371
Tensile loading, 349, 361
Temperature
    Arrhenius, 152
    Vogel, 213
Theory
    Arrhenius activation, 83
    crack deflection, 446
    dislocation, 23
    Doolittle free volume, 83
    Fourier transformation, 277
    Mark-Curro, 521
    rotational isomeric state, 520
    Williams-Landell-Ferry (WLF), 83
    Wu's, 441
    Wu's percolation, 443
Thermoplastics, 343, 345, 356, 371
    flexible-chain, 186
    semicrystalline, 361
Time
    reptation relaxation, 253
    terminal relaxation, 253
Time-temperature superposition, 82
    principle, 311
Titania, 317
Titanium dioxide, 141
Toughening, 332
    effect, 95
Toughness, 341, 347, 361, 362, 363, 393, 397, 404, 411, 427
    composites, 456
    enhancements, 400
    modifiers, 452
    tensile, 347, 349, 363

Transcrystallization, 396, 398, 438
Transition
  brittle-ductile, 400
  insulator-conductor, 159
Tube, 213
Twinning, 21

Unit cell, 551

Vickers indentation, 491

Weld line(s), 496, 502, 504, 505, 507, 508, 511